TABLE 6.1 Summary of Basic Op Amp Circuits

Name	Circuit Schematic	Input-Output Relation
Inverting Amplifier		$v_{out} = -\dfrac{R_f}{R_1} v_{in}$
Noninverting Amplifier		$v_{out} = \left(1 + \dfrac{R_f}{R_1}\right) v_{in}$
Voltage Follower (also known as a Unity Gain Amplifier)		$v_{out} = v_{in}$
Summing Amplifier		$v_{out} = -\dfrac{R_f}{R}(v_1 + v_2 + v_3)$
Difference Amplifier		$v_{out} = v_2 - v_1$

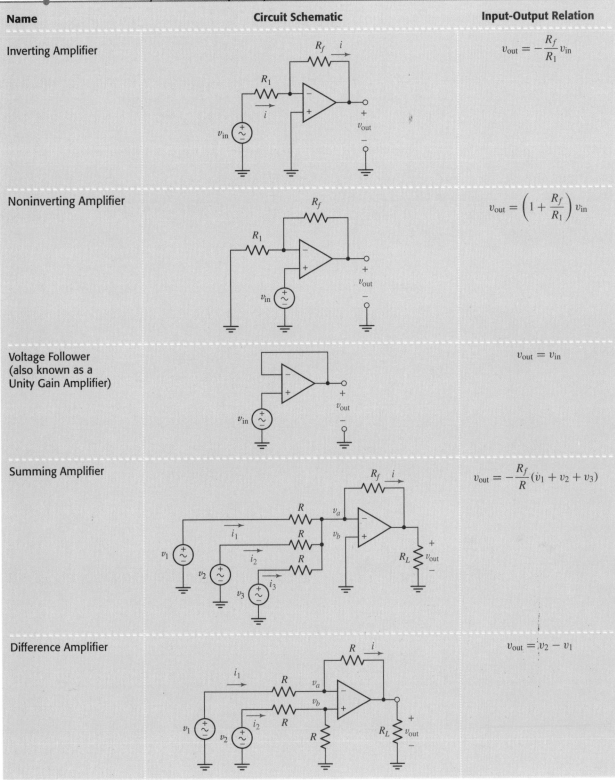

ENGINEERING CIRCUIT ANALYSIS

ENGINEERING CIRCUIT ANALYSIS

EIGHTH EDITION

William H. Hayt, Jr. (deceased)
Purdue University

Jack E. Kemmerly (deceased)
California State University

Steven M. Durbin
University at Buffalo
The State University of New York

Mc Graw Hill

Connect
Learn
Succeed™

ENGINEERING CIRCUIT ANALYSIS, EIGHTH EDITION

Published by McGraw-Hill, a business unit of The McGraw-Hill Companies, Inc., 1221 Avenue of the Americas, New York, NY 10020. Copyright © 2012 by The McGraw-Hill Companies, Inc. All rights reserved. Previous editions © 2007, 2002, and 1993. Printed in the United States of America.

Some ancillaries, including electronic and print components, may not be available to customers outside the United States.

This book is printed on acid-free paper.

1 2 3 4 5 6 7 8 9 0 DOW/DOW 1 0 9 8 7 6 5 4 3 2 1

ISBN   978-0-07-352957-8
MHID   0-07-352957-5

Vice President & Editor-in-Chief: *Marty Lange*
Vice President & Director of Specialized Publishing: *Janice M. Roerig-Blong*
Editorial Director: *Michael Lange*
Global Publisher: *Raghothaman Srinivasan*
Senior Marketing Manager: *Curt Reynolds*
Developmental Editor: *Darlene M. Schueller*
Lead Project Manager: *Jane Mohr*
Buyer: *Kara Kudronowicz*
Design Coordinator: *Brenda A. Rolwes*
Senior Photo Research Coordinator: *John C. Leland*
Senior Media Project Manager: *Tammy Juran*
Compositor: *MPS Limited, a Macmillan Company*
Typeface: *10/12 Times Roman*
Printer: *R. R. Donnelley*

Cover Image: © *Getty Images*
Cover Designer: *Studio Montage, St. Louis, Missouri*

MATLAB is a registered trademark of The MathWorks, Inc.
PSpice is a registered trademark of Cadence Design Systems, Inc.

The following photos are courtesy of Steve Durbin: Page 5, Fig. 2.22a, 2.24a–c, 5.34, 6.1a, 7.2a–c, 7.11a–b, 13.15, 17.29

Library of Congress Cataloging-in-Publication Data

Hayt, William Hart, 1920–1999
  Engineering circuit analysis / William H. Hayt, Jr., Jack E. Kemmerly, Steven M. Durbin. — 8th ed.
    p. cm.
  Includes index.
  ISBN 978-0-07-352957-8
  1. Electric circuit analysis. 2. Electric network analysis. I. Kemmerly, Jack E. (Jack Ellsworth), 1924–1998
II. Durbin, Steven M. III. Title.

  TK454.H4   2012
  621.319'2—dc22   2011009912

www.mhhe.com

To Sean and Kristi.
The best part of every day.

BRIEF CONTENTS

CONTENTS

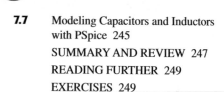

The target audience colors everything about a book, being a major factor in decisions big and small, particularly both the pace and the overall writing style. Consequently it is important to note that the authors have made the conscious decision to write this book to the **student**, and not to the instructor. Our underlying philosophy is that reading the book should be enjoyable, despite the level of technical detail that it must incorporate. When we look back to the very first edition of *Engineering Circuit Analysis*, it's clear that it was developed specifically to be more of a conversation than a dry, dull discourse on a prescribed set of fundamental topics. To keep it conversational, we've had to work hard at updating the book so that it continues to speak to the increasingly diverse group of students using it all over the world.

Although in many engineering programs the introductory circuits course is preceded or accompanied by an introductory physics course in which electricity and magnetism are introduced (typically from a fields perspective), this is not required to use this book. After finishing the course, many students find themselves truly amazed that such a broad set of analytical tools have been derived from **only three simple scientific laws**—Ohm's law and Kirchhoff's voltage and current laws. The first six chapters assume only a familiarity with algebra and simultaneous equations; subsequent chapters assume a first course in calculus (derivatives and integrals) is being taken in tandem. Beyond that, we have tried to incorporate sufficient details to allow the book to be read on its own.

So, what key features have been designed into this book with the student in mind? First, individual chapters are organized into relatively short subsections, each having a single primary topic. The language has been updated to remain informal and to flow smoothly. Color is used to highlight important information as opposed to merely improve the aesthetics of the page layout, and white space is provided for jotting down short notes and questions. New terms are defined as they are introduced, and examples are placed strategically to demonstrate not only basic concepts, but problem-solving approaches as well. Practice problems relevant to the examples are placed in proximity so that students can try out the techniques for themselves before attempting the end-of-chapter exercises. The exercises represent a broad range of difficulties, generally ordered from simpler to more complex, and grouped according to the relevant section of each chapter. Answers to selected odd-numbered end-of-chapter exercises are posted on the book's website at www.mhhe.com/haytdurbin8e.

Engineering is an intensive subject to study, and students often find themselves faced with deadlines and serious workloads. This does not mean that textbooks have to be dry and pompous, however, or that coursework should never contain any element of fun. In fact, successfully solving a problem often *is* fun, and learning how to do that can be fun as well. Determining how to best accomplish this within the context of a textbook is an ongoing process. The authors have always relied on the often very candid feedback received from our own students at Purdue University; the California State University, Fullerton; Fort Lewis College in Durango, the joint engineering program at Florida A&M University and Florida State University, the University of Canterbury (New Zealand) and the University at Buffalo. We also rely on comments, corrections, and suggestions from instructors and students worldwide, and for this edition, consideration has been given to a new source of comments, namely, semianonymous postings on various websites.

The first edition of *Engineering Circuit Analysis* was written by Bill Hayt and Jack Kemmerly, two engineering professors who very much enjoyed teaching, interacting with their students, and training generations of future engineers. It was well received due to its compact structure, "to the point" informal writing style, and logical organization. There is no timidity when it comes to presenting the theory underlying a specific topic, or pulling punches when developing mathematical expressions. Everything, however, was carefully designed to assist students in their learning, present things in a straightforward fashion, and leave theory for theory's sake to other books. They clearly put a great deal of thought into writing the book, and their enthusiasm for the subject comes across to the reader.

KEY FEATURES OF THE EIGHTH EDITION

We have taken great care to retain key features from the seventh edition which were clearly working well. These include the general layout and sequence of chapters, the basic style of both the text and line drawings, the use of four-color printing where appropriate, numerous worked examples and related practice problems, and grouping of end-of-chapter exercises according to section. Transformers continue to merit their own chapter, and complex frequency is briefly introduced through a student-friendly extension of the phasor technique, instead of indirectly by merely stating the Laplace transform integral. We also have retained the use of icons, an idea first introduced in the sixth edition:

 Provides a heads-up to common mistakes;

 Indicates a point that's worth noting;

 Denotes a design problem to which there is no unique answer;

 Indicates a problem which requires computer-aided analysis.

The introduction of engineering-oriented analysis and design software in the book has been done with the mind-set that it should assist, not replace, the learning process. Consequently, the computer icon denotes problems that are typically phrased such that the software is used to *verify* answers, and not simply provide them. Both MATLAB® and PSpice® are used in this context.

SPECIFIC CHANGES FOR THE EIGHTH EDITION INCLUDE:

- A new section in Chapter 16 on the analysis and design of multistage Butterworth filters
- Over 1000 new and revised end-of-chapter exercises
- A new overarching philosophy on end-of-chapter exercises, with each section containing problems similar to those solved in worked examples and practice problems, before proceeding to more complex problems to test the reader's skills
- Introduction of Chapter-Integrating Exercises at the end of each chapter. For the convenience of instructors and students, end-of-chapter exercises are grouped by section. To provide the opportunity for assigning exercises with less emphasis on an explicit solution method (for example, mesh or nodal analysis), as well as to give a broader perspective on key topics within each chapter, a select number of Chapter-Integrating Exercises appear at the end of each chapter.
- New photos, many in full color, to provide connection to the real world
- Updated screen captures and text descriptions of computer-aided analysis software
- New worked examples and practice problems
- Updates to the Practical Application feature, introduced to help students connect material in each chapter to broader concepts in engineering. Topics include distortion in amplifiers, modeling automotive suspension systems, practical aspects of grounding, the relationship of poles to stability, resistivity, and the memristor, sometimes called "the missing element"
- Streamlining of text, especially in the worked examples, to get to the point faster
- Answers to selected odd-numbered end-of-chapter exercises are posted on the book's website at www.mhhe.com/haytdurbin8e.

I joined the book in 1999, and sadly never had the opportunity to speak to either Bill or Jack about the revision process, although I count myself lucky to have taken a circuits course from Bill Hayt while I was a student at Purdue. It is a distinct privilege to serve as a coauthor to *Engineering Circuit Analysis*, and in working on this book I give its fundamental philosophy and target audience the highest priority. I greatly appreciate the many people who have already provided feedback—both positive and negative— on aspects of previous editions, and welcome others to do so as well, either through the publishers (McGraw-Hill Higher Education) or to me directly (durbin@ieee.org).

Of course, this project has been a team effort, as is the case with every modern textbook. In particular I would like to thank Raghu Srinivasan (Global Publisher), Peter Massar (Sponsoring Editor), Curt Reynolds (Marketing Manager), Jane Mohr (Project Manager), Brittney-Corrigan-McElroy (Project Manager), Brenda Rolwes (Designer), Tammy Juran (Media Project Manager), and most importantly, Developmental Editor Darlene Schueller, who helped me with many, many details, issues, deadlines, and questions. She is absolutely the best, and I'm very grateful for all the support from the team at McGraw-Hill. I would also like to thank various McGraw-Hill representatives, especially Nazier Hassan, who dropped by whenever on campus to just say hello and ask how things were going. Special thanks are also due to Catherine Shultz and Michael Hackett, former editors who continue to keep in contact. Cadence® and The MathWorks kindly provided assistance with software-aided analysis software, which was much appreciated. Several colleagues have generously supplied or helped with photographs and technical details, for which I'm very grateful: Prof. Masakazu Kobayashi of Waseda University; Dr. Wade Enright, Prof. Pat Bodger, Prof. Rick Millane, Mr. Gary Turner, and Prof. Richard Blaikie of the University of Canterbury; and Prof. Reginald Perry and Prof. Jim Zheng of Florida A&M University and the Florida State University. For the eighth edition, the following individuals deserve acknowledgment and a debt of gratitude for taking the time to review various versions of the manuscript:

Chong Koo An, *The University of Ulsan*

Mark S. Andersland, *The University of Iowa*

Marc Cahay, *University of Cincinnati*

Claudio Canizares, *University of Waterloo*

Teerapon Dachokiatawan, *King Mongkut's University of Technology North Bangkok*

John Durkin, *The University of Akron*

Lauren M. Fuentes, *Durham College*

Lalit Goel, *Nanyang Technological University*

Rudy Hofer, *Conestoga College ITAL*

Mark Jerabek, *West Virginia University*

Michael Kelley, *Cornell University*

Hua Lee, *University of California, Santa Barbara*

Georges Livanos, *Humber College Institute of Technology*

Ahmad Nafisi, *Cal Poly State University*

Arnost Neugroschel, *University of Florida*

Pravin Patel, *Durham College*

Jamie Phillips, *The University of Michigan*

Daryl Reynolds, *West Virginia University*

G.V.K.R. Sastry, *Andhra University*

Michael Scordilis, *University of Miami*

Yu Sun, *University of Toronto, Canada*

Chanchana Tangwongsan, *Chulalongkorn University*

Edward Wheeler, *Rose-Hulman Institute of Technology*

Xiao-Bang Xu, *Clemson University*

Tianyu Yang, *Embry-Riddle Aeronautical University*

Zivan Zabar, *Polytechnic Institute of NYU*

I would also like to thank Susan Lord, *University of San Diego,* Archie L. Holmes, Jr., *University of Virginia,* Arnost Neugroschel, *University of Florida,* and Michael Scordilis, *University of Miami,* for their assistance in accuracy checking answers to selected end-of-chapter exercises.

Finally, I would like to briefly thank a number of other people who have contributed both directly and indirectly to the eighth edition. First and foremost, my wife, Kristi, and our son, Sean, for their patience, understanding, support, welcome distractions, and helpful advice. Throughout the day it has always been a pleasure to talk to friends and colleagues about what should be taught, how it should be taught, and how to measure learning. In particular, Martin Allen, Richard Blaikie, Alex Cartwright, Peter Cottrell, Wade Enright, Jeff Gray, Mike Hayes, Bill Kennedy, Susan Lord, Philippa Martin, Theresa Mayer, Chris McConville, Reginald Perry, Joan Redwing, Roger Reeves, Dick Schwartz, Leonard Tung, Jim Zheng, and many others have provided me with many useful insights, as has my father, Jesse Durbin, an electrical engineering graduate of the Indiana Institute of Technology.

Steven M. Durbin
Buffalo, New York

McGRAW-HILL DIGITAL OFFERINGS INCLUDE:

McGraw-Hill Connect™ Engineering

McGraw-Hill Connect Engineering is a web-based assignment and assessment platform that gives students the means to better connect with their coursework, with their instructors, and with the important concepts that they will need to know for success now and in the future. With Connect Engineering, instructors can deliver assignments, quizzes, and tests easily online. Students can practice important skills at their own pace and on their own schedule. Ask your McGraw-Hill representative for more details and check it out at www.mcgrawhillconnect.com/engineering.

Instructor support materials are available from the book's website at www.mhhe.com/haytdurbin8e. The materials include a password-protected solutions manual and image library. Instructors can also benefit from McGraw-Hill's Complete Online Solutions Manual Organization System (COSMOS). COSMOS enables instructors to generate a limitless supply of problem material for assignments, as well as transfer and integrate their own problems into the software. Contact your McGraw-Hill sales representative for additional information.

McGraw-Hill Create™

Craft your teaching resources to match the way you teach! With McGraw-Hill Create, www.mcgrawhillcreate.com, you can easily rearrange chapters, combine material from other content sources, and quickly upload content you have written such as your course syllabus or teaching notes. Find the content you need in Create by searching through thousands of leading McGraw-Hill textbooks. Arrange your book to fit your teaching style. Create even allows you to personalize your book's appearance by selecting the cover and adding your name, school, and course information. Order a Create book and you'll receive a complimentary print review copy in 3–5 business days or a complimentary electronic review copy (eComp) via email in minutes. Go to www.mcgrawhillcreate.com today and register to experience how McGraw-Hill Create empowers you to teach *your* students *your* way.

McGraw-Hill Higher Education and Blackboard® Have Teamed Up.

Blackboard, the web-based course management system, has partnered with McGraw-Hill to better allow students and faculty to use online materials and activities to complement face-to-face teaching. Blackboard features exciting social learning and teaching tools that foster more logical, visually impactful, and active learning opportunities for students. You'll transform your closed-door classrooms into communities where students remain connected to their educational experience 24 hours a day.

This partnership allows you and your students access to McGraw-Hill's Connect™ and Create™ right from within your Blackboard course—all with one single sign-on.

Not only do you get single sign-on with Connect and Create, you also get deep integration of McGraw-Hill content and content engines right in Blackboard. Whether you're choosing a book for your course or building Connect assignments, all the tools you need are right where you want them—inside of Blackboard.

Gradebooks are now seamless. When a student completes an integrated Connect assignment, the grade for that assignment automatically (and instantly) feeds your Blackboard grade center.

McGraw-Hill and Blackboard can now offer you easy access to industry leading technology and content, whether your campus hosts it or we do. Be sure to ask your local McGraw-Hill representative for details.

Electronic Textbook Option

This text is offered through CourseSmart for both instructors and students. CourseSmart is an online resource where students can purchase the complete text online at almost half the cost of a traditional text. Purchasing the eTextbook allows students to take advantage of CourseSmart's web tools for learning, which include full text search, notes and highlighting, and email tools for sharing notes between classmates. To learn more about CourseSmart options, contact your sales representative or visit www.CourseSmart.com.

Introduction

KEY CONCEPTS

Linear versus Nonlinear
Circuits

Four Main Categories of
Circuit Analysis:

• DC Analysis

• Transient Analysis

• Sinusoidal Analysis

• Frequency Response

Circuit Analysis Beyond
Circuits

Analysis and Design

Use of Engineering Software

A Problem-Solving Strategy

PREAMBLE

Although there are clear specialties within the field of engineering,
all engineers share a considerable amount of common ground,
particularly when it comes to problem solving. In fact, many practicing engineers find it is possible to work in a large variety of
settings and even outside their traditional specialty, as their skill set
is often transferrable to other environments. Today's engineering
graduates are employed in a broad range of jobs, from design of
individual components and systems, to assisting in solving socio-
economic problems such as air and water pollution, urban planning,
communication, mass transportation, power generation and distribution, and efficient use and conservation of natural resources.

Circuit analysis has long been a traditional introduction to **the
art of problem solving** from an engineering perspective, even for
those whose interests lie outside electrical engineering. There are
many reasons for this, but one of the best is that in today's world
it's extremely unlikely for any engineer to encounter a system that
does not in some way include electrical circuitry. As circuits become smaller and require less power, and power sources become
smaller and cheaper, embedded circuits are seemingly everywhere.
Since most engineering situations require a team effort at some
stage, having a working knowledge of circuit analysis therefore
helps to provide everyone on a project with the background needed
for effective communication.

Consequently, this book is not just about "circuit analysis" from
an engineering perspective, but is also about developing basic
problem-solving skills as they apply to situations an engineer is
likely to encounter. As part of this, we also find that we're developing an intuitive understanding at a general level, and often we can

1

Not all electrical engineers routinely make use of circuit analysis, but they often bring to bear analytical and problem-solving skills learned early on in their careers. A circuit analysis course is one of the first exposures to such concepts. (*Solar Mirrors:* © *Corbis; Skyline:* © *Getty Images/PhotoLink; Oil Rig:* © *Getty Images; Dish:* © *Getty Images/J. Luke/PhotoLink*)

understand a complex system by its analogy to an electrical circuit. Before launching into all this, however, we'll begin with a quick preview of the topics found in the remainder of the book, pausing briefly to ponder the difference between analysis and design, and the evolving role computer tools play in modern engineering.

1.1　OVERVIEW OF TEXT

The fundamental subject of this text is **_linear circuit analysis_**, which sometimes prompts a few readers to ask,

> "Is there ever any *nonlinear* circuit analysis?"

Sure! We encounter nonlinear circuits every day: they capture and decode signals for our TVs and radios, perform calculations millions of times a second inside microprocessors, convert speech into electrical signals for transmission over phone lines, and execute many other functions outside our field of view. In designing, testing, and implementing such nonlinear circuits, detailed analysis is unavoidable.

> "Then why study *linear* circuit analysis?"

Television sets include many nonlinear circuits. A great deal of them, however, can be understood and analyzed with the assistance of linear models. (© *Sony Electronics, Inc.*)

you might ask. An excellent question. The simple fact of the matter is that no physical system (including electrical circuits) is ever perfectly linear. Fortunately for us, however, a great many systems behave in a reasonably linear fashion over a limited range—allowing us to model them as linear systems if we keep the range limitations in mind.

For example, consider the common function

$$f(x) = e^x$$

A linear approximation to this function is

$$f(x) \approx 1 + x$$

Let's test this out. Table 1.1 shows both the exact value and the approximate value of $f(x)$ for a range of x. Interestingly, the linear approximation is exceptionally accurate up to about $x = 0.1$, when the relative error is still less than 1%. Although many engineers are rather quick on a calculator, it's hard to argue that any approach is faster than just adding 1.

TABLE **1.1** Comparison of a Linear Model for e^x to Exact Value

x	f(x)*	1 + x	Relative error**
0.0001	1.0001	1.0001	0.0000005%
0.001	1.0010	1.001	0.00005%
0.01	1.0101	1.01	0.005%
0.1	1.1052	1.1	0.5%
1.0	2.7183	2.0	26%

*Quoted to four significant figures.

**Relative error $\triangleq \left| 100 \times \dfrac{e^x - (1 + x)}{e^x} \right|$

Linear problems are inherently more easily solved than their nonlinear counterparts. For this reason, we often seek reasonably accurate linear approximations (or *models*) to physical situations. Furthermore, the linear models are more easily manipulated and understood—making design a more straightforward process.

The circuits we will encounter in subsequent chapters all represent linear approximations to physical electric circuits. Where appropriate, brief discussions of potential inaccuracies or limitations to these models are provided, but generally speaking we find them to be suitably accurate for most applications. When greater accuracy is required in practice, nonlinear models are employed, but with a considerable increase in solution complexity. A detailed discussion of what constitutes a *linear electric circuit* can be found in Chap. 2.

Linear circuit analysis can be separated into four broad categories: (1) *dc analysis,* where the energy sources do not change with time; (2) *transient analysis,* where things often change quickly; (3) *sinusoidal analysis,* which applies to both ac power and signals; and (4) *frequency response,* which is the most general of the four categories, but typically assumes something is changing with time. We begin our journey with the topic of resistive circuits, which may include simple examples such as a flashlight or a toaster. This provides us with a perfect opportunity to learn a number of very powerful engineering circuit analysis techniques, such as *nodal analysis, mesh analysis, superposition, source transformation, Thévenin's theorem, Norton's*

Modern trains are powered by electric motors. Their electrical systems are best analyzed using ac or phasor analysis techniques. (*Used with permission. Image copyright © 2010 M. Kobayashi. All rights reserved.*)

Frequency-dependent circuits lie at the heart of many electronic devices, and they can be a great deal of fun to design. (© *The McGraw-Hill Companies, Inc.*)

theorem, and several methods for simplifying networks of components connected in series or parallel. The single most redeeming feature of resistive circuits is that the time dependence of any quantity of interest does not affect our analysis procedure. In other words, if asked for an electrical quantity of a resistive circuit at several specific instants in time, we do not need to analyze the circuit more than once. As a result, we will spend most of our effort early on considering only dc circuits—those circuits whose electrical parameters do not vary with time.

Although dc circuits such as flashlights or automotive rear window defoggers are undeniably important in everyday life, things are often much more interesting when something happens suddenly. In circuit analysis parlance, we refer to *transient analysis* as the suite of techniques used to study circuits which are suddenly energized or de-energized. To make such circuits interesting, we need to add elements that respond to the rate of change of electrical quantities, leading to circuit equations which include derivatives and integrals. Fortunately, we can obtain such equations using the simple techniques learned in the first part of our study.

Still, not all time-varying circuits are turned on and off suddenly. Air conditioners, fans, and fluorescent lights are only a few of the many examples we may see daily. In such situations, a calculus-based approach for every analysis can become tedious and time-consuming. Fortunately, there is a better alternative for situations where equipment has been allowed to run long enough for transient effects to die out, and this is commonly referred to as ac or sinusoidal analysis, or sometimes *phasor analysis.*

The final leg of our journey deals with a subject known as *frequency response.* Working directly with the differential equations obtained in time-domain analysis helps us develop an intuitive understanding of the operation of circuits containing energy storage elements (e.g., capacitors and inductors). As we shall see, however, circuits with even a relatively small number of components can be somewhat onerous to analyze, and so much more straightforward methods have been developed. These methods, which include Laplace and Fourier analysis, allow us to transform differential equations into algebraic equations. Such methods also enable us to design circuits to respond in specific ways to particular frequencies. We make use of frequency-dependent circuits every day when we dial a telephone, select our favorite radio station, or connect to the Internet.

1.2 • RELATIONSHIP OF CIRCUIT ANALYSIS TO ENGINEERING

Whether we intend to pursue further circuit analysis at the completion of this course or not, it is worth noting that there are several layers to the concepts under study. Beyond the nuts and bolts of circuit analysis techniques lies the opportunity to develop a methodical approach to problem solving, the ability to determine the goal or goals of a particular problem, skill at collecting the information needed to effect a solution, and, perhaps equally importantly, opportunities for practice at verifying solution accuracy.

Students familiar with the study of other engineering topics such as fluid flow, automotive suspension systems, bridge design, supply chain management, or process control will recognize the general form of many of the

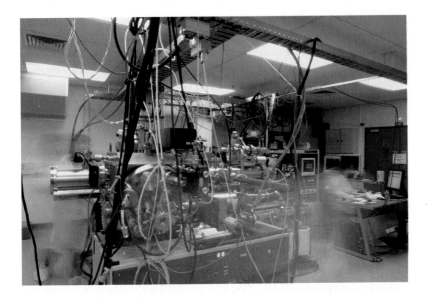

A molecular beam epitaxy crystal growth facility. The equations governing its operation closely resemble those used to describe simple linear circuits.

equations we develop to describe the behavior of various circuits. We simply need to learn how to "translate" the relevant variables (for example, replacing *voltage* with *force, charge* with *distance, resistance* with *friction coefficient*, etc.) to find that we already know how to work a new type of problem. Very often, if we have previous experience in solving a similar or related problem, our intuition can guide us through the solution of a totally new problem.

What we are about to learn regarding linear circuit analysis forms the basis for many subsequent electrical engineering courses. The study of electronics relies on the analysis of circuits with devices known as diodes and transistors, which are used to construct power supplies, amplifiers, and digital circuits. The skills which we will develop are typically applied in a rapid, methodical fashion by electronics engineers, who sometimes can analyze a complicated circuit without even reaching for a pencil! The time-domain and frequency-domain chapters of this text lead directly into discussions of signal processing, power transmission, control theory, and communications. We find that frequency-domain analysis in particular is an extremely powerful technique, easily applied to any physical system subjected to time-varying excitation, and particularly helpful in the design of filters.

1.3 ANALYSIS AND DESIGN

Engineers take a fundamental understanding of scientific principles, combine this with practical knowledge often expressed in mathematical terms, and (frequently with considerable creativity) arrive at a solution to a given problem. *Analysis* is the process through which we determine the scope of a problem, obtain the information required to understand it, and compute the parameters of interest. *Design* is the process by which we synthesize something new as part of the solution to a problem. Generally speaking, there is an expectation that a problem requiring design will have no unique solution, whereas the analysis phase typically will. Thus, the last step in designing is always analyzing the result to see if it meets specifications.

An example of a robotic manipulator. The feedback control system can be modeled using linear circuit elements to determine situations in which the operation may become unstable. (*NASA Marshall Space Flight Center.*)

This text is focused on developing our ability to analyze and solve problems because it is the starting point in every engineering situation. The philosophy of this book is that we need clear explanations, well-placed examples, and plenty of practice to develop such an ability. Therefore, elements of design are integrated into end-of-chapter problems and later chapters so as to be enjoyable rather than distracting.

1.4 COMPUTER-AIDED ANALYSIS

Solving the types of equations that result from circuit analysis can often become notably cumbersome for even moderately complex circuits. This of course introduces an increased probability that errors will be made, in addition to considerable time in performing the calculations. The desire to find a tool to help with this process actually predates electronic computers, with purely mechanical computers such as the Analytical Engine designed by Charles Babbage in the 1880s proposed as possible solutions. Perhaps the earliest successful electronic computer designed for solution of differential equations was the 1940s-era ENIAC, whose vacuum tubes filled a large room. With the advent of low-cost desktop computers, however, computer-aided circuit analysis has developed into an invaluable everyday tool which has become an integral part of not only analysis but design as well.

One of the most powerful aspects of computer-aided design is the relatively recent integration of multiple programs in a fashion transparent to the user. This allows the circuit to be drawn schematically on the screen, reduced automatically to the format required by an analysis program (such as SPICE, introduced in Chap. 4), and the resulting output smoothly transferred to a third program capable of plotting various electrical quantities of

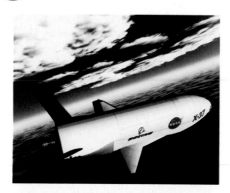

Two proposed designs for a next-generation space shuttle. Although both contain similar elements, each is unique. (*NASA Dryden Flight Research Center.*)

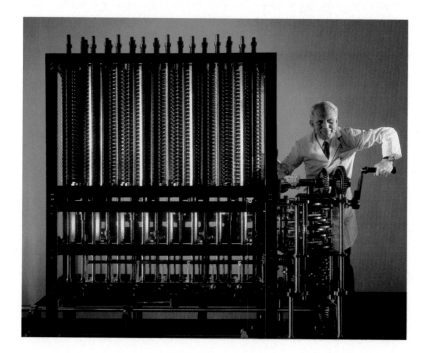

Charles Babbage's "Difference Engine Number 2," as completed by the Science Museum (London) in 1991. (© *Science Museum/Science & Society Picture Library.*)

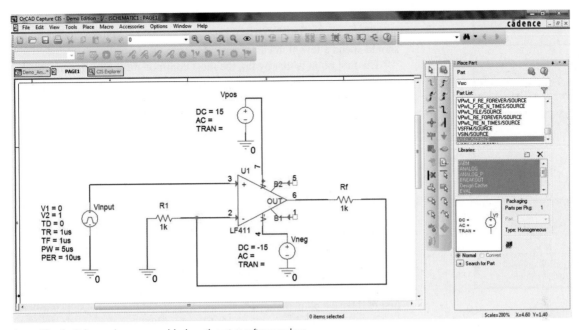

An amplifier circuit drawn using a commercial schematic capture software package.

interest that describe the operation of the circuit. Once the engineer is satisfied with the simulated performance of the design, the same software can generate the printed circuit board layout using geometrical parameters in the components library. This level of integration is continually increasing, to the point where soon an engineer will be able to draw a schematic, click a few buttons, and walk to the other side of the table to pick up a manufactured version of the circuit, ready to test!

The reader should be wary, however, of one thing. Circuit analysis software, although fun to use, is by no means a replacement for good old-fashioned paper-and-pencil analysis. We need to have a solid understanding of how circuits work in order to develop an ability to design them. Simply going through the motions of running a particular software package is a little like playing the lottery: with user-generated entry errors, hidden default parameters in the myriad of menu choices, and the occasional shortcoming of human-written code, there is no substitute for having at least an approximate idea of the expected behavior of a circuit. Then, if the simulation result does not agree with expectations, we can find the error early, rather than after it's too late.

Still, computer-aided analysis is a powerful tool. It allows us to vary parameter values and evaluate the change in circuit performance, and to consider several variations to a design in a straightforward manner. The result is a reduction of repetitive tasks, and more time to concentrate on engineering details.

1.5 SUCCESSFUL PROBLEM-SOLVING STRATEGIES

As the reader might have picked up, this book is just as much about problem solving as it is about circuit analysis. As a result, the expectation is that during your time as an engineering student, you are learning how to solve problems—so just at this moment, those skills are not yet fully developed. As you proceed

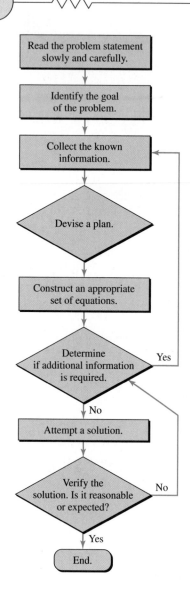

through your course of study, you will pick up techniques that work for you, and likely continue to do so as a practicing engineer. At this stage, then, we should spend a few moments discussing some basic points.

The first point is that by far, the most common difficulty encountered by engineering students is *not knowing how to start* a problem. This improves with experience, but early on that's of no help. The best advice we can give is to adopt a methodical approach, beginning with reading the problem statement slowly and carefully (and more than once, if needed). Since experience usually gives us some type of insight into how to deal with a specific problem, worked examples appear throughout the book. Rather than just read them, however, it might be helpful to work through them with a pencil and a piece of paper.

Once we've read through the problem, and feel we might have some useful experience, the next step is to identify the goal of the problem—perhaps to calculate a voltage or a power, or to select a component value. Knowing where we're going is a big help. The next step is to collect as much information as we can, and to organize it somehow.

At this point *we're still not ready to reach for the calculator.* It's best first to devise a plan, perhaps based on experience, perhaps based simply on our intuition. Sometimes plans work, and sometimes they don't. Starting with our initial plan, it's time to construct an initial set of equations. If they appear complete, we can solve them. If not, we need to either locate more information, modify our plan, or both.

Once we have what appears to be a working solution, we should not stop, even if exhausted and ready for a break. **No engineering problem is solved unless the solution is tested somehow.** We might do this by performing a computer simulation, or solving the problem a different way, or perhaps even just estimating what answer might be reasonable.

Since not everyone likes to read to learn, these steps are summarized in the adjacent flowchart. This is just one particular problem-solving strategy, and the reader of course should feel free to modify it as necessary. The real key, however, is to try and learn in a relaxed, low-stress environment free of distractions. Experience is the best teacher, and learning from our own mistakes will always be part of the process of becoming a skilled engineer.

READING FURTHER

This relatively inexpensive, best-selling book teaches the reader how to develop winning strategies in the face of seemingly impossible problems:

G. Polya, *How to Solve It.* Princeton, N.J.: Princeton University Press, 1971.

Basic Components and Electric Circuits

KEY CONCEPTS

Basic Electrical Quantities and Associated Units: Charge, Current, Voltage, and Power

Current Direction and Voltage Polarity

The Passive Sign Convention for Calculating Power

Ideal Voltage and Current Sources

Dependent Sources

Resistance and Ohm's Law

INTRODUCTION

In conducting circuit analysis, we often find ourselves seeking specific *currents, voltages,* or *powers,* so here we begin with a brief description of these quantities. In terms of components that can be used to build electrical circuits, we have quite a few from which to choose. We initially focus on the *resistor,* a simple passive component, and a range of idealized active sources of voltage and current. As we move forward, new components will be added to the inventory to allow more complex (and useful) circuits to be considered.

A quick word of advice before we begin: Pay close attention to the role of "+" and "−" signs when labeling voltages, and the significance of the arrow in defining current; they often make the difference between wrong and right answers.

2.1 UNITS AND SCALES

In order to state the value of some measurable quantity, we must give both a *number* and a *unit,* such as "3 meters." Fortunately, we all use the same number system. This is not true for units, and a little time must be spent in becoming familiar with a suitable system. We must agree on a standard unit and be assured of its permanence and its general acceptability. The standard unit of length, for example, should not be defined in terms of the distance between two marks on a certain rubber band; this is not permanent, and furthermore everybody else is using another standard.

The most frequently used system of units is the one adopted by the National Bureau of Standards in 1964; it is used by all major professional engineering societies and is the language in which today's textbooks are written. This is the International System of Units (abbreviated *SI* in all languages), adopted by the General

Conference on Weights and Measures in 1960. Modified several times since, the SI is built upon seven basic units: the *meter, kilogram, second, ampere, kelvin, mole,* and *candela* (see Table 2.1). This is a "metric system," some form of which is now in common use in most countries of the world, although it is not yet widely used in the United States. Units for other quantities such as volume, force, energy, etc., are derived from these seven base units.

TABLE 2.1 SI Base Units

Base Quantity	Name	Symbol
length	meter	m
mass	kilogram	kg
time	second	s
electric current	ampere	A
thermodynamic temperature	kelvin	K
amount of substance	mole	mol
luminous intensity	candela	cd

There is some inconsistency regarding whether units named after a person should be capitalized. Here, we will adopt the most contemporary convention,[1,2] where such units are written out in lowercase (e.g., watt, joule), but abbreviated with an uppercase symbol (e.g., W, J).

(1) H. Barrell, *Nature* **220**, 1968, p. 651.
(2) V. N. Krutikov, T. K. Kanishcheva, S. A. Kononogov, L. K. Isaev, and N. I. Khanov, *Measurement Techniques* **51**, 2008, p. 1045.

The "calorie" used with food, drink, and exercise is really a kilocalorie, 4.187 J.

The fundamental unit of work or energy is the *joule* (J). One joule (a kg m^2 s^{-2} in SI base units) is equivalent to 0.7376 foot pound-force (ft · lbf). Other energy units include the calorie (cal), equal to 4.187 J; the British thermal unit (Btu), which is 1055 J; and the kilowatthour (kWh), equal to 3.6×10^6 J. Power is defined as the *rate* at which work is done or energy is expended. The fundamental unit of power is the *watt* (W), defined as 1 J/s. One watt is equivalent to 0.7376 ft · lbf/s or, equivalently, 1/745.7 horsepower (hp).

The SI uses the decimal system to relate larger and smaller units to the basic unit, and employs prefixes to signify the various powers of 10. A list of prefixes and their symbols is given in Table 2.2; the ones most commonly encountered in engineering are highlighted.

TABLE 2.2 SI Prefixes

Factor	Name	Symbol	Factor	Name	Symbol
10^{-24}	yocto	y	10^{24}	yotta	Y
10^{-21}	zepto	z	10^{21}	zetta	Z
10^{-18}	atto	a	10^{18}	exa	E
10^{-15}	femto	f	10^{15}	peta	P
10^{-12}	pico	p	10^{12}	tera	T
10^{-9}	nano	n	10^{9}	giga	G
10^{-6}	micro	μ	10^{6}	mega	M
10^{-3}	milli	m	10^{3}	kilo	k
10^{-2}	centi	c	10^{2}	hecto	h
10^{-1}	deci	d	10^{1}	deka	da

These prefixes are worth memorizing, for they will appear often both in this text and in other technical work. Combinations of several prefixes, such as the millimicrosecond, are unacceptable. It is worth noting that in terms of distance, it is common to see "micron (μm)" as opposed to "microme-ter," and often the angstrom (Å) is used for 10^{-10} meter. Also, in circuit analysis and engineering in general, it is fairly common to see numbers expressed in what are frequently termed "engineering units." In engineering notation, a quantity is represented by a number between 1 and 999 and an appropriate metric unit using a power divisible by 3. So, for example, it is preferable to express the quantity 0.048 W as 48 mW, instead of 4.8 cW, 4.8×10^{-2} W, or 48,000 μW.

PRACTICE

2.1 A krypton fluoride laser emits light at a wavelength of 248 nm. This is the same as: (*a*) 0.0248 mm; (*b*) 2.48 μm; (*c*) 0.248 μm; (*d*) 24,800 Å.

2.2 A single logic gate in a prototype integrated circuit is found to be capable of switching from the "on" state to the "off" state in 12 ps. This corresponds to: (*a*) 1.2 ns; (*b*) 120 ns; (*c*) 1200 ns; (*d*) 12,000 ns.

2.3 A typical incandescent reading lamp runs at 60 W. If it is left on constantly, how much energy (J) is consumed per day, and what is the weekly cost if energy is charged at a rate of 12.5 cents per kilowatthour?

Ans: 2.1 (c); 2.2 (d); 2.3 5.18 MJ, $1.26.

2.2 CHARGE, CURRENT, VOLTAGE, AND POWER

Charge

One of the most fundamental concepts in electric circuit analysis is that of charge conservation. We know from basic physics that there are two types of charge: positive (corresponding to a proton) and negative (corresponding to an electron). For the most part, this text is concerned with circuits in which only electron flow is relevant. There are many devices (such as batteries, diodes, and transistors) in which positive charge motion is important to understanding internal operation, but external to the device we typically concentrate on the electrons which flow through the connecting wires. Although we continuously transfer charges between different parts of a circuit, we do nothing to change the total amount of charge. In other words, we neither create nor destroy electrons (or protons) when running electric circuits.[1] Charge in motion represents a *current*.

In the SI system, the fundamental unit of charge is the **coulomb** (C). It is defined in terms of the **ampere** by counting the total charge that passes through an arbitrary cross section of a wire during an interval of one second; one coulomb is measured each second for a wire carrying a current of 1 ampere (Fig. 2.1). In this system of units, a single electron has a charge of -1.602×10^{-19} C and a single proton has a charge of $+1.602 \times 10^{-19}$ C.

(1) Although the occasional appearance of smoke may seem to suggest otherwise. . .

As seen in Table 2.1, the base units of the SI are not derived from fundamental physical quantities. Instead, they represent historically agreed upon measurements, leading to definitions which occasionally seem backward. For example, it would make more sense physically to define the ampere based on electronic charge.

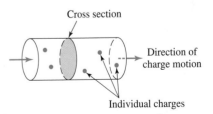

■ **FIGURE 2.1** The definition of current illustrated using current flowing through a wire; 1 ampere corresponds to 1 coulomb of charge passing through the arbitrarily chosen cross section in 1 second.

A quantity of charge that does not change with time is typically represented by Q. The instantaneous amount of charge (which may or may not be time-invariant) is commonly represented by $q(t)$, or simply q. This convention is used throughout the remainder of the text: capital letters are reserved for constant (time-invariant) quantities, whereas lowercase letters represent the more general case. Thus, a constant charge may be represented by *either* Q or q, but an amount of charge that changes over time *must* be represented by the lowercase letter q.

Current

The idea of "transfer of charge" or "charge in motion" is of vital importance to us in studying electric circuits because, in moving a charge from place to place, we may also transfer energy from one point to another. The familiar cross-country power-transmission line is a practical example of a device that transfers energy. Of equal importance is the possibility of varying the rate at which the charge is transferred in order to communicate or transfer information. This process is the basis of communication systems such as radio, television, and telemetry.

The current present in a discrete path, such as a metallic wire, has both a *numerical value* and a *direction* associated with it; it is a measure of the rate at which charge is moving past a given reference point in a specified direction.

Once we have specified a reference direction, we may then let $q(t)$ be the total charge that has passed the reference point since an arbitrary time $t = 0$, moving in the defined direction. A contribution to this total charge will be negative if negative charge is moving in the reference direction, or if positive charge is moving in the opposite direction. As an example, Fig. 2.2 shows a history of the total charge $q(t)$ that has passed a given reference point in a wire (such as the one shown in Fig. 2.1).

We define the current at a specific point and flowing in a specified direction as the instantaneous rate at which net positive charge is moving past that point in the specified direction. This, unfortunately, is the historical definition, which came into popular use before it was appreciated that current in wires is actually due to negative, not positive, charge motion. Current is symbolized by I or i, and so

$$i = \frac{dq}{dt} \qquad [1]$$

The unit of current is the ampere (A), named after A. M. Ampère, a French physicist. It is commonly abbreviated as an "amp," although this is unofficial and somewhat informal. One ampere equals 1 coulomb per second.

Using Eq. [1], we compute the instantaneous current and obtain Fig. 2.3. The use of the lowercase letter i is again to be associated with an instantaneous value; an uppercase I would denote a constant (i.e., time-invariant) quantity.

The charge transferred between time t_0 and t may be expressed as a definite integral:

$$\int_{q(t_0)}^{q(t)} dq = \int_{t_0}^{t} i \, dt'$$

The total charge transferred over all time is thus given by

$$q(t) = \int_{t_0}^{t} i \, dt' + q(t_0) \qquad [2]$$

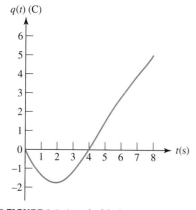

■ **FIGURE 2.2** A graph of the instantaneous value of the total charge $q(t)$ that has passed a given reference point since $t = 0$.

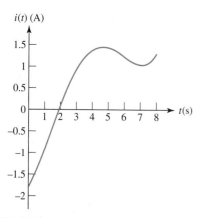

■ **FIGURE 2.3** The instantaneous current $i = dq/dt$, where q is given in Fig. 2.2.

Several different types of current are illustrated in Fig. 2.4. A current that is constant in time is termed a direct current, or simply dc, and is shown by Fig. 2.4a. We will find many practical examples of currents that vary sinusoidally with time (Fig. 2.4b); currents of this form are present in normal household circuits. Such a current is often referred to as alternating current, or ac. Exponential currents and damped sinusoidal currents (Fig. 2.4c and d) will also be encountered later.

We create a graphical symbol for current by placing an arrow next to the conductor. Thus, in Fig. 2.5a the direction of the arrow and the value 3 A indicate either that a net positive charge of 3 C/s is moving to the right or that a net negative charge of −3 C/s is moving to the left each second. In Fig. 2.5b there are again two possibilities: either −3 A is flowing to the left or +3 A is flowing to the right. All four statements and both figures represent currents that are equivalent in their electrical effects, and we say that they are equal. A nonelectrical analogy that may be easier to visualize is to think in terms of a personal savings account: e.g., a deposit can be viewed as either a *negative* cash flow *out of* your account or a *positive* flow *into* your account.

It is convenient to think of current as the motion of positive charge, even though it is known that current flow in metallic conductors results from electron motion. In ionized gases, in electrolytic solutions, and in some semiconductor materials, however, positive charges in motion constitute part or all of the current. Thus, any definition of current can agree with the physical nature of conduction only part of the time. The definition and symbolism we have adopted are standard.

It is essential that we realize that the current arrow does not indicate the "actual" direction of current flow but is simply part of a convention that allows us to talk about "the current in the wire" in an unambiguous manner. The arrow is a fundamental part of the definition of a current! Thus, to talk about the value of a current $i_1(t)$ without specifying the arrow is to discuss an undefined entity. For example, Fig. 2.6a and b are meaningless representations of $i_1(t)$, whereas Fig. 2.6c is complete.

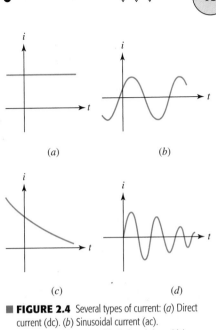

FIGURE 2.4 Several types of current: (a) Direct current (dc). (b) Sinusoidal current (ac). (c) Exponential current. (d) Damped sinusoidal current.

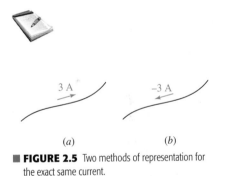

(a)   (b)

FIGURE 2.5 Two methods of representation for the exact same current.

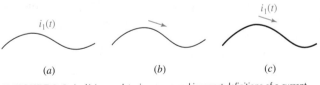

(a)   (b)   (c)

FIGURE 2.6 (a, b) Incomplete, improper, and incorrect definitions of a current. (c) The correct definition of $i_1(t)$.

PRACTICE

2.4 In the wire of Fig. 2.7, electrons are moving *left* to *right* to create a current of 1 mA. Determine I_1 and I_2.

FIGURE 2.7

Ans: $I_1 = -1$ mA; $I_2 = +1$ mA.

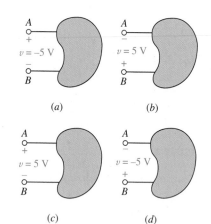

■ **FIGURE 2.8** A general two-terminal circuit element.

■ **FIGURE 2.9** (*a, b*) Terminal *B* is 5 V positive with respect to terminal *A*; (*c, d*) terminal *A* is 5 V positive with respect to terminal *B*.

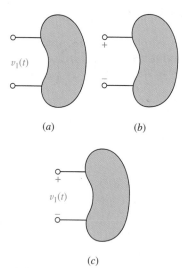

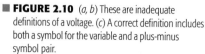

■ **FIGURE 2.10** (*a, b*) These are inadequate definitions of a voltage. (*c*) A correct definition includes both a symbol for the variable and a plus-minus symbol pair.

Voltage

We must now begin to refer to a circuit element, something best defined in general terms to begin with. Such electrical devices as fuses, light bulbs, resistors, batteries, capacitors, generators, and spark coils can be represented by combinations of simple circuit elements. We begin by showing a very general circuit element as a shapeless object possessing two terminals at which connections to other elements may be made (Fig. 2.8).

There are two paths by which current may enter or leave the element. In subsequent discussions we will define particular circuit elements by describing the electrical characteristics that may be observed at their terminals.

In Fig. 2.8, let us suppose that a dc current is sent into terminal *A*, through the general element, and back out of terminal *B*. Let us also assume that pushing charge through the element requires an expenditure of energy. We then say that an electrical voltage (or a *potential difference*) exists between the two terminals, or that there is a voltage "across" the element. Thus, the voltage across a terminal pair is a measure of the work required to move charge through the element. The unit of voltage is the volt,[2] and 1 volt is the same as 1 J/C. Voltage is represented by V or v.

A voltage can exist between a pair of electrical terminals whether a current is flowing or not. An automobile battery, for example, has a voltage of 12 V across its terminals even if nothing whatsoever is connected to the terminals.

According to the principle of conservation of energy, the energy that is expended in forcing charge through the element must appear somewhere else. When we later meet specific circuit elements, we will note whether that energy is stored in some form that is readily available as electric energy or whether it changes irreversibly into heat, acoustic energy, or some other nonelectrical form.

We must now establish a convention by which we can distinguish between energy supplied *to* an element and energy that is supplied *by* the element itself. We do this by our choice of sign for the voltage of terminal *A* with respect to terminal *B*. If a positive current is entering terminal *A* of the element and an external source must expend energy to establish this current, then terminal *A* is positive with respect to terminal *B*. (Alternatively, we may say that terminal *B* is negative with respect to terminal *A*.)

The sense of the voltage is indicated by a plus-minus pair of algebraic signs. In Fig. 2.9*a*, for example, the placement of the + sign at terminal *A* indicates that terminal *A* is v volts positive with respect to terminal *B*. If we later find that v happens to have a numerical value of −5 V, then we may say either that *A* is −5 V positive with respect to *B* or that *B* is 5 V positive with respect to *A*. Other cases are shown in Fig. 2.9*b, c,* and *d*.

Just as we noted in our definition of current, it is essential to realize that the plus-minus pair of algebraic signs does not indicate the "actual" polarity of the voltage but is simply part of a convention that enables us to talk unambiguously about "the voltage across the terminal pair." *The definition of any voltage must include a plus-minus sign pair!* Using a quantity $v_1(t)$ without specifying the location of the plus-minus sign pair is using an undefined term. Figure 2.10*a* and *b* do *not* serve as definitions of $v_1(t)$; Fig. 2.10*c* does.

---

(2) We are probably fortunate that the full name of the 18th century Italian physicist, *Alessandro Giuseppe Antonio Anastasio Volta,* is not used for our unit of potential difference!

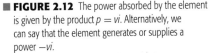

PRACTICE

2.5 For the element in Fig. 2.11, $v_1 = 17$ V. Determine v_2.

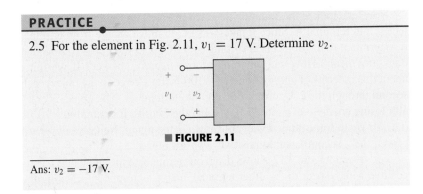

■ **FIGURE 2.11**

Ans: $v_2 = -17$ V.

Power

We have already defined power, and we will represent it by P or p. If one joule of energy is expended in transferring one coulomb of charge through the device in one second, then the rate of energy transfer is one watt. The absorbed power must be proportional both to the number of coulombs transferred per second (current) and to the energy needed to transfer one coulomb through the element (voltage). Thus,

$$p = vi \qquad [3]$$

Dimensionally, the right side of this equation is the product of joules per coulomb and coulombs per second, which produces the expected dimension of joules per second, or watts. The conventions for current, voltage, and power are shown in Fig. 2.12.

We now have an expression for the power being absorbed by a circuit element in terms of a voltage across it and current through it. Voltage was defined in terms of an energy expenditure, and power is the rate at which energy is expended. However, no statement can be made concerning energy transfer in any of the four cases shown in Fig. 2.9, for example, until the direction of the current is specified. Let us imagine that a current arrow is placed alongside each upper lead, directed to the right, and labeled "+2 A." First, consider the case shown in Fig. 2.9c. Terminal A is 5 V positive with respect to terminal B, which means that 5 J of energy is required to move each coulomb of positive charge into terminal A, through the object, and out terminal B. Since we are injecting $+2$ A (a current of 2 coulombs of positive charge per second) into terminal A, we are doing (5 J/C) $\times$ (2 C/s) $= 10$ J of work per second on the object. In other words, the object is absorbing 10 W of power from whatever is injecting the current.

We know from an earlier discussion that there is no difference between Fig. 2.9c and Fig. 2.9d, so we expect the object depicted in Fig. 2.9d to also be absorbing 10 W. We can check this easily enough: we are injecting $+2$ A into terminal A of the object, so $+2$ A flows out of terminal B. Another way of saying this is that we are injecting -2 A of current into terminal B. It takes -5 J/C to move charge from terminal B to terminal A, so the object is absorbing $(-5$ J/C) $\times (-2$ C/s) $= +10$ W as expected. The only difficulty in describing this particular case is keeping the minus signs straight, but with a bit of care we see the correct answer can be obtained regardless of our choice of positive reference terminal (terminal A in Fig. 2.9c, and terminal B in Fig. 2.9d).

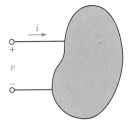

■ **FIGURE 2.12** The power absorbed by the element is given by the product $p = vi$. Alternatively, we can say that the element generates or supplies a power $-vi$.

Now let's look at the situation depicted in Fig. 2.9*a*, again with +2 A injected into terminal *A*. Since it takes −5 J/C to move charge from terminal *A* to terminal *B*, the object is absorbing (−5 J/C) × (2 C/s) = −10 W. What does this mean? How can anything absorb *negative* power? If we think about this in terms of energy transfer, −10 J is transferred to the object each second through the 2 A current flowing into terminal *A*. The object is actually losing energy—at a rate of 10 J/s. In other words, it is supplying 10 J/s (i.e., 10 W) to some other object not shown in the figure. Negative *absorbed* power, then, is equivalent to positive *supplied* power.

Let's recap. Figure 2.12 shows that if one terminal of the element is *v* volts positive with respect to the other terminal, and if a current *i* is entering the element through that terminal, then a power $p = vi$ is being *absorbed* by the element; it is also correct to say that a power $p = vi$ is being *delivered* to the element. When the current arrow is directed into the element at the plus-marked terminal, we satisfy the ***passive sign convention.*** This convention should be studied carefully, understood, and memorized. In other words, it says that if the current arrow and the voltage polarity signs are placed such that the current enters that end of the element marked with the positive sign, then the power *absorbed* by the element can be expressed by the product of the specified current and voltage variables. If the numerical value of the product is negative, then we say that the element is absorbing negative power, or that it is actually generating power and delivering it to some external element. For example, in Fig. 2.12 with $v = 5$ V and $i = -4$ A, the element may be described as either absorbing −20 W or generating 20 W.

Conventions are only required when there is more than one way to do something, and confusion may result when two different groups try to communicate. For example, it is rather arbitrary to always place "North" at the top of a map; compass needles don't point "up," anyway. Still, if we were talking to people who had secretly chosen the opposite convention of placing "South" at the top of their maps, imagine the confusion that could result! In the same fashion, there is a general convention that always draws the current arrows pointing into the positive voltage terminal, regardless of whether the element supplies or absorbs power. This convention is not incorrect but sometimes results in counterintuitive currents labeled on circuit schematics. The reason for this is that it simply seems more natural to refer to positive current flowing out of a voltage or current source that is supplying positive power to one or more circuit elements.

> If the current arrow is directed into the "+" marked terminal of an element, then $p = vi$ yields the *absorbed* power. A negative value indicates that power is actually being generated by the element.

> If the current arrow is directed out of the "+" terminal of an element, then $p = vi$ yields the *supplied* power. A negative value in this case indicates that power is being absorbed.

EXAMPLE 2.1

Compute the power absorbed by each part in Fig. 2.13.

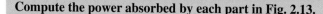
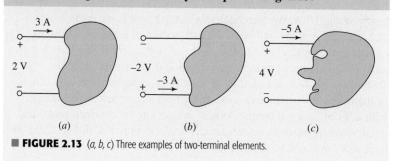

(*a*)   (*b*)   (*c*)

■ FIGURE 2.13 (*a, b, c*) Three examples of two-terminal elements.

In Fig. 2.13a, we see that the reference current is defined consistent with the passive sign convention, which assumes that the element is absorbing power. With +3 A flowing into the positive reference terminal, we compute

$$P = (2 \text{ V})(3 \text{ A}) = 6 \text{ W}$$

of power absorbed by the element.

Figure 2.13b shows a slightly different picture. Now, we have a current of −3 A flowing into the positive reference terminal. This gives us an absorbed power

$$P = (-2 \text{ V})(-3 \text{ A}) = 6 \text{ W}$$

Thus, we see that the two cases are actually equivalent: A current of +3 A flowing into the top terminal is the same as a current of +3 A flowing out of the bottom terminal, or, equivalently, a current of −3 A flowing into the bottom terminal.

Referring to Fig. 2.13c, we again apply the passive sign convention rules and compute an absorbed power

$$P = (4 \text{ V})(-5 \text{ A}) = -20 \text{ W}$$

Since we computed a negative *absorbed* power, this tells us that the element in Fig. 2.13c is actually *supplying* +20 W (i.e., it's a source of energy).

PRACTICE

2.6 Determine the power being absorbed by the circuit element in Fig. 2.14a.

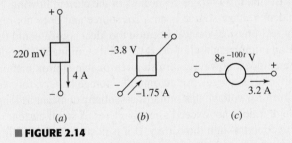

(a)   *(b)*   *(c)*

■ **FIGURE 2.14**

2.7 Determine the power being generated by the circuit element in Fig. 2.14b.

2.8 Determine the power being delivered to the circuit element in Fig. 2.14c at $t = 5$ ms.

Ans: 880 mW; 6.65 W; −15.53 W.

2.3 VOLTAGE AND CURRENT SOURCES

Using the concepts of current and voltage, it is now possible to be more specific in defining a *circuit element.*

In so doing, it is important to differentiate between the physical device itself and the mathematical model which we will use to analyze its behavior in a circuit. The model is only an approximation.

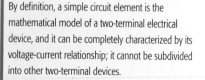

Let us agree that we will use the expression *circuit element* to refer to the mathematical model. The choice of a particular model for any real device must be made on the basis of experimental data or experience; we will usually assume that this choice has already been made. For simplicity, we initially consider circuits with idealized components represented by simple models.

All the simple circuit elements that we will consider can be classified according to the relationship of the current through the element to the voltage across the element. For example, if the voltage across the element is linearly proportional to the current through it, we will call the element a resistor. Other types of simple circuit elements have terminal voltages which are proportional to the *derivative* of the current with respect to time (an inductor), or to the *integral* of the current with respect to time (a capacitor). There are also elements in which the voltage is completely independent of the current, or the current is completely independent of the voltage; these are termed *independent sources*. Furthermore, we will need to define special kinds of sources for which either the source voltage or current depends upon a current or voltage elsewhere in the circuit; such sources are referred to as *dependent sources*. Dependent sources are used a great deal in electronics to model both dc and ac behavior of transistors, especially in amplifier circuits.

Independent Voltage Sources

The first element we will consider is the ***independent voltage source.*** The circuit symbol is shown in Fig. 2.15*a*; the subscript *s* merely identifies the voltage as a "source" voltage, and is common but not required. *An independent voltage source is characterized by a terminal voltage which is completely independent of the current through it.* Thus, if we are given an independent voltage source and are notified that the terminal voltage is 12 V, then we always assume this voltage, regardless of the current flowing.

The independent voltage source is an *ideal* source and does not represent exactly any real physical device, because the ideal source could theoretically deliver an infinite amount of energy from its terminals. This idealized voltage source does, however, furnish a reasonable approximation to several practical voltage sources. An automobile storage battery, for example, has a 12 V terminal voltage that remains essentially constant as long as the current through it does not exceed a few amperes. A small current may flow in either direction through the battery. If it is positive and flowing out of the positively marked terminal, then the battery is furnishing power to the headlights, for example; if the current is positive and flowing into the positive terminal, then the battery is charging by absorbing energy from the alternator.[3] An ordinary household electrical outlet also approximates an independent voltage source, providing a voltage $v_s = 115\sqrt{2} \cos 2\pi 60t$ V; this representation is valid for currents less than 20 A or so.

A point worth repeating here is that the presence of the plus sign at the upper end of the symbol for the independent voltage source in Fig. 2.15*a* does not necessarily mean that the upper terminal is numerically positive with respect to the lower terminal. Instead, it means that the upper terminal is v_s volts positive with respect to the lower. If at some instant v_s happens to be negative, then the upper terminal is actually negative with respect to the lower at that instant.

By definition, a simple circuit element is the mathematical model of a two-terminal electrical device, and it can be completely characterized by its voltage-current relationship; it cannot be subdivided into other two-terminal devices.

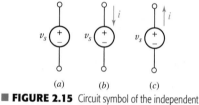

■ FIGURE 2.15 Circuit symbol of the independent voltage source.

If you've ever noticed the room lights dim when an air conditioner kicks on, it's because the sudden large current demand temporarily led to a voltage drop. After the motor starts moving, it takes less current to keep it in motion. At that point, the current demand is reduced, the voltage returns to its original value, and the wall outlet again provides a reasonable approximation of an ideal voltage source.

(3) Or the battery of a friend's car, if you accidentally left your headlights on. . .

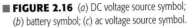

Consider a current arrow labeled "*i*" placed adjacent to the upper conductor of the source as in Fig. 2.15*b*. The current *i* is entering the terminal at which the positive sign is located, the passive sign convention is satisfied, and the source thus *absorbs* power $p = v_s i$. More often than not, a source is expected to deliver power to a network and not to absorb it. Consequently, we might choose to direct the arrow as in Fig. 2.15*c* so that $v_s i$ will represent the power *delivered* by the source. Technically, either arrow direction may be chosen; whenever possible, we will adopt the convention of Fig. 2.15*c* in this text for voltage and current sources, which are not usually considered passive devices.

An independent voltage source with a constant terminal voltage is often termed an independent dc voltage source and can be represented by either of the symbols shown in Fig. 2.16*a* and *b*. Note in Fig. 2.16*b* that when the physical plate structure of the battery is suggested, the longer plate is placed at the positive terminal; the plus and minus signs then represent redundant notation, but they are usually included anyway. For the sake of completeness, the symbol for an independent ac voltage source is shown in Fig. 2.16*c*.

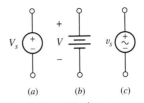

FIGURE 2.16 (*a*) DC voltage source symbol; (*b*) battery symbol; (*c*) ac voltage source symbol.

Terms like dc voltage source and dc current source are commonly used. Literally, they mean "direct-current voltage source" and "direct-current current source," respectively. Although these terms may seem a little odd or even redundant, the terminology is so widely used there's no point in fighting it.

Independent Current Sources

Another ideal source which we will need is the ***independent current source.*** Here, the current through the element is completely independent of the voltage across it. The symbol for an independent current source is shown in Fig. 2.17. If i_s is constant, we call the source an independent dc current source. An ac current source is often drawn with a tilde through the arrow, similar to the ac voltage source shown in Fig. 2.16*c*.

Like the independent voltage source, the independent current source is at best a reasonable approximation for a physical element. In theory it can deliver infinite power from its terminals because it produces the same finite current for any voltage across it, no matter how large that voltage may be. It is, however, a good approximation for many practical sources, particularly in electronic circuits.

Although most students seem happy enough with an independent voltage source providing a fixed voltage but essentially any current, *it is a common mistake* to view an independent current source as having zero voltage across its terminals while providing a fixed current. In fact, we do not know a priori what the voltage across a current source will be—it depends entirely on the circuit to which it is connected.

FIGURE 2.17 Circuit symbol for the independent current source.

Dependent Sources

The two types of ideal sources that we have discussed up to now are called *independent* sources because the value of the source quantity is not affected in any way by activities in the remainder of the circuit. This is in contrast with yet another kind of ideal source, the *dependent,* or *controlled,* source, in which the source quantity is determined by a voltage or current existing at some other location in the system being analyzed. Sources such as these appear in the equivalent electrical models for many electronic devices, such as transistors, operational amplifiers, and integrated circuits. To distinguish between dependent and independent sources, we introduce the diamond symbols shown in Fig. 2.18. In Fig. 2.18*a* and *c*, *K* is a dimensionless scaling constant. In Fig.2.18*b*, *g* is a scaling factor with units of A/V; in Fig. 2.18*d*, *r* is a scaling factor with units of V/A. The controlling current i_x and the controlling voltage v_x must be defined in the circuit.

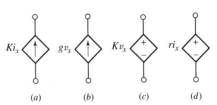

FIGURE 2.18 The four different types of dependent sources: (*a*) current-controlled current source; (*b*) voltage-controlled current source; (*c*) voltage-controlled voltage source; (*d*) current-controlled voltage source.

It does seem odd at first to have a current source whose value depends on a voltage, or a voltage source which is controlled by a current flowing through some other element. Even a voltage source depending on a remote voltage can appear strange. Such sources are invaluable for modeling complex systems, however, making the analysis algebraically straightforward. Examples include the drain current of a field effect transistor as a function of the gate voltage, or the output voltage of an analog integrated circuit as a function of differential input voltage. When encountered during circuit analysis, we write down the entire controlling expression for the dependent source just as we would if it was a numerical value attached to an independent source. This often results in the need for an additional equation to complete the analysis, unless the controlling voltage or current is already one of the specified unknowns in our system of equations.

EXAMPLE 2.2

In the circuit of Fig. 2.19a, if v_2 is known to be 3 V, find v_L.

We have been provided with a partially labeled circuit diagram and the additional information that $v_2 = 3$ V. This is probably worth adding to our diagram, as shown in Fig. 2.19b.

Next we step back and look at the information collected. In examining the circuit diagram, we notice that the desired voltage v_L is the same as the voltage across the dependent source. Thus,

$$v_L = 5v_2$$

At this point, we would be done with the problem if only we knew v_2! Returning to our diagram, we see that we actually do know v_2—it was specified as 3 V. We therefore write

$$v_2 = 3$$

We now have two (simple) equations in two unknowns, and solve to find $v_L = 15$ V.

An important lesson at this early stage of the game is that *the time it takes to completely label a circuit diagram is always a good investment.* As a final step, we should go back and check over our work to ensure that the result is correct.

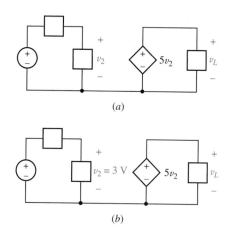

(a)

(b)

■ **FIGURE 2.19** (a) An example circuit containing a voltage-controlled voltage source. (b) The additional information provided is included on the diagram.

PRACTICE

2.9 Find the power *absorbed* by each element in the circuit in Fig. 2.20.

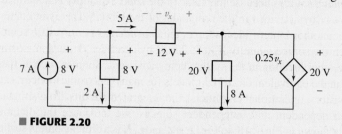

■ **FIGURE 2.20**

Ans: (left to right) −56 W; 16 W; −60 W; 160 W; −60 W.

Dependent and independent voltage and current sources are *active* elements; they are capable of delivering power to some external device. For the present we will think of a *passive* element as one which is capable only of receiving power. However, we will later see that several passive elements are able to store finite amounts of energy and then return that energy later to various external devices; since we still wish to call such elements passive, it will be necessary to improve upon our two definitions a little later.

Networks and Circuits

The interconnection of two or more simple circuit elements forms an electrical **network.** If the network contains at least one closed path, it is also an electric **circuit.** Note: Every circuit is a network, but not all networks are circuits (see Fig. 2.21)!

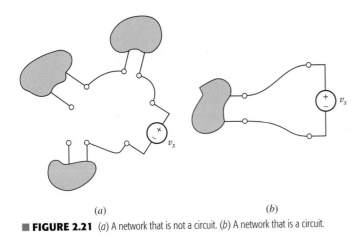

(a)   (b)

■ **FIGURE 2.21** (*a*) A network that is not a circuit. (*b*) A network that is a circuit.

A network that contains at least one active element, such as an independent voltage or current source, is an active network. A network that does not contain any active elements is a passive network.

We have now defined what we mean by the term **circuit element,** and we have presented the definitions of several specific circuit elements, the independent and dependent voltage and current sources. Throughout the remainder of the book we will define only five additional circuit elements: the resistor, inductor, capacitor, transformer, and the ideal operational amplifier ("op amp," for short). These are all ideal elements. They are important because we may combine them into networks and circuits that represent real devices as accurately as we require. Thus, the transistor shown in Fig. 2.22*a* and *b* may be modeled by the voltage terminals designated v_{gs} and the single dependent current source of Fig. 2.22*c*. Note that the dependent current source produces a current that depends on a voltage elsewhere in the circuit. The parameter g_m, commonly referred to as the transconductance, is calculated using transistor-specific details as well as the operating point determined by the circuit connected to the transistor. It is generally a small number, on the order of 10^{-2} to perhaps 10 A/V. This model works pretty well as long as the frequency of any sinusoidal source is neither very large nor very small; the model can be modified to account for frequency-dependent

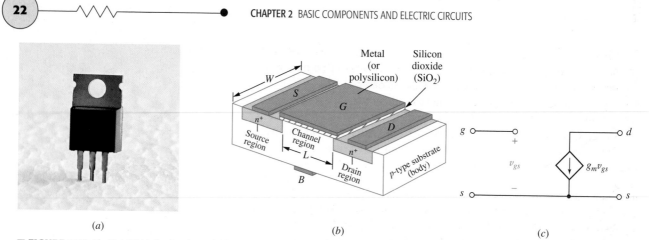

■ **FIGURE 2.22** The Metal Oxide Semiconductor Field Effect Transistor (MOSFET). (*a*) An IRF540 N-channel power MOSFET in a TO-220 package, rated at 100 V and 22 A; (*b*) cross-sectional view of a basic MOSFET (*R. Jaeger, Microelectronic Circuit Design, McGraw-Hill, 1997*); (*c*) equivalent circuit model for use in ac circuit analysis.

effects by including additional ideal circuit elements such as resistors and capacitors.

Similar (but much smaller) transistors typically constitute only one small part of an integrated circuit that may be less than 2 mm × 2 mm square and 200 μm thick and yet contains several thousand transistors plus various resistors and capacitors. Thus, we may have a physical device that is about the size of one letter on this page but requires a model composed of ten thousand ideal simple circuit elements. We use this concept of "circuit modeling" in a number of electrical engineering topics covered in other courses, including electronics, energy conversion, and antennas.

2.4 OHM'S LAW

So far, we have been introduced to both dependent and independent voltage and current sources and were cautioned that they were *idealized* active elements that could only be approximated in a real circuit. We are now ready to meet another idealized element, the linear resistor. The resistor is the simplest passive element, and we begin our discussion by considering the work of an obscure German physicist, Georg Simon Ohm, who published a pamphlet in 1827 that described the results of one of the first efforts to measure currents and voltages, and to describe and relate them mathematically. One result was a statement of the fundamental relationship we now call ***Ohm's law,*** even though it has since been shown that this result was discovered 46 years earlier in England by Henry Cavendish, a brilliant semirecluse.

Ohm's law states that the voltage across conducting materials is directly proportional to the current flowing through the material, or

$$v = Ri \qquad [4]$$

where the constant of proportionality R is called the ***resistance.*** The unit of resistance is the *ohm,* which is 1 V/A and customarily abbreviated by a capital omega, Ω.

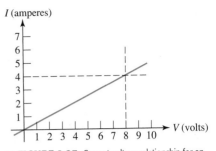

When this equation is plotted on *i*-versus-*v* axes, the graph is a straight line passing through the origin (Fig. 2.23). Equation [4] is a linear equation, and we will consider it as the definition of a *linear resistor*. Resistance is normally considered to be a positive quantity, although negative resistances may be simulated with special circuitry.

Again, it must be emphasized that the linear resistor is an idealized circuit element; it is only a mathematical model of a real, physical device. "Resistors" may be easily purchased or manufactured, but it is soon found that the voltage-current ratios of these physical devices are reasonably constant only within certain ranges of current, voltage, or power, and depend also on temperature and other environmental factors. We usually refer to a linear resistor as simply a resistor; any resistor that is nonlinear will always be described as such. Nonlinear resistors should not necessarily be considered undesirable elements. Although it is true that their presence complicates an analysis, the performance of the device may depend on or be greatly improved by the nonlinearity. For example, fuses for overcurrent protection and Zener diodes for voltage regulation are very nonlinear in nature, a fact that is exploited when using them in circuit design.

■ FIGURE 2.23 Current-voltage relationship for an example 2 Ω linear resistor. Note the slope of the line is 0.5 A/V, or 500 mΩ^{-1}.

Power Absorption

Figure 2.24 shows several different resistor packages, as well as the most common circuit symbol used for a resistor. In accordance with the voltage, current, and power conventions already adopted, the product of *v* and *i* gives the power absorbed by the resistor. That is, *v* and *i* are selected to satisfy the passive sign convention. The absorbed power appears physically

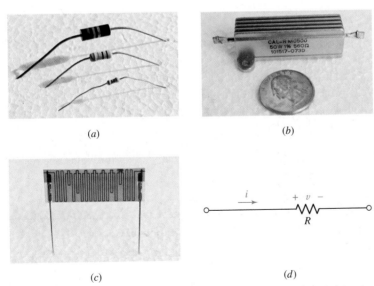

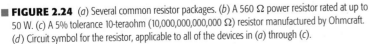

■ FIGURE 2.24 (*a*) Several common resistor packages. (*b*) A 560 Ω power resistor rated at up to 50 W. (*c*) A 5% tolerance 10-teraohm (10,000,000,000,000 Ω) resistor manufactured by Ohmcraft. (*d*) Circuit symbol for the resistor, applicable to all of the devices in (*a*) through (*c*).

as heat and/or light and is always positive; a (positive) resistor is a passive element that cannot deliver power or store energy. Alternative expressions for the absorbed power are

$$p = vi = i^2R = v^2/R \qquad [5]$$

One of the authors (who shall remain anonymous) had the unfortunate experience of inadvertently connecting a 100 Ω, 2 W carbon resistor across a 110 V source. The ensuing flame, smoke, and fragmentation were rather disconcerting, demonstrating clearly that a practical resistor has definite limits to its ability to behave like the ideal linear model. In this case, the unfortunate resistor was called upon to absorb 121 W; since it was designed to handle only 2 W, its reaction was understandably violent.

EXAMPLE 2.3

The 560 Ω resistor shown in Fig. 2.24b is connected to a circuit which causes a current of 42.4 mA to flow through it. Calculate the voltage across the resistor and the power it is dissipating.

The voltage across the resistor is given by Ohm's law:

$$v = Ri = (560)(0.0424) = 23.7 \text{ V}$$

The dissipated power can be calculated in several different ways. For instance,

$$p = vi = (23.7)(0.0424) = 1.005 \text{ W}$$

Alternatively,

$$p = v^2/R = (23.7)^2/560 = 1.003 \text{ W}$$

or

$$p = i^2R = (0.0424)^2(560) = 1.007 \text{ W}$$

We note several things.

First, we calculated the power in three different ways, and we seem to have obtained *three different answers*!

In reality, however, we rounded our voltage to three significant digits, which will impact the accuracy of any subsequent quantity we calculate with it. With this in mind, we see that the answers show reasonable agreement (within 1%).

The other point worth noting is that the resistor is rated to 50 W— since we are only dissipating approximately 2% of this value, the resistor is in no danger of overheating.

PRACTICE

With reference to Fig. 2.25, compute the following:

2.10 R if $i = -2 \mu$A and $v = -44$ V.

2.11 The power absorbed by the resistor if $v = 1$ V and $R = 2$ kΩ.

2.12 The power absorbed by the resistor if $i = 3$ nA and $R = 4.7$ MΩ.

Ans: 22 MΩ; 500 μW; 42.3 pW.

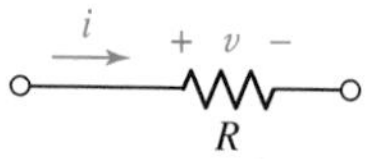

■ **FIGURE 2.25**

PRACTICAL APPLICATION

Wire Gauge

Technically speaking, any material (except for a superconductor) will provide resistance to current flow. As in all introductory circuits texts, however, we tacitly assume that wires appearing in circuit diagrams have zero resistance. This implies that there is no potential difference between the ends of a wire, and hence no power absorbed or heat generated. Although usually not an unreasonable assumption, it does neglect practical considerations when choosing the appropriate wire diameter for a specific application.

Resistance is determined by (1) the inherent resistivity of a material and (2) the device geometry. ***Resistivity,*** represented by the symbol ρ, is a measure of the ease with which electrons can travel through a certain material. Since it is the ratio of the electric field (V/m) to the areal density of current flowing in the material (A/m^2), the general unit of ρ is an $\Omega \cdot$ m, although metric prefixes are often employed. Every material has a different inherent resistivity, which depends on temperature. Some examples are shown in Table 2.3; as can be seen, there is a small variation between different types of copper (less than 1%) but a very large difference between different metals. In particular, although physically stronger than copper, steel wire is several times more resistive. In some technical discussions, it is more common to see the conductivity (symbolized by σ) of a

material quoted, which is simply the reciprocal of the resistivity.

The resistance of a particular object is obtained by multiplying the resistivity by the length ℓ of the resistor and dividing by the cross-sectional area (A) as in Eq. [6]; these parameters are illustrated in Fig. 2.26.

$$R = \rho \frac{\ell}{A} \qquad [6]$$

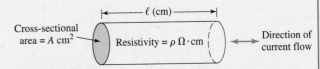

■ **FIGURE 2.26** Definition of geometrical parameters used to compute the resistance of a wire. The resistivity of the material is assumed to be spatially uniform.

We determine the resistivity when we select the material from which to fabricate a wire and measure the temperature of the application environment. Since a finite amount of power is absorbed by the wire due to its resistance, current flow leads to the production of heat. Thicker wires have lower resistance and also dissipate heat more easily but are heavier, take up a larger volume, and are more expensive. Thus, we are motivated by practical considerations to choose the smallest wire that

TABLE 2.3 Common Electrical Wire Materials and Resistivities*

ASTM Specification**	Temper and Shape	Resistivity at 20°C ($\mu\Omega \cdot$ cm)
B33	Copper, tinned soft, round	1.7654
B75	Copper, tube, soft, OF copper	1.7241
B188	Copper, hard bus tube, rectangular or square	1.7521
B189	Copper, lead-coated soft, round	1.7654
B230	Aluminum, hard, round	2.8625
B227	Copper-clad steel, hard, round, grade 40 HS	4.3971
B355	Copper, nickel-coated soft, round Class 10	1.9592
B415	Aluminum-clad steel, hard, round	8.4805

* C. B. Rawlins, "Conductor materials," *Standard Handbook for Electrical Engineering,* 13th ed., D. G. Fink and H. W. Beaty, eds. New York: McGraw-Hill, 1993, pp. 4-4 to 4-8.

**,American Society of Testing and Materials.

(Continued on next page)

can safely do the job, rather than simply choosing the largest-diameter wire available in an effort to minimize resistive losses. The American Wire Gauge (AWG) is a standard system of specifying wire size. In selecting a wire gauge, smaller AWG corresponds to a larger wire diameter; an abbreviated table of common gauges is given in Table 2.4. Local fire and electrical safety codes typically dictate the required gauge for specific wiring applications, based on the maximum current expected as well as where the wires will be located.

TABLE 2.4 Some Common Wire Gauges and the Resistance of (Soft) Solid Copper Wire*

Conductor Size (AWG)	Cross-Sectional Area (mm²)	Ohms per 1000 ft at 20°C
28	0.0804	65.3
24	0.205	25.7
22	0.324	16.2
18	0.823	6.39
14	2.08	2.52
12	3.31	1.59
6	13.3	0.3952
4	21.1	0.2485
2	33.6	0.1563

* C. B. Rawlins et al., *Standard Handbook for Electrical Engineering*, 13th ed., D. G. Fink and H. W. Beaty, eds. New York: McGraw-Hill, 1993, p. 4-47.

EXAMPLE 2.4

A dc power link is to be made between two islands separated by a distance of 24 miles. The operating voltage is 500 kV and the system capacity is 600 MW. Calculate the maximum dc current flow, and estimate the resistivity of the cable, assuming a diameter of 2.5 cm and a solid (not stranded) wire.

Dividing the maximum power (600 MW, or 600×10^6 W) by the operating voltage (500 kV, or 500×10^3 V) yields a maximum current of

$$\frac{600 \times 10^6}{500 \times 10^3} = 1200 \text{ A}$$

The cable resistance is simply the ratio of the voltage to the current, or

$$R_{cable} = \frac{500 \times 10^3}{1200} = 417 \text{ }\Omega$$

We know the length:

$$\ell = (24 \text{ miles}) \left(\frac{5280 \text{ ft}}{1 \text{ mile}} \right) \left(\frac{12 \text{ in}}{1 \text{ ft}} \right) \left(\frac{2.54 \text{ cm}}{1 \text{ in}} \right) = 3{,}862{,}426 \text{ cm}$$

Given that most of our information appears to be valid to only two significant figures, we round this to 3.9×10^6 cm.

With the cable diameter specified as 2.5 cm, we know its cross-sectional area is 4.9 cm^2.

Thus, $\rho_{\text{cable}} = R_{\text{cable}} \dfrac{A}{\ell} = 417 \left(\dfrac{4.9}{3.9 \times 10^6} \right) = 520 \ \mu\Omega \cdot \text{cm}$

PRACTICE

2.13 A 500 ft long 24 AWG soft copper wire is carrying a current of 100 mA. What is the voltage dropped across the wire?

Ans: 3.26 V.

Conductance

For a linear resistor the ratio of current to voltage is also a constant

$$\frac{i}{v} = \frac{1}{R} = G \tag{7}$$

where G is called the *conductance*. The SI unit of conductance is the siemens (S), 1 A/V. An older, unofficial unit for conductance is the mho, which was often abbreviated as ℧ and is still occasionally written as Ω^{-1}. You will occasionally see it used on some circuit diagrams, as well as in catalogs and texts. The same circuit symbol (Fig. 2.24d) is used to represent both resistance and conductance. The absorbed power is again necessarily positive and may be expressed in terms of the conductance by

$$p = vi = v^2 G = \frac{i^2}{G} \tag{8}$$

Thus a 2 Ω resistor has a conductance of $\frac{1}{2}$ S, and if a current of 5 A is flowing through it, then a voltage of 10 V is present across the terminals and a power of 50 W is being absorbed.

All the expressions given so far in this section were written in terms of instantaneous current, voltage, and power, such as $v = iR$ and $p = vi$. We should recall that this is a shorthand notation for $v(t) = Ri(t)$ and $p(t) = v(t)\,i(t)$. The current through and voltage across a resistor must both vary with time in the same manner. Thus, if $R = 10\ \Omega$ and $v = 2 \sin 100t$ V, then $i = 0.2 \sin 100t$ A. Note that the power is given by $0.4 \sin^2 100t$ W, and a simple sketch will illustrate the different nature of its variation with time. Although the current and voltage are each negative during certain time intervals, the absorbed power is *never* negative!

Resistance may be used as the basis for defining two commonly used terms, *short circuit* and *open circuit*. We define a short circuit as a resistance of zero ohms; then, since $v = iR$, the voltage across a short circuit must be zero, although the current may have any value. In an analogous manner, we define an open circuit as an infinite resistance. It follows from Ohm's law that the current must be zero, regardless of the voltage across the open circuit. Although real wires have a small resistance associated with them, we always assume them to have zero resistance unless otherwise specified. Thus, in all of our circuit schematics, wires are taken to be perfect short circuits.

SUMMARY AND REVIEW

In this chapter, we introduced the topic of units – specifically those relevant to electrical circuits—and their relationship to fundamental (SI) units. We also discussed current and current sources, voltage and voltage sources, and the fact that the product of voltage and current yields power (the rate of energy consumption or generation). Since power can be either positive or negative depending on the current direction and voltage polarity, the passive sign convention was described to ensure we always know if an element is *absorbing* or *supplying* energy to the rest of the circuit. Four additional sources were introduced, forming a general class known as dependent sources. They are often used to model complex systems and electrical components, but the actual value of voltage or current supplied is typically unknown until the entire circuit is analyzed. We concluded the chapter with the resistor—by far the most common circuit element—whose voltage and current are linearly related (described by Ohm's law). Whereas the *resistivity* of a material is one of its fundamental properties (measured in $\Omega \cdot cm$), *resistance* describes a device property (measured in Ω) and hence depends not only on resistivity but on the device geometry (i.e., length and area) as well.

We conclude with key points of this chapter to review, along with appropriate examples.

❑ The system of units most commonly used in electrical engineering is the SI.

❑ The direction in which positive charges are moving is the direction of positive current flow; alternatively, positive current flow is in the direction opposite that of moving electrons.

❑ To define a current, both a value and a direction must be given. Currents are typically denoted by the uppercase letter "I" for constant (dc) values, and either $i(t)$ or simply i otherwise.

❑ To define a voltage across an element, it is necessary to label the terminals with "+" and "−" signs as well as to provide a value (either an algebraic symbol or a numerical value).

❑ Any element is said to supply positive power if positive current flows out of the positive voltage terminal. Any element absorbs positive power if positive current flows into the positive voltage terminal. (Example 2.1)

❑ There are six sources: the independent voltage source, the independent current source, the current-controlled dependent current source, the voltage-controlled dependent current source, the voltage-controlled dependent voltage source, and the current-controlled dependent voltage source. (Example 2.2)

Note that a current represented by i or $i(t)$ can be constant (dc) or time-varying, but currents represented by the symbol I must be non-time-varying.

❑ Ohm's law states that the voltage across a linear resistor is directly proportional to the current flowing through it; i.e., $v = Ri$. (Example 2.3)

❑ The power dissipated by a resistor (which leads to the production of heat) is given by $p = vi = i^2 R = v^2/R$. (Example 2.3)

❑ Wires are typically assumed to have zero resistance in circuit analysis. When selecting a wire gauge for a specific application, however, local electrical and fire codes must be consulted. (Example 2.4)

READING FURTHER

A good book that discusses the properties and manufacture of resistors in considerable depth:

Felix Zandman, Paul-René Simon, and Joseph Szwarc, *Resistor Theory and Technology.* Raleigh, N.C.: SciTech Publishing, 2002.

A good all-purpose electrical engineering handbook:

Donald G. Fink and H. Wayne Beaty, *Standard Handbook for Electrical Engineers,* 13th ed., New York: McGraw-Hill, 1993.

In particular, pp. 1-1 to 1-51, 2-8 to 2-10, and 4-2 to 4-207 provide an in-depth treatment of topics related to those discussed in this chapter.

A detailed reference for the SI is available on the Web from the National Institute of Standards:

Ambler Thompson and Barry N. Taylor, *Guide for the Use of the International System of Units (SI),* NIST Special Publication 811, 2008 edition, www.nist.gov.

EXERCISES

2.1 Units and Scales

1. Convert the following to engineering notation:
   - (*a*) 0.045 W
   - (*b*) 2000 pJ
   - (*c*) 0.1 ns
   - (*d*) 39,212 as
   - (*e*) 3 Ω
   - (*f*) 18,000 m
   - (*g*) 2,500,000,000,000 bits
   - (*h*) 10^{15} atoms/cm^3

2. Convert the following to engineering notation:
   - (*a*) 1230 fs
   - (*b*) 0.0001 decimeter
   - (*c*) 1400 mK
   - (*d*) 32 nm
   - (*e*) 13,560 kHz
   - (*f*) 2021 micromoles
   - (*g*) 13 deciliters
   - (*h*) 1 hectometer

3. Express the following in engineering units:
   - (*a*) 1212 mV
   - (*b*) 10^{11} pA
   - (*c*) 1000 yoctoseconds
   - (*d*) 33.9997 zeptoseconds
   - (*e*) 13,100 attoseconds
   - (*f*) 10^{-14} zettasecond
   - (*g*) 10^{-5} second
   - (*h*) 10^{-9} Gs

4. Expand the following distances in simple meters:
   - (*a*) 1 Zm
   - (*b*) 1 Em
   - (*c*) 1 Pm
   - (*d*) 1 Tm
   - (*e*) 1 Gm
   - (*f*) 1 Mm

5. Convert the following to SI units, taking care to employ proper engineering notation:

   (*a*) 212°F   (*b*) 0°F   (*c*) 0 K

   (*d*) 200 hp   (*e*) 1 yard   (*f*) 1 mile

6. Convert the following to SI units, taking care to employ proper engineering notation:

   (*a*) 100°C   (*b*) 0°C   (*c*) 4.2 K

   (*d*) 150 hp   (*e*) 500 Btu   (*f*) 100 J/s

7. A certain krypton fluoride laser generates 15 ns long pulses, each of which contains 550 mJ of energy. (*a*) Calculate the peak instantaneous output power of the laser. (*b*) If up to 100 pulses can be generated per second, calculate the maximum average power output of the laser.

8. When operated at a wavelength of 750 nm, a certain Ti:sapphire laser is capable of producing pulses as short as 50 fs, each with an energy content of 500 μJ. (*a*) Calculate the instantaneous output power of the laser. (*b*) If the laser is capable of a pulse repetition rate of 80 MHz, calculate the maximum average output power that can be achieved.

9. An electric vehicle is driven by a single motor rated at 40 hp. If the motor is run continuously for 3 h at maximum output, calculate the electrical energy consumed. Express your answer in SI units using engineering notation.

10. Under insolation conditions of 500 W/m^2 (direct sunlight), and 10% solar cell efficiency (defined as the ratio of electrical output power to incident solar power), calculate the area required for a photovoltaic (solar cell) array capable of running the vehicle in Exer. 9 at half power.

11. A certain metal oxide nanowire piezoelectricity generator is capable of producing 100 pW of usable electricity from the type of motion obtained from a person jogging at a moderate pace. (*a*) How many nanowire devices are required to operate a personal MP3 player which draws 1 W of power? (*b*) If the nanowires can be produced with a density of 5 devices per square micron directly onto a piece of fabric, what area is required, and would it be practical?

12. A particular electric utility charges customers different rates depending on their daily rate of energy consumption: $0.05/kWh up to 20 kWh, and $0.10/kWh for all energy usage above 20 kWh in any 24 hour period. (*a*) Calculate how many 100 W light bulbs can be run continuously for less than $10 per week. (*b*) Calculate the daily energy cost if 2000 kW of power is used continuously.

13. The Tilting Windmill Electrical Cooperative LLC Inc. has instituted a differential pricing scheme aimed at encouraging customers to conserve electricity use during daylight hours, when local business demand is at its highest. If the price per kilowatthour is $0.033 between the hours of 9 p.m. and 6 a.m., and $0.057 for all other times, how much does it cost to run a 2.5 kW portable heater continuously for 30 days?

14. Assuming a global population of 9 billion people, each using approximately 100 W of power continuously throughout the day, calculate the total land area that would have to be set aside for photovoltaic power generation, assuming 800 W/m^2 of incident solar power and a conversion efficiency (sunlight to electricity) of 10%.

2.2 Charge, Current, Voltage, and Power

15. The total charge flowing out of one end of a small copper wire and into an unknown device is determined to follow the relationship $q(t) = 5e^{-t/2}$ C, where t is expressed in seconds. Calculate the current flowing into the device, taking note of the sign.

16. The current flowing into the collector lead of a certain bipolar junction transistor (BJT) is measured to be 1 nA. If no charge was transferred in or out of the collector lead prior to $t = 0$, and the current flows for 1 min, calculate the total charge which crosses into the collector.

17. The total charge stored on a 1 cm diameter insulating plate is -10^{13} C. (a) How many electrons are on the plate? (b) What is the areal density of electrons (number of electrons per square meter)? (c) If additional electrons are added to the plate from an external source at the rate of 10^6 electrons per second, what is the magnitude of the current flowing between the source and the plate?

18. A mysterious device found in a forgotten laboratory accumulates charge at a rate specified by the expression $q(t) = 9 - 10t$ C from the moment it is switched on. (a) Calculate the total charge contained in the device at $t = 0$. (b) Calculate the total charge contained at $t = 1$ s. (c) Determine the current flowing into the device at $t = 1$ s, 3 s, and 10 s.

19. A new type of device appears to accumulate charge according to the expression $q(t) = 10t^2 - 22t$ mC (t in s). (a) In the interval $0 \le t < 5$ s, at what time does the current flowing into the device equal zero? (b) Sketch $q(t)$ and $i(t)$ over the interval $0 \le t < 5$ s.

20. The current flowing through a tungsten-filament light bulb is determined to follow $i(t) = 114\sin(100\pi t)$ A. (a) Over the interval defined by $t = 0$ and $t = 2$ s, how many times does the current equal zero amperes? (b) How much charge is transported through the light bulb in the first second?

21. The current waveform depicted in Fig. 2.27 is characterized by a period of 8 s. (a) What is the average value of the current over a single period? (b) If $q(0) = 0$, sketch $q(t), 0 < t < 20$ s.

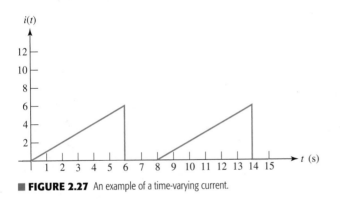

■ FIGURE 2.27 An example of a time-varying current.

22. The current waveform depicted in Fig. 2.28 is characterized by a period of 4 s. (a) What is the average value of the current over a single period? (b) Compute the average current over the interval $1 < t < 3$ s. (c) If $q(0) = 1$ C, sketch $q(t), 0 < t < 4$ s.

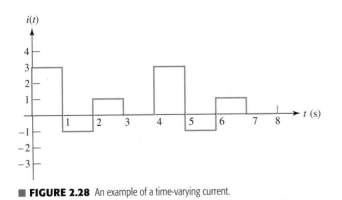

■ FIGURE 2.28 An example of a time-varying current.

23. A path around a certain electric circuit has discrete points labeled A, B, C, and D. To move an electron from points A to C requires 5 pJ. To move an electron from B to C requires 3 pJ. To move an electron from A to D requires 8 pJ. (a) What is the potential difference (in volts) between points B and C, assuming a "+" reference at C? (b) What is the potential difference (in volts) between points B and D, assuming a "+" reference at D? (c) What is the potential difference (in volts) between points A and B (again, in volts), assuming a "+" reference at B?

24. Two metallic terminals protrude from a device. The terminal on the left is the positive reference for a voltage called v_x (the other terminal is the negative reference). The terminal on the right is the positive reference for a voltage called v_y (the other terminal being the negative reference). If it takes 1 mJ of energy to push a single electron into the left terminal, determine the voltages v_x and v_y.

25. The convention for voltmeters is to use a black wire for the negative reference terminal and a red wire for the positive reference terminal. (a) Explain why two wires are required to measure a voltage. (b) If it is dark and the wires into the voltmeter are swapped by accident, what will happen during the next measurement?

26. Determine the power absorbed by each of the elements in Fig. 2.29.

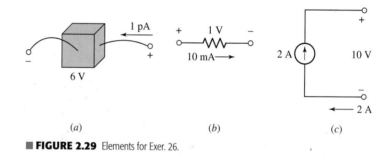

(a)   (b)   (c)

■ **FIGURE 2.29** Elements for Exer. 26.

27. Determine the power absorbed by each of the elements in Fig. 2.30.

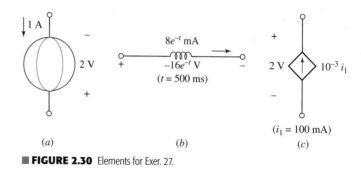

(a)   (b)   (c)

■ **FIGURE 2.30** Elements for Exer. 27.

28. A constant current of 1 ampere is measured flowing into the positive reference terminal of a pair of leads whose voltage we'll call v_p. Calculate the absorbed power at $t = 1$ s if $v_p(t)$ equals (a) $+1$ V; (b) -1 V; (c) $2 + 5\cos(5t)$ V; (d) $4e^{-2t}$ V, (e) Explain the significance of a negative value for absorbed power.

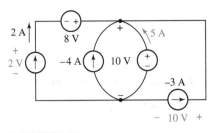

29. Determine the power supplied by the leftmost element in the circuit of Fig. 2.31.

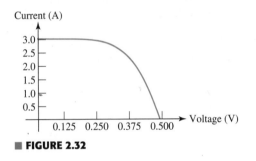

■ **FIGURE 2.31**

30. The current-voltage characteristic of a silicon solar cell exposed to direct sunlight at noon in Florida during midsummer is given in Fig. 2.32. It is obtained by placing different-sized resistors across the two terminals of the device and measuring the resulting currents and voltages.
    (a) What is the value of the short-circuit current? – no voltage
    (b) What is the value of the voltage at open circuit? – no current
    (c) Estimate the maximum power that can be obtained from the device.

Current (A)

3.0
2.5
2.0
1.5
1.0
0.5

0.125   0.250   0.375   0.500   Voltage (V)

■ **FIGURE 2.32**

2.3 Voltage and Current Sources

31. Some of the ideal sources in the circuit of Fig. 2.31 are supplying positive power, and others are absorbing positive power. Determine which are which, and show that the algebraic sum of the power absorbed by each element (taking care to preserve signs) is equal to zero.

32. By careful measurements it is determined that a benchtop argon ion laser is consuming (absorbing) 1.5 kW of electric power from the wall outlet, but only producing 5 W of optical power. Where is the remaining power going? Doesn't conservation of energy require the two quantities to be equal?

33. Refer to the circuit represented in Fig. 2.33, while noting that the same current flows through each element. The voltage-controlled dependent source provides a current which is 5 times as large as the voltage V_x. (a) For $V_R = 10$ V and $V_x = 2$ V, determine the power absorbed by each element. (b) Is element A likely a passive or active source? Explain.

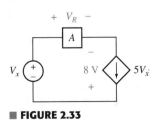

■ **FIGURE 2.33**

34. Refer to the circuit represented in Fig. 2.33, while noting that the same current flows through each element. The voltage-controlled dependent source provides a current which is 5 times as large as the voltage V_x. (*a*) For $V_R = 100$ V and $V_x = 92$ V, determine the power supplied by each element. (*b*) Verify that the algebraic sum of the supplied powers is equal to zero.

35. The circuit depicted in Fig. 2.34 contains a dependent current source; the magnitude and direction of the current it supplies are directly determined by the voltage labeled v_1. Note that therefore $i_2 = -3v_1$. Determine the voltage v_1 if $v_2 = 33i_2$ and $i_2 = 100$ mA. *A current in opp. direction so,*

$$i_2 = -3v_1 \text{ A}$$

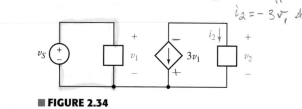

$v_1 = v_S$ so,

$3v_1 = 3v_S$

■ **FIGURE 2.34**

36. To protect an expensive circuit component from being delivered too much power, you decide to incorporate a fast-blowing fuse into the design. Knowing that the circuit component is connected to 12 V, its minimum power consumption is 12 W, and the maximum power it can safely dissipate is 100 W, which of the three available fuse ratings should you select: 1 A, 4 A, or 10 A? Explain your answer.

37. The dependent source in the circuit of Fig. 2.35 provides a voltage whose value depends on the current i_x. What value of i_x is required for the dependent source to be supplying 1 W?

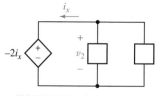

■ **FIGURE 2.35**

2.4 Ohm's Law

38. Determine the magnitude of the current flowing through a 4.7 kΩ resistor if the voltage across it is (*a*) 1 mV; (*b*) 10 V; (*c*) $4e^{-t}$ V; (*d*) 100 cos(5*t*) V; (*e*) −7 V.

39. Real resistors can only be manufactured to a specific tolerance, so that in effect the value of the resistance is uncertain. For example, a 1 Ω resistor specified as 5% tolerance could in practice be found to have a value anywhere in the range of 0.95 to 1.05 Ω. Calculate the voltage across a 2.2 kΩ 10% tolerance resistor if the current flowing through the element is (*a*) 1 mA; (*b*) 4 sin 44*t* mA.

40. (*a*) Sketch the current-voltage relationship (current on the *y*-axis) of a 2 kΩ resistor over the voltage range of −10 V ≤ V_{resistor} ≤ +10 V. Be sure to label both axes appropriately. (*b*) What is the numerical value of the slope (express your answer in siemens)?

41. Sketch the voltage across a 33 Ω resistor over the range 0 < *t* < 2π s, if the current is given by 2.8 cos(*t*) A. Assume both the current and voltage are defined according to the passive sign convention.

42. Figure 2.36 depicts the current-voltage characteristic of three different resistive elements. Determine the resistance of each, assuming the voltage and current are defined in accordance with the passive sign convention.

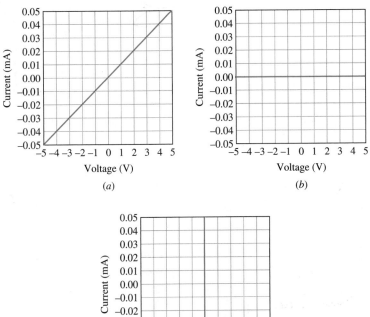

FIGURE 2.36

43. Determine the conductance (in siemens) of the following: (*a*) 0 Ω; (*b*) 100 MΩ; (*c*) 200 mΩ.

44. Determine the magnitude of the current flowing through a 10 mS conductance if the voltage across it is (*a*) 2 mV; (*b*) −1 V; (*c*) $100e^{-2t}$ V; (*d*) 5 sin(5*t*) V; (*e*) 0 V.

45. A 1% tolerance 1 kΩ resistor may in reality have a value anywhere in the range of 990 to 1010 Ω. Assuming a voltage of 9 V is applied across it, determine (*a*) the corresponding range of current and (*b*) the corresponding range of absorbed power. (*c*) If the resistor is replaced with a 10% tolerance 1 kΩ resistor, repeat parts (*a*) and (*b*).

46. The following experimental data is acquired for an unmarked resistor, using a variable-voltage power supply and a current meter. The current meter readout is somewhat unstable, unfortunately, which introduces error into the measurement.

Voltage (V)	Current (mA)
−2.0	−0.89
−1.2	−0.47
0.0	0.01
1.0	0.44
1.5	0.70

(*a*) Plot the measured current-versus-voltage characteristic.

(*b*) Using a best-fit line, estimate the value of the resistance.

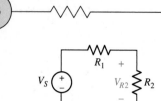

■ **FIGURE 2.37**

47. Utilize the fact that in the circuit of Fig. 2.37, the total power supplied by the voltage source must equal the total power absorbed by the two resistors to show that

$$V_{R_2} = V_S \frac{R_2}{R_1 + R_2}$$

You may assume the same current flows through each element (a requirement of charge conservation).

48. For each of the circuits in Fig. 2.38, find the current I and compute the power absorbed by the resistor.

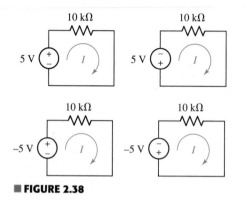

■ **FIGURE 2.38**

49. Sketch the power absorbed by a 100 Ω resistor as a function of voltage over the range $-2\ \text{V} \leq V_{\text{resistor}} \leq +2\ \text{V}$.

Chapter-Integrating Exercises

50. So-called "n-type" silicon has a resistivity given by $\rho = (-qN_D\mu_n)^{-1}$, where N_D is the volume density of phosphorus atoms (atoms/cm^3), μ_n is the electron mobility (cm^2/V $\cdot$ s), and $q = -1.602 \times 10^{-19}$ C is the charge of each electron. Conveniently, a relationship exists between mobility and N_D, as shown in Fig. 2.39. Assume an 8 inch diameter silicon wafer (disk) having a thickness of 300 μm. Design a 10 Ω resistor by specifying a phosphorus concentration in the range of 2×10^{15} cm$^{-3} \leq N_D \leq 2 \times 10^{17}$ cm^{-3}, along with a suitable geometry (the wafer may be cut, but not thinned).

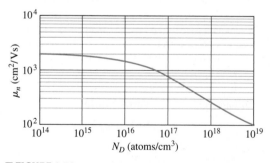

■ **FIGURE 2.39**

51. Figure 2.39 depicts the relationship between electron mobility μ_n and dopant density N_D for n-type silicon. With the knowledge that resistivity in this material is given by $\rho = N_D\mu_n/q$, plot resistivity as a function of dopant density over the range 10^{14} cm$^{-3} \leq N_D \leq 10^{19}$ cm^{-3}.

DP 52. Referring to the data of Table 2.4, design a resistor whose value can be varied mechanically in the range of 100 to 500 Ω (assume operation at 20°C).

53. A 250 ft long span separates a dc power supply from a lamp which draws 25 A of current. If 14 AWG wire is used (note that two wires are needed for a total of 500 ft), calculate the amount of power wasted in the wire.

54. The resistance values in Table 2.4 are calibrated for operation at 20°C. They may be corrected for operation at other temperatures using the relationship[4]

$$\frac{R_2}{R_1} = \frac{234.5 + T_2}{234.5 + T_1}$$

where T_1 = reference temperature (20°C in present case)
T_2 = desired operating temperature
R_1 = resistance at T_1
R_2 = resistance at T_2

A piece of equipment relies on an external wire made of 28 AWG soft copper, which has a resistance of 50.0 Ω at 20°C. Unfortunately, the operating environment has changed, and it is now 110.5°F. (*a*) Calculate the length of the original wire. (*b*) Determine by how much the wire should be shortened so that it is once again 50.0 Ω.

DP 55. Your favorite meter contains a precision (1% tolerance) 10 Ω resistor. Unfortunately, the last person who borrowed this meter somehow blew the resistor, and it needs to be replaced. Design a suitable replacement, assuming at least 1000 ft of each of the wire gauges listed in Table 2.4 is readily available to you.

56. At a new installation, you specified that all wiring should conform to the ASTM B33 specification (see Table 2.3). Unfortunately the subcontractor misread your instructions and installed B415 wiring instead (but the same gauge). Assuming the operating voltage is unchanged, (*a*) by how much will the current be reduced, and (*b*) how much additional power will be wasted in the lines? (*Express both answers in terms of percentage.*)

57. If 1 mA of current is forced through a 1 mm diameter, 2.3 meter long piece of hard, round, aluminum-clad steel (B415) wire, how much power is wasted as a result of resistive losses? If instead wire of the same dimensions but conforming to B75 specifications is used, by how much will the power wasted due to resistive losses be reduced?

58. The network shown in Fig. 2.40 can be used to accurately model the behavior of a bipolar junction transistor provided that it is operating in the forward active mode. The parameter β is known as the current gain. If for this device

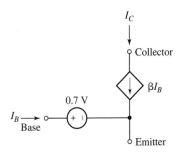

■ **FIGURE 2.40** DC model for a bipolar junction transistor operating in forward active mode.

(4) D. G. Fink and H. W. Beaty, *Standard Handbook for Electrical Engineers,* 13th ed. New York: McGraw-Hill, 1993, p. 2–9.

$\beta = 100$, and I_B is determined to be 100 μA, calculate (a) I_C, the current flowing into the collector terminal; and (b) the power dissipated by the base-emitter region.

59. A 100 W tungsten filament light bulb functions by taking advantage of resistive losses in the filament, absorbing 100 joules each second of energy from the wall socket. How much *optical* energy per second do you expect it to produce, and does this violate the principle of energy conservation?

60. Batteries come in a wide variety of types and sizes. Two of the most common are called "AA" and "AAA." A single battery of either type is rated to produce a terminal voltage of 1.5 V when fully charged. So what are the differences between the two, other than size? (*Hint:* Think about energy.)

Voltage and Current Laws

INTRODUCTION

In Chap. 2 we were introduced to independent voltage and current sources, dependent sources, and resistors. We discovered that *dependent* sources come in four varieties, and are controlled by a voltage or current which exists elsewhere. Once we know the voltage across a resistor, we know its current (and vice versa); this is not the case for sources, however. In general, circuits must be analyzed to determine a complete set of voltages and currents. This turns out to be reasonably straightforward, and only two simple laws are needed in addition to Ohm's law. These new laws are Kirchhoff's current law (KCL) and Kirchhoff's voltage law (KVL), and they are simply restatements of charge and energy conservation, respectively. They apply to any circuit we will ever encounter, although in later chapters we will learn more efficient techniques for specific types of situations.

3.1 • NODES, PATHS, LOOPS, AND BRANCHES

We now focus our attention on the current-voltage relationships in simple networks of two or more circuit elements. The elements will be connected by wires (sometimes referred to as "leads"), which have zero resistance. Since the network then appears as a number of simple elements and a set of connecting leads, it is called a ***lumped-parameter network.*** A more difficult analysis problem arises when we are faced with a ***distributed-parameter network,*** which contains an essentially infinite number of vanishingly small elements. We will concentrate on lumped-parameter networks in this text.

⚠️

In circuits assembled in the real world, the wires will always have finite resistance. However, this resistance is typically so small compared to other resistances in the circuit that we can neglect it without introducing significant error. In our idealized circuits, we will therefore refer to "zero resistance" wires from now on.

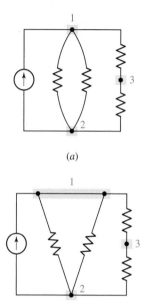

(a)

(b)

■ **FIGURE 3.1** (*a*) A circuit containing three nodes and five branches. (*b*) Node 1 is redrawn to look like two nodes; it is still one node.

A point at which two or more elements have a common connection is called a ***node.*** For example, Fig. 3.1*a* shows a circuit containing three nodes. Sometimes networks are drawn so as to trap an unwary student into believing that there are more nodes present than is actually the case. This occurs when a node, such as node 1 in Fig. 3.1*a*, is shown as two separate junctions connected by a (zero-resistance) conductor, as in Fig. 3.1*b*. However, all that has been done is to spread the common point out into a common zero-resistance line. Thus, we must necessarily consider all of the perfectly conducting leads or portions of leads attached to the node as part of the node. Note also that every element has a node at each of its ends.

Suppose that we start at one node in a network and move through a simple element to the node at the other end. We then continue from that node through a different element to the next node, and continue this movement until we have gone through as many elements as we wish. If no node was encountered more than once, then the set of nodes and elements that we have passed through is defined as a ***path.*** If the node at which we started is the same as the node on which we ended, then the path is, by definition, a closed path or a ***loop.***

For example, in Fig. 3.1*a*, if we move from node 2 through the current source to node 1, and then through the upper right resistor to node 3, we have established a path; since we have not continued on to node 2 again, we have not made a loop. If we proceeded from node 2 through the current source to node 1, down through the left resistor to node 2, and then up through the central resistor to node 1 again, we do not have a path, since a node (actually two nodes) was encountered more than once; we also do not have a loop, because a loop must be a path.

Another term whose use will prove convenient is ***branch.*** We define a branch as a single path in a network, composed of one simple element and the node at each end of that element. Thus, a path is a particular collection of branches. The circuit shown in Fig. 3.1*a* and *b* contains five branches.

3.2 • KIRCHHOFF'S CURRENT LAW

We are now ready to consider the first of the two laws named for Gustav Robert Kirchhoff (two *h*'s and two *f*'s), a German university professor who was born about the time Ohm was doing his experimental work. This axiomatic law is called Kirchhoff's current law (abbreviated KCL), and it simply states that

> The algebraic sum of the currents entering any node is zero.

This law represents a mathematical statement of the fact that charge cannot accumulate at a node. *A node is not a circuit element,* and it certainly cannot store, destroy, or generate charge. Hence, the currents must sum to zero. A hydraulic analogy is sometimes useful here: for example, consider three water pipes joined in the shape of a Y. We define three "currents" as flowing *into* each of the three pipes. If we insist that water is always flowing, then obviously we cannot have three positive water currents, or the pipes would burst. This is a result of our defining currents independent of

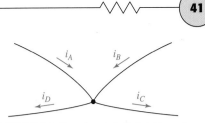
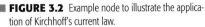

the direction that water is actually flowing. Therefore, the value of either one or two of the currents as defined must be negative.

Consider the node shown in Fig. 3.2. The algebraic sum of the four currents entering the node must be zero:

$$i_A + i_B + (-i_C) + (-i_D) = 0$$

However, the law could be equally well applied to the algebraic sum of the currents *leaving* the node:

$$(-i_A) + (-i_B) + i_C + i_D = 0$$

We might also wish to equate the sum of the currents having reference arrows directed into the node to the sum of those directed out of the node:

$$i_A + i_B = i_C + i_D$$

which simply states that the sum of the currents going in must equal the sum of the currents going out.

■ FIGURE 3.2 Example node to illustrate the application of Kirchhoff's current law.

EXAMPLE 3.1

For the circuit in Fig. 3.3a, compute the current through resistor R_3 if it is known that the voltage source supplies a current of 3 A.

▶ **Identify the goal of the problem.**
The current through resistor R_3, labeled as i on the circuit diagram.

▶ **Collect the known information.**
The node at the top of R_3 is connected to four branches.
    Two of these currents are clearly labeled: 2 A flows out of the node into R_2, and 5 A flows into the node from the current source. We are told the current out of the 10 V source is 3 A.

▶ **Devise a plan.**
If we label the current through R_1 (Fig. 3.3b), we may write a KCL equation at the top node of resistors R_2 and R_3.

▶ **Construct an appropriate set of equations.**
Summing the currents flowing into the node:

$$i_{R_1} - 2 - i + 5 = 0$$

The currents flowing into this node are shown in the expanded diagram of Fig. 3.3c for clarity.

▶ **Determine if additional information is required.**
We have one equation but two unknowns, which means we need to obtain an additional equation. At this point, the fact that we know the 10 V source is supplying 3 A comes in handy: KCL shows us that this is also the current i_{R_1}.

▶ **Attempt a solution.**
Substituting, we find that $i = 3 - 2 + 5 = 6$ A.

▶ **Verify the solution. Is it reasonable or expected?**
It is always worth the effort to recheck our work. Also, we can attempt to evaluate whether at least the magnitude of the solution is

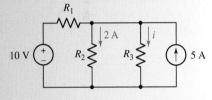

(a)

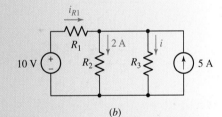

(b)

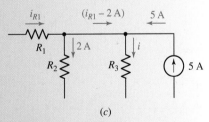

(c)

■ FIGURE 3.3 (a) Simple circuit for which the current through resistor R_3 is desired. (b) The current through resistor R_1 is labeled so that a KCL equation can be written. (c) The currents into the top node of R_3 are redrawn for clarity.

(Continued on next page)

reasonable. In this case, we have two sources—one supplies 5 A, and the other supplies 3 A. There are no other sources, independent or dependent. Thus, we would not expect to find any current in the circuit in excess of 8 A.

PRACTICE

3.1 Count the number of branches and nodes in the circuit in Fig. 3.4. If $i_x = 3$ A and the 18 V source delivers 8 A of current, what is the value of R_A? (*Hint:* You need Ohm's law as well as KCL.)

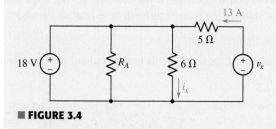

■ **FIGURE 3.4**

Ans: 5 branches, 3 nodes, 1Ω.

A compact expression for Kirchhoff's current law is

$$\sum_{n=1}^{N} i_n = 0 \qquad [1]$$

which is just a shorthand statement for

$$i_1 + i_2 + i_3 + \cdots + i_N = 0 \qquad [2]$$

When Eq. [1] or Eq. [2] is used, it is understood that the N current arrows are either all directed toward the node in question, or are all directed away from it.

3.3 KIRCHHOFF'S VOLTAGE LAW

Current is related to the charge flowing *through* a circuit element, whereas voltage is a measure of potential energy difference *across* the element. There is a single unique value for any voltage in circuit theory. Thus, the energy required to move a unit charge from point A to point B in a circuit must have a value independent of the path chosen to get from A to B (there is often more than one such path). We may assert this fact through Kirchhoff's voltage law (abbreviated **KVL**):

The algebraic sum of the voltages around any closed path is zero.

In Fig. 3.5, if we carry a charge of 1 C from A to B through element 1, the reference polarity signs for v_1 show that we do v_1 joules of work.[1] Now

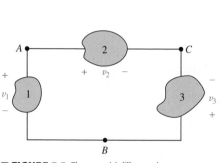

■ **FIGURE 3.5** The potential difference between points A and B is independent of the path selected.

(1) Note that we chose a 1 C charge for the sake of numerical convenience: therefore, we did (1 C)(v_1 J/C) = v_1 joules of work.

if, instead, we choose to proceed from A to B via node C, then we expend $(v_2 - v_3)$ joules of energy. The work done, however, is independent of the path in a circuit, and so any route must lead to the same value for the voltage. In other words,

$$v_1 = v_2 - v_3 \qquad [3]$$

It follows that if we trace out a closed path, the algebraic sum of the voltages across the individual elements around it must be zero. Thus, we may write

$$v_1 + v_2 + v_3 + \cdots + v_N = 0$$

or, more compactly,

$$\sum_{n=1}^{N} v_n = 0 \qquad [4]$$

We can apply KVL to a circuit in several different ways. One method that leads to fewer equation-writing errors than others consists of moving mentally around the closed path in a clockwise direction and writing down directly the voltage of each element whose $(+)$ terminal is entered, and writing down the negative of every voltage first met at the $(-)$ sign. Applying this to the single loop of Fig. 3.5, we have

$$-v_1 + v_2 - v_3 = 0$$

which agrees with our previous result, Eq. [3].

EXAMPLE 3.2

In the circuit of Fig. 3.6, find v_x and i_x.

We know the voltage across two of the three elements in the circuit. Thus, KVL can be applied immediately to obtain v_x.

Beginning with the bottom node of the 5 V source, we apply KVL clockwise around the loop:

$$-5 - 7 + v_x = 0$$

so $v_x = 12$ V.

KCL applies to this circuit, but only tells us that the same current (i_x) flows through all three elements. We now know the voltage across the 100 Ω resistor, however.

Invoking Ohm's law,

$$i_x = \frac{v_x}{100} = \frac{12}{100}\ \text{A} = 120\ \text{mA}$$

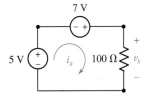

■ **FIGURE 3.6** A simple circuit with two voltage sources and a single resistor.

PRACTICE

3.2 Determine i_x and v_x in the circuit of Fig. 3.7.

Ans: $v_x = -4$ V; $i_x = -400$ mA.

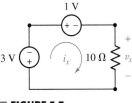

■ **FIGURE 3.7**

EXAMPLE 3.3

In the circuit of Fig. 3.8 there are eight circuit elements. Find v_{R2} (the voltage across R_2) and the voltage labeled v_x.

The best approach for finding v_{R2} is to look for a loop to which we can apply KVL. There are several options, but the leftmost loop offers a straightforward route, as two of the voltages are clearly specified. Thus, we find v_{R2} by writing a KVL equation around the loop on the left, starting at point c:

$$4 - 36 + v_{R2} = 0$$

which leads to $v_{R2} = 32$ V.

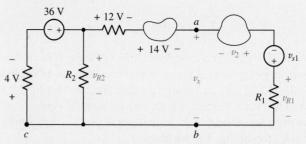

■ **FIGURE 3.8** A circuit with eight elements for which we desire v_{R2} and v_x.

To find v_x, we might think of this as the (algebraic) sum of the voltages across the three elements on the right. However, since we do not have values for these quantities, such an approach would not lead to a numerical answer. Instead, we apply KVL beginning at point c, moving up and across the top to a, through v_x to b, and through the conducting lead to the starting point:

$$+4 - 36 + 12 + 14 + v_x = 0$$

so that

$$v_x = 6 \text{ V}$$

An alternative approach: Knowing v_{R2}, we might have taken the shortcut through R_2:

$$-32 + 12 + 14 + v_x = 0$$

yielding $v_x = 6$ V once again.

Points b and c, as well as the wire between them, are all part of the same node.

PRACTICE

3.3 For the circuit of Fig. 3.9, determine (*a*) v_{R2} and (*b*) v_2, if $v_{R1} = 1$ V.

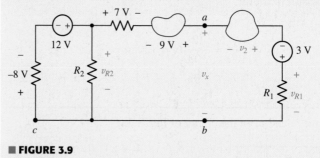

■ **FIGURE 3.9**

Ans: (*a*) 4 V; (*b*) −8 V.

As we have just seen, the key to correctly analyzing a circuit is to first methodically label all voltages and currents on the diagram. This way, carefully written KCL or KVL equations will yield correct relationships, and Ohm's law can be applied as necessary if more unknowns than equations are obtained initially. We illustrate these principles with a more detailed example.

EXAMPLE **3.4**

Determine v_x in the circuit of Fig. 3.10a.

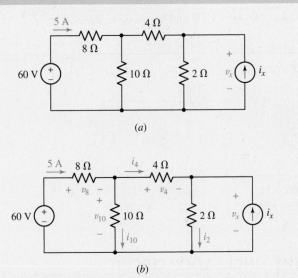

(a)

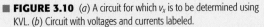

(b)

■ **FIGURE 3.10** (a) A circuit for which v_x is to be determined using KVL. (b) Circuit with voltages and currents labeled.

We begin by labeling voltages and currents on the rest of the elements in the circuit (Fig. 3.10b). Note that v_x appears across the 2 Ω resistor and the source i_x as well.

If we can obtain the current through the 2 Ω resistor, Ohm's law will yield v_x. Writing the appropriate KCL equation, we see that

$$i_2 = i_4 + i_x$$

Unfortunately, we do not have values for any of these three quantities. Our solution has (temporarily) stalled.

Since we were given the current flowing from the 60 V source, perhaps we should consider starting from that side of the circuit. Instead of finding v_x using i_2, it might be possible to find v_x directly using KVL. We can write the following KVL equations:

$$-60 + v_8 + v_{10} = 0$$

and

$$-v_{10} + v_4 + v_x = 0 \qquad [5]$$

This is progress: we now have two equations in four unknowns, an improvement over one equation in which *all* terms were unknown. In fact, we know that $v_8 = 40$ V through Ohm's law, as we were told that 5 A flows through the 8 Ω resistor. Thus, $v_{10} = 0 + 60 - 40 = 20$ V,

(Continued on next page)

so Eq. [5] reduces to

$$v_x = 20 - v_4$$

If we can determine v_4, the problem is solved.

The best route to finding a numerical value for the voltage v_4 in this case is to employ Ohm's law, which requires a value for i_4. From KCL, we see that

$$i_4 = 5 - i_{10} = 5 - \frac{v_{10}}{10} = 5 - \frac{20}{10} = 3$$

so that $v_4 = (4)(3) = 12$ V and hence $v_x = 20 - 12 = 8$ V.

PRACTICE

3.4 Determine v_x in the circuit of Fig. 3.11.

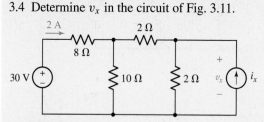

■ **FIGURE 3.11**

Ans: $v_x = 12.8$ V.

3.4 THE SINGLE-LOOP CIRCUIT

We have seen that repeated use of KCL and KVL in conjunction with Ohm's law can be applied to nontrivial circuits containing several loops and a number of different elements. Before proceeding further, this is a good time to focus on the concept of series (and, in the next section, parallel) circuits, as they form the basis of any network we will encounter in the future.

All of the elements in a circuit that carry the same current are said to be connected in **series.** As an example, consider the circuit of Fig. 3.10. The 60 V source is in series with the 8 Ω resistor; they carry the same 5 A current. However, the 8 Ω resistor is not in series with the 4 Ω resistor; they carry different currents. Note that elements may carry equal currents and not be in series; two 100 W light bulbs in neighboring houses may very well carry equal currents, but they certainly do not carry the same current and are *not* connected in series.

Figure 3.12*a* shows a simple circuit consisting of two batteries and two resistors. Each terminal, connecting lead, and solder glob is assumed to have zero resistance; together they constitute an individual node of the circuit diagram in Fig. 3.12*b*. Both batteries are modeled by ideal voltage sources; any internal resistances they may have are assumed to be small enough to neglect. The two resistors are assumed to be ideal (linear) resistors.

We seek the current *through* each element, the voltage *across* each element, and the power *absorbed* by each element. Our first step in the analysis is the assumption of reference directions for the unknown currents. Arbitrarily, let us select a clockwise current i which flows out of the upper terminal of the voltage source on the left. This choice is indicated by an arrow labeled i at that point in the circuit, as shown in Fig. 3.12*c*. A trivial

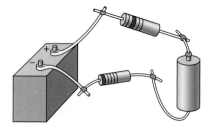

(a)

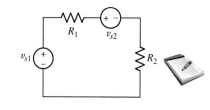

(b)

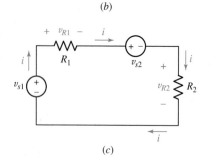

(c)

■ **FIGURE 3.12** (*a*) A single-loop circuit with four elements. (*b*) The circuit model with source voltages and resistance values given. (*c*) Current and voltage reference signs have been added to the circuit.

application of Kirchhoff's current law assures us that this same current must also flow through every other element in the circuit; we emphasize this fact this one time by placing several other current symbols about the circuit.

Our second step in the analysis is a choice of the voltage reference for each of the two resistors. The passive sign convention requires that the resistor current and voltage variables be defined so that the current enters the terminal at which the positive voltage reference is located. Since we already (arbitrarily) selected the current direction, v_{R1} and v_{R2} are defined as in Fig. 3.12c.

The third step is the application of Kirchhoff's voltage law to the only closed path. Let us decide to move around the circuit in the clockwise direction, beginning at the lower left corner, and to write down directly every voltage first met at its positive reference, and to write down the negative of every voltage encountered at the negative terminal. Thus,

$$-v_{s1} + v_{R1} + v_{s2} + v_{R2} = 0 \qquad [6]$$

We then apply Ohm's law to the resistive elements:

$$v_{R1} = R_1 i \quad \text{and} \quad v_{R2} = R_2 i$$

Substituting into Eq. [6] yields

$$-v_{s1} + R_1 i + v_{s2} + R_2 i = 0$$

Since i is the only unknown, we find that

$$i = \frac{v_{s1} - v_{s2}}{R_1 + R_2}$$

The voltage or power associated with any element may now be obtained by applying $v = Ri$, $p = vi$, or $p = i^2 R$.

PRACTICE

3.5 In the circuit of Fig. 3.12b, $v_{s1} = 120$ V, $v_{s2} = 30$ V, $R_1 = 30\ \Omega$, and $R_2 = 15\ \Omega$. Compute the power absorbed by each element.

Ans: $p_{120V} = -240$ W; $p_{30V} = +60$ W; $p_{30\Omega} = 120$ W; $p_{15\Omega} = 60$ W.

EXAMPLE **3.5**

Compute the power absorbed in each element for the circuit shown in Fig. 3.13a.

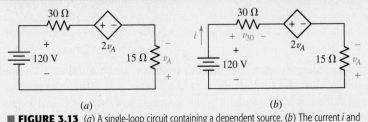

(a)   (b)

■ **FIGURE 3.13** (a) A single-loop circuit containing a dependent source. (b) The current i and voltage v_{30} are assigned.

(Continued on next page)

We first assign a reference direction for the current i and a reference polarity for the voltage v_{30} as shown in Fig. 3.13b. There is no need to assign a voltage to the 15 Ω resistor, since the controlling voltage v_A for the dependent source is already available. (It is worth noting, however, that the reference signs for v_A are reversed from those we would have assigned based on the passive sign convention.)

This circuit contains a dependent voltage source, the value of which remains unknown until we determine v_A. However, its algebraic value $2v_A$ can be used in the same fashion as if a numerical value were available. Thus, applying KVL around the loop:

$$-120 + v_{30} + 2v_A - v_A = 0 \qquad [7]$$

Using Ohm's law to introduce the known resistor values:

$$v_{30} = 30i \quad \text{and} \quad v_A = -15i$$

Note that the negative sign is required since i flows into the negative terminal of v_A.

Substituting into Eq. [7] yields

$$-120 + 30i - 30i + 15i = 0$$

and so we find that

$$i = 8 \text{ A}$$

Computing the power *absorbed* by each element:

$$
\begin{aligned}
p_{120V} &= (120)(-8) = -960 \text{ W} \\
p_{30\Omega} &= (8)^2(30) \quad = 1920 \text{ W} \\
p_{\text{dep}} &= (2v_A)(8) \quad = 2[(-15)(8)](8) \\
&\qquad\qquad\quad = -1920 \text{ W} \\
p_{15\Omega} &= (8)^2(15) \quad = 960 \text{ W}
\end{aligned}
$$

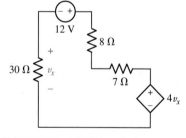

■ **FIGURE 3.14** A simple loop circuit.

PRACTICE

3.6 In the circuit of Fig. 3.14, find the power absorbed by each of the five elements in the circuit.

Ans: (CW from left) 0.768 W, 1.92 W, 0.2048 W, 0.1792 W, −3.072 W.

In the preceding example and practice problem, we were asked to compute the power absorbed by each element of a circuit. It is difficult to think of a situation, however, in which *all* of the absorbed power quantities of a circuit would be positive, for the simple reason that the energy must come from somewhere. Thus, from simple conservation of energy, we expect that ***the sum of the absorbed power for each element of a circuit should be zero.*** In other words, at least one of the quantities should be negative (neglecting the trivial case where the circuit is not operating). Stated another way, the sum of the supplied power for each element should be zero. More pragmatically, *the sum of the absorbed power equals the sum of the supplied power,* which seems reasonable enough at face value.

Let's test this with the circuit of Fig. 3.13 from Example 3.5, which consists of two sources (one dependent and one independent) and two resistors. Adding the power absorbed by each element, we find

$$\sum_{\text{all elements}} p_{\text{absorbed}} = -960 + 1920 - 1920 + 960 = 0$$

In reality (our indication is the sign associated with the absorbed power) the 120 V source *supplies* $+960$ W, and the dependent source supplies $+1920$ W. Thus, the sources supply a total of $960 + 1920 = 2880$ W. The resistors are expected to absorb positive power, which in this case sums to a total of $1920 + 960 = 2880$ W. Thus, if we take into account each element of the circuit,

$$\sum p_{\text{absorbed}} = \sum p_{\text{supplied}}$$

as we expect.

Turning our attention to Practice Problem 3.6, the solution to which the reader might want to verify, we see that the absorbed powers sum to $0.768 + 1.92 + 0.2048 + 0.1792 - 3.072 = 0$. Interestingly enough, the 12 V independent voltage source is absorbing $+1.92$ W, which means it is *dissipating* power, not supplying it. Instead, the dependent voltage source appears to be supplying all the power in this particular circuit. Is such a thing possible? We usually expect a source to supply positive power, but since we are employing idealized sources in our circuits, it is in fact possible to have a net power flow into any source. If the circuit is changed in some way, the same source might then be found to supply positive power. The result is not known until a circuit analysis has been completed.

3.5 THE SINGLE-NODE-PAIR CIRCUIT

The companion of the single-loop circuit discussed in Sec. 3.4 is the single-node-pair circuit, in which any number of simple elements are connected between the same pair of nodes. An example of such a circuit is shown in Fig. 3.15a. KVL forces us to recognize that the voltage across each branch is the same as that across any other branch. *Elements in a circuit having a common voltage across them are said to be connected in* **parallel.**

EXAMPLE **3.6**

Find the voltage, current, and power associated with each element in the circuit of Fig. 3.15a.

We first define a voltage v and arbitrarily select its polarity as shown in Fig. 3.15b. Two currents, flowing in the resistors, are selected in conformance with the passive sign convention, as shown in Fig. 3.15b.

(Continued on next page)

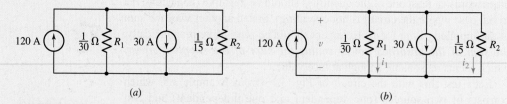

(a)   (b)

■ **FIGURE 3.15** (a) A single-node-pair circuit. (b) A voltage and two currents are assigned.

Determining either current i_1 or i_2 will enable us to obtain a value for v. Thus, our next step is to apply KCL to either of the two nodes in the circuit. Equating the algebraic sum of the currents leaving the upper node to zero:

$$-120 + i_1 + 30 + i_2 = 0$$

Writing both currents in terms of the voltage v using Ohm's law

$$i_1 = 30v \quad \text{and} \quad i_2 = 15v$$

we obtain

$$-120 + 30v + 30 + 15v = 0$$

Solving this equation for v results in

$$v = 2 \text{ V}$$

and invoking Ohm's law then gives

$$i_1 = 60 \text{ A} \quad \text{and} \quad i_2 = 30 \text{ A}$$

The absorbed power in each element can now be computed. In the two resistors,

$$p_{R1} = 30(2)^2 = 120 \text{ W} \quad \text{and} \quad p_{R2} = 15(2)^2 = 60 \text{ W}$$

and for the two sources,

$$p_{120A} = 120(-2) = -240 \text{ W} \quad \text{and} \quad p_{30A} = 30(2) = 60 \text{ W}$$

Since the 120 A source absorbs negative 240 W, it is actually *supplying* power to the other elements in the circuit. In a similar fashion, we find that the 30 A source is actually *absorbing* power rather than *supplying* it.

PRACTICE

3.7 Determine v in the circuit of Fig. 3.16.

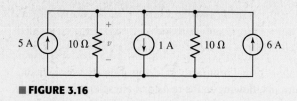

■ **FIGURE 3.16**

Ans: 50 V.

EXAMPLE **3.7**

Determine the value of v and the power supplied by the independent current source in Fig. 3.17.

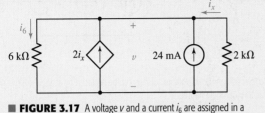

■ **FIGURE 3.17** A voltage v and a current i_6 are assigned in a single-node-pair circuit containing a dependent source.

By KCL, the sum of the currents leaving the upper node must be zero, so that

$$i_6 - 2i_x - 0.024 - i_x = 0$$

Again, note that the value of the dependent source ($2i_x$) is treated the same as any other current would be, even though its exact value is not known until the circuit has been analyzed.

We next apply Ohm's law to each resistor:

$$i_6 = \frac{v}{6000} \quad \text{and} \quad i_x = \frac{-v}{2000}$$

Therefore,

$$\frac{v}{6000} - 2\left(\frac{-v}{2000}\right) - 0.024 - \left(\frac{-v}{2000}\right) = 0$$

and so $v = (600)(0.024) = 14.4$ V.

Any other information we may want to find for this circuit is now easily obtained, usually in a single step. For example, the power supplied by the independent source is $p_{24} = 14.4(0.024) = 0.3456$ W (345.6 mW).

PRACTICE

3.8 For the single-node-pair circuit of Fig. 3.18, find i_A, i_B, and i_C.

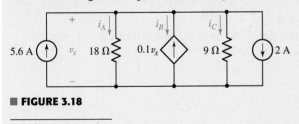

■ **FIGURE 3.18**

Ans: 3 A; −5.4 A; 6 A.

3.6 ● SERIES AND PARALLEL CONNECTED SOURCES

It turns out that some of the equation writing that we have been doing for series and parallel circuits can be avoided by combining sources. Note, however, that all the current, voltage, and power relationships in the remainder of the circuit will be unchanged. For example, several voltage

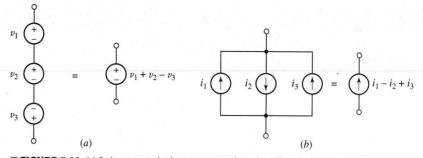

FIGURE 3.19 (*a*) Series-connected voltage sources can be replaced by a single source. (*b*) Parallel current sources can be replaced by a single source.

sources in series may be replaced by an equivalent voltage source having a voltage equal to the algebraic sum of the individual sources (Fig. 3.19*a*). Parallel current sources may also be combined by algebraically adding the individual currents, and the order of the parallel elements may be rearranged as desired (Fig. 3.19*b*).

EXAMPLE 3.8

Determine the current i in the circuit of Fig. 3.20*a* by first combining the sources into a single equivalent voltage source.

To be able to combine the voltage sources, they must be in series. Since the same current (i) flows through each, this condition is satisfied.

Starting from the bottom left-hand corner and proceeding clockwise,

$$-3 - 9 - 5 + 1 = -16 \text{ V}$$

so we may replace the four voltage sources with a single 16 V source having its negative reference as shown in Fig. 3.20*b*.

KVL combined with Ohm's law then yields

$$-16 + 100i + 220i = 0$$

or

$$i = \frac{16}{320} = 50 \text{ mA}$$

We should note that the circuit in Fig. 3.20*c* is also equivalent, a fact easily verified by computing i.

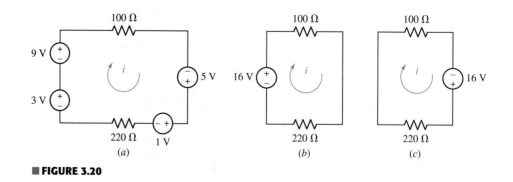

(*a*)   (*b*)   (*c*)

FIGURE 3.20

PRACTICE

3.9 Determine the current i in the circuit of Fig. 3.21 after first replacing the four sources with a single equivalent source.

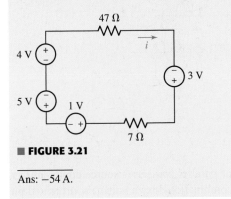

■ **FIGURE 3.21**

Ans: -54 A.

EXAMPLE **3.9**

Determine the voltage v in the circuit of Fig. 3.22a by first combining the sources into a single equivalent current source.

The sources may be combined if the same voltage appears across each one, which we can easily verify is the case. Thus, we create a new source, arrow pointing upward into the top node, by adding the currents that flow into that node:

$$2.5 - 2.5 - 3 = -3 \text{ A}$$

One equivalent circuit is shown in Fig. 3.22b.

KCL then allows us to write

$$-3 + \frac{v}{5} + \frac{v}{5} = 0$$

Solving, we find $v = 7.5$ V.

Another equivalent circuit is shown in Fig. 3.22c.

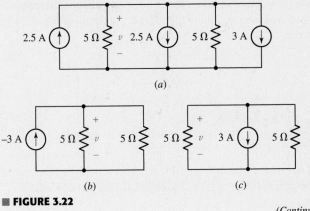

(a)

(b)   (c)

■ **FIGURE 3.22**

(Continued on next page)

PRACTICE

3.10 Determine the voltage v in the circuit of Fig. 3.23 after first replacing the three sources with a single equivalent source.

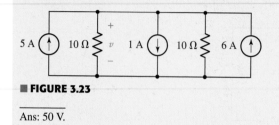

■ **FIGURE 3.23**

Ans: 50 V.

To conclude the discussion of parallel and series source combinations, we should consider the parallel combination of two voltage sources and the series combination of two current sources. For instance, what is the equivalent of a 5 V source in parallel with a 10 V source? By the definition of a voltage source, the voltage across the source cannot change; by Kirchhoff's voltage law, then, 5 equals 10 and we have hypothesized a physical impossibility. Thus, *ideal* voltage sources in parallel are permissible only when each has the same terminal voltage at every instant. In a similar way, two current sources may not be placed in series unless each has the same current, including sign, for every instant of time.

EXAMPLE 3.10

Determine which of the circuits of Fig. 3.24 are valid.

The circuit of Fig. 3.24a consists of two voltage sources in parallel. The value of each source is different, so this circuit violates KVL. For example, if a resistor is placed in parallel with the 5 V source, it is also in parallel with the 10 V source. The actual voltage across it is therefore ambiguous, and clearly the circuit cannot be constructed as indicated. If we attempt to build such a circuit in real life, we will find it impossible to locate "ideal" voltage sources—all real-world sources have an internal resistance. The presence of such resistance allows a voltage difference between the two *real* sources. Along these lines, the circuit of Fig. 3.24b is perfectly valid.

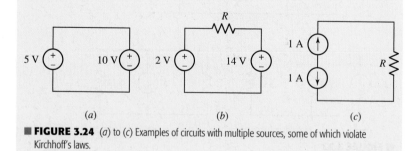

       (a)   (b)   (c)

■ **FIGURE 3.24** (a) to (c) Examples of circuits with multiple sources, some of which violate Kirchhoff's laws.

The circuit of Fig. 3.24*c* violates KCL: it is unclear what current actually flows through the resistor *R*.

PRACTICE

3.11 Determine whether the circuit of Fig. 3.25 violates either of Kirchhoff's laws.

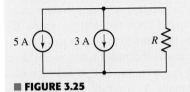

■ **FIGURE 3.25**

Ans: No. If the resistor were removed, however, the resulting circuit would.

3.7 RESISTORS IN SERIES AND PARALLEL

It is often possible to replace relatively complicated resistor combinations with a single equivalent resistor. This is useful when we are not specifically interested in the current, voltage, or power associated with any of the individual resistors in the combinations. *All the current, voltage, and power relationships in the remainder of the circuit will be unchanged.*

Consider the series combination of *N* resistors shown in Fig. 3.26*a*. We want to simplify the circuit with replacing the *N* resistors with a single resistor R_{eq} so that the remainder of the circuit, in this case only the voltage source, does not realize that any change has been made. The current, voltage, and power of the source must be the same before and after the replacement.

First, apply KVL:

$$v_s = v_1 + v_2 + \cdots + v_N$$

and then Ohm's law:

$$v_s = R_1 i + R_2 i + \cdots + R_N i = (R_1 + R_2 + \cdots + R_N)i$$

Now compare this result with the simple equation applying to the equivalent circuit shown in Fig. 3.26*b*:

$$v_s = R_{eq} i$$

> Helpful Tip: Inspection of the KVL equation for any series circuit will show that *the order* in which elements are placed in such a circuit *makes no difference*.

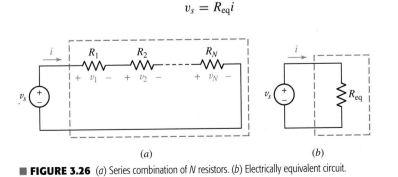

(a) (b)

■ **FIGURE 3.26** (*a*) Series combination of *N* resistors. (*b*) Electrically equivalent circuit.

Thus, the value of the equivalent resistance for N series resistors is

$$R_{eq} = R_1 + R_2 + \cdots + R_N \qquad [8]$$

We are therefore able to replace a two-terminal network consisting of N series resistors with a single two-terminal element R_{eq} that has the same v-i relationship.

It should be emphasized again that we might be interested in the current, voltage, or power of one of the original elements. For example, the voltage of a dependent voltage source may depend upon the voltage across R_3. Once R_3 is combined with several series resistors to form an equivalent resistance, then it is gone and the voltage across it cannot be determined until R_3 is identified by removing it from the combination. In that case, it would have been better to look ahead and not make R_3 a part of the combination initially.

EXAMPLE 3.11

Use resistance and source combinations to determine the current i in Fig. 3.27a and the power delivered by the 80 V source.

We first interchange the element positions in the circuit, being careful to preserve the proper sense of the sources, as shown in Fig. 3.27b. The

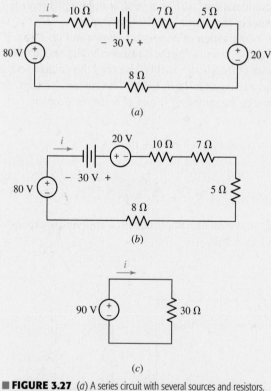

FIGURE 3.27 (a) A series circuit with several sources and resistors. (b) The elements are rearranged for the sake of clarity. (c) A simpler equivalent.

next step is to then combine the three voltage sources into an equivalent 90 V source, and the four resistors into an equivalent 30 Ω resistance, as in Fig. 3.27c. Thus, instead of writing

$$-80 + 10i - 30 + 7i + 5i + 20 + 8i = 0$$

we have simply

$$-90 + 30i = 0$$

and so we find that

$$i = 3 \text{ A}$$

In order to calculate the power delivered to the circuit by the 80 V source appearing in the given circuit, it is necessary to return to Fig. 3.27a with the knowledge that the current is 3 A. The desired power is then 80 V $\times$ 3 A $= 240$ W.

It is interesting to note that no element of the original circuit remains in the equivalent circuit.

PRACTICE

3.12 Determine i in the circuit of Fig. 3.28.

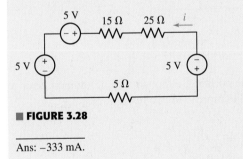

■ **FIGURE 3.28**

Ans: -333 mA.

Similar simplifications can be applied to parallel circuits. A circuit containing N resistors in parallel, as in Fig. 3.29a, leads to the KCL equation

$$i_s = i_1 + i_2 + \cdots + i_N$$

or

$$i_s = \frac{v}{R_1} + \frac{v}{R_2} + \cdots + \frac{v}{R_N}$$

$$= \frac{v}{R_{eq}}$$

Thus,

$$\boxed{\frac{1}{R_{eq}} = \frac{1}{R_1} + \frac{1}{R_2} + \cdots + \frac{1}{R_N}}$$ [9]

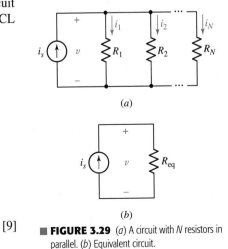

■ **FIGURE 3.29** (a) A circuit with N resistors in parallel. (b) Equivalent circuit.

which can be written as

$$R_{eq}^{-1} = R_1^{-1} + R_2^{-1} + \cdots + R_N^{-1}$$

or, in terms of conductances, as

$$G_{eq} = G_1 + G_2 + \cdots + G_N$$

The simplified (equivalent) circuit is shown in Fig. 3.29b.

A parallel combination is routinely indicated by the following shorthand notation:

$$R_{eq} = R_1 \| R_2 \| R_3$$

The special case of only two parallel resistors is encountered fairly often, and is given by

$$R_{eq} = R_1 \| R_2$$

$$= \frac{1}{\dfrac{1}{R_1} + \dfrac{1}{R_2}}$$

Or, more simply,

$$\boxed{R_{eq} = \frac{R_1 R_2}{R_1 + R_2}} \qquad [10]$$

The last form is worth memorizing, although it is a common error to attempt to generalize Eq. [10] to more than two resistors, e.g.,

$$R_{eq} \neq \frac{R_1 R_2 R_3}{R_1 + R_2 + R_3}$$

A quick look at the units of this equation will immediately show that the expression cannot possibly be correct.

PRACTICE

3.13 Determine v in the circuit of Fig. 3.30 by first combining the three current sources, and then the two 10 Ω resistors.

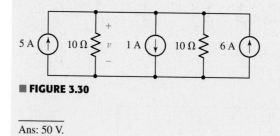

■ **FIGURE 3.30**

Ans: 50 V.

EXAMPLE 3.12

Calculate the power and voltage of the dependent source in Fig. 3.31a.

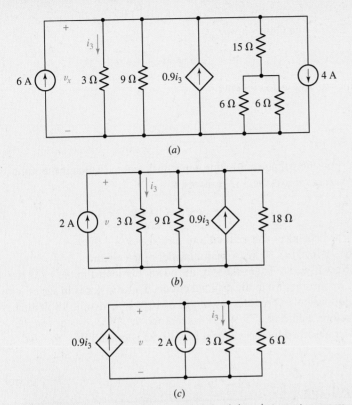

(a)

(b)

(c)

■ **FIGURE 3.31** (a) A multinode circuit. (b) The two independent current sources are combined into a 2 A source, and the 15 Ω resistor in series with the two parallel 6 Ω resistors are replaced with a single 18 Ω resistor. (c) A simplified equivalent circuit.

We will seek to simplify the circuit before analyzing it, but take care not to include the dependent source since its voltage and power characteristics are of interest.

Despite not being drawn adjacent to one another, the two independent current sources are in fact in parallel, so we replace them with a 2 A source.

The two 6 Ω resistors are in parallel and can be replaced with a single 3 Ω resistor in series with the 15 Ω resistor. Thus, the two 6 Ω resistors and the 15 Ω resistor are replaced by an 18 Ω resistor (Fig. 3.31b).

No matter how tempting, *we should not combine the remaining three resistors;* the controlling variable i_3 depends on the 3 Ω resistor and so that resistor must remain untouched. The only further simplification, then, is 9 Ω∥18 Ω = 6 Ω, as shown in Fig. 3.31c.

(Continued on next page)

Applying KCL at the top node of Fig. 3.31c, we have

$$-0.9i_3 - 2 + i_3 + \frac{v}{6} = 0$$

Employing Ohm's law,

$$v = 3i_3$$

which allows us to compute

$$i_3 = \frac{10}{3} \text{ A}$$

Thus, the voltage across the dependent source (which is the same as the voltage across the 3 Ω resistor) is

$$v = 3i_3 = 10 \text{ V}$$

The dependent source therefore furnishes $v \times 0.9i_3 = 10(0.9)(10/3) = 30$ W to the remainder of the circuit.

Now if we are later asked for the power dissipated in the 15 Ω resistor, we must return to the original circuit. This resistor is in series with an equivalent 3 Ω resistor; a voltage of 10 V is across the 18 Ω total; therefore, a current of 5/9 A flows through the 15 Ω resistor and the power absorbed by this element is $(5/9)^2(15)$ or 4.63 W.

PRACTICE

3.14 For the circuit of Fig. 3.32, calculate the voltage v_x.

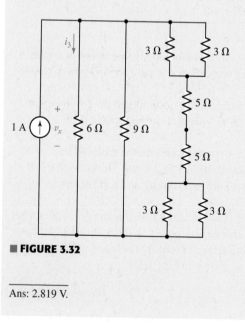

■ FIGURE 3.32

Ans: 2.819 V.

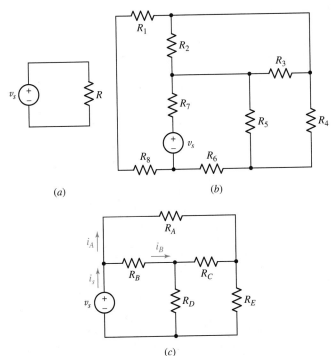

■ FIGURE 3.33 These two circuit elements are both in series and in parallel.
(b) R_2 and R_3 are in parallel, and R_1 and R_8 are in series. (c) There are no circuit
elements either in series or in parallel with one another.

Three final comments on series and parallel combinations might be helpful. The first is illustrated by referring to Fig. 3.33a and asking, "*Are v_s and R in series or in parallel?*" The answer is "*Both.*" The two elements carry the same current and are therefore in series; they also enjoy the same voltage and consequently are in parallel.

The second comment is a word of caution. Circuits can be drawn in such a way as to make series or parallel combinations difficult to spot. In Fig. 3.33b, for example, the only two resistors in parallel are R_2 and R_3, while the only two in series are R_1 and R_8.

The final comment is simply that a simple circuit element need not be in series or parallel with any other simple circuit element in a circuit. For example, R_4 and R_5 in Fig. 3.33b are not in series or parallel with any other simple circuit element, and there are no simple circuit elements in Fig. 3.33c that are in series or parallel with any other simple circuit element. In other words, we cannot simplify that circuit further using any of the techniques discussed in this chapter.

3.8 • VOLTAGE AND CURRENT DIVISION

By combining resistances and sources, we have found one method of shortening the work of analyzing a circuit. Another useful shortcut is the application of the ideas of voltage and current division. Voltage division is used to express the voltage across one of several series resistors in terms of the

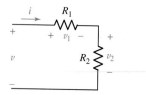

FIGURE 3.34 An illustration of voltage division.

voltage across the combination. In Fig. 3.34, the voltage across R_2 is found via KVL and Ohm's law:

$$v = v_1 + v_2 = iR_1 + iR_2 = i(R_1 + R_2)$$

so

$$i = \frac{v}{R_1 + R_2}$$

Thus,

$$v_2 = iR_2 = \left(\frac{v}{R_1 + R_2}\right)R_2$$

or

$$v_2 = \frac{R_2}{R_1 + R_2}v$$

and the voltage across R_1 is, similarly,

$$v_1 = \frac{R_1}{R_1 + R_2}v$$

If the network of Fig. 3.34 is generalized by removing R_2 and replacing it with the series combination of $R_2, R_3, \ldots, R_N$, then we have the general result for voltage division across a string of N series resistors

$$v_k = \frac{R_k}{R_1 + R_2 + \cdots + R_N}v \qquad [11]$$

which allows us to compute the voltage v_k that appears across an arbitrary resistor R_k of the series.

EXAMPLE 3.13

Determine v_x in the circuit of Fig. 3.35a.

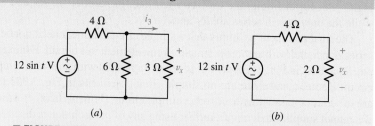

(a)   (b)

FIGURE 3.35 A numerical example illustrating resistance combination and voltage division. (a) Original circuit. (b) Simplified circuit.

We first combine the 6 Ω and 3 Ω resistors, replacing them with $(6)(3)/(6 + 3) = 2$ Ω.

Since v_x appears across the parallel combination, our simplification has not lost this quantity. However, further simplification of the circuit by replacing the series combination of the 4 Ω resistor with our new 2 Ω resistor would.

Thus, we proceed by simply applying voltage division to the circuit in Fig. 3.35b:

$$v_x = (12 \sin t) \frac{2}{4+2} = 4 \sin t \qquad \text{volts}$$

PRACTICE

3.15 Use voltage division to determine v_x in the circuit of Fig. 3.36.

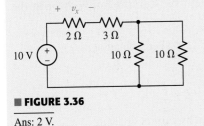

■ **FIGURE 3.36**

Ans: 2 V.

The dual[2] of voltage division is current division. We are now given a total current supplied to several parallel resistors, as shown in the circuit of Fig. 3.37.

The current flowing through R_2 is

$$i_2 = \frac{v}{R_2} = \frac{i(R_1 \| R_2)}{R_2} = \frac{i}{R_2} \frac{R_1 R_2}{R_1 + R_2}$$

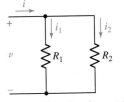

■ **FIGURE 3.37** An illustration of current division.

or

$$\boxed{i_2 = i \frac{R_1}{R_1 + R_2}} \qquad [12]$$

and, similarly,

$$\boxed{i_1 = i \frac{R_2}{R_1 + R_2}} \qquad [13]$$

Nature has not smiled on us here, for these last two equations have a factor which differs subtly from the factor used with voltage division, and some effort is going to be needed to avoid errors. Many students look on the expression for voltage division as "obvious" and that for current division as being "different." It helps to realize that the larger of two parallel resistors always carries the smaller current.

For a parallel combination of N resistors, the current through resistor R_k is

$$\boxed{i_k = i \frac{\dfrac{1}{R_k}}{\dfrac{1}{R_1} + \dfrac{1}{R_2} + \cdots + \dfrac{1}{R_N}}} \qquad [14]$$

(2) The principle of duality is encountered often in engineering. We will consider the topic briefly in Chap. 7 when we compare inductors and capacitors.

Written in terms of conductances,

$$i_k = i \frac{G_k}{G_1 + G_2 + \cdots + G_N}$$

which strongly resembles Eq. [11] for voltage division.

EXAMPLE **3.14**

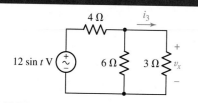

FIGURE 3.38 A circuit used as an example of current division. The wavy line in the voltage source symbol indicates a sinusoidal variation with time.

Write an expression for the current through the 3 Ω resistor in the circuit of Fig. 3.38.

The total current flowing into the 3 Ω–6 Ω combination is

$$i(t) = \frac{12 \sin t}{4 + 3 \| 6} = \frac{12 \sin t}{4 + 2} = 2 \sin t \quad \text{A}$$

and thus the desired current is given by current division:

$$i_3(t) = (2 \sin t) \left(\frac{6}{6 + 3} \right) = \frac{4}{3} \sin t \quad \text{A}$$

Unfortunately, current division is sometimes applied when it is not applicable. As one example, let us consider again the circuit shown in Fig. 3.33c, a circuit that we have already agreed contains no circuit elements that are in series or in parallel. Without parallel resistors, there is no way that current division can be applied. Even so, there are too many students who take a quick look at resistors R_A and R_B and try to apply current division, writing an incorrect equation such as

$$i_A \neq i_S \frac{R_B}{R_A + R_B}$$

Remember, *parallel resistors must be branches between the same pair of nodes*.

PRACTICE

3.16 In the circuit of Fig. 3.39, use resistance combination methods and current division to find i_1, i_2, and v_3.

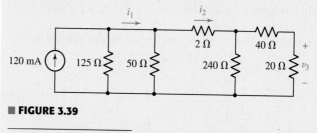

FIGURE 3.39

Ans: 100 mA; 50 mA; 0.8 V.

PRACTICAL APPLICATION

Not the Earth Ground from Geology

Up to now, we have been drawing circuit schematics in a fashion similar to that of the one shown in Fig. 3.40, where voltages are defined across two clearly marked terminals. Special care was taken to emphasize the fact that voltage cannot be defined at a single point—it is by definition the *difference* in potential between *two* points. However, many schematics make use of the convention of taking the earth as defining zero volts, so that all other voltages are implicitly referenced to this potential. The concept is often referred to as **earth ground,** and is fundamentally tied to safety regulations designed to prevent fires, fatal electrical shocks, and related mayhem. The symbol for earth ground is shown in Fig. 3.41*a*.

Since earth ground is defined as zero volts, it is often convenient to use this as a common terminal in schematics. The circuit of Fig. 3.40 is shown redrawn in this fashion in Fig. 3.42, where the earth ground symbol represents a common node. It is important to note that the two circuits are equivalent in terms of our value for v_a (4.5 V in either case), but are no longer exactly the same. The circuit in Fig. 3.40 is said to be "floating" in that it could for all practical purposes be installed on a circuit board of a satellite in geosynchronous orbit (or on its way to Pluto). The circuit in Fig. 3.42, however, is somehow physically connected to the ground through a conducting path. For this reason, there are two other symbols that are occasionally used to denote a common terminal. Figure 3.41*b* shows what is commonly referred to as **signal ground;** there can be (and often is) a large voltage between earth ground and any terminal tied to signal ground.

The fact that the common terminal of a circuit may or may not be connected by some low-resistance pathway to earth ground can lead to potentially dangerous situations. Consider the diagram of Fig. 3.43*a*, which depicts an innocent bystander about to touch a piece of equipment powered by an ac outlet. Only two terminals have been used from the wall socket; the round ground pin

of the receptacle was left unconnected. The common terminal of every circuit in the equipment has been tied together and electrically connected to the conducting equipment chassis; this terminal is often denoted using the **chassis ground** symbol of Fig. 3.41*c*. Unfortunately, a wiring fault exists, due to either poor manufacturing or perhaps just wear and tear. At any rate, the chassis is not "grounded," so there is a very large resistance between chassis ground and earth ground. A pseudo-schematic (some liberty was taken with the person's equivalent resistance symbol) of the situation is shown in Fig. 3.43*b*. The electrical path between the conducting chassis and ground may in fact be the table, which could represent a resistance of hundreds of megaohms or more. The resistance of the person, however, is many orders of magnitude lower. Once the person taps on the equipment to see why it isn't working properly . . . well, let's just say not all stories have happy endings.

The fact that "ground" is not always "earth ground" can cause a wide range of safety and electrical noise problems. One example is occasionally encountered in older buildings, where plumbing originally consisted of electrically conducting copper pipes. In such buildings, any water pipe was often treated as a low-resistance path to earth ground, and therefore used in many electrical connections. However, when corroded pipes are replaced with more modern and cost-effective

(*a*)   (*b*)   (*c*)

■ **FIGURE 3.41** Three different symbols used to represent a ground or common terminal: (*a*) earth ground; (*b*) signal ground; (*c*) chassis ground.

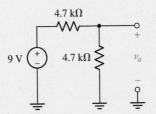

■ **FIGURE 3.42** The circuit of Fig. 3.40, redrawn using the earth ground symbol. The rightmost ground symbol is redundant; it is only necessary to label the positive terminal of v_a; the negative reference is then implicitly ground, or zero volts.

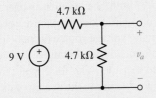

■ **FIGURE 3.40** A simple circuit with a voltage v_a defined between two terminals.

(Continued on next page)
(Continued on next page)

nonconducting PVC piping, the low-resistance path to earth ground no longer exists. A related problem occurs when the composition of the earth varies greatly over a particular region. In such situations, it is possible to actually have two separated buildings in which the two "earth grounds" are not equal, and current can flow as a result.

Within this text, the earth ground symbol will be used exclusively. It is worth remembering, however, that not all grounds are created equal in practice.

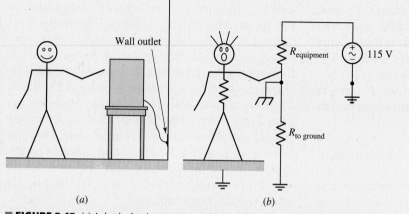

■ FIGURE 3.43 (*a*) A sketch of an innocent person about to touch an improperly grounded piece of equipment. It's not going to be pretty. (*b*) A schematic of an equivalent circuit for the situation as it is about to unfold; the person has been represented by an equivalent resistance, as has the equipment. A resistor has been used to represent the nonhuman path to ground.

SUMMARY AND REVIEW

We began this chapter by discussing connections of circuit elements, and introducing the terms *node*, *path*, *loop*, and *branch*. The next two topics could be considered the two most important in the entire textbook, namely, Kirchhoff's current law (KCL) and Kirchhoff's voltage law. The first is derived from conservation of charge, and can be thought of in terms of "what goes in (current) must come out." The second is based on conservation of energy, and can be viewed as "what goes up (potential) must come down." These two laws allow us to analyze *any* circuit, linear or otherwise, provided we have a way of relating the voltage and current associated with passive elements (e.g., Ohm's law for the resistor). In the case of a single-loop circuit, the elements are connected in *series* and hence each carries the same current. The single-node-pair circuit, in which elements are connected in *parallel* with one another, is characterized by a single voltage common to each element. Extending these concepts allowed us to develop a means of simplifying voltage sources connected in series, or current sources in parallel; subsequently we obtained classic expressions for series and parallel connected resistors. The final topic, that of voltage and current division, finds considerable use in the design of circuits where a specific voltage or current is required but our choice of source is limited.

Let's conclude with key points of this chapter to review, highlighting appropriate examples.

- Kirchhoff's current law (KCL) states that the algebraic sum of the currents entering any node is zero. (Examples 3.1, 3.4)
- Kirchhoff's voltage law (KVL) states that the algebraic sum of the voltages around any closed path in a circuit is zero. (Examples 3.2, 3.3)
- All elements in a circuit that carry the same current are said to be connected in series. (Example 3.5)
- Elements in a circuit having a common voltage across them are said to be connected in parallel. (Examples 3.6, 3.7)
- Voltage sources in series can be replaced by a single source, provided care is taken to note the individual polarity of each source. (Examples 3.8, 3.10)
- Current sources in parallel can be replaced by a single source, provided care is taken to note the direction of each current arrow. (Examples 3.9, 3.10)
- A series combination of N resistors can be replaced by a single resistor having the value $R_{eq} = R_1 + R_2 + \cdots + R_N$. (Example 3.11)
- A parallel combination of N resistors can be replaced by a single resistor having the value

$$\frac{1}{R_{eq}} = \frac{1}{R_1} + \frac{1}{R_2} + \cdots + \frac{1}{R_N}$$

(Example 3.12)

- Voltage division allows us to calculate what fraction of the total voltage across a series string of resistors is dropped across any one resistor (or group of resistors). (Example 3.13)
- Current division allows us to calculate what fraction of the total current into a parallel string of resistors flows through any one of the resistors. (Example 3.14)

READING FURTHER

A discussion of the principles of conservation of energy and conservation of charge, as well as Kirchhoff's laws, can be found in

R. Feynman, R. B. Leighton, and M. L. Sands, *The Feynman Lectures on Physics*. Reading, Mass.: Addison-Wesley, 1989, pp. 4-1, 4-7, and 25-9.

Detailed discussions of numerous aspects of grounding practices consistent with the 2008 National Electrical Code® can be found throughout

J. E. McPartland, B. J. McPartland, and F. P. Hartwell, *McGraw-Hill's National Electrical Code® 2008 Handbook*, 26th ed. New York, McGraw-Hill, 2008.

EXERCISES

3.1 Nodes, Paths, Loops, and Branches

1. Referring to the circuit depicted in Fig. 3.44, count the number of (*a*) nodes; (*b*) elements; (*c*) branches.

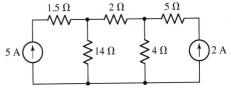

■ **FIGURE 3.44**

2. Referring to the circuit depicted in Fig. 3.45, count the number of (*a*) nodes; (*b*) elements; (*c*) branches.

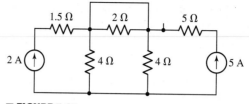

■ **FIGURE 3.45**

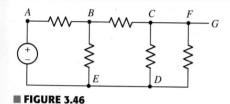

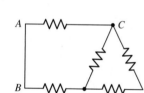

■ **FIGURE 3.46**

3. For the circuit of Fig. 3.46:
   (*a*) Count the number of nodes.
   (*b*) In moving from *A* to *B*, have we formed a path? Have we formed a loop?
   (*c*) In moving from *C* to *F* to *G*, have we formed a path? Have we formed a loop?

4. For the circuit of Fig. 3.46:
   (*a*) Count the number of circuit elements.
   (*b*) If we move from *B* to *C* to *D*, have we formed a path? Have we formed a loop?
   (*c*) If we move from *E* to *D* to *C* to *B*, have we formed a path? Have we formed a loop?

5. Refer to the circuit of Fig. 3.47, and answer the following:
   (*a*) How many distinct nodes are contained in the circuit?
   (*b*) How many elements are contained in the circuit?
   (*c*) How many branches does the circuit have?
   (*d*) Determine if each of the following represents a path, a loop, both, or neither:
      (*i*)   *A* to *B*
      (*ii*) *B* to *D* to *C* to *E*
      (*iii*) *C* to *E* to *D* to *B* to *A* to *C*
      (*iv*) *C* to *D* to *B* to *A* to *C* to *E*

■ **FIGURE 3.47**

3.2 Kirchhoff's Current Law

6. A local restaurant has a neon sign constructed from 12 separate bulbs; when a bulb fails, it appears as an infinite resistance and cannot conduct current. In wiring the sign, the manufacturer offers two options (Fig. 3.48). From what you've learned about KCL, which one should the restaurant owner select? Explain.

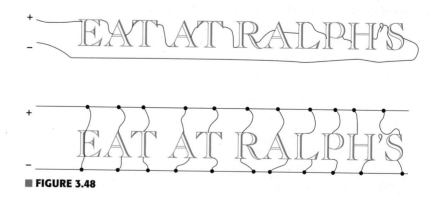

■ **FIGURE 3.48**

7. Referring to the single node diagram of Fig. 3.49, compute:
   (a) i_B, if $i_A = 1$ A, $i_D = -2$ A, $i_C = 3$ A, and $i_E = 0$;
   (b) i_E, if $i_A = -1$ A, $i_B = -1$ A, $i_C = -1$ A, and $i_D = -1$ A.

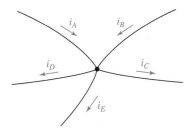

■ **FIGURE 3.49**

8. Determine the current labeled I in each of the circuits of Fig. 3.50.

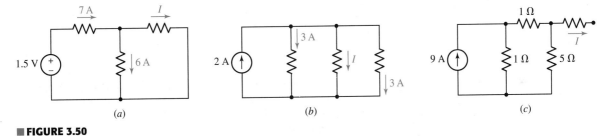

(a)   (b)   (c)

■ **FIGURE 3.50**

9. In the circuit shown in Fig. 3.51, the resistor values are unknown, but the 2 V source is known to be supplying a current of 7 A to the rest of the circuit. Calculate the current labeled i_2.

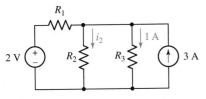

■ **FIGURE 3.51**

10. The voltage source in the circuit of Fig. 3.52 has a current of 1 A flowing out of its positive terminal into resistor R_1. Calculate the current labeled i_2.

11. In the circuit depicted in Fig. 3.53, i_x is determined to be 1.5 A, and the 9 V source supplies a current of 7.6 A (that is, a current of 7.6 A leaves the positive reference terminal of the 9 V source). Determine the value of resistor R_A.

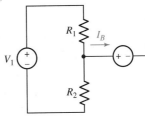

■ **FIGURE 3.52**

12. For the circuit of Fig. 3.54 (which is a model for the dc operation of a bipolar junction transistor biased in forward active region), I_B is measured to be 100 μA. Determine I_C and I_E.

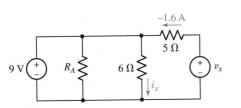

■ **FIGURE 3.53**

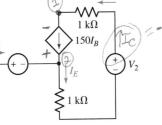

■ **FIGURE 3.54**

13. Determine the current labeled I_3 in the circuit of Fig. 3.55.

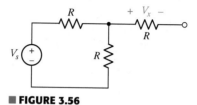

■ **FIGURE 3.55**

14. Study the circuit depicted in Fig. 3.56, and explain (in terms of KCL) why the voltage labeled V_x must be zero.

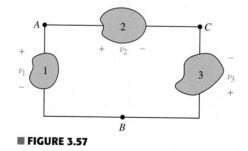

■ **FIGURE 3.56**

15. In many households, multiple electrical outlets within a given room are often all part of the same circuit. Draw the circuit for a four-walled room which has a single electrical outlet per wall, with a lamp (represented by a 1 Ω resistor) connected to each outlet.

3.3 Kirchoff's Voltage Law

16. For the circuit of Fig. 3.57:

   (a) Determine the voltage v_1 if $v_2 = 0$ and $v_3 = -17$ V.
   (b) Determine the voltage v_1 if $v_2 = -2$ V and $v_3 = +2$ V.
   (c) Determine the voltage v_2 if $v_1 = 7$ V and $v_3 = 9$ V.
   (d) Determine the voltage v_3 if $v_1 = -2.33$ V and $v_2 = -1.70$ V.

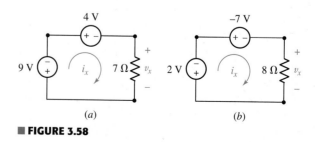

■ **FIGURE 3.57**

17. For each of the circuits in Fig. 3.58, determine the voltage v_x and the current i_x.

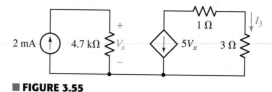

(a)   (b)

■ **FIGURE 3.58**

18. Use KVL to obtain a numerical value for the current labeled i in each circuit depicted in Fig. 3.59.

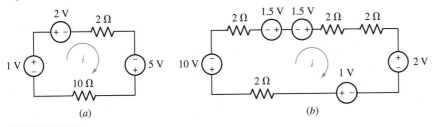

■ **FIGURE 3.59**

19. In the circuit of Fig. 3.60, it is determined that $v_1 = 3$ V and $v_3 = 1.5$ V. Calculate v_R and v_2.

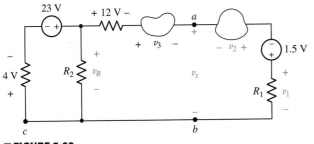

■ **FIGURE 3.60**

20. In the circuit of Fig. 3.60, a voltmeter is used to measure the following: $v_1 = 2$ V and $v_3 = -1.5$ V. Calculate v_x.

21. Determine the value of v_x as labeled in the circuit of Fig. 3.61.

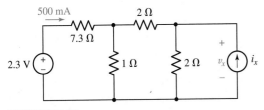

■ **FIGURE 3.61**

22. Consider the simple circuit shown in Fig. 3.62. Using KVL, derive the expressions

$$v_1 = v_s \frac{R_1}{R_1 + R_2} \quad \text{and} \quad v_2 = v_s \frac{R_2}{R_1 + R_2}$$

23. (*a*) Determine a numerical value for each current and voltage (i_1, v_1, etc.) in the circuit of Fig. 3.63. (*b*) Calculate the power absorbed by each element and verify that they sum to zero.

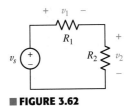

■ **FIGURE 3.62**

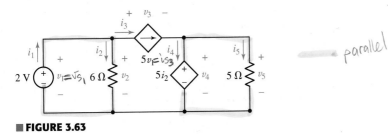

■ **FIGURE 3.63**

24. The circuit shown in Fig. 3.64 includes a device known as an op amp. This device has two unusual properties in the circuit shown: (1) $V_d = 0$ V, and (2) no current can flow into either input terminal (marked "−" and "+" inside the symbol), but it *can* flow through the output terminal (marked "OUT"). This seemingly impossible situation—in direct conflict with KCL—is a result of power leads to the device that are not included in the symbol. Based on this information, calculate V_{out}. (*Hint*: two KVL equations are required, both involving the 5 V source.)

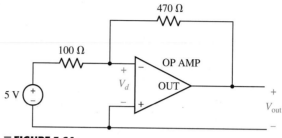

■ FIGURE 3.64

3.4 The Single-Loop Circuit

25. The circuit of Fig. 3.12*b* is constructed with the following: $v_{s1} = -8$ V, $R_1 = 1 \, \Omega$, $v_{s2} = 16$ V, and $R_2 = 4.7 \, \Omega$. Calculate the power absorbed by each element. Verify that the absorbed powers sum to zero.

26. Obtain a numerical value for the power absorbed by each element in the circuit shown in Fig. 3.65.

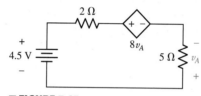

■ FIGURE 3.65

27. Compute the power absorbed by each element of the circuit of Fig. 3.66.

28. Compute the power absorbed by each element in the circuit of Fig. 3.67 if the mysterious element X is (*a*) a 13 Ω resistor; (*b*) a dependent voltage source labeled $4v_1$, "+" reference on top; (*c*) a dependent voltage source labeled $4i_x$, "+" reference on top.

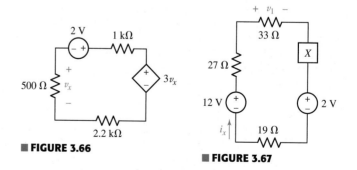

■ FIGURE 3.66

■ FIGURE 3.67

29. Kirchhoff's laws apply whether or not Ohm's law applies to a particular element. The *I-V* characteristic of a diode, for example, is given by

$$I_D = I_S \left(e^{V_D/V_T} - 1 \right)$$

where $V_T = 27$ mV at room temperature and I_S can vary from 10^{-12} to 10^{-3} A. In the circuit of Fig. 3.68, use KVL/KCL to obtain V_D if $I_S = 29$ pA. (*Note: This problem results in a transcendental equation, requiring an iterative approach to obtaining a numerical solution. Most scientific calculators will perform such a function.*)

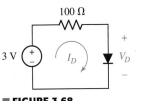

■ **FIGURE 3.68**

3.5 The Single-Node-Pair Circuit

30. Referring to the circuit of Fig. 3.69, (*a*) determine the two currents i_1 and i_2; (*b*) compute the power absorbed by each element.

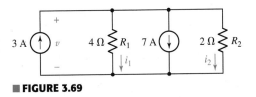

■ **FIGURE 3.69**

31. Determine a value for the voltage v as labeled in the circuit of Fig. 3.70, and compute the power supplied by the two current sources.

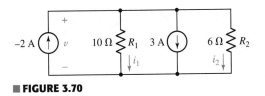

■ **FIGURE 3.70**

32. Referring to the circuit depicted in Fig. 3.71, determine the value of the voltage v.

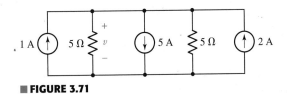

■ **FIGURE 3.71**

33. Determine the voltage v as labeled in Fig. 3.72, and calculate the power supplied by each current source.

34. Although drawn so that it may not appear obvious at first glance, the circuit of Fig. 3.73 is in fact a single-node-pair circuit. (*a*) Determine the power absorbed by each resistor. (*b*) Determine the power supplied by each current source. (*c*) Show that the sum of the absorbed power calculated in (*a*) is equal to the sum of the supplied power calculated in (*c*).

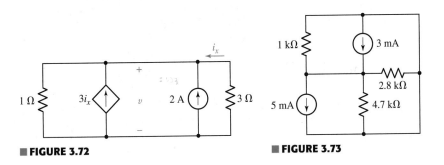

■ **FIGURE 3.72**   ■ **FIGURE 3.73**

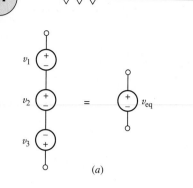

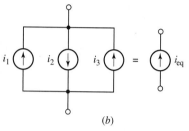

(a)

(b)

■ **FIGURE 3.74**

3.6 Series and Parallel Connected Sources

35. Determine the numerical value for v_{eq} in Fig. 3.74a, if (a) $v_1 = 0$, $v_2 = -3$ V, and $v_3 = +3$ V; (b) $v_1 = v_2 = v_3 = 1$ V; (c) $v_1 = -9$ V, $v_2 = 4.5$ V, $v_3 = 1$ V.

36. Determine the numerical value for i_{eq} in Fig. 3.74b, if (a) $i_1 = 0$, $i_2 = -3$ A, and $i_3 = +3$ A; (b) $i_1 = i_2 = i_3 = 1$ A; (c) $i_1 = -9$ A, $i_2 = 4.5$ A, $i_3 = 1$ A.

37. For the circuit presented in Fig. 3.75, determine the current labeled i by first combining the four sources into a single equivalent source.

38. Determine the value of v_1 required to obtain a zero value for the current labeled i in the circuit of Fig. 3.76.

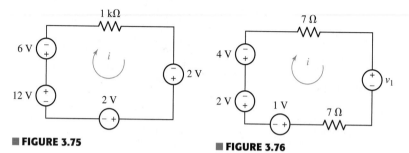

■ **FIGURE 3.75**   ■ **FIGURE 3.76**

39. (a) For the circuit of Fig. 3.77, determine the value for the voltage labeled v, after first simplifying the circuit to a single current source in parallel with two resistors. (b) Verify that the power supplied by your equivalent source is equal to the sum of the supplied powers of the individual sources in the original circuit.

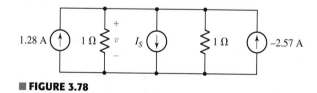

■ **FIGURE 3.77**

40. What value of I_S in the circuit of Fig. 3.78 will result in a zero voltage v?

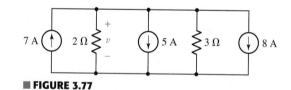

■ **FIGURE 3.78**

41. (a) Determine the values for I_X and V_Y in the circuit shown in Fig. 3.79. (b) Are those values necessarily unique for that circuit? Explain. (c) Simplify the circuit of Fig. 3.79 as much as possible and still maintain the values for v and i. (Your circuit must contain the 1 Ω resistor.)

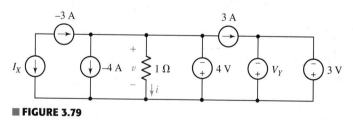

■ **FIGURE 3.79**

3.7 Resistors in Series and Parallel

42. Determine the equivalent resistance of each of the networks shown in Fig. 3.80.

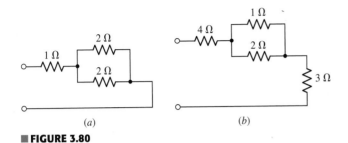

(a)   (b)

■ **FIGURE 3.80**

43. For each network depicted in Fig. 3.81, determine a single equivalent resistance.

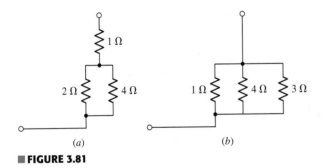

(a)   (b)

■ **FIGURE 3.81**

44. (a) Simplify the circuit of Fig. 3.82 as much as possible by using source and resistor combinations. (b) Calculate i, using your simplified circuit. (c) To what voltage should the 1 V source be changed to reduce i to zero? (d) Calculate the power absorbed by the 5 Ω resistor.

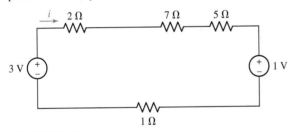

■ **FIGURE 3.82**

45. (a) Simplify the circuit of Fig. 3.83, using appropriate source and resistor combinations. (b) Determine the voltage labeled v, using your simplified circuit. (c) Calculate the power provided by the 2 A source to the rest of the circuit.

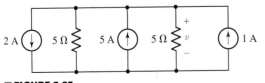

■ **FIGURE 3.83**

46. Making appropriate use of resistor combination techniques, calculate i_3 in the circuit of Fig. 3.84 and the power provided to the circuit by the single current source.

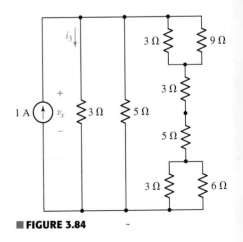

■ **FIGURE 3.84**

47. Calculate the voltage labeled v_x in the circuit of Fig. 3.85 after first simplifying, using appropriate source and resistor combinations.

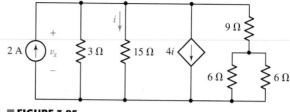

■ FIGURE 3.85

48. Determine the power absorbed by the 15 Ω resistor in the circuit of Fig. 3.86.

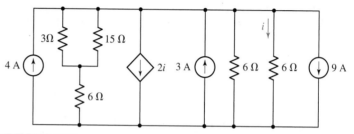

■ FIGURE 3.86

49. Calculate the equivalent resistance R_{eq} of the network shown in Fig. 3.87 if
$R_1 = 2R_2 = 3R_3 = 4R_4$ etc. and $R_{11} = 3$ Ω.

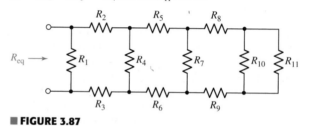

■ FIGURE 3.87

50. Show how to combine four 100 Ω resistors to obtain an equivalent resistance
of (a) 25 Ω; (b) 60 Ω; (c) 40 Ω.

3.8 Voltage and Current Division

51. In the voltage divider network of Fig. 3.88, calculate (a) v_2 if $v = 9.2$ V and
$v_1 = 3$ V; (b) v_1 if $v_2 = 1$ V and $v = 2$ V; (c) v if $v_1 = 3$ V and $v_2 = 6$ V;
(d) R_1/R_2 if $v_1 = v_2$; (e) v_2 if $v = 3.5$ V and $R_1 = 2R_2$; (f) v_1 if $v = 1.8$ V,
$R_1 = 1$ kΩ, and $R_2 = 4.7$ kΩ.

52. In the current divider network represented in Fig. 3.89, calculate (a) i_1 if
$i = 8$ A and $i_2 = 1$ A; (b) v if $R_1 = 100$ kΩ, $R_2 = 100$ kΩ, and $i = 1$ mA;
(c) i_2 if $i = 20$ mA, $R_1 = 1$ Ω, and $R_2 = 4$ Ω; (d) i_1 if $i = 10$ A, $R_1 = R_2 = 9$ Ω;
(e) i_2 if $i = 10$ A, $R_1 = 100$ MΩ, and $R_2 = 1$Ω.

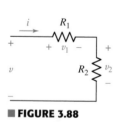

■ FIGURE 3.88

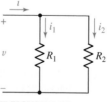

■ FIGURE 3.89

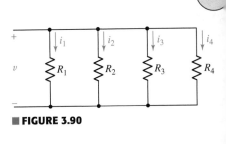

53. Choose a voltage $v < 2.5$ V and values for the resistors R_1, R_2, R_3, and R_4 in the circuit of Fig. 3.90 so that $i_1 = 1$ A, $i_2 = 1.2$ A, $i_3 = 8$ A, and $i_4 = 3.1$ A.

54. Employ voltage division to assist in the calculation of the voltage labeled v_x in the circuit of Fig. 3.91.

55. A network is constructed from a series connection of five resistors having values 1 Ω, 3 Ω, 5 Ω, 7 Ω, and 9 Ω. If 9 V is connected across the terminals of the network, employ voltage division to calculate the voltage across the 3 Ω resistor, and the voltage across the 7 Ω resistor.

56. Employing resistance combination and current division as appropriate, determine values for i_1, i_2, and v_3 in the circuit of Fig. 3.92.

■ **FIGURE 3.90**

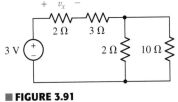

■ **FIGURE 3.91**

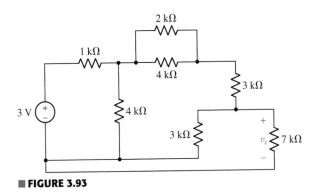

■ **FIGURE 3.92**

57. In the circuit of Fig. 3.93, only the voltage v_x is of interest. Simplify the circuit using appropriate resistor combinations and iteratively employ voltage division to determine v_x.

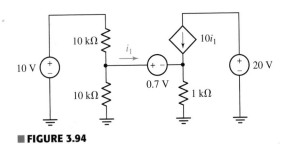

■ **FIGURE 3.93**

Chapter-Integrating Exercises

58. The circuit shown in Fig. 3.94 is a linear model of a bipolar junction transistor biased in the forward active region of operation. Explain why voltage division is not a valid approach for determining the voltage across either 10 kΩ resistor.

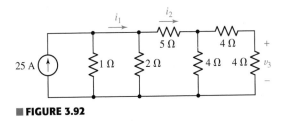

■ **FIGURE 3.94**

59. A common midfrequency model for a field effect–based amplifier circuit is shown in Fig. 3.95. If the controlling parameter g_m (known as the *transconductance*) is equal to 1.2 mS, employ current division to obtain the current through the 1 kΩ resistor, and then calculate the amplifier output voltage v_{out}.

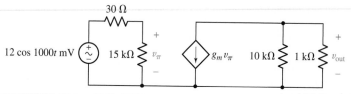
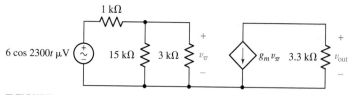

■ FIGURE 3.95

60. The circuit depicted in Fig. 3.96 is routinely employed to model the midfrequency operation of a bipolar junction transistor–based amplifier. Calculate the amplifier output v_{out} if the transconductance g_m is equal to 322 mS.

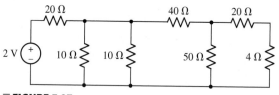

■ FIGURE 3.96

61. With regard to the circuit shown in Fig. 3.97, compute (*a*) the voltage across the two 10 Ω resistors, assuming the top terminal is the positive reference; (*b*) the power dissipated by the 4 Ω resistor.

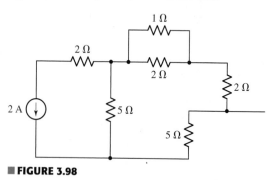

■ FIGURE 3.97

62. Delete the leftmost 10 Ω resistor in the circuit of Fig. 3.97, and compute (*a*) the current flowing into the left-hand terminal of the 40 Ω resistor; (*b*) the power supplied by the 2 V source; (*c*) the power dissipated by the 4 Ω resistor.

63. Consider the seven-element circuit depicted in Fig. 3.98. (*a*) How many nodes, loops, and branches does it contain? (*b*) Calculate the current flowing through each resistor. (*c*) Determine the voltage across the current source, assuming the top terminal is the positive reference terminal.

■ FIGURE 3.98

Basic Nodal and Mesh Analysis

INTRODUCTION

Armed with the trio of Ohm's and Kirchhoff's laws, analyzing a simple linear circuit to obtain useful information such as the current, voltage, or power associated with a particular element is perhaps starting to seem a straightforward enough venture. Still, for the moment at least, every circuit seems unique, requiring (to some degree) a measure of creativity in approaching the analysis. In this chapter, we learn two basic circuit analysis techniques— *nodal analysis* and *mesh analysis*—both of which allow us to investigate many different circuits with a consistent, methodical approach. The result is a streamlined analysis, a more uniform level of complexity in our equations, fewer errors and, perhaps most importantly, a reduced occurrence of "*I don't know how to even start!*"

Most of the circuits we have seen up to now have been rather simple and (to be honest) of questionable practical use. Such circuits are valuable, however, in helping us to learn to apply fundamental techniques. Although the more complex circuits appearing in this chapter may represent a variety of electrical systems including control circuits, communication networks, motors, or integrated circuits, as well as electric circuit models of nonelectrical systems, we believe it best not to dwell on such specifics at this early stage. Rather, it is important to initially focus on the *methodology of problem solving* that we will continue to develop throughout the book.

4.1 NODAL ANALYSIS

We begin our study of general methods for methodical circuit analysis by considering a powerful method based on KCL, namely *nodal analysis.* In Chap. 3 we considered the analysis of a simple circuit containing only two nodes. We found that the major step of the analysis was obtaining a single equation in terms of a single unknown quantity—the voltage between the pair of nodes.

We will now let the number of nodes increase and correspondingly provide one additional unknown quantity and one additional equation for each added node. Thus, a three-node circuit should have two unknown voltages and two equations; a 10-node circuit will have nine unknown voltages and nine equations; an N-node circuit will need $(N - 1)$ voltages and $(N - 1)$ equations. Each equation is a simple KCL equation.

To illustrate the basic technique, consider the three-node circuit shown in Fig. 4.1a, redrawn in Fig. 4.1b to emphasize the fact that there are only three nodes, numbered accordingly. Our goal will be to determine the voltage across each element, and the next step in the analysis is critical. We designate one node as a *reference node;* it will be the negative terminal of our $N - 1 = 2$ nodal voltages, as shown in Fig. 4.1c.

A little simplification in the resultant equations is obtained if the node connected to the greatest number of branches is identified as the reference node. If there is a ground node, it is usually most convenient to select it as the reference node, although many people seem to prefer selecting the bottom node of a circuit as the reference, especially if no explicit ground is noted.

The voltage of node 1 *relative to the reference node* is named v_1, and v_2 is defined as the voltage of node 2 with respect to the reference node. These

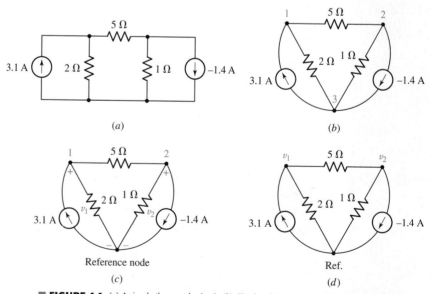

(a)

(b)

(c)

Reference node

(d)

Ref.

■ **FIGURE 4.1** (*a*) A simple three-node circuit. (*b*) Circuit redrawn to emphasize nodes. (*c*) Reference node selected and voltages assigned. (*d*) Shorthand voltage references. If desired, an appropriate ground symbol may be substituted for "Ref."

two voltages are all we need, as the voltage between any other pair of nodes may be found in terms of them. For example, the voltage of node 1 with respect to node 2 is $v_1 - v_2$. The voltages v_1 and v_2 and their reference signs are shown in Fig. 4.1c. It is common practice once a reference node has been labeled to omit the reference signs for the sake of clarity; the node labeled with the voltage is taken to be the positive terminal (Fig. 4.1d). This is understood to be a type of shorthand voltage notation.

We now apply KCL to nodes 1 and 2. We do this by equating the total current leaving the node through the several resistors to the total source current entering the node. Thus,

$$\frac{v_1}{2} + \frac{v_1 - v_2}{5} = 3.1 \qquad [1]$$

or

$$0.7v_1 - 0.2v_2 = 3.1 \qquad [2]$$

At node 2 we obtain

$$\frac{v_2}{1} + \frac{v_2 - v_1}{5} = -(-1.4) \qquad [3]$$

or

$$-0.2v_1 + 1.2v_2 = 1.4 \qquad [4]$$

Equations [2] and [4] are the desired two equations in two unknowns, and they may be solved easily. The results are $v_1 = 5$ V and $v_2 = 2$ V.

From this, it is straightforward to determine the voltage across the 5 Ω resistor: $v_{5\Omega} = v_1 - v_2 = 3$ V. The currents and absorbed powers may also be computed in one step.

We should note at this point that there is more than one way to write the KCL equations for nodal analysis. For example, the reader may prefer to sum all the currents entering a given node and set this quantity to zero. Thus, for node 1 we might have written

$$3.1 - \frac{v_1}{2} - \frac{v_1 - v_2}{5} = 0$$

or

$$3.1 + \frac{-v_1}{2} + \frac{v_2 - v_1}{5} = 0$$

either of which is equivalent to Eq. [1].

Is one way better than any other? Every instructor and every student develop a personal preference, and at the end of the day the most important thing is to be consistent. The authors prefer constructing KCL equations for nodal analysis in such a way as to end up with all current source terms on one side and all resistor terms on the other. Specifically,

$\sum$ currents entering the node from current sources
$= \sum$ currents leaving the node through resistors

There are several advantages to such an approach. First, there is never any confusion regarding whether a term should be "$v_1 - v_2$" or "$v_2 - v_1$;" the

The reference node in a schematic is implicitly defined as zero volts. However, it is important to remember that any terminal can be designated as the reference terminal. Thus, the reference node is at zero volts with respect to the other defined nodal voltages, and not necessarily with respect to *earth* ground.

first voltage in every resistor current expression corresponds to the node for which a KCL equation is being written, as seen in Eqs. [1] and [3]. Second, it allows a quick check that a term has not been accidentally omitted. Simply count the current sources connected to a node and then the resistors; grouping them in the stated fashion makes the comparison a little easier.

EXAMPLE 4.1

Determine the current flowing left to right through the 15 Ω resistor of Fig. 4.2a.

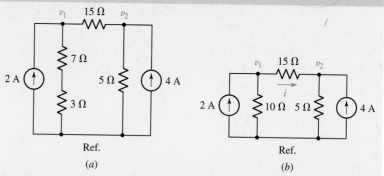

■ **FIGURE 4.2** (*a*) A four-node circuit containing two independent current sources. (*b*) The two resistors in series are replaced with a single 10 Ω resistor, reducing the circuit to three nodes.

Nodal analysis will directly yield numerical values for the nodal voltages v_1 and v_2, and the desired current is given by $i = (v_1 - v_2)/15$.

Before launching into nodal analysis, however, we first note that no details regarding either the 7 Ω resistor or the 3 Ω resistor are of interest. Thus, we may replace their series combination with a 10 Ω resistor as in Fig. 4.2*b*. The result is a reduction in the number of equations to solve.

Writing an appropriate KCL equation for node 1,

$$2 = \frac{v_1}{10} + \frac{v_1 - v_2}{15} \qquad [5]$$

and for node 2,

$$4 = \frac{v_2}{5} + \frac{v_2 - v_1}{15} \qquad [6]$$

Rearranging, we obtain

$$5v_1 - 2v_2 = 60$$

and

$$-v_1 + 4v_2 = 60$$

Solving, we find that $v_1 = 20$ V and $v_2 = 20$ V so that $v_1 - v_2 = 0$. In other words, ***zero current*** is flowing through the 15 Ω resistor in this circuit!

PRACTICE

4.1 For the circuit of Fig. 4.3, determine the nodal voltages v_1 and v_2.

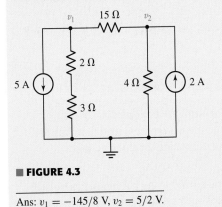

■ **FIGURE 4.3**

Ans: $v_1 = -145/8$ V, $v_2 = 5/2$ V.

Now let us increase the number of nodes so that we may use this technique to work a slightly more difficult problem.

EXAMPLE **4.2**

Determine the nodal voltages for the circuit of Fig. 4.4a, as referenced to the bottom node.

▶ *Identify the goal of the problem.*
There are four nodes in this circuit. With the bottom node as our reference, we label the other three nodes as shown in Fig. 4.4b. The circuit has been redrawn for clarity, taking care to identify the two relevant nodes for the 4 Ω resistor.

▶ *Collect the known information.*
We have three unknown voltages, v_1, v_2, and v_3. All current sources and resistors have designated values, which are marked on the schematic.

▶ *Devise a plan.*
This problem is well suited to nodal analysis, as three independent KCL equations may be written in terms of the current sources and the current through each resistor.

▶ *Construct an appropriate set of equations.*
We begin by writing a KCL equation for node 1:

$$-8 - 3 = \frac{v_1 - v_2}{3} + \frac{v_1 - v_3}{4}$$

or

$$0.5833v_1 - 0.3333v_2 - 0.25v_3 = -11 \qquad [7]$$

At node 2:

$$-(-3) = \frac{v_2 - v_1}{3} + \frac{v_2}{1} + \frac{v_2 - v_3}{7}$$

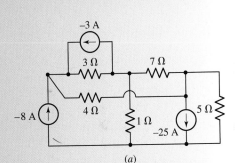

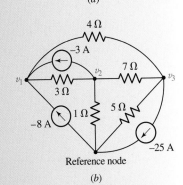

■ **FIGURE 4.4** (*a*) A four-node circuit. (*b*) Redrawn circuit with reference node chosen and voltages labeled.

(*Continued on next page*)

or

$$-0.3333v_1 + 1.4762v_2 - 0.1429v_3 = 3 \qquad [8]$$

And, at node 3:

$$-(-25) = \frac{v_3}{5} + \frac{v_3 - v_2}{7} + \frac{v_3 - v_1}{4}$$

or, more simply,

$$-0.25v_1 - 0.1429v_2 + 0.5929v_3 = 25 \qquad [9]$$

▶ **Determine if additional information is required.**
We have three equations in three unknowns. Provided that they are independent, this is sufficient to determine the three voltages.

▶ **Attempt a solution.**
Equations [7] through [9] can be solved using a scientific calculator (Appendix 5), software packages such as MATLAB, or more traditional "plug-and-chug" techniques such as elimination of variables, matrix methods, or Cramer's rule. Using the latter method, described in Appendix 2, we have

$$v_1 = \frac{\begin{vmatrix} -11 & -0.3333 & -0.2500 \\ 3 & 1.4762 & -0.1429 \\ 25 & -0.1429 & 0.5929 \end{vmatrix}}{\begin{vmatrix} 0.5833 & -0.3333 & -0.2500 \\ -0.3333 & 1.4762 & -0.1429 \\ -0.2500 & -0.1429 & 0.5929 \end{vmatrix}} = \frac{1.714}{0.3167} = 5.412 \text{ V}$$

Similarly,

$$v_2 = \frac{\begin{vmatrix} 0.5833 & -11 & -0.2500 \\ -0.3333 & 3 & -0.1429 \\ -0.2500 & 25 & 0.5929 \end{vmatrix}}{0.3167} = \frac{2.450}{0.3167} = 7.736 \text{ V}$$

and

$$v_3 = \frac{\begin{vmatrix} 0.5833 & -0.3333 & -11 \\ -0.3333 & 1.4762 & 3 \\ -0.2500 & -0.1429 & 25 \end{vmatrix}}{0.3167} = \frac{14.67}{0.3167} = 46.32 \text{ V}$$

▶ **Verify the solution. Is it reasonable or expected?**
Substituting the nodal voltages into any of our three nodal equations is sufficient to ensure we made no computational errors. Beyond that, is it possible to determine whether these voltages are "reasonable" values? We have a maximum possible current of $3 + 8 + 25 = 36$ amperes anywhere in the circuit. The largest resistor is 7 Ω, so we do not expect any voltage magnitude greater than $7 \times 36 = 252$ V.

There are, of course, numerous methods available for the solution of linear systems of equations, and we describe several in Appendix 2 in detail. Prior to the advent of the scientific calculator, Cramer's rule as seen in Example 4.2 was very common in circuit analysis, although occasionally tedious to implement. It is, however, straightforward to use on a simple four-function calculator, and so an awareness of the technique can be valuable. MATLAB, on the other hand, although not likely to be available during an examination, is a powerful software package that can greatly simplify the solution process; a brief tutorial on getting started is provided in Appendix 6.

For the situation encountered in Example 4.2, there are several options available through MATLAB. First, we can represent Eqs. [7] to [9] in *matrix form:*

$$
\begin{bmatrix} 0.5833 & -0.3333 & -0.25 \\ -0.3333 & 1.4762 & -0.1429 \\ -0.25 & -0.1429 & 0.5929 \end{bmatrix} \begin{bmatrix} v_1 \\ v_2 \\ v_3 \end{bmatrix} = \begin{bmatrix} -11 \\ 3 \\ 25 \end{bmatrix}
$$

so that

$$
\begin{bmatrix} v_1 \\ v_2 \\ v_3 \end{bmatrix} = \begin{bmatrix} 0.5833 & -0.3333 & -0.25 \\ -0.3333 & 1.4762 & -0.1429 \\ -0.25 & -0.1429 & 0.5929 \end{bmatrix}^{-1} \begin{bmatrix} -11 \\ 3 \\ 25 \end{bmatrix}
$$

In MATLAB, we write

```
>> a = [0.5833 -0.3333 -0.25; -0.3333 1.4762 -0.1429; -0.25 -0.1429 0.5929];
>> c = [-11; 3; 25];
>> b = a^-1 * c

b =

   5.4124
   7.7375
  46.3127

>>
```

where spaces separate elements along rows, and a semicolon separates rows. The matrix named **b**, which can also be referred to as a *vector* as it has only one column, is our solution. Thus, $v_1 = 5.412$ V, $v_2 = 7.738$ V, and $v_3 = 46.31$ V (some rounding error has been incurred).

We could also use the KCL equations as we wrote them initially if we employ the symbolic processor of MATLAB.

```
>> eqn1 = '-8 -3 = (v1 - v2)/ 3 + (v1 - v3)/ 4';
>> eqn2 = '-(-3) = (v2 - v1)/ 3 + v2/ 1 + (v2 - v3)/ 7';
>> eqn3 = '-(-25) = v3/ 5 + (v3 - v2)/ 7 + (v3 - v1)/ 4';
>> answer = solve(eqn1, eqn2, eqn3, 'v1', 'v2', 'v3');
>> answer.v1

ans =

720/133

>> answer.v2

ans =

147/19

>> answer.v3

ans =

880/19

>>
```

which results in exact answers, with no rounding errors. The *solve()* routine is invoked with the list of symbolic equations we named eqn1, eqn2, and eqn3, but the variables v1, v2 and v3 must also be specified. If *solve()* is called with fewer variables than equations, an algebraic solution is returned. The form of the solution is worth a quick comment; it is returned in what is referred to in programming parlance as a *structure;* in this case, we called our structure "answer." Each component of the structure is accessed separately by name as shown.

PRACTICE

4.2 For the circuit of Fig. 4.5, compute the voltage across each current source.

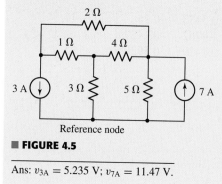

FIGURE 4.5

Ans: $v_{3A} = 5.235$ V; $v_{7A} = 11.47$ V.

The previous examples have demonstrated the basic approach to nodal analysis, but it is worth considering what happens if dependent sources are present as well.

EXAMPLE 4.3

Determine the power supplied by the dependent source of Fig. 4.6a.

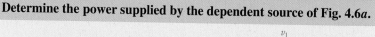
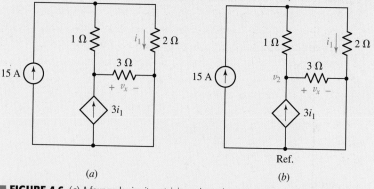

FIGURE 4.6 (*a*) A four-node circuit containing a dependent current source. (*b*) Circuit labeled for nodal analysis.

We choose the bottom node as our reference, since it has a large number of branch connections, and proceed to label the nodal voltages v_1 and v_2 as shown in Fig. 4.6b. The quantity labeled v_x is actually equal to v_2.

At node 1, we write

$$15 = \frac{v_1 - v_2}{1} + \frac{v_1}{2} \qquad [10]$$

and at node 2

$$3i_1 = \frac{v_2 - v_1}{1} + \frac{v_2}{3} \qquad [11]$$

Unfortunately, we have only two equations but three unknowns; *this is a direct result of the presence of the dependent current source, since it is not controlled by a nodal voltage.* Thus, we need an additional equation that relates i_1 to one or more nodal voltages.

In this case, we find that

$$i_1 = \frac{v_1}{2} \qquad [12]$$

which upon substitution into Eq. [11] yields (with a little rearranging)

$$3v_1 - 2v_2 = 30 \qquad [13]$$

and Eq. [10] simplifies to

$$-15v_1 + 8v_2 = 0 \qquad [14]$$

Solving, we find that $v_1 = -40$ V, $v_2 = -75$ V, and $i_1 = 0.5v_1 = -20$ A. Thus, the power supplied by the dependent source is equal to $(3i_1)(v_2) = (-60)(-75) = 4.5$ kW.

We see that the presence of a dependent source will create the need for an additional equation in our analysis if the controlling quantity is not a nodal voltage. Now let's look at the same circuit, but with the controlling variable of the dependent current source changed to a different quantity—the voltage across the 3 Ω resistor, which is in fact a nodal voltage. We will find that only *two* equations are required to complete the analysis.

EXAMPLE 4.4

Determine the power supplied by the dependent source of Fig. 4.7a.

We select the bottom node as our reference and label the nodal voltages as shown in Fig. 4.7b. We have labeled the nodal voltage v_x explicitly for clarity. Note that our choice of reference node is important in this case; it led to the quantity v_x being a nodal voltage.

Our KCL equation for node 1 is

$$15 = \frac{v_1 - v_x}{1} + \frac{v_1}{2} \qquad [15]$$

(Continued on next page)

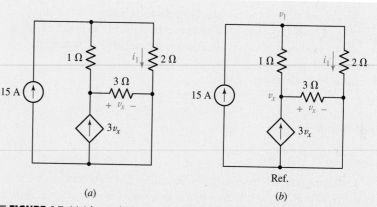

(a)   (b)

■ **FIGURE 4.7** (a) A four-node circuit containing a dependent current source. (b) Circuit labeled for nodal analysis.

and for node x is

$$3v_x = \frac{v_x - v_1}{1} + \frac{v_2}{3} \qquad [16]$$

Grouping terms and solving, we find that $v_1 = \frac{50}{7}$ V and $v_x = -\frac{30}{7}$ V. Thus, the dependent source in this circuit generates $(3v_x)(v_x) = 55.1$ W.

PRACTICE

4.3 For the circuit of Fig. 4.8, determine the nodal voltage v_1 if A is (a) $2i_1$; (b) $2v_1$.

Ans: (a) $\frac{70}{9}$ V; (b) -10 V.

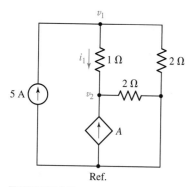

■ **FIGURE 4.8**

Summary of Basic Nodal Analysis Procedure

1. **Count the number of nodes** (N).

2. **Designate a reference node.** The number of terms in your nodal equations can be minimized by selecting the node with the greatest number of branches connected to it.

3. **Label the nodal voltages** (there are $N - 1$ of them).

4. **Write a KCL equation for each of the nonreference nodes.** Sum the currents flowing *into* a node from sources on one side of the equation. On the other side, sum the currents flowing *out of* the node through resistors. Pay close attention to "−" signs.

5. **Express any additional unknowns such as currents or voltages other than nodal voltages in terms of appropriate nodal voltages.** This situation can occur if voltage sources or dependent sources appear in our circuit.

6. **Organize the equations.** Group terms according to nodal voltages.

7. **Solve the system of equations for the nodal voltages** (there will be $N - 1$ of them).

These seven basic steps will work on any circuit we ever encounter, although the presence of voltage sources will require extra care. Such situations are discussed next.

4.2 THE SUPERNODE

As an example of how voltage sources are best handled when performing nodal analysis, consider the circuit shown in Fig. 4.9*a*. The original four-node circuit of Fig. 4.4 has been changed by replacing the 7 Ω resistor between nodes 2 and 3 with a 22 V voltage source. We still assign the same node-to-reference voltages v_1, v_2, and v_3. Previously, the next step was the application of KCL at each of the three nonreference nodes. If we try to do that once again, we see that we will run into some difficulty at both nodes 2 and 3, for we do not know what the current is in the branch with the voltage source. There is no way by which we can express the current as a function of the voltage, for the definition of a voltage source is exactly that the voltage is independent of the current.

There are two ways out of this dilemma. The more difficult approach is to assign an unknown current to the branch which contains the voltage source, proceed to apply KCL three times, and then apply KVL ($v_3 - v_2 = 22$) once between nodes 2 and 3; the result is then four equations in four unknowns.

The easier method is to treat node 2, node 3, and the voltage source together as a sort of ***supernode*** and apply KCL to both nodes at the same time; the supernode is indicated by the region enclosed by the broken line in Fig. 4.9*a*. This is okay because if the total current leaving node 2 is zero and the total current leaving node 3 is zero, then the total current leaving the combination of the two nodes is zero. This concept is represented graphically in the expanded view of Fig. 4.9*b*.

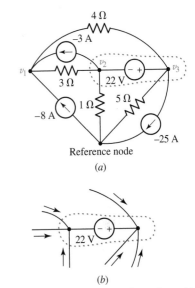

■ **FIGURE 4.9** (*a*) The circuit of Example 4.2 with a 22 V source in place of the 7 Ω resistor. (*b*) Expanded view of the region defined as a supernode; KCL requires that all currents flowing into the region sum to zero, or we would pile up or run out of electrons.

EXAMPLE 4.5

Determine the value of the unknown node voltage v_1 in the circuit of Fig. 4.9*a*.

The KCL equation at node 1 is unchanged from Example 4.2:

$$-8 - 3 = \frac{v_1 - v_2}{3} + \frac{v_1 - v_3}{4}$$

or

$$0.5833v_1 - 0.3333v_2 - 0.2500v_3 = -11 \qquad [17]$$

Next we consider the 2-3 supernode. Two current sources are connected, and four resistors. Thus,

$$3 + 25 = \frac{v_2 - v_1}{3} + \frac{v_3 - v_1}{4} + \frac{v_3}{5} + \frac{v_2}{1}$$

or

$$-0.5833v_1 + 1.3333v_2 + 0.45v_3 = 28 \qquad [18]$$

(Continued on next page)

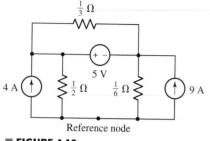

■ FIGURE 4.10

Since we have three unknowns, we need one additional equation, and it must utilize the fact that there is a 22 V voltage source between nodes 2 and 3:

$$v_2 - v_3 = -22 \qquad [19]$$

Solving Eqs. [17] to [19], the solution for v_1 is 1.071 V.

PRACTICE

4.4 For the circuit of Fig. 4.10, compute the voltage across each current source.

Ans: 5.375 V, 375 mV.

The presence of a voltage source thus reduces by 1 the number of nonreference nodes at which we must apply KCL, regardless of whether the voltage source extends between two nonreference nodes or is connected between a node and the reference. We should be careful in analyzing circuits such as that of Practice Problem 4.4. Since both ends of the resistor are part of the supernode, there must technically be *two* corresponding current terms in the KCL equation, but they cancel each other out. We can summarize the supernode method as follows:

Summary of Supernode Analysis Procedure

1. **Count the number of nodes** (N).
2. **Designate a reference node.** The number of terms in your nodal equations can be minimized by selecting the node with the greatest number of branches connected to it.
3. **Label the nodal voltages** (there are $N - 1$ of them).
4. **If the circuit contains voltage sources, form a supernode about each one.** This is done by enclosing the source, its two terminals, and any other elements connected between the two terminals within a broken-line enclosure.
5. **Write a KCL equation for each nonreference node and for each supernode *that does not contain the reference node.*** Sum the currents flowing *into* a node/supernode from current sources on one side of the equation. On the other side, sum the currents flowing *out* of the node/supernode through resistors. Pay close attention to "−" signs.
6. **Relate the voltage across each voltage source to nodal voltages.** This is accomplished by simple application of KVL; one such equation is needed for each supernode defined.
7. **Express any additional unknowns (i.e., currents or voltages other than nodal voltages) in terms of appropriate nodal voltages.** This situation can occur if dependent sources appear in our circuit.
8. **Organize the equations.** Group terms according to nodal voltages.
9. **Solve the system of equations for the nodal voltages** (there will be $N - 1$ of them).

We see that we have added two additional steps from our general nodal analysis procedure. In reality, however, application of the supernode technique to a circuit containing voltage sources not connected to the reference node will result in a reduction in the number of KCL equations required. With this in mind, let's consider the circuit of Fig. 4.11, which contains all four types of sources and has five nodes.

EXAMPLE 4.6

Determine the node-to-reference voltages in the circuit of Fig. 4.11.

After establishing a supernode about each *voltage* source, we see that we need to write KCL equations only at node 2 and at the supernode containing the dependent voltage source. By inspection, it is clear that $v_1 = -12$ V.

At node 2,

$$\frac{v_2 - v_1}{0.5} + \frac{v_2 - v_3}{2} = 14 \qquad [20]$$

while at the 3-4 supernode,

$$0.5v_x = \frac{v_3 - v_2}{2} + \frac{v_4}{1} + \frac{v_4 - v_1}{2.5} \qquad [21]$$

We next relate the source voltages to the node voltages:

$$v_3 - v_4 = 0.2v_y \qquad [22]$$

and

$$0.2v_y = 0.2(v_4 - v_1) \qquad [23]$$

Finally, we express the dependent current source in terms of the assigned variables:

$$0.5v_x = 0.5(v_2 - v_1) \qquad [24]$$

Five nodes requires *four* KCL equations in general nodal analysis, but we have reduced this requirement to *only two,* as we formed two separate supernodes. Each supernode required a KVL equation (Eq. [22] and $v_1 = -12$, the latter written by inspection). Neither dependent source was controlled by a nodal voltage, so two additional equations were needed as a result.

With this done, we can now eliminate v_x and v_y to obtain a set of four equations in the four node voltages:

$$
\begin{aligned}
-2v_1 + 2.5v_2 - 0.5v_3 &= 14 \\
0.1v_1 - v_2 + 0.5v_3 + 1.4v_4 &= 0 \\
v_1 &= -12 \\
0.2v_1 + v_3 - 1.2v_4 &= 0
\end{aligned}
$$

Solving, $v_1 = -12$ V, $v_2 = -4$ V, $v_3 = 0$ V, and $v_4 = -2$ V.

PRACTICE

4.5 Determine the nodal voltages in the circuit of Fig. 4.12.

Ans: $v_1 = 3$ V, $v_2 = -2.33$ V, $v_3 = -1.91$ V, $v_4 = 0.945$ V.

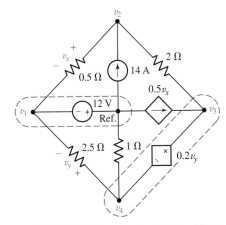

■ **FIGURE 4.11** A five-node circuit with four different types of sources.

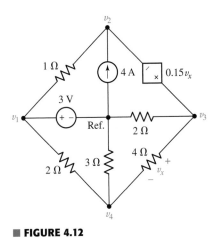

■ **FIGURE 4.12**

4.3 • MESH ANALYSIS

As we have seen, nodal analysis is a straightforward analysis technique when only current sources are present, and voltage sources are easily accommodated with the supernode concept. Still, nodal analysis is based on KCL, and the reader might at some point wonder if there isn't a similar approach based on KVL. There is—it's known as **mesh analysis**—and although only strictly speaking applicable to what we will shortly define as a planar circuit, it can in many cases prove simpler to apply than nodal analysis.

If it is possible to draw the diagram of a circuit on a plane surface in such a way that no branch passes over or under any other branch, then that circuit is said to be a **planar circuit.** Thus, Fig. 4.13a shows a planar network, Fig. 4.13b shows a nonplanar network, and Fig. 4.13c also shows a planar network, although it is drawn in such a way as to make it appear nonplanar at first glance.

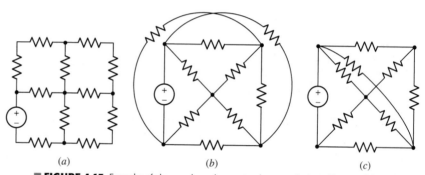

(a)   (b)   (c)

■ **FIGURE 4.13** Examples of planar and nonplanar networks; crossed wires without a solid dot are not in physical contact with each other.

In Sec. 3.1, the terms **path, closed path,** and **loop** were defined. Before we define a mesh, let us consider the sets of branches drawn with heavy lines in Fig. 4.14. The first set of branches is not a path, since four branches are connected to the center node, and it is of course also not a loop. The second set of branches does not constitute a path, since it is traversed only by passing through the central node twice. The remaining four paths are all loops. The circuit contains 11 branches.

The mesh is a property of a planar circuit and is undefined for a nonplanar circuit. We define a **mesh** as a loop that does not contain any other loops within it. Thus, the loops indicated in Fig. 4.14c and d are not meshes, whereas those of parts e and f are meshes. Once a circuit has been drawn neatly in planar form, it often has the appearance of a multipaned window; the boundary of each pane in the window may be considered to be a mesh.

If a network is planar, mesh analysis can be used to accomplish the analysis. This technique involves the concept of a **mesh current,** which we introduce by considering the analysis of the two-mesh circuit of Fig. 4.15a.

As we did in the single-loop circuit, we will begin by defining a current through one of the branches. Let us call the current flowing to the right through the 6 Ω resistor i_1. We will apply KVL around each of the two meshes, and the two resulting equations are sufficient to determine two unknown currents. We next define a second current i_2 flowing to the right in

We should mention that mesh-type analysis can be applied to nonplanar circuits, but since it is not possible to define a complete set of unique meshes for such circuits, assignment of unique mesh currents is not possible.

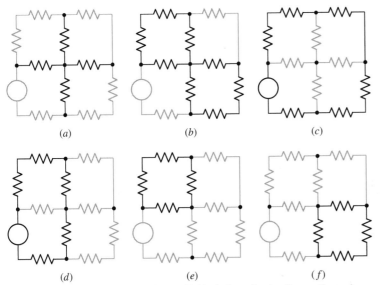

■ FIGURE 4.14 (a) The set of branches identified by the heavy lines is neither a path nor a loop. (b) The set of branches here is not a path, since it can be traversed only by passing through the central node twice. (c) This path is a loop but not a mesh, since it encloses other loops. (d) This path is also a loop but not a mesh. (e, f) Each of these paths is both a loop and a mesh.

the 4 Ω resistor. We might also choose to call the current flowing downward through the central branch i_3, but it is evident from KCL that i_3 may be expressed in terms of the two previously assumed currents as $(i_1 - i_2)$. The assumed currents are shown in Fig. 4.15b.

Following the method of solution for the single-loop circuit, we now apply KVL to the left-hand mesh,

$$-42 + 6i_1 + 3(i_1 - i_2) = 0$$

or

$$9i_1 - 3i_2 = 42 \qquad [25]$$

Applying KVL to the right-hand mesh,

$$-3(i_1 - i_2) + 4i_2 - 10 = 0$$

or

$$-3i_1 + 7i_2 = 10 \qquad [26]$$

Equations [25] and [26] are independent equations; one cannot be derived from the other. With two equations and two unknowns, the solution is easily obtained:

$$i_1 = 6 \text{ A} \qquad i_2 = 4 \text{ A} \qquad \text{and} \qquad (i_1 - i_2) = 2 \text{ A}$$

If our circuit contains M meshes, then we expect to have M mesh currents and therefore will be required to write M independent equations.

Now let us consider this same problem in a slightly different manner by using mesh currents. We define a ***mesh current*** as a current that flows only around the perimeter of a mesh. One of the greatest advantages in the use of mesh currents is the fact that Kirchhoff's current law is automatically satisfied. If a mesh current flows *into* a given node, it flows *out* of it also.

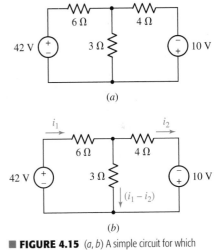

■ FIGURE 4.15 (a, b) A simple circuit for which currents are required.

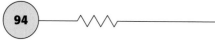

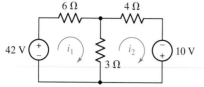

■ **FIGURE 4.16** The same circuit considered in Fig. 4.15*b*, but viewed a slightly different way.

A mesh current may often be identified as a branch current, as i_1 and i_2 have been identified in this example. This is not always true, however, for consideration of a square nine-mesh network soon shows that the central mesh current cannot be identified as the current in any branch.

If we call the left-hand mesh of our problem mesh 1, then we may establish a mesh current i_1 flowing in a clockwise direction about this mesh. A mesh current is indicated by a curved arrow that almost closes on itself and is drawn inside the appropriate mesh, as shown in Fig. 4.16. The mesh current i_2 is established in the remaining mesh, again in a clockwise direction. Although the directions are arbitrary, we will always choose clockwise mesh currents because a certain error-minimizing symmetry then results in the equations.

We no longer have a current or current arrow shown directly on each branch in the circuit. The current through any branch must be determined by considering the mesh currents flowing in every mesh in which that branch appears. This is not difficult, because no branch can appear in more than two meshes. For example, the 3 Ω resistor appears in both meshes, and the current flowing downward through it is $i_1 - i_2$. The 6 Ω resistor appears only in mesh 1, and the current flowing to the right in that branch is equal to the mesh current i_1.

For the left-hand mesh,

$$-42 + 6i_1 + 3(i_1 - i_2) = 0$$

while for the right-hand mesh,

$$3(i_2 - i_1) + 4i_2 - 10 = 0$$

and these two equations are equivalent to Eqs. [25] and [26].

EXAMPLE 4.7

Determine the power supplied by the 2 V source of Fig. 4.17a.

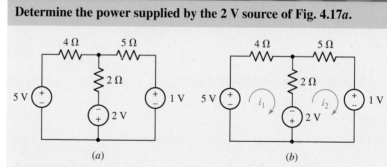

(a)   (b)

■ **FIGURE 4.17** (*a*) A two-mesh circuit containing three sources. (*b*) Circuit labeled for mesh analysis.

We first define two clockwise mesh currents as shown in Fig. 4.17*b*.

Beginning at the bottom left node of mesh 1, we write the following KVL equation as we proceed clockwise through the branches:

$$-5 + 4i_1 + 2(i_1 - i_2) - 2 = 0$$

Doing the same for mesh 2, we write

$$+2 + 2(i_2 - i_1) + 5i_2 + 1 = 0$$

Rearranging and grouping terms,

$$6i_1 - 2i_2 = 7$$

and

$$-2i_1 + 7i_2 = -3$$

Solving, $i_1 = \dfrac{43}{38} = 1.132$ A and $i_2 = -\dfrac{2}{19} = -0.1053$ A.

The current flowing out of the positive reference terminal of the 2 V source is $i_1 - i_2$. Thus, the 2 V source supplies $(2)(1.237) = 2.474$ W.

PRACTICE

4.6 Determine i_1 and i_2 in the circuit in Fig. 4.18.

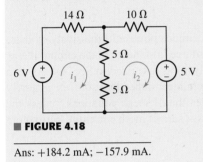

■ **FIGURE 4.18**

Ans: $+184.2$ mA; -157.9 mA.

Let us next consider the five-node, seven-branch, three-mesh circuit shown in Fig. 4.19. This is a slightly more complicated problem because of the additional mesh.

EXAMPLE 4.8

Use mesh analysis to determine the three mesh currents in the circuit of Fig. 4.19.

The three required mesh currents are assigned as indicated in Fig. 4.19, and we methodically apply KVL about each mesh:

$$-7 + 1(i_1 - i_2) + 6 + 2(i_1 - i_3) = 0$$
$$1(i_2 - i_1) + 2i_2 + 3(i_2 - i_3) = 0$$
$$2(i_3 - i_1) - 6 + 3(i_3 - i_2) + 1i_3 = 0$$

Simplifying,

$$3i_1 - i_2 - 2i_3 = 1$$
$$-i_1 + 6i_2 - 3i_3 = 0$$
$$-2i_1 - 3i_2 + 6i_3 = 6$$

and solving, we obtain $i_1 = 3$ A, $i_2 = 2$ A, and $i_3 = 3$ A.

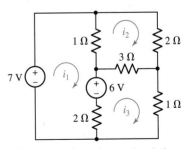

■ **FIGURE 4.19** A five-node, seven-branch, three-mesh circuit.

PRACTICE

4.7 Determine i_1 and i_2 in the circuit of Fig 4.20.

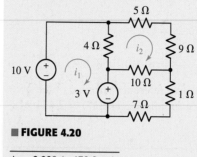

■ **FIGURE 4.20**

Ans: 2.220 A, 470.0 mA.

The previous examples dealt with circuits powered exclusively by independent voltage sources. If a current source is included in the circuit, it may either simplify or complicate the analysis, as discussed in Sec. 4.4. As seen in our study of the nodal analysis technique, dependent sources generally require an additional equation besides the M mesh equations, unless the controlling variable is a mesh current (or sum of mesh currents). We explore this in the following example.

EXAMPLE 4.9

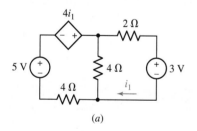

(a)

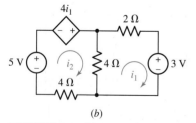

(b)

■ **FIGURE 4.21** (a) A two-mesh circuit containing a dependent source. (b) Circuit labeled for mesh analysis.

Determine the current i_1 in the circuit of Fig. 4.21a.

The current i_1 is actually a mesh current, so rather than redefine it we label the rightmost mesh current i_1 and define a clockwise mesh current i_2 for the left mesh, as shown in Fig. 4.21b.

For the left mesh, KVL yields

$$-5 - 4i_1 + 4(i_2 - i_1) + 4i_2 = 0 \qquad [27]$$

and for the right mesh we find

$$4(i_1 - i_2) + 2i_1 + 3 = 0 \qquad [28]$$

Grouping terms, these equations may be written more compactly as

$$-8i_1 + 8i_2 = 5$$

and

$$6i_1 - 4i_2 = -3$$

Solving, $i_2 = 375$ mA, so $i_1 = -250$ mA.

Since the dependent source of Fig. 4.21 is controlled by a mesh current (i_1), only two equations—Eqs. [27] and [28]—were required to analyze the two-mesh circuit. In the following example, we explore the situation that arises if the controlling variable is *not* a mesh current.

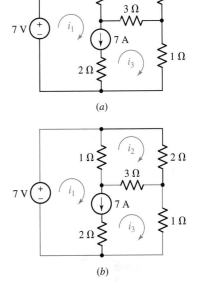

is that of meshes 1 and 3 as shown in Fig. 4.24b. Applying KVL about this loop,

$$-7 + 1(i_1 - i_2) + 3(i_3 - i_2) + 1i_3 = 0$$

or

$$i_1 - 4i_2 + 4i_3 = 7 \qquad [32]$$

and around mesh 2,

$$1(i_2 - i_1) + 2i_2 + 3(i_2 - i_3) = 0$$

or

$$-i_1 + 6i_2 - 3i_3 = 0 \qquad [33]$$

Finally, the independent source current is related to the mesh currents,

$$i_1 - i_3 = 7 \qquad [34]$$

Solving Eqs. [32] through [34], we find $i_1 = 9$ A, $i_2 = 2.5$ A, and $i_3 = 2$ A.

PRACTICE

4.9 Determine the current i_1 in the circuit of Fig. 4.25.

Ans: −1.93 A.

■ **FIGURE 4.24** (a) A three-mesh circuit with an independent current source. (b) A supermesh is defined by the colored line.

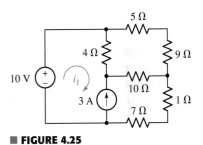

■ **FIGURE 4.25**

The presence of one or more dependent sources merely requires each of these source quantities and the variable on which it depends to be expressed in terms of the assigned mesh currents. In Fig. 4.26, for example, we note that both a dependent and an independent current source are included in the network. Let's see how their presence affects the analysis of the circuit and actually simplifies it.

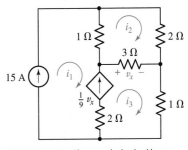

■ **FIGURE 4.26** A three-mesh circuit with one dependent and one independent current source.

EXAMPLE 4.12

Evaluate the three unknown currents in the circuit of Fig. 4.26.

The current sources appear in meshes 1 and 3. Since the 15 A source is located on the perimeter of the circuit, we may eliminate mesh 1 from consideration—it is clear that $i_1 = 15$ A.

We find that because we now know one of the two mesh currents relevant to the dependent current source, there is no need to write a supermesh equation about meshes 1 and 3. Instead, we simply relate i_1 and i_3 to the current from the dependent source using KCL:

$$\frac{v_x}{9} = i_3 - i_1 = \frac{3(i_3 - i_2)}{9}$$

(Continued on next page)

which can be written more compactly as

$$-i_1 + \frac{1}{3}i_2 + \frac{2}{3}i_3 = 0 \quad \text{or} \quad \frac{1}{3}i_2 + \frac{2}{3}i_3 = 15 \qquad [35]$$

With one equation in two unknowns, all that remains is to write a KVL equation about mesh 2:

$$1(i_2 - i_1) + 2i_2 + 3(i_2 - i_3) = 0$$

or

$$6i_2 - 3i_3 = 15 \qquad [36]$$

Solving Eqs. [35] and [36], we find that $i_2 = 11$ A and $i_3 = 17$ A; we already determined that $i_1 = 15$ A by inspection.

PRACTICE

4.10 Determine v_3 in the circuit of Fig. 4.27.

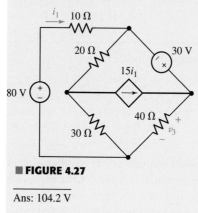

■ **FIGURE 4.27**

Ans: 104.2 V

We can now summarize the general approach to writing mesh equations, whether or not dependent sources, voltage sources, and/or current sources are present, provided that the circuit can be drawn as a planar circuit:

Summary of Supermesh Analysis Procedure

1. **Determine if the circuit is a planar circuit.** If not, perform nodal analysis instead.
2. **Count the number of meshes (M).** Redraw the circuit if necessary.
3. **Label each of the M mesh currents.** Generally, defining all mesh currents to flow clockwise results in a simpler analysis.
4. **If the circuit contains current sources shared by two meshes, form a supermesh to enclose both meshes.** A highlighted enclosure helps when writing KVL equations.
5. **Write a KVL equation around each mesh/supermesh.** Begin with a convenient node and proceed in the direction of the mesh current. Pay close attention to "−" signs. If a current source lies

on the periphery of a mesh, no KVL equation is needed and the mesh current is determined by inspection.

6. **Relate the current flowing from each current source to mesh currents.** This is accomplished by simple application of KCL; one such equation is needed for each supermesh defined.

7. **Express any additional unknowns such as voltages or currents other than mesh currents in terms of appropriate mesh currents.** This situation can occur if dependent sources appear in our circuit.

8. **Organize the equations.** Group terms according to nodal voltages.

9. **Solve the system of equations for the mesh currents** (there will be M of them).

4.5 NODAL VS. MESH ANALYSIS: A COMPARISON

Now that we have examined two distinctly different approaches to circuit analysis, it seems logical to ask if there is ever any advantage to using one over the other. If the circuit is nonplanar, then there is no choice: only nodal analysis may be applied.

Provided that we are indeed considering the analysis of a *planar* circuit, however, there are situations where one technique has a small advantage over the other. If we plan to use nodal analysis, then a circuit with N nodes will lead to at most $(N - 1)$ KCL equations. Each supernode defined will further reduce this number by 1. If the same circuit has M distinct meshes, then we will obtain at most M KVL equations; each supermesh will reduce this number by 1. Based on these facts, we should select the approach that will result in the smaller number of simultaneous equations.

If one or more dependent sources are included in the circuit, then each controlling quantity may influence our choice of nodal or mesh analysis. For example, a dependent voltage source controlled by a nodal voltage does not require an additional equation when we perform nodal analysis. Likewise, a dependent current source controlled by a mesh current does not require an additional equation when we perform mesh analysis. *What about the situation where a dependent voltage source is controlled by a current? Or the converse, where a dependent current source is controlled by a voltage?* Provided that the controlling quantity can be easily related to mesh currents, we might expect mesh analysis to be the more straightforward option. Likewise, if the controlling quantity can be easily related to nodal voltages, nodal analysis may be preferable. One final point in this regard is to keep in mind the *location* of the source; current sources which lie on the periphery of a mesh, whether dependent or independent, are easily treated in mesh analysis; voltage sources connected to the reference terminal are easily treated in nodal analysis.

When either method results in essentially the same number of equations, it may be worthwhile to also consider what quantities are being sought. Nodal analysis results in direct calculation of nodal voltages, whereas mesh analysis provides currents. If we are asked to find currents through a set of resistors, for example, after performing nodal analysis, we must still invoke Ohm's law at each resistor to determine the current.

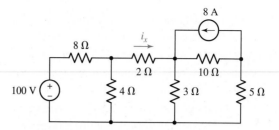

■ **FIGURE 4.28** A planar circuit with five nodes and four meshes.

As an example, consider the circuit in Fig. 4.28. We wish to determine the current i_x.

We choose the bottom node as the reference node, and note that there are four nonreference nodes. Although this means that we can write four distinct equations, there is no need to label the node between the 100 V source and the 8 Ω resistor, since that node voltage is clearly 100 V. Thus, we label the remaining node voltages v_1, v_2, and v_3 as in Fig. 4.29.

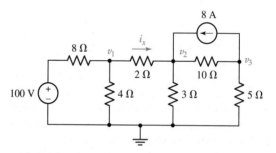

■ **FIGURE 4.29** The circuit of Fig. 4.28 with node voltages labeled. Note that an earth ground symbol was chosen to designate the reference terminal.

We write the following three equations:

$$\frac{v_1 - 100}{8} + \frac{v_1}{4} + \frac{v_1 - v_2}{2} = 0 \qquad \text{or} \quad 0.875v_1 - 0.5v_2 \qquad\qquad = 12.5 \quad [37]$$

$$\frac{v_2 - v_1}{2} + \frac{v_2}{3} + \frac{v_2 - v_3}{10} - 8 = 0 \quad \text{or} \quad -0.5v_1 - 0.9333v_2 - 0.1v_3 = 8 \quad [38]$$

$$\frac{v_3 - v_2}{10} + \frac{v_3}{5} + 8 = 0 \qquad\qquad \text{or} \qquad\qquad -0.1v_2 \quad + 0.3v_3 = -8 \quad [39]$$

Solving, we find that $v_1 = 25.89$ V and $v_2 = 20.31$ V. We determine the current i_x by application of Ohm's law:

$$i_x = \frac{v_1 - v_2}{2} = 2.79 \text{ A} \qquad\qquad [40]$$

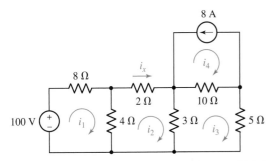

■ **FIGURE 4.30** The circuit of Fig. 4.28 with mesh currents labeled.

Next, we consider the same circuit using mesh analysis. We see in Fig. 4.30 that we have four distinct meshes, although it is obvious that $i_4 = -8$ A; we therefore need to write three distinct equations.

Writing a KVL equation for meshes 1, 2, and 3:

$$-100 + 8i_1 + 4(i_1 - i_2) = 0 \quad \text{or} \quad 12i_1 - 4i_2 \qquad = 100 \qquad [41]$$

$$4(i_2 - i_1) + 2i_2 + 3(i_2 - i_3) = 0 \quad \text{or} \quad -4i_1 + 9i_2 - 3i_3 = 0 \qquad [42]$$

$$3(i_3 - i_2) + 10(i_3 + 8) + 5i_3 = 0 \quad \text{or} \qquad -3i_2 + 18i_3 = -80 \qquad [43]$$

Solving, we find that $i_2 (= i_x) = 2.79$ A. For this particular problem, mesh analysis proved to be simpler. Since either method is valid, however, working the same problem both ways can also serve as a means to check our answers.

4.6 • COMPUTER-AIDED CIRCUIT ANALYSIS

We have seen that it does not take many components at all to create a circuit of respectable complexity. As we continue to examine even more complex circuits, it will become obvious rather quickly that it is easy to make errors during the analysis, and verifying solutions by hand can be time-consuming. A powerful computer software package known as PSpice is commonly employed for rapid analysis of circuits, and the schematic capture tools are typically integrated with either a printed circuit board or integrated circuit layout tool. Originally developed in the early 1970s at the University of California at Berkeley, SPICE (*Simulation Program with Integrated Circuit Emphasis*) is now an industry standard. MicroSim Corporation introduced PSpice in 1984, which built intuitive graphical interfaces around the core SPICE program. Depending on the type of circuit application being considered, there are now several companies offering variations of the basic SPICE package.

Although computer-aided analysis is a relatively quick means of determining voltages and currents in a circuit, we should be careful not to allow simulation packages to completely replace traditional "paper and pencil" analysis. There are several reasons for this. First, in order to design we must be able to analyze. Overreliance on software tools can inhibit the development of necessary analytical skills, similar to introducing calculators too early in grade school. Second, it is virtually impossible to use a complicated software package over a long period of time without making some type of data-entry error. If we have no basic intuition as to what type of answer to

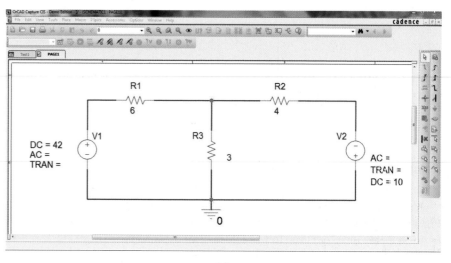

(a)

(b)

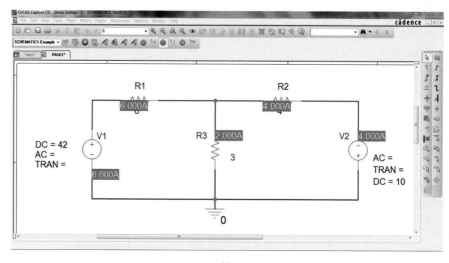

(c)

■ **FIGURE 4.31** (a) Circuit of Fig. 4.15a drawn using Orcad schematic capture software. (b) Current, voltage, and power display buttons. (c) Circuit after simulation run, with current display enabled.

expect from a simulation, then there is no way to determine whether or not it is valid. Thus, the generic name really is a fairly accurate description: computer-*aided* analysis. Human brains are not obsolete. Not yet, anyway.

As an example, consider the circuit of Fig. 4.15b, which includes two dc voltage sources and three resistors. We wish to simulate this circuit using PSpice so that we may determine the currents i_1 and i_2. Figure 4.31a shows the circuit as drawn using a schematic capture program.[1]

(1) Refer to Appendix 4 for a brief tutorial on PSpice and schematic capture.

In order to determine the mesh currents, we need only run a bias point simulation. Under **PSpice,** select **New Simulation Profile,** type in a name (such as Example), and click on **Create**. Under the **Analysis type**: pull-down menu, select **Bias Point**, then click on **OK**. Returning to the original schematic window, under **PSpice** select **Run** (or use either of the two shortcuts: pressing the F11 key or clicking on the blue "Play" symbol). To see the currents calculated by PSpice, make sure the current button is selected (Fig. 4.31*b*). The results of our simulation are shown in Fig. 4.31*c*. We see that the two currents i_1 and i_2 are 6 A and 4 A, respectively, as we found previously.

As a further example, consider the circuit shown in Fig. 4.32*a*. It contains a dc voltage source, a dc current source, and a voltage-controlled current source. We are interested in the three nodal voltages, which from either nodal or mesh analysis are found to be 82.91 V, 69.9 V, and 59.9 V, respectively, as we move from left to right across the top of the circuit. Figure 4.32*b* shows this circuit after the simulation was performed. The three nodal voltages are indicated directly on the schematic. Note that in drawing a dependent source using the schematic capture tool, we must *explicitly* link two terminals of the source to the controlling voltage or current.

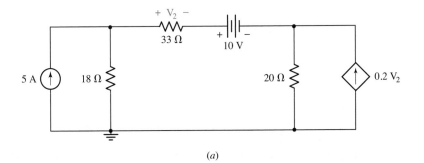

(a)

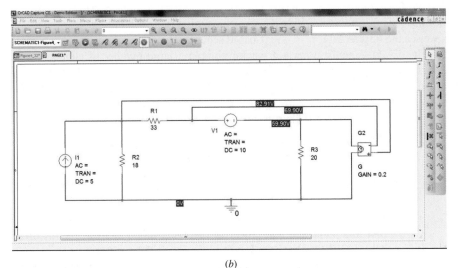

(b)

■ **FIGURE 4.32** (*a*) Circuit with dependent current source. (*b*) Circuit drawn using a schematic capture tool, with simulation results presented directly on the schematic.

PRACTICAL APPLICATION

Node-Based PSpice Schematic Creation

The most common method of describing a circuit in conjunction with computer-aided circuit analysis is with some type of graphical schematic drawing package, an example output of which was shown in Fig. 4.32. SPICE, however, was written before the advent of such software, and as such requires circuits to be described in a specific text-based format. The format has its roots in the syntax used for punch cards, which gives it a somewhat distinct appearance. The basis for circuit description is the definition of elements, each terminal of which is assigned a node number. So, although we have just studied two different generalized circuit analysis methods—the nodal and mesh techniques—it is interesting that SPICE and PSpice were written using a clearly defined nodal analysis approach.

Even though modern circuit analysis is largely done using graphics-oriented interactive software, when errors are generated (usually due to a mistake in drawing the schematic or in selecting a combination of analysis options), the ability to read the text-based "input deck" generated by the schematic capture tool can be invaluable in tracking down the specific problem. The easiest way to develop such an ability is to learn how to run PSpice directly from a user-written input deck.

Consider, for example, the sample input deck below (lines beginning with an asterisk are comments, and are skipped by SPICE).

```
* Example SPICE input deck for simple voltage divider circuit.

.OP                         (Requests dc operating point)

R1  1 2 1k                  (Locates R1 between nodes 1 and 2; value is 1 kΩ)
R2  2 0 1k                  (Locates R2 between nodes 2 and 0; also 1 kΩ)
V1  1 0 DC 5                (Locates 5 V source between nodes 1 and 0)

* End of input deck.
```

We can create the input deck by using the Notepad program from Windows or our favorite text editor. Saving the file under the name example.cir, we next invoke PSpice A/D (see Appendix. 4). Under **File,** we choose **Open,** locate the directory in which we saved our file example.cir, and for **Files of type:** select **Circuit Files (*.cir).** After selecting our file and clicking **Open,** we see the PSpice A/D window with our circuit file loaded (Fig. 4.33a). A netlist such as this, containing instructions for the simulation to be performed, can be created by schematic capture software or created manually as in this example.

We run the simulation by either clicking the green "play" symbol at the top right, or selecting **Run** under **Simulation.**

To view the results, we select **Output File** from under the **View** menu, which provides the window shown in Fig. 4.33b. Here it is worth noting that the output provides the expected nodal voltages (5 V at node 1, 2.5 V across resistor R2), but the current is quoted using the passive sign convention (i.e., -2.5 mA).

Text-based schematic entry is reasonably straightforward, but for complex (large number of elements) circuits, it can quickly become cumbersome. It is also easy to misnumber nodes, an error that can be difficult to isolate. However, reading the input and output files is often helpful when running simulations, so some experience with this format is useful.

At this point, the real power of computer-aided analysis begins to be apparent: Once you have the circuit drawn in the schematic capture program, it is easy to experiment by simply changing component values and observing the effect on currents and voltages. To gain a little experience at this point, try simulating any of the circuits shown in previous examples and practice problems.

■ **FIGURE 4.33** (a) PSpice A/D window after the input deck describing our voltage divider is loaded. (b) Output window, showing nodal voltages and current from the source (but quoted using the passive sign convention). Note that the voltage across R1 requires post-simulation subtraction.

SUMMARY AND REVIEW

Although Chap. 3 introduced KCL and KVL, both of which are sufficient to enable us to analyze any circuit, a more methodical approach proves helpful in everyday situations. Thus, in this chapter we developed the nodal analysis technique based on KCL, which results in a voltage at each node (with respect to some designated "reference" node). We generally need to solve a system of simultaneous equations, unless voltage sources are connected so that they automatically provide nodal voltages. The controlling quantity of a *dependent* source is written down just as we would write down the numerical value of an "independent" source. Typically an additional equation is then required, unless the dependent source is controlled by a nodal voltage. When a voltage source bridges two nodes, the basic technique can be extended by creating a *supernode*; KCL dictates that the sum of the currents flowing into a group of connections so defined is equal to the sum of the currents flowing out.

As an alternative to nodal analysis, the mesh analysis technique was developed through application of KVL; it yields the complete set of *mesh* currents, which do not always represent the *net* current flowing through any particular element (for example, if an element is shared by two meshes). The presence of a current source will simplify the analysis if it lies on the periphery of a mesh; if the source is shared, then the *supermesh* technique is best. In that case, we write a KVL equation around a path that avoids the shared current source, then algebraically link the two corresponding mesh currents using the source.

A common question is: "*Which analysis technique should I use?*" We discussed some of the issues that might go into choosing a technique for a given circuit. These included whether or not the circuit is planar, what types of sources are present and how they are connected, and also what specific information is required (i.e., a voltage, current, or power). For complex circuits, it may take a greater effort than it is worth to determine the "optimum" approach, in which case most people will opt for the method with which they feel most comfortable. We concluded the chapter by introducing PSpice, a common circuit simulation tool, which is very useful for checking our results.

At this point we wrap up by identifying key points of this chapter to review, along with relevant example(s).

❑ Start each analysis with a neat, simple circuit diagram. Indicate all element and source values. (Example 4.1)

❑ For nodal analysis,

  ❑ Choose one node as the reference node. Then label the node voltages $v_1, v_2, \ldots, v_{N-1}$. Each is understood to be measured with respect to the reference node. (Examples 4.1, 4.2)

  ❑ If the circuit contains only current sources, apply KCL at each nonreference node. (Examples 4.1, 4.2)

  ❑ If the circuit contains voltage sources, form a supernode about each one, and then apply KCL at all nonreference nodes and supernodes. (Examples 4.5, 4.6)

❑ For mesh analysis, first make certain that the network is a planar network.

  ❑ Assign a clockwise mesh current in each mesh: $i_1, i_2, \ldots, i_M$. (Example 4.7)

  ❑ If the circuit contains only voltage sources, apply KVL around each mesh. (Examples 4.7, 4.8, 4.9)

  ❑ If the circuit contains current sources, create a supermesh for each one that is common to two meshes, and then apply KVL around each mesh and supermesh. (Examples 4.11, 4.12)

- Dependent sources will add an additional equation to nodal analysis if the controlling variable is a current, but not if the controlling variable is a nodal voltage. (Conversely, a dependent source will add an additional equation to mesh analysis if the controlling variable is a voltage, but not if the controlling variable is a mesh current). (Examples 4.3, 4.4, 4.6, 4.9, 4.10, 4.12)

- In deciding whether to use nodal or mesh analysis for a planar circuit, a circuit with fewer nodes/supernodes than meshes/supermeshes will result in fewer equations using nodal analysis.

- Computer-aided analysis is useful for checking results and analyzing circuits with large numbers of elements. However, common sense must be used to check simulation results.

READING FURTHER

A detailed treatment of nodal and mesh analysis can be found in:

R. A. DeCarlo and P. M. Lin, *Linear Circuit Analysis,* 2nd ed. New York: Oxford University Press, 2001.

A solid guide to SPICE is

P. Tuinenga, *SPICE: A Guide to Circuit Simulation and Analysis Using PSPICE,* 3rd ed. Upper Saddle River, N.J.: Prentice-Hall, 1995.

EXERCISES

4.1 Nodal Analysis

1. Solve the following systems of equations:
   (a) $2v_2 - 4v_1 = 9$ and $v_1 - 5v_2 = -4$;
   (b) $-v_1 + 2v_3 = 8$; $2v_1 + v_2 - 5v_3 = -7$; $4v_1 + 5v_2 + 8v_3 = 6$.

2. Evaluate the following determinants:
   (a) $\begin{vmatrix} 2 & 1 \\ -4 & 3 \end{vmatrix}$ (b) $\begin{vmatrix} 0 & 2 & 11 \\ 6 & 4 & 1 \\ 3 & -1 & 5 \end{vmatrix}$.

3. Employ Cramer's rule to solve for v_2 in each part of Exercise 1.

4. (a) Solve the following system of equations:

$$3 = \frac{v_1}{5} - \frac{v_2 - v_1}{22} + \frac{v_1 - v_3}{3}$$

$$2 - 1 = \frac{v_2 - v_1}{22} + \frac{v_2 - v_3}{14}$$

$$0 = \frac{v_3}{10} + \frac{v_3 - v_1}{3} + \frac{v_3 - v_2}{14}$$

   (b) Verify your solution using MATLAB.

5. (a) Solve the following system of equations:

$$7 = \frac{v_1}{2} - \frac{v_2 - v_1}{12} + \frac{v_1 - v_3}{19}$$

$$15 = \frac{v_2 - v_1}{12} + \frac{v_2 - v_3}{2}$$

$$4 = \frac{v_3}{7} + \frac{v_3 - v_1}{19} + \frac{v_3 - v_2}{2}$$

   (b) Verify your solution using MATLAB.

6. Correct (and verify by running) the following MATLAB code:

```
>> e1 = '3 = v/7 - (v2 - v1)/2 + (v1 - v3)/3;
>> e2 = '2 = (v2 - v1)/2 + (v2 - v3)/14';
>> e  '0 = v3/10 + (v3 - v1)/3 + (v3 - v2)/14';
>>
>> a = sove(e e2 e3, 'v1', v2, 'v3')
```

7. Identify the obvious errors in the following complete set of nodal equations if the last equation is known to be correct:

$$7 = \frac{v_1}{4} - \frac{v_2 - v}{1} + \frac{v_1 - v_3}{9}$$

$$0 = \frac{v_2 - v_1}{2} + \frac{v_2 - v_3}{2}$$

$$4 = \frac{v_3}{7} + \frac{v_3 - v_1}{19} + \frac{v_3 - v_2}{2}$$

8. In the circuit of Fig. 4.34, determine the current labeled i with the assistance of nodal analysis techniques.

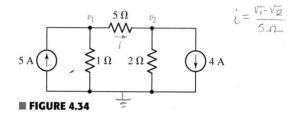

$$i = \frac{v_1 - v_2}{5\,\Omega}$$

■ **FIGURE 4.34**

9. Calculate the power dissipated in the 1 Ω resistor of Fig. 4.35.

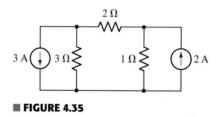

■ **FIGURE 4.35**

10. With the assistance of nodal analysis, determine $v_1 - v_2$ in the circuit shown in Fig. 4.36.

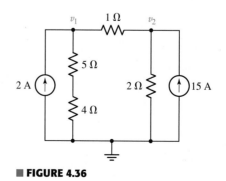

■ **FIGURE 4.36**

11. For the circuit of Fig. 4.37, determine the value of the voltage labeled v_1 and the current labeled i_1.

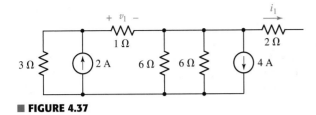

■ FIGURE 4.37

12. Use nodal analysis to find v_P in the circuit shown in Fig. 4.38.

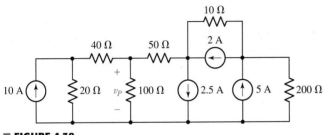

■ FIGURE 4.38

13. Using the bottom node as reference, determine the voltage across the 5 Ω resistor in the circuit of Fig. 4.39, and calculate the power dissipated by the 7 Ω resistor.

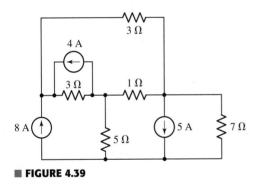

■ FIGURE 4.39

14. For the circuit of Fig. 4.40, use nodal analysis to determine the current i_5.

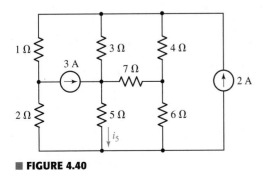

■ FIGURE 4.40

15. Determine a numerical value for each nodal voltage in the circuit of Fig. 4.41.

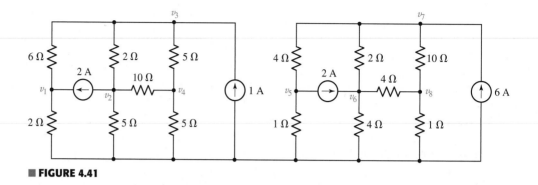

■ FIGURE 4.41

16. Determine the current i_2 as labeled in the circuit of Fig. 4.42, with the assistance of nodal analysis.

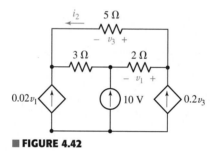

■ FIGURE 4.42

17. Using nodal analysis as appropriate, determine the current labeled i_1 in the circuit of Fig. 4.43.

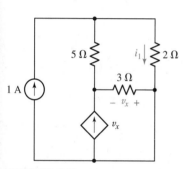

■ FIGURE 4.43

4.2 The Supernode

18. Determine the nodal voltages as labeled in Fig. 4.44, making use of the supernode technique as appropriate.

19. For the circuit shown in Fig. 4.45, determine a numerical value for the voltage labeled v_1.

20. For the circuit of Fig. 4.46, determine all four nodal voltages.

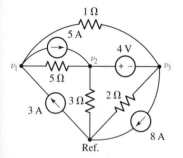

■ FIGURE 4.44

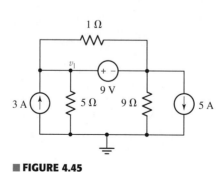

■ FIGURE 4.45

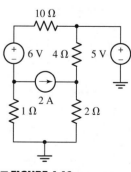

■ FIGURE 4.46

21. Employing supernode/nodal analysis techniques as appropriate, determine the power dissipated by the 1 Ω resistor in the circuit of Fig. 4.47.

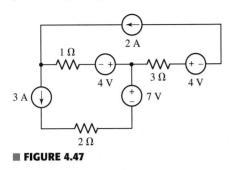

■ **FIGURE 4.47**

22. Referring to the circuit of Fig. 4.48, obtain a numerical value for the power supplied by the 1 V source.

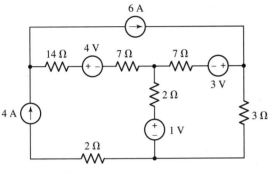

■ **FIGURE 4.48**

23. Determine the voltage labeled v in the circuit of Fig. 4.49.
24. Determine the voltage v_x in the circuit of Fig. 4.50, and the power supplied by the 1 A source.

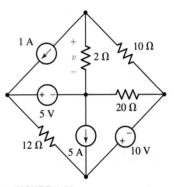

■ **FIGURE 4.49**

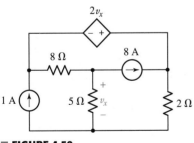

■ **FIGURE 4.50**

25. Consider the circuit of Fig. 4.51. Determine the current labeled i_1.

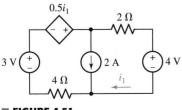

■ **FIGURE 4.51**

26. Determine the value of k that will result in v_x being equal to zero in the circuit of Fig. 4.52.

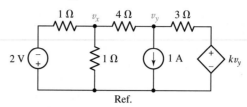

■ FIGURE 4.52

27. For the circuit depicted in Fig. 4.53, determine the voltage labeled v_1 across the 3 Ω resistor.

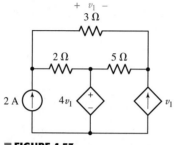

■ FIGURE 4.53

28. For the circuit of Fig. 4.54, determine all four nodal voltages.

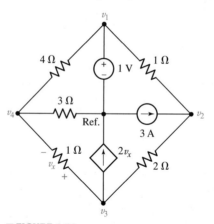

■ FIGURE 4.54

4.3 Mesh Analysis

29. Determine the currents flowing out of the positive terminal of each voltage source in the circuit of Fig. 4.55.

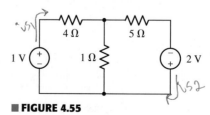

■ FIGURE 4.55

30. Obtain numerical values for the two mesh currents i_1 and i_2 in the circuit shown in Fig. 4.56.

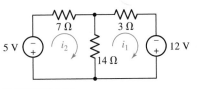

■ **FIGURE 4.56**

31. Use mesh analysis as appropriate to determine the two mesh currents labeled in Fig. 4.57.

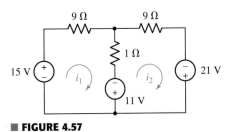

■ **FIGURE 4.57**

32. Determine numerical values for each of the three mesh currents as labeled in the circuit diagram of Fig. 4.58.

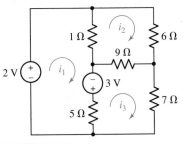

■ **FIGURE 4.58**

33. Calculate the power dissipated by each resistor in the circuit of Fig. 4.58.

34. Employing mesh analysis as appropriate, obtain (*a*) a value for the current i_y and (*b*) the power dissipated by the 220 Ω resistor in the circuit of Fig. 4.59.

35. Choose nonzero values for the three voltage sources of Fig. 4.60 so that no current flows through any resistor in the circuit.

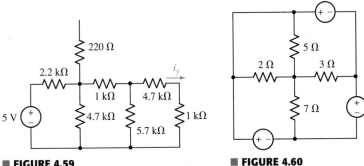

■ **FIGURE 4.59**   ■ **FIGURE 4.60**

36. Calculate the current i_x in the circuit of Fig. 4.61.

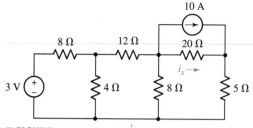

■ **FIGURE 4.61**

37. Employing mesh analysis procedures, obtain a value for the current labeled i in the circuit represented by Fig. 4.62.

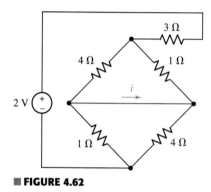

■ **FIGURE 4.62**

38. Determine the power dissipated in the 4 Ω resistor of the circuit shown in Fig. 4.63.

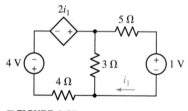

■ **FIGURE 4.63**

39. (*a*) Employ mesh analysis to determine the power dissipated by the 1 Ω resistor in the circuit represented schematically by Fig. 4.64. (*b*) Check your answer using nodal analysis.

40. Define three clockwise mesh currents for the circuit of Fig. 4.65, and employ mesh analysis to obtain a value for each.

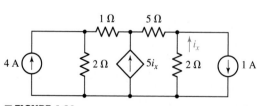

■ **FIGURE 4.64**

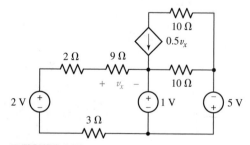

■ **FIGURE 4.65**

41. Employ mesh analysis to obtain values for i_x and v_a in the circuit of Fig. 4.66.

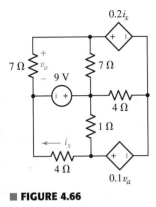

FIGURE 4.66

4.4 The Supermesh

42. Determine values for the three mesh currents of Fig. 4.67.

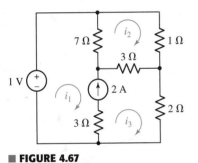

FIGURE 4.67

43. Through appropriate application of the supermesh technique, obtain a numerical value for the mesh current i_3 in the circuit of Fig. 4.68, and calculate the power dissipated by the 1 Ω resistor.

44. For the circuit of Fig. 4.69, determine the mesh current i_1 and the power dissipated by the 1 Ω resistor.

45. Calculate the three mesh currents labeled in the circuit diagram of Fig. 4.70.

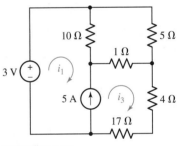

FIGURE 4.68

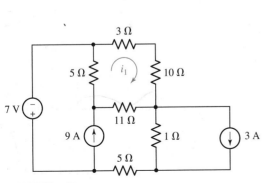

FIGURE 4.69

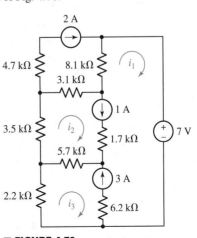

FIGURE 4.70

46. Employing the supermesh technique to best advantage, obtain numerical values for each of the mesh currents identified in the circuit depicted in Fig. 4.71.

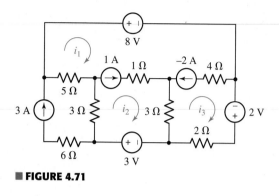

■ **FIGURE 4.71**

47. Through careful application of the supermesh technique, obtain values for all three mesh currents as labeled in Fig. 4.72.

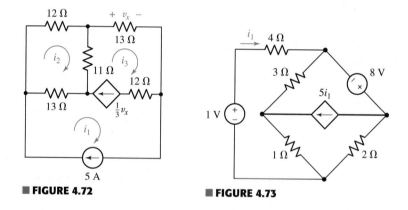

■ **FIGURE 4.72**   ■ **FIGURE 4.73**

48. Determine the power supplied by the 1 V source in Fig. 4.73.
49. Define three clockwise mesh currents for the circuit of Fig. 4.74, and employ the supermesh technique to obtain a numerical value for each.
50. Determine the power absorbed by the 10 Ω resistor in Fig. 4.75.

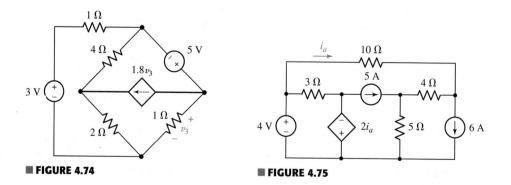

■ **FIGURE 4.74**   ■ **FIGURE 4.75**

4.5 Nodal vs. Mesh Analysis: A Comparison

51. For the circuit represented schematically in Fig. 4.76: (a) How many nodal equations would be required to determine i_5? (b) Alternatively, how many mesh equations would be required? (c) Would your preferred analysis method change if only the voltage across the 7 Ω resistor were needed? *Explain.*

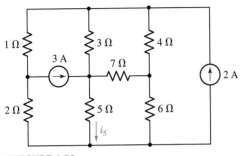

■ **FIGURE 4.76**

52. The circuit of Fig. 4.76 is modified such that the 3 A source is replaced by a 3 V source whose positive reference terminal is connected to the 7 Ω resistor. (a) Determine the number of nodal equations required to determine i_5. (b) Alternatively, how many mesh equations would be required? (c) Would your preferred analysis method change if only the voltage across the 7 Ω resistor were needed? *Explain.*

53. The circuit of Fig. 4.77 contains three sources. (a) As presently drawn, would nodal or mesh analysis result in fewer equations to determine the voltages v_1 and v_2? Explain. (b) If the voltage source were replaced with current sources, and the current source replaced with a voltage source, would your answer to part (a) change? *Explain?*

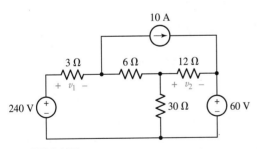

■ **FIGURE 4.77**

54. Solve for the voltage v_x as labeled in the circuit of Fig. 4.78 using (a) mesh analysis. (b) Repeat using nodal analysis. (c) Which approach was easier, and why?

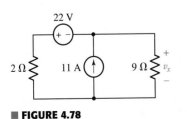

■ **FIGURE 4.78**

55. Consider the five-source circuit of Fig. 4.79. Determine the total number of simultaneous equations that must be solved in order to determine v_1 using (*a*) nodal analysis; (*b*) mesh analysis. (*c*) Which method is preferred, and does it depend on which side of the 40 Ω resistor is chosen as the reference node? *Explain your answer.*

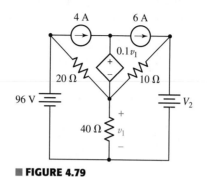

■ **FIGURE 4.79**

56. Replace the dependent voltage source in the circuit of Fig. 4.79 with a dependent current source oriented such that the arrow points upward. The controlling expression 0.1 v_1 remains unchanged. The value V_2 is zero. (*a*) Determine the total number of simultaneous equations required to obtain the power dissipated by the 40 Ω resistor if nodal analysis is employed. (*b*) Is mesh analysis preferred instead? *Explain.*

57. After studying the circuit of Fig. 4.80, determine the total number of simultaneous equations that must be solved to determine voltages v_1 and v_3 using (*a*) nodal analysis; (*b*) mesh analysis.

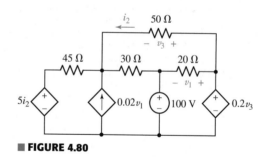

■ **FIGURE 4.80**

 58. From the perspective of determining voltages and currents associated with all components, (*a*) design a five-node, four-mesh circuit that is analyzed more easily using nodal techniques. (*b*) Modify your circuit by replacing only one component such that it is now more easily analyzed using mesh techniques.

4.6 Computer-Aided Circuit Analysis

 59. Employ PSpice (or similar CAD tool) to verify the solution of Exercise 8. Submit a printout of a properly labeled schematic with the answer highlighted, along with your hand calculations.

 60. Employ PSpice (or similar CAD tool) to verify the solution of Exercise 10. Submit a printout of a properly labeled schematic with the two nodal voltages highlighted, along with your hand calculations solving for the same quantities.

 61. Employ PSpice (or similar CAD tool) to verify the voltage across the 5 Ω resistor in the circuit of Exercise 13. Submit a printout of a properly labeled schematic with the answer highlighted, along with your hand calculations.

62. Verify numerical values for each nodal voltage in Exercise 15 by employing PSpice or a similar CAD tool. Submit a printout of an appropriately labeled schematic with the nodal voltages highlighted, along with your hand calculations.

63. Verify the numerical values for i_1 and v_x as indicated in the circuit accompanying Exercise 17, using PSpice or a similar CAD tool. Submit a printout of a properly labeled schematic with the answers highlighted, along with hand calculations.

64. (a) Generate an input deck for SPICE to determine the voltage v_9 as labeled in Fig. 4.81. Submit a printout of the output file with the solution highlighted. (b) Verify your answer by hand.

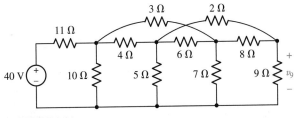

■ **FIGURE 4.81**

Chapter-Integrating Exercises

65. (a) Design a circuit employing only 9 V batteries and standard 5% tolerance value resistors that provide voltages of 1.5 V, 4.5 V, and 5 V and at least one mesh current of 1 mA. (b) Verify your design using PSpice or similar CAD tool.

66. A decorative string of multicolored outdoor lights is installed on a home in a quiet residential area. After plugging the 12 V ac adapter into the electrical socket, the homeowner immediately notes that two bulbs are burned out. (a) Are the individual lights connected in series or parallel? *Explain.* (b) Simulate the string by writing a SPICE input deck, assuming 44 lights, 12 V dc power supply, 24 AWG soft solid copper wire, and individual bulbs rated at 10 mW each. Submit a printout of the output file, with the power supplied by the 12 V supply highlighted. (c) Verify your simulation with hand calculations.

67. Consider the circuit depicted in Fig. 4.82. Employ either nodal or mesh analysis as a design tool to obtain a value of 200 mA for i_1, if elements A, B, C, D, E, and F must be either current or voltage sources with nonzero values.

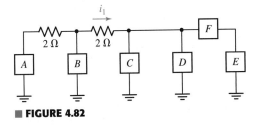

■ **FIGURE 4.82**

68. (a) Under what circumstances does the presence of an independent voltage source greatly simplify nodal analysis? *Explain.* (b) Under what circumstances does the presence of an independent current source significantly simplify mesh analysis? *Explain.* (c) On which fundamental physical principle do we base nodal analysis? (d) On which fundamental physical principle do we base mesh analysis?

69. Referring to Fig. 4.83, (*a*) determine whether nodal or mesh analysis is more appropriate in determining i_2 if element A is replaced with a short circuit, then carry out the analysis. (*b*) Verify your answer with an appropriate PSpice simulation. Submit a properly labeled schematic along with the answer highlighted.

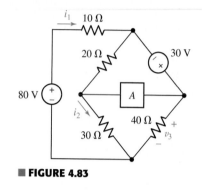

■ **FIGURE 4.83**

70. The element marked A in the circuit of Fig. 4.83 is replaced by a 2.5 V independent voltage source with the positive reference terminal connected to the common node of the 20 Ω and 30 Ω resistors. (*a*) Determine whether mesh or nodal analysis is more straightforward for determining the voltage marked v_3. (*b*) Verify your answer using PSpice. (*c*) Would your conclusion for part (*a*) change if the current i_2 were required as well? *Explain.*

to a complicated amplifier circuit in order to obtain a zero output voltage. We also notice that there are several other types of parameter sweeps that we can perform, including a dc voltage sweep. The ability to vary temperature is useful only when dealing with component models that have a temperature parameter built in, such as diodes and transistors.

Unfortunately, it usually turns out that little if any time is saved in analyzing a circuit containing one or more dependent sources by use of the superposition principle, for there must always be at least two sources in operation: one independent source and all the dependent sources.

We must constantly be aware of the limitations of superposition. It is applicable only to linear responses, and thus the most common nonlinear response—power—is not subject to superposition. For example, consider two 1 V batteries in series with a 1 Ω resistor. The power delivered to the resistor is 4 W, but if we mistakenly try to apply superposition, we might say that each battery alone furnished 1 W and thus the calculated power is only 2 W. This is incorrect, but a surprisingly easy mistake to make.

5.2 SOURCE TRANSFORMATIONS

Practical Voltage Sources

So far, we've only worked with *ideal* sources—elements whose terminal voltage is independent of the current flowing through them. To see the relevance of this fact, consider a simple independent ("ideal") 9 V source connected to a 1 Ω resistor. The 9 volt source will force a current of 9 amperes through the 1 Ω resistor (perhaps this seems reasonable enough), but the same source would apparently force 9,000,000 amperes through a 1 mΩ resistor (which hopefully does not seem reasonable). On paper, there's nothing to stop us from reducing the resistor value all the way to 0 Ω ... but that would lead to a contradiction, as the source would be "trying" to maintain 9 V across a dead short, which Ohm's law tells us can't happen ($V = 9 = RI = 0$?).

What happens in the real world when we do this type of experiment? For example, if we try to start a car with the headlights already on, we most likely notice the headlights dim as the battery is asked to supply a large (∼100 A or more) starter current in parallel with the current running to the headlights. If we model the 12 V battery with an ideal 12 V source as in Fig. 5.11*a*, our observation cannot be explained. Another way of saying this is that our model breaks down when the load draws a large current from the source.

To better approximate the behavior of a real device, the ideal voltage source must be modified to account for the lowering of its terminal voltage when large currents are drawn from it. Let us suppose that we observe experimentally that our car battery has a terminal voltage of 12 V when no current is flowing through it, and a reduced voltage of 11 V when 100 A is flowing. How could we model this behavior? Well, a more accurate model might be an ideal voltage source of 12 V in series with a resistor across which 1 V appears when 100 A flows through it. A quick calculation shows that the resistor must be 1 V/100 A = 0.01 Ω, and the ideal voltage source and this series resistor constitute a ***practical voltage source*** (Fig. 5.11*b*).

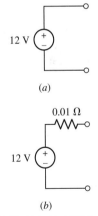

(a)

(b)

■ **FIGURE 5.11** (*a*) An ideal 12 V dc voltage source used to model a car battery. (*b*) A more accurate model that accounts for the observed reduction in terminal voltage at large currents.

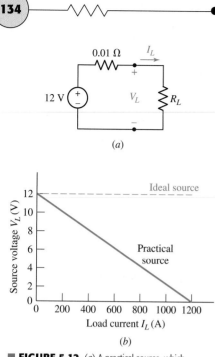

(a)

(b)

■ **FIGURE 5.12** *(a)* A practical source, which approximates the behavior of a certain 12 V automobile battery, is shown connected to a load resistor R_L. *(b)* The relationship between I_L and V_L is linear.

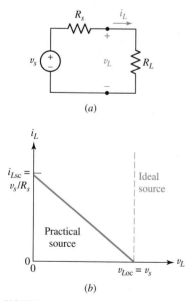

(a)

(b)

■ **FIGURE 5.13** *(a)* A general practical voltage source connected to a load resistor R_L. *(b)* The terminal voltage of a practical voltage source decreases as i_L increases and $R_L = v_L/i_L$ decreases. The terminal voltage of an ideal voltage source (also plotted) remains the same for any current delivered to a load.

Thus, we are using the series combination of two ideal circuit elements, an independent voltage source and a resistor, to model a real device.

We do not expect to find such an arrangement of ideal elements inside our car battery, of course. Any real device is characterized by a certain current-voltage relationship at its terminals, and our problem is to develop some combination of ideal elements that can furnish a similar current-voltage characteristic, at least over some useful range of current, voltage, or power.

In Fig. 5.12*a*, we show our two-piece practical model of the car battery now connected to some load resistor R_L. The terminal voltage of the practical source is the same as the voltage across R_L and is marked[2] V_L. Figure 5.12*b* shows a plot of load voltage V_L as a function of the load current I_L for this practical source. The KVL equation for the circuit of Fig. 5.12*a* may be written in terms of I_L and V_L:

$$12 = 0.01I_L + V_L$$

and thus

$$V_L = -0.01I_L + 12$$

This is a linear equation in I_L and V_L, and the plot in Fig. 5.12*b* is a straight line. Each point on the line corresponds to a different value of R_L. For example, the midpoint of the straight line is obtained when the load resistance is equal to the internal resistance of the practical source, or $R_L = 0.01 \, \Omega$. Here, the load voltage is exactly one-half the ideal source voltage.

When $R_L = \infty$ and no current whatsoever is being drawn by the load, the practical source is open-circuited and the terminal voltage, or open-circuit voltage, is $V_{Loc} = 12$ V. If, on the other hand, $R_L = 0$, thereby short-circuiting the load terminals, then a load current or short-circuit current, $I_{Lsc} = 1200$ A, would flow. (*In practice, such an experiment would probably result in the destruction of the short circuit, the battery, and any measuring instruments incorporated in the circuit!*)

Since the plot of V_L versus I_L is a straight line for this practical voltage source, we should note that the values of V_{Loc} and I_{Lsc} uniquely determine the entire V_L–I_L curve.

The horizontal broken line of Fig. 5.12*b* represents the V_L–I_L plot for an *ideal* voltage source; the terminal voltage remains constant for any value of load current. For the practical voltage source, the terminal voltage has a value near that of the ideal source only when the load current is relatively small.

Let us now consider a *general* practical voltage source, as shown in Fig. 5.13*a*. The voltage of the ideal source is v_s, and a resistance R_s, called an *internal resistance* or *output resistance*, is placed in series with it. Again, we must note that the resistor is not really present as a separate component but merely serves to account for a terminal voltage that decreases as the load current increases. Its presence enables us to model the behavior of a physical voltage source more closely.

The linear relationship between v_L and i_L is

$$v_L = v_s - R_s i_L \qquad [9]$$

---

(2) From this point on we will endeavor to adhere to the standard convention of referring to strictly dc quantities using capital letters, whereas lowercase letters denote a quantity that we know to possess some time-varying component. However, in describing general theorems which apply to either dc or ac, we will continue to use lowercase to emphasize the general nature of the concept.

and this is plotted in Fig. 5.13*b*. The open-circuit voltage ($R_L = \infty$, so $i_L = 0$) is

$$v_{L\text{oc}} = v_s \qquad [10]$$

and the short-circuit current ($R_L = 0$, so $v_L = 0$) is

$$i_{L\text{sc}} = \frac{v_s}{R_s} \qquad [11]$$

Once again, these values are the intercepts for the straight line in Fig. 5.13*b*, and they serve to define it completely.

Practical Current Sources

An ideal current source is also nonexistent in the real world; there is no physical device that will deliver a constant current regardless of the load resistance to which it is connected or the voltage across its terminals. Certain transistor circuits will deliver a constant current to a wide range of load resistances, but the load resistance can always be made sufficiently large that the current through it becomes very small. Infinite power is simply never available (unfortunately).

A practical current source is defined as an ideal current source in parallel with an internal resistance R_p. Such a source is shown in Fig. 5.14*a*, and the current i_L and voltage v_L associated with a load resistance R_L are indicated. Application of KCL yields

$$i_L = i_s - \frac{v_L}{R_p} \qquad [12]$$

which is again a linear relationship. The open-circuit voltage and the short-circuit current are

$$v_{L\text{oc}} = R_p i_s \qquad [13]$$

and

$$i_{L\text{sc}} = i_s \qquad [14]$$

The variation of load current with changing load voltage may be investigated by changing the value of R_L as shown in Fig. 5.14*b*. The straight line is traversed from the short-circuit, or "northwest," end to the open-circuit termination at the "southeast" end by increasing R_L from zero to infinite ohms. The midpoint occurs for $R_L = R_p$. The load current i_L and the ideal source current are approximately equal only for small values of load voltage, which are obtained with values of R_L that are small compared to R_p.

Equivalent Practical Sources

It may be no surprise that we can improve upon models to increase their accuracy; at this point we now have a practical voltage source model and also a practical current source model. Before we proceed, however, let's take a moment to compare Fig. 5.13*b* and Fig. 5.14*b*. One is for a circuit with a voltage source and the other, with a current source, *but the graphs are indistinguishable!*

It turns out that this is no coincidence. In fact, we are about to show that a practical voltage source *can be* electrically equivalent to a practical current source—meaning that a load resistor R_L connected to either will have

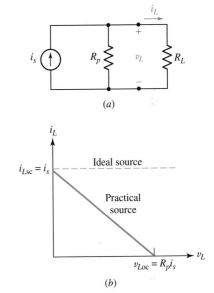

(a)

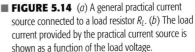

(b)

■ FIGURE 5.14 (*a*) A general practical current source connected to a load resistor R_L. (*b*) The load current provided by the practical current source is shown as a function of the load voltage.

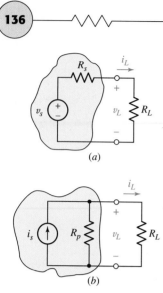

FIGURE 5.15 (a) A given practical voltage source connected to a load R_L. (b) The equivalent practical current source connected to the same load.

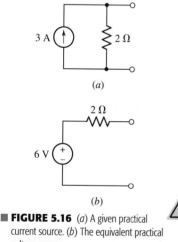

FIGURE 5.16 (a) A given practical current source. (b) The equivalent practical voltage source.

the same v_L and i_L. This means we can replace one practical source with the other and the rest of the circuit will not know the difference.

Consider the practical voltage source and resistor R_L shown in Fig. 5.15a, and the circuit composed of a practical current source and resistor R_L shown in Fig. 5.15b. A simple calculation shows that the voltage across the load R_L of Fig. 5.15a is

$$v_L = v_s \frac{R_L}{R_s + R_L} \qquad [15]$$

A similar calculation shows that the voltage across the load R_L in Fig. 5.15b is

$$v_L = \left(i_s \frac{R_p}{R_p + R_L} \right) \cdot R_L$$

The two practical sources are electrically equivalent, then, if

$$R_s = R_p \qquad [16]$$

and

$$v_s = R_p i_s = R_s i_s \qquad [17]$$

where we now let R_s represent the internal resistance of either practical source, which is the conventional notation.

Let's try this with the practical current source shown in Fig. 5.16a. Since its internal resistance is 2 Ω, the internal resistance of the equivalent practical voltage source is also 2 Ω; the voltage of the ideal voltage source contained within the practical voltage source is $(2)(3) = 6$ V. The equivalent practical voltage source is shown in Fig. 5.16b.

To check the equivalence, let us visualize a 4 Ω resistor connected to each source. In both cases a current of 1 A, a voltage of 4 V, and a power of 4 W are associated with the 4 Ω load. However, we should note very carefully that the ideal current source is delivering a total power of 12 W, while the ideal voltage source is delivering only 6 W. Furthermore, the internal resistance of the practical current source is absorbing 8 W, whereas the internal resistance of the practical voltage source is absorbing only 2 W. Thus we see that the two practical sources are equivalent only with respect to what transpires at the load terminals; they are *not* equivalent internally!

EXAMPLE 5.4

Compute the current through the 4.7 kΩ resistor in Fig. 5.17a after transforming the 9 mA source into an equivalent voltage source.

It's not just the 9 mA source at issue, but also the resistance in parallel with it (5 kΩ). We remove these components, leaving two terminals "dangling." We then replace them with a voltage source in series with a 5 kΩ resistor. The value of the voltage source must be $(0.09)(5000) = 45$ V.

Redrawing the circuit as in Fig. 5.17b, we can write a simple KVL equation

$$-45 + 5000I + 4700I + 3000I + 3 = 0$$

which is easily solved to yield $I = 3.307$ mA.

We can check our answer of course by analyzing the circuit of Fig. 5.17a using either nodal or mesh techniques.

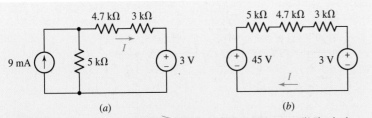

■ **FIGURE 5.17** (*a*) A circuit with both a voltage source and a current source. (*b*) The circuit after the 9 mA source is transformed into an equivalent voltage source.

PRACTICE

5.3 For the circuit of Fig. 5.18, compute the current I_X through the 47 kΩ resistor after performing a source transformation on the voltage source.

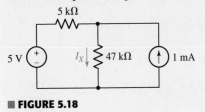

■ **FIGURE 5.18**

Ans: 192 μA.

EXAMPLE **5.5**

Calculate the current through the 2 Ω resistor in Fig. 5.19a by making use of source transformations to first simplify the circuit.

We begin by transforming each current source into a voltage source (Fig. 5.19*b*), the strategy being to convert the circuit into a simple loop.

We must be careful to retain the 2 Ω resistor for two reasons: first, the dependent source controlling variable appears across it, and second, we desire the current flowing through it. However, we can combine the 17 Ω and 9 Ω resistors, since they appear in series. We also see that the 3 Ω and 4 Ω resistors may be combined into a single 7 Ω resistor, which can then be used to transform the 15 V source into a 15/7 A source as in Fig. 5.19*c*.

Finally, we note that the two 7 Ω resistors can be combined into a single 3.5 Ω resistor, which may be used to transform the 15/7 A current source into a 7.5 V voltage source. The result is a simple loop circuit, shown in Fig. 5.19*d*.

The current I can now be found using KVL:

$$-7.5 + 3.5I - 51V_x + 28I + 9 = 0$$

where

$$V_x = 2I$$

Thus,

$$I = 21.28 \text{ mA}$$

(*Continued on next page*)

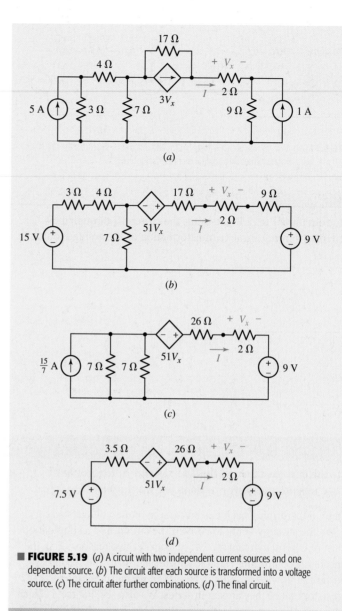

FIGURE 5.19 (*a*) A circuit with two independent current sources and one dependent source. (*b*) The circuit after each source is transformed into a voltage source. (*c*) The circuit after further combinations. (*d*) The final circuit.

PRACTICE

5.4 For the circuit of Fig. 5.20, compute the voltage V across the 1 MΩ resistor using repeated source transformations.

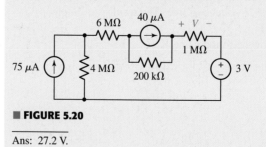

FIGURE 5.20

Ans: 27.2 V.

Several Key Points

We conclude our discussion of practical sources and source transformations with a few observations. First, when we transform a voltage source, we must be sure that the source is in fact *in series* with the resistor under consideration. For example, in the circuit of Fig. 5.21, it is perfectly valid to perform a source transformation on the voltage source using the 10 Ω resistor, as they are in series. However, it would be incorrect to attempt a source transformation using the 60 V source and the 30 Ω resistor—a very common type of error.

In a similar fashion, when we transform a current source and resistor combination, we must be sure that they are in fact *in parallel.* Consider the current source shown in Fig. 5.22a. We may perform a source transformation including the 3 Ω resistor, as they are in parallel, but after the transformation there may be some ambiguity as to where to place the resistor. In such circumstances, it is helpful to first redraw the components to be transformed as in Fig. 5.22b. Then the transformation to a voltage source in series with a resistor may be drawn correctly as shown in Fig. 5.22c; the resistor may in fact be drawn above or below the voltage source.

It is also worthwhile to consider the unusual case of a current source in series with a resistor, and its dual, the case of a voltage source in parallel

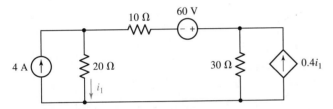

■ **FIGURE 5.21** An example circuit to illustrate how to determine if a source transformation can be performed.

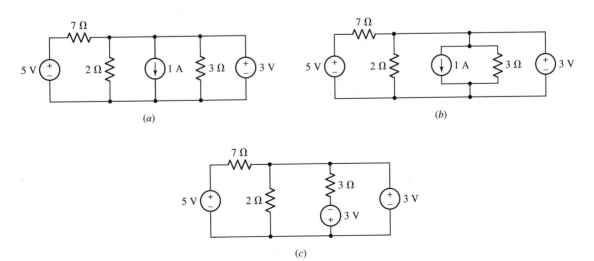

(a)

(b)

(c)

■ **FIGURE 5.22** (a) A circuit with a current source to be transformed to a voltage source. (b) Circuit redrawn so as to avoid errors. (c) Transformed source/resistor combination.

with a resistor. Let's start with the simple circuit of Fig. 5.23a, where we are interested only in the voltage across the resistor marked R_2. We note that regardless of the value of resistor R_1, $V_{R_2} = I_x R_2$. Although we might be tempted to perform an inappropriate source transformation on such a circuit, in fact *we may simply omit resistor R_1* (provided that it is of no interest to us itself). A similar situation arises with a voltage source in parallel with a resistor, as depicted in Fig. 5.23b. Again, if we are only interested in some quantity regarding resistor R_2, we may find ourselves tempted to perform some strange (and incorrect) source transformation on the voltage source and resistor R_1. In reality, we may omit resistor R_1 from our circuit as far as resistor R_2 is concerned—its presence does not alter the voltage across, the current through, or the power dissipated by resistor R_2.

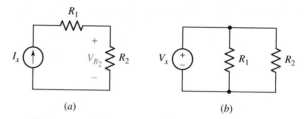

■ **FIGURE 5.23** (*a*) Circuit with a resistor R_1 in series with a current source. (*b*) A voltage source in parallel with two resistors.

Summary of Source Transformation

1. **A common goal in source transformation is to end up with either all current sources or all voltage sources in the circuit.** This is especially true if it makes nodal or mesh analysis easier.

2. **Repeated source transformations can be used to simplify a circuit by allowing resistors and sources to eventually be combined.**

3. **The resistor value does not change during a source transformation, but it is not the same resistor.** This means that currents or voltages associated with the original resistor are irretrievably lost when we perform a source transformation.

4. **If the voltage or current associated with a particular resistor is used as a controlling variable for a dependent source, it should not be included in any source transformation.** The original resistor must be retained in the final circuit, untouched.

5. **If the voltage or current associated with a particular element is of interest, that element should not be included in any source transformation.** The original element must be retained in the final circuit, untouched.

6. **In a source transformation, the head of the current source arrow corresponds to the "+" terminal of the voltage source.**

7. **A source transformation on a current source and resistor requires that the two elements be in parallel.**

8. **A source transformation on a voltage source and resistor requires that the two elements be in series.**

24. With regard to the circuit represented in Fig. 5.68, first transform both voltage sources to current sources, reduce the number of elements as much as possible, and determine the voltage v_3.

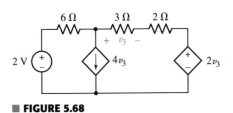

■ **FIGURE 5.68**

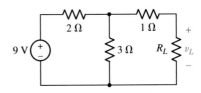

5.3 Thévenin and Norton Equivalent Circuits

25. Referring to Fig. 5.69, determine the Thévenin equivalent of the network connected to R_L. (b) Determine v_L for $R_L = 1\ \Omega, 3.5\ \Omega, 6.257\ \Omega$, and $9.8\ \Omega$.

■ **FIGURE 5.69**

26. (a) With respect to the circuit depicted in Fig. 5.69, obtain the Norton equivalent of the network connected to R_L. (b) Plot the power dissipated in resistor R_L as a function of i_L corresponding to the range of $0 < R_L < 5\ \Omega$. (c) Using your graph, estimate at what value of R_L does the dissipated power reach its maximum value.

27. (a) Obtain the Norton equivalent of the network connected to R_L in Fig. 5.70. (b) Obtain the Thévenin equivalent of the same network. (c) Use either to calculate i_L for $R_L = 0\ \Omega, 1\ \Omega, 4.923\ \Omega$, and $8.107\ \Omega$.

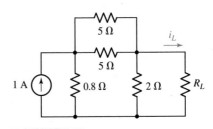

■ **FIGURE 5.70**

28. (a) Determine the Thévenin equivalent of the circuit depicted in Fig. 5.71 by first finding V_{oc} and I_{sc} (defined as flowing into the positive reference terminal of V_{oc}). (b) Connect a 4.7 kΩ resistor to the open terminals of your new network and calculate the power it dissipates.

29. Referring to the circuit of Fig. 5.71: (a) Determine the Norton equivalent of the circuit by first finding V_{oc} and I_{sc} (defined as flowing into the positive reference terminal of V_{oc}). (b) Connect a 1.7 kΩ resistor to the open terminals of your new network and calculate the power supplied to that resistor.

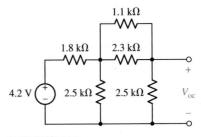

30. (a) Employ Thévenin's theorem to obtain a simple two-component equivalent of the circuit shown in Fig. 5.72. (b) Use your equivalent circuit to determine the power delivered to a 100 Ω resistor connected to the open terminals. (c) Verify your solution by analyzing the original circuit with the same 100 Ω resistor connected across the open terminals.

■ **FIGURE 5.71**

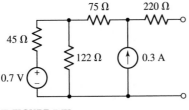

■ **FIGURE 5.72**

 31. (a) Employ Thévenin's theorem to obtain a two-component equivalent for the network shown in Fig. 5.73. (b) Determine the power supplied to a 1 MΩ resistor connected to the network if $i_1 = 19\ \mu A$, $R_1 = R_2 = 1.6$ MΩ, $R_2 = 3$ MΩ, and $R_4 = R_5 = 1.2$ MΩ. (c) Verify your solution by simulating both circuits with PSpice or another appropriate CAD tool.

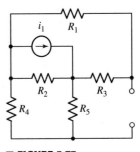

■ **FIGURE 5.73**

32. Determine the Thévenin equivalent of the network shown in Fig. 5.74 as seen looking into the two open terminals.

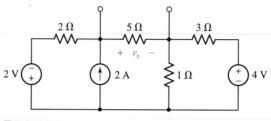

■ **FIGURE 5.74**

33. (*a*) Determine the Norton equivalent of the circuit depicted in Fig. 5.74 as seen looking into the two open terminals. (*b*) Compute power dissipated in a 5 Ω resistor connected in parallel with the existing 5 Ω resistor. (*c*) Compute the current flowing through a short circuit connecting the two terminals.

34. For the circuit of Fig. 5.75: (*a*) Employ Norton's theorem to reduce the network connected to R_L to only two components. (*b*) Calculate the downward-directed current flowing through R_L if it is a 3.3 kΩ resistor. (*c*) Verify your answer by simulating both circuits with PSpice or a comparable CAD tool.

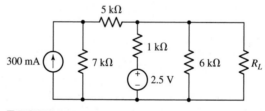

■ **FIGURE 5.75**

35. (*a*) Obtain a value for the Thévenin equivalent resistance seen looking into the open terminals of the circuit in Fig. 5.76 by first finding V_{oc} and I_{sc}. (*b*) Connect a 1 A test source to the open terminals of the original circuit after shorting the voltage source, and use this to obtain R_{TH}. (*c*) Connect a 1 V test source to the open terminals of the original circuit after again zeroing the 2 V source, and use this now to obtain R_{TH}.

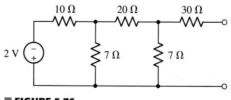

■ **FIGURE 5.76**

36. Refer to the circuit depicted in Fig. 5.77. (*a*) Obtain a value for the Thévenin equivalent resistance seen looking into the open terminals by first finding V_{oc} and I_{sc}. (*b*) Connect a 1 A test source to the open terminals of the original

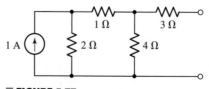

■ **FIGURE 5.77**

circuit after deactivating the other current source, and use this to obtain R_{TH}.
(c) Connect a 1 V test source to the open terminals of the original circuit, once
again zeroing out the original source, and use this now to obtain R_{TH}.

37. Obtain a value for the Thévenin equivalent resistance seen looking into the
open terminals of the circuit in Fig. 5.78 by (a) finding V_{oc} and I_{sc}, and then
taking their ratio; (b) setting all independent sources to zero and using resistor
combination techniques; (c) connecting an unknown current source to the
terminals, deactivating (zero out) all other sources, finding an algebraic expression for the voltage that develops across the source, and taking the ratio of the
two quantities.

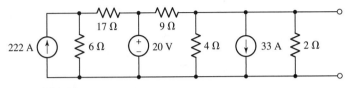

■ FIGURE 5.78

38. With regard to the network depicted in Fig. 5.79, determine the Thévenin
equivalent as seen by an element connected to terminals (a) a and b; (b) a
and c; (c) b and c. (d) Verify your answers using PSpice or other suitable CAD
tool. (*Hint: Connect a test source to the terminals of interest.*)

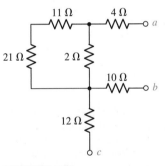

■ FIGURE 5.79

39. Determine the Thévenin and Norton equivalents of the circuit represented in
Fig. 5.80 from the perspective of the open terminals. (There should be no
dependent sources in your answer.)

40. Determine the Norton equivalent of the circuit drawn in Fig. 5.81 as seen by
terminals a and b. (There should be no dependent sources in your answer.)

41. With regard to the circuit of Fig. 5.82, determine the power dissipated by (a) a
1 kΩ resistor connected between a and b; (b) a 4.7 kΩ resistor connected
between a and b; (c) a 10.54 kΩ resistor connected between a and b.

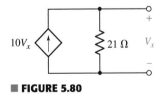

■ FIGURE 5.80

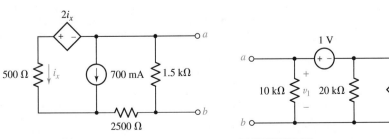

■ FIGURE 5.81   **■ FIGURE 5.82**

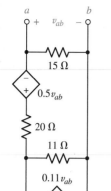

■ **FIGURE 5.83**

42. Determine the Thévenin and Norton equivalents of the circuit shown in Fig. 5.83, as seen by an unspecified element connected between terminals a and b.

43. Referring to the circuit of Fig. 5.84, determine the Thévenin equivalent resistance of the circuit to the right of the dashed line. This circuit is a common-source transistor amplifier, and you are calculating its input resistance.

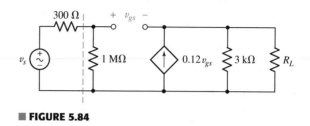

■ **FIGURE 5.84**

44. Referring to the circuit of Fig. 5.85, determine the Thévenin equivalent resistance of the circuit to the right of the dashed line. This circuit is a common-collector transistor amplifier, and you are calculating its input resistance.

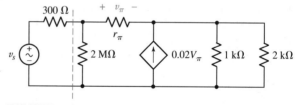

■ **FIGURE 5.85**

45. The circuit shown in Fig. 5.86 is a reasonably accurate model of an operational amplifier. In cases where R_i and A are very large and $R_o \sim 0$, a resistive load (such as a speaker) connected between ground and the terminal labeled v_{out} will see a voltage $-R_f/R_1$ times larger than the input signal v_{in}. Find the Thévenin equivalent of the circuit, taking care to label v_{out}.

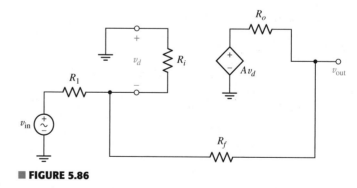

■ **FIGURE 5.86**

5.4 Maximum Power Transfer

46. (*a*) For the simple circuit of Fig. 5.87, graph the power dissipated by the resistor R as a function of R/R_S, if $0 \leq R \leq 3000 \ \Omega$. (*b*) Graph the first derivative of the power versus R/R_S, and verify that maximum power is transferred to R when it is equal to R_S.

■ **FIGURE 5.87**

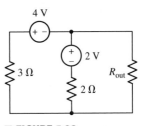

47. For the circuit drawn in Fig. 5.88, (*a*) determine the Thévenin equivalent connected to R_{out}. (*b*) Choose R_{out} such that maximum power is delivered to it.

48. Study the circuit of Fig. 5.89. (*a*) Determine the Norton equivalent connected to resistor R_{out}. (*b*) Select a value for R_{out} such that maximum power will be delivered to it.

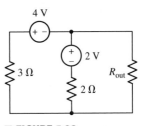

■ **FIGURE 5.88**

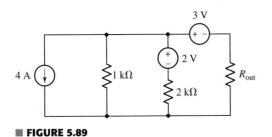

■ **FIGURE 5.89**

49. Assuming that we can determine the Thévenin equivalent resistance of our wall socket, why don't toaster, microwave oven, and TV manufacturers match each appliance's Thévenin equivalent resistance to this value? Wouldn't it permit maximum power transfer from the utility company to our household appliances?

50. For the circuit of Fig. 5.90, what value of R_L will ensure it absorbs the maximum possible amount of power?

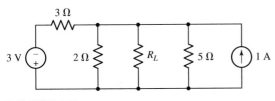

■ **FIGURE 5.90**

51. With reference to the circuit of Fig. 5.91, (*a*) calculate the power absorbed by the 9 Ω resistor; (*b*) adjust the size of the 5 Ω resistor so that the new network delivers maximum power to the 9 Ω resistor.

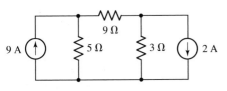

■ **FIGURE 5.91**

52. Referring to the circuit of Fig. 5.92, (*a*) determine the power absorbed by the 3.3 Ω resistor; (*b*) replace the 3.3 Ω resistor with another resistor such that it absorbs maximum power from the rest of the circuit.

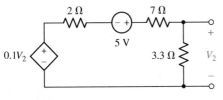

■ **FIGURE 5.92**

53. Select a value for R_L in Fig. 5.93 such that it is ensured to absorb maximum power from the circuit.

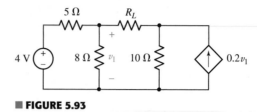

■ **FIGURE 5.93**

54. Determine what value of resistance would absorb maximum power from the circuit of Fig. 5.94 when connected across terminals a and b.

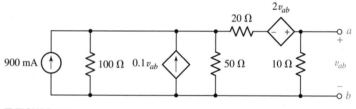

■ **FIGURE 5.94**

5.5 Delta-Wye Conversion

55. Derive the equations required to convert from a Y-connected network to a Δ-connected network.

56. Convert the Δ- (or "Π-") connected networks in Fig. 5.95 to Y-connected networks.

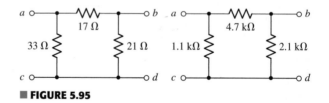

■ **FIGURE 5.95**

57. Convert the Y- (or "T-") connected networks in Fig. 5.96 to Δ-connected networks.

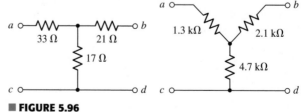

■ **FIGURE 5.96**

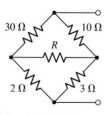

■ **FIGURE 5.97**

58. For the network of Fig. 5.97, select a value of R such that the network has an equivalent resistance of 9 Ω. Round your answer to two significant figures.

59. For the network of Fig. 5.98, select a value of R such that the network has an equivalent resistance of 70.6 Ω.

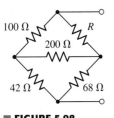

■ FIGURE 5.98

60. Determine the effective resistance R_{in} of the network exhibited in Fig. 5.99.

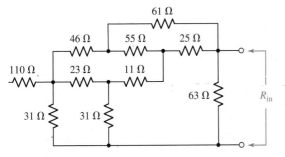

Each R is 2.2 kΩ

■ FIGURE 5.99

61. Calculate R_{in} as indicated in Fig. 5.100.

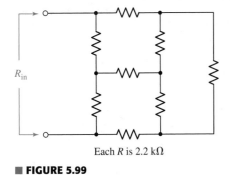

■ FIGURE 5.100

62. Employ Δ/Y conversion techniques as appropriate to determine R_{in} as labeled in Fig. 5.101.

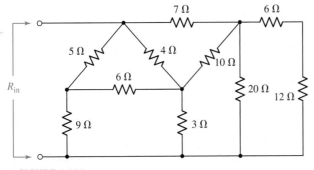

■ FIGURE 5.101

63. (*a*) Determine the two-component Thévenin equivalent of the network in Fig. 5.102. (*b*) Calculate the power dissipated by a 1 Ω resistor connected between the open terminals.

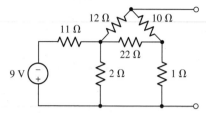

■ **FIGURE 5.102**

64. (*a*) Use appropriate techniques to obtain both the Thévenin and Norton equivalents of the network drawn in Fig. 5.103. (*b*) Verify your answers by simulating each of the three circuits connected to a 1 Ω resistor.

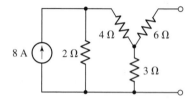

■ **FIGURE 5.103**

65. (*a*) Replace the network in Fig. 5.104 with an equivalent three-resistor Δ network.

(*b*) Perform a PSpice analysis to verify that your answer is in fact equivalent. (*Hint:* Try adding a load resistor.)

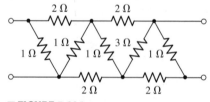

■ **FIGURE 5.104**

5.6 Selecting an Approach: A Summary of Various Techniques

66. Determine the power absorbed by a resistor connected between the open terminal of the circuit shown in Fig. 5.105 if it has a value of (*a*) 1 Ω; (*b*) 100 Ω; (*c*) 2.65 kΩ; (*d*) 1.13 MΩ.

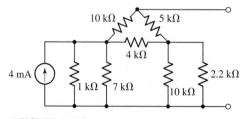

■ **FIGURE 5.105**

67. It is known that a load resistor of some type will be connected between terminals *a* and *b* of the network of Fig. 5.106. (*a*) Change the value of the 25 V source such that both voltage sources contribute equally to the power delivered to the load resistor, assuming its value is chosen such that it absorbs maximum power. (*b*) Calculate the value of the load resistor.

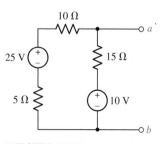

■ **FIGURE 5.106**

68. A 2.57 Ω load is connected between terminals a and b of the network drawn in Fig. 5.106. Unfortunately, the power delivered to the load is only 50% of the required amount. Altering only voltage sources, modify the circuit so that the required power is delivered and both sources contribute equally.

69. A load resistor is connected across the open terminals of the circuit shown in Fig. 5.107, and its value was chosen carefully to ensure maximum power transfer from the rest of the circuit. (*a*) What is the value of the resistor? (*b*) If the power absorbed by the load resistor is three times as large as required, modify the circuit so that it performs as desired, without losing the maximum power transfer condition already enjoyed.

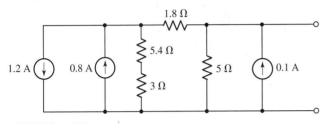

■ **FIGURE 5.107**

70. A backup is required for the circuit depicted in Fig. 5.107. It is unknown what will be connected to the open terminals, or whether it will be purely linear. If a simple battery is to be used, what no-load ("open circuit") voltage should it have, and what is the maximum tolerable internal resistance?

Chapter-Integrating Exercises

71. Three 45 W light bulbs originally wired in a Y network configuration with a 120 V ac source connected across each port are rewired as a Δ network. The neutral, or center, connection is not used. If the intensity of each light is proportional to the power it draws, design a new 120 V ac power circuit so that the three lights have the same intensity in the Δ configuration as they did when connected in a Y configuration. Verify your design using PSpice by comparing the power drawn by each light in your circuit (modeled as an appropriately chosen resistor value) with the power each would draw in the original Y-connected circuit.

72. (*a*) Explain in general terms how source transformation can be used to simplify a circuit prior to analysis. (*b*) Even if source transformations can greatly simplify a particular circuit, when might it not be worth the effort? (*c*) Multiplying all the independent sources in a circuit by the same scaling factor results in all other voltages and currents being scaled by the same amount. Explain why we don't scale the dependent sources as well. (*d*) In a general circuit, if we set an independent voltage source to zero, what current can flow through it? (*e*) In a general circuit, if we set an independent current source to zero, what voltage can be sustained across its terminals?

73. The load resistor in Fig. 5.108 can safely dissipate up to 1 W before overheating and bursting into flame. The lamp can be treated as a 10.6 Ω resistor if less than 1 A flows through it and a 15 Ω resistor if more than 1 A flows through it. What is the maximum permissible value of I_s? Verify your answer with PSpice.

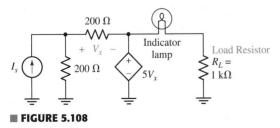

■ **FIGURE 5.108**

74. A certain red LED has a maximum current rating of 35 mA, and if this value is exceeded, overheating and catastrophic failure will result. The resistance of the LED is a nonlinear function of its current, but the manufacturer warrants a minimum resistance of 47 Ω and a maximum resistance of 117 Ω. Only 9 V batteries are available to power the LED. Design a suitable circuit to deliver the maximum power possible to the LED without damaging it. Use only combinations of the standard resistor values given in the inside front cover.

 75. As part of a security system, a very thin 100 Ω wire is attached to a window using nonconducting epoxy. Given only a box of 12 rechargeable 1.5 V AAA batteries, one thousand 1 Ω resistors, and a 2900 Hz piezo buzzer that draws 15 mA at 6 V, design a circuit with no moving parts that will set off the buzzer if the window is broken (and hence the thin wire as well). Note that the buzzer requires a dc voltage of at least 6 V (maximum 28 V) to operate.

CHAPTER

6

The Operational Amplifier

INTRODUCTION

At this point we have a good set of circuit analysis tools at our disposal, but have focused primarily on somewhat general circuits composed of only sources and resistors. In this chapter, we introduce a new component which, although technically nonlinear, can be treated effectively with linear models. This element, known as the *operational amplifier* or *op amp* for short, finds daily usage in a large variety of electronic applications. It also provides us a new element to use in building circuits, and another opportunity to test out our developing analytical skills.

6.1 BACKGROUND

The origins of the operational amplifier date to the 1940s, when basic circuits were constructed using vacuum tubes to perform mathematical operations such as addition, subtraction, multiplication, division, differentiation, and integration. This enabled the construction of analog (as opposed to digital) computers tasked with the solution of complex differential equations. The first commercially available op amp *device* is generally considered to be the K2-W, manufactured by Philbrick Researches, Inc. of Boston from about 1952 through the early 1970s (Fig. 6.1*a*). These early vacuum tube devices weighed 3 oz (85 g), measured $1^{33}/_{64}$ in $\times$ $2^9/_{64}$ in $\times$ $4^7/_{64}$ in (3.8 cm $\times$ 5.4 cm $\times$ 10.4 cm), and sold for about US$22. In contrast, integrated circuit (IC) op amps such as the Fairchild KA741 weigh less than 500 mg, measure 5.7 mm $\times$ 4.9 mm $\times$ 1.8 mm, and sell for approximately US$0.22.

Compared to op amps based on vacuum tubes, modern IC op amps are constructed using perhaps 25 or more transistors all on the same silicon "chip," as well as resistors and capacitors needed to obtain the desired performance characteristics. As a result, they run at

KEY CONCEPTS

Characteristics of Ideal Op Amps

Inverting and Noninverting Amplifiers

Summing and Difference Amplifier Circuits

Cascaded Op Amp Stages

Using Op Amps to Build Voltage and Current Sources

Nonideal Characteristics of Op Amps

Voltage Gain and Feedback

Basic Comparator and Instrumentation Amplifier Circuits

175

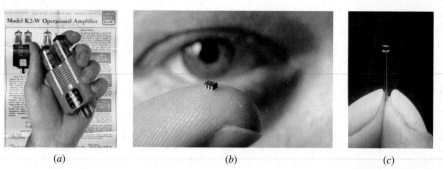

(a)   (b)   (c)

■ **FIGURE 6.1** (a) A Philbrick K2-W op amp, based on a matched pair of 12AX7A vacuum tubes.
(b) LMV321 op amp, used in a variety of phone and game applications. (c) LMC6035 operational amplifier,
which packs 114 transistors into a package so small that it fits on the head of a pin.

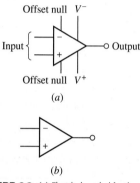

■ **FIGURE 6.2** (a) Electrical symbol for the op amp.
(b) Minimum required connections to be shown on a
circuit schematic.

much lower dc supply voltages (± 18 V, for example, as opposed to ± 300 V for the K2-W), are more reliable, and considerably smaller (Fig. 6.1b,c). In some cases, the IC may contain several op amps. In addition to the output pin and the two inputs, other pins enable power to be supplied to run the transistors, and for external adjustments to be made to balance and compensate the op amp. The symbol commonly used for an op amp is shown in Fig. 6.2a. At this point, we are not concerned with the internal circuitry of the op amp or the IC, but only with the voltage and current relationships that exist between the input and output terminals. Thus, for the time being we will use a simpler electrical symbol, shown in Fig. 6.2b. Two input terminals are shown on the left, and a single output terminal appears at the right. The terminal marked by a "+" is referred to as the **noninverting input**, and the "−" marked terminal is called the **inverting input**.

6.2 THE IDEAL OP AMP: A CORDIAL INTRODUCTION

In practice, we find that most op amps perform so well that we can often make the assumption that we are dealing with an "ideal" op amp. The characteristics of an **ideal op amp** form the basis for two fundamental rules that at first may seem somewhat unusual:

Ideal Op Amp Rules
1. No current ever flows into either input terminal.
2. There is no voltage difference between the two input terminals.

In a real op amp, a very small leakage current will flow into the input (sometimes as low as 40 femtoamperes). It is also possible to obtain a very small voltage across the two input terminals. However, compared to other voltages and currents in most circuits, such values are so small that including them in the analysis does not typically affect our calculations.

When analyzing op amp circuits, we should keep one other point in mind. As opposed to the circuits that we have studied so far, an op amp circuit always has an *output* that depends on some type of *input*. Therefore, we will analyze op amp circuits with the goal of obtaining an expression for the output in terms of the input quantities. *We will find that it is usually a good idea to begin the analysis of an op amp circuit at the input, and proceed from there.*

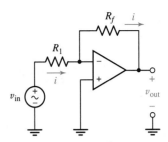

177

The circuit shown in Fig. 6.3 is known as an ***inverting amplifier.*** We choose to analyze this circuit using KVL, beginning with the input voltage source. The current labeled i flows only through the two resistors R_1 and R_f; ideal op amp rule 1 states that no current flows into the inverting input terminal. Thus, we can write

$$-v_{in} + R_1 i + R_f i + v_{out} = 0$$

which can be rearranged to obtain an equation that relates the output to the input:

$$v_{out} = v_{in} - (R_1 + R_f)i \qquad [1]$$

Given $v_{in} = 5 \sin 3t$ mV, $R_1 = 4.7$ kΩ, and $R_f = 47$ kΩ, we require one additional equation that expresses i only in terms of v_{out}, v_{in}, R_1, and/or R_f.

This is a good time to mention that we have not yet made use of ideal op amp rule 2. Since the noninverting input is grounded, it is at zero volts. By ideal op amp rule 2, the inverting input is therefore also at zero volts! *This does not mean that the two inputs are physically shorted together, and we should be careful not to make such an assumption.* Rather, the two input voltages simply track each other: if we try to change the voltage at one pin, the other pin will be driven by internal circuitry to the same value. Thus, we can write one more KVL equation:

$$-v_{in} + R_1 i + 0 = 0$$

or

$$i = \frac{v_{in}}{R_1} \qquad [2]$$

Combining Eq. [2] with Eq. [1], we obtain an expression for v_{out} in terms of v_{in}:

$$v_{out} = -\frac{R_f}{R_1} v_{in} \qquad [3]$$

Substituting $v_{in} = 5 \sin 3t$ mV, $R_1 = 4.7$ kΩ, and $R_f = 47$ kΩ,

$$v_{out} = -50 \sin 3t \qquad \text{mV}$$

Since $R_f > R_1$, this circuit amplifies the input voltage signal v_{in}. If we choose $R_f < R_1$, the signal will be attenuated instead. We also note that the output voltage has the opposite sign of the input voltage,[1] hence the name "inverting amplifier." The output is sketched in Fig. 6.4, along with the input waveform for comparison.

At this point, it is worth mentioning that the ideal op amp seems to be violating KCL. Specifically, in the above circuit no current flows into or out of either input terminal, but somehow current is able to flow into the output pin! This would imply that the op amp is somehow able to either create electrons out of nowhere or store them forever (depending on the direction of current flow). Obviously, this is not possible. The conflict arises because we have been treating the op amp the same way we treated passive elements

FIGURE 6.3 An op amp used to construct an inverting amplifier circuit. The current i flows to ground through the output pin of the op amp.

The fact that the inverting input terminal finds itself at zero volts in this type of circuit configuration leads to what is often referred to as a "virtual ground." This does not mean that the pin is actually grounded, which is sometimes a source of confusion for students. The op amp makes whatever internal adjustments are necessary to prevent a voltage difference between the input terminals. The input terminals are not shorted together.

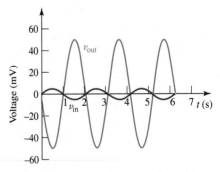

FIGURE 6.4 Input and output waveforms of the inverting amplifier circuit.

---

(1) Or, "*the output is 180° out of phase with the input*," which sounds more impressive.

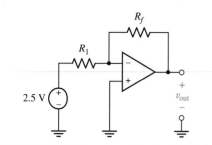

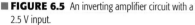

■ **FIGURE 6.5** An inverting amplifier circuit with a 2.5 V input.

such as the resistor. In reality, however, the op amp cannot function unless it is connected to external power sources. It is through those power sources that we can direct current flow through the output terminal.

Although we have shown that the inverting amplifier circuit of Fig. 6.3 can amplify an ac signal (a sine wave in this case having a frequency of 3 rad/s and an amplitude of 5 mV), it works just as well with dc inputs. We consider this type of situation in Fig. 6.5, where values for R_1 and R_f are to be selected to obtain an output voltage of -10 V.

This is the same circuit as shown in Fig. 6.3, but with a 2.5 V dc input. Since no other change has been made, the expression we presented as Eq. [3] is valid for this circuit as well. To obtain the desired output, we seek a ratio of R_f to R_1 of 10/2.5, or 4. Since it is only the ratio that is important here, we simply need to pick a convenient value for one resistor, and the other resistor value is then fixed at the same time. For example, we could choose $R_1 = 100\ \Omega$ (so $R_f = 400\ \Omega$), or even $R_f = 8\ \text{M}\Omega$ (so $R_1 = 2\ \text{M}\Omega$). In practice, other constraints (such as bias current) may limit our choices.

This circuit configuration therefore acts as a convenient type of voltage amplifier (or ***attenuator,*** if the ratio of R_f to R_1 is less than 1), but does have the sometimes inconvenient property of inverting the sign of the input. There is an alternative, however, which is analyzed just as easily—the non-inverting amplifier shown in Fig. 6.6. We examine such a circuit in the following example.

EXAMPLE 6.1

Sketch the output waveform of the noninverting amplifier circuit in Fig. 6.6a. Use $v_{in} = 5 \sin 3t$ mV, $R_1 = 4.7$ kΩ, and $R_f = 47$ kΩ.

▶ **Identify the goal of the problem.**
We require an expression for v_{out} that only depends on the known quantities v_{in}, R_1, and R_f.

▶ **Collect the known information.**
Since values have been specified for the resistors and the input waveform, we begin by labeling the current i and the two input voltages as shown in Fig. 6.6b. We will assume that the op amp is an ideal op amp.

▶ **Devise a plan.**
Although mesh analysis is a favorite technique of students, it turns out to be more practical in most op amp circuits to apply nodal analysis, since there is no direct way to determine the current flowing out of the op amp output.

▶ **Construct an appropriate set of equations.**
Note that we are using ideal op amp rule 1 implicitly by defining the same current through both resistors: no current flows into the inverting input terminal. Employing nodal analysis to obtain our expression for v_{out} in terms of v_{in}, we thus find that

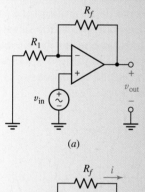

(a)

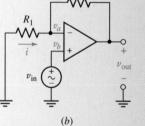

(b)

■ **FIGURE 6.6** (a) An op amp used to construct a noninverting amplifier circuit. (b) Circuit with the current through R_1 and R_f defined, as well as both input voltages labeled.

At node a:

$$0 = \frac{v_a}{R_1} + \frac{v_a - v_{out}}{R_f} \qquad [4]$$

At node b:

$$v_b = v_{in} \qquad [5]$$

▶ **Determine if additional information is required.**

Our goal is to obtain a single expression that relates the input and output voltages, although neither Eq. [4] nor Eq. [5] appears to do so. However, we have not yet employed ideal op amp rule 2, and we will find that in almost every op amp circuit *both* rules need to be invoked in order to obtain such an expression.

Thus, we recognize that $v_a = v_b = v_{in}$, and Eq. [4] becomes

$$0 = \frac{v_{in}}{R_1} + \frac{v_{in} - v_{out}}{R_f}$$

▶ **Attempt a solution.**

Rearranging, we obtain an expression for the output voltage in terms of the input voltage v_{in}:

$$v_{out} = \left(1 + \frac{R_f}{R_1}\right) v_{in} = 11 v_{in} = 55 \sin 3t \quad mV$$

▶ **Verify the solution. Is it reasonable or expected?**

The output waveform is sketched in Fig. 6.7, along with the input waveform for comparison. In contrast to the output waveform of the inverting amplifier circuit, we note that the input and output are in phase for the noninverting amplifier. This should not be entirely unexpected: it is implicit in the name "noninverting amplifier."

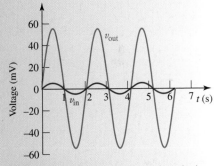

■ **FIGURE 6.7** Input and output waveforms for the noninverting amplifier circuit.

PRACTICE

6.1 Derive an expression for v_{out} in terms of v_{in} for the circuit shown in Fig. 6.8.

Ans: $v_{out} = v_{in}$. The circuit is known as a "*voltage follower*," since the output voltage tracks or "*follows*" the input voltage.

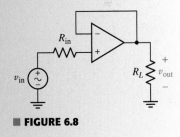

■ **FIGURE 6.8**

Just like the inverting amplifier, the noninverting amplifier works with dc as well as ac inputs, but has a voltage gain of $v_{out}/v_{in} = 1 + (R_f/R_1)$. Thus, if we set $R_f = 9\ \Omega$ and $R_1 = 1\ \Omega$, we obtain an output v_{out} which is 10 times larger than the input voltage v_{in}. In contrast to the inverting amplifier, the output and input of the noninverting amplifier always have the same sign, and the output voltage cannot be less than the input; the minimum gain is 1. Which amplifier we choose depends on the application we are considering. In the special case of the voltage follower circuit shown in Fig. 6.8, which represents a noninverting amplifier with R_1 set to ∞ and R_f set to zero, the output is identical to the input in both sign *and* magnitude. This may seem rather pointless as a general type of circuit, but we should keep in mind that *the voltage follower draws no current from the input* (in the ideal case)—it therefore can act as a ***buffer*** between the voltage v_{in} and some resistive load R_L connected to the output of the op amp.

We mentioned earlier that the name "operational amplifier" originates from using such devices to perform arithmetical operations on analog (i.e., nondigitized, real-time, real-world) signals. As we see in the following two circuits, this includes both addition and subtraction of input voltage signals.

EXAMPLE 6.2

Obtain an expression for v_{out} in terms of v_1, v_2, and v_3 for the op amp circuit in Fig. 6.9, also known as a *summing amplifier*.

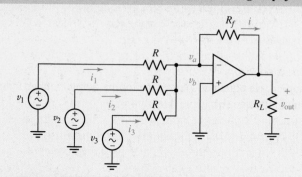

■ **FIGURE 6.9** Basic summing amplifier circuit with three inputs.

We first note that this circuit is similar to the inverting amplifier circuit of Fig. 6.3. Again, the goal is to obtain an expression for v_{out} (which in this case appears across a load resistor R_L) in terms of the inputs (v_1, v_2, and v_3).

Since no current can flow into the inverting input terminal, we can write

$$i = i_1 + i_2 + i_3$$

Therefore, we can write the following equation at the node labeled v_a:

$$0 = \frac{v_a - v_{out}}{R_f} + \frac{v_a - v_1}{R} + \frac{v_a - v_2}{R} + \frac{v_a - v_3}{R}$$

This equation contains both v_{out} and the input voltages, but unfortunately it also contains the nodal voltage v_a. To remove this unknown quantity from our expression, we need to write an additional equation that relates v_a to v_{out}, the input voltages, R_f, and/or R. At this point, we remember that we have not yet used ideal op amp rule 2, and that we will almost certainly require the use of both rules when analyzing an op amp circuit. Thus, since $v_a = v_b = 0$, we can write the following:

$$0 = \frac{v_{out}}{R_f} + \frac{v_1}{R} + \frac{v_2}{R} + \frac{v_3}{R}$$

which describes the input-output characteristics of the circuit shown in Fig. 6.15. We may not always be able to reduce such a circuit to familiar stages, however, so it is worth seeing how the two-stage circuit of Fig. 6.15 can be analyzed as a whole.

When analyzing cascaded circuits, it is sometimes helpful to begin with the last stage and work backward toward the input stage. Referring to ideal op amp rule 1, the same current flows through R_1 and R_2. Writing the appropriate nodal equation at the node labeled v_c yields

$$0 = \frac{v_c - v_x}{R_1} + \frac{v_c - v_{out}}{R_2} \qquad [10]$$

Applying ideal op amp rule 2, we can set $v_c = 0$ in Eq. [10], resulting in

$$0 = \frac{v_x}{R_1} + \frac{v_{out}}{R_2} \qquad [11]$$

Since our goal is an expression for v_{out} in terms of v_1 and v_2, we proceed to the first op amp in order to obtain an expression for v_x in terms of the two input quantities.

Applying ideal op amp rule 1 at the inverting input of the first op amp,

$$0 = \frac{v_a - v_x}{R_f} + \frac{v_a - v_1}{R} + \frac{v_a - v_2}{R} \qquad [12]$$

Ideal op amp rule 2 allows us to replace v_a in Eq. [12] with zero, since $v_a = v_b = 0$. Thus, Eq. [12] becomes

$$0 = \frac{v_x}{R_f} + \frac{v_1}{R} + \frac{v_2}{R} \qquad [13]$$

We now have an equation for v_{out} in terms of v_x (Eq. [11]), and an equation for v_x in terms of v_1 and v_2 (Eq. [13]). These equations are identical to Eqs. [7] and [8], respectively, which means that cascading the two separate circuits as in Fig. 6.15 did not affect the input-output relationship of either stage. Combining Eqs. [11] and [13], we find that the input-output relationship for the cascaded op amp circuit is

$$v_{out} = \frac{R_2}{R_1} \frac{R_f}{R} (v_1 + v_2) \qquad [14]$$

which is identical to Eq. [9].

Thus, the cascaded circuit acts as a summing amplifier, but without a phase reversal between the input and output. By choosing the resistor values carefully, we can either amplify or attenuate the sum of the two input voltages. If we select $R_2 = R_1$ and $R_f = R$, we can also obtain an amplifier circuit where $v_{out} = v_1 + v_2$, if desired.

EXAMPLE 6.3

A multiple-tank gas propellant fuel system is installed in a small lunar orbit runabout. The amount of fuel in any tank is monitored by measuring the tank pressure (in psia).[2] Technical details for tank capacity as well as sensor pressure and voltage range are given in Table 6.2. Design a circuit which provides a positive dc voltage signal proportional to the total fuel remaining, such that 1 V = 100 percent.

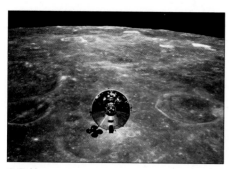

© Corbis

TABLE 6.2 Technical Data for Tank Pressure Monitoring System

Tank 1 Capacity	10,000 psia
Tank 2 Capacity	10,000 psia
Tank 3 Capacity	2000 psia
Sensor Pressure Range	0 to 12,500 psia
Sensor Voltage Output	0 to 5 Vdc

We see from Table 6.2 that the system has three separate gas tanks, requiring three separate sensors. Each sensor is rated up to 12,500 psia, with a corresponding output of 5 V. Thus, when tank 1 is full, its sensor will provide a voltage signal of $5 \times (10{,}000/12{,}500) = 4$ V; the same is true for the sensor monitoring tank 2. The sensor connected to tank 3, however, will only provide a maximum voltage signal of $5 \times (2000/12{,}500) = 800$ mV.

One possible solution is the circuit shown in Fig. 6.16a, which employs a summing amplifier stage with v_1, v_2, and v_3 representing the three sensor outputs, followed by an inverting amplifier to adjust the voltage sign and magnitude. Since we are not told the output resistance of the sensor, we employ a buffer for each one as shown in Fig. 6.16b; the result is (in the ideal case) no current flow from the sensor.

To keep the design as simple as possible, we begin by choosing R_1, R_2, R_3, and R_4 to be 1 kΩ; any value will do as long as all four resistors are equal. Thus, the output of the summing stage is

$$v_x = -(v_1 + v_2 + v_3)$$

The final stage must invert this voltage and scale it such that the output voltage is 1 V when all three tanks are full. The full condition results in $v_x = -(4 + 4 + 0.8) = -8.8$ V. Thus, the final stage needs a voltage ratio of $R_6/R_5 = 1/8.8$. Arbitrarily choosing $R_6 = 1$ kΩ, we find that a value of 8.8 kΩ for R_5 completes the design.

---

(2) Pounds per square inch, absolute. This is a differential pressure measurement relative to a vacuum reference.

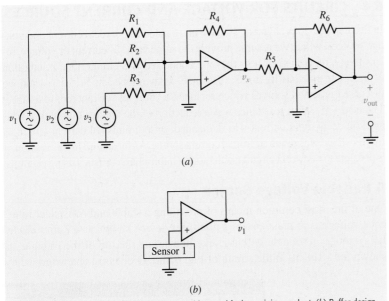

(a)

(b)

■ FIGURE 6.16 (a) A proposed circuit to provide a total fuel remaining readout. (b) Buffer design to avoid errors associated with the internal resistance of the sensor and limitations on its ability to provide current. One such buffer is used for each sensor, providing the inputs v_1, v_2, and v_3 to the summing amplifier stage.

PRACTICE

6.3 An historic bridge is showing signs of deterioration. Until renovations can be performed, it is decided that only cars weighing less than 1600 kg will be allowed across. To monitor this, a four-pad weighing system is designed. There are four independent voltage signals, one from each wheel pad, with 1 mV = 1 kg. Design a circuit to provide a positive voltage signal to be displayed on a DMM (digital multimeter) that represents the total weight of a vehicle, such that 1 mV = 1 kg. You may assume there is no need to buffer the wheel pad voltage signals.

Ans: See Fig. 6.17.

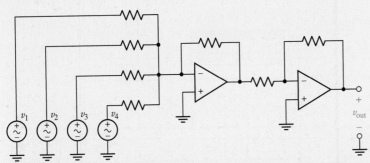

■ FIGURE 6.17 One possible solution to Practice Problem 6.3; all resistors are 10 kΩ (although any value will do as long as they are all equal). Input voltages v_1, v_2, v_3, and v_4 represent the voltage signals from the four wheel pad sensors, and v_{out} is the output signal to be connected to the positive input terminal of the DMM. All five voltages are referenced to ground, and the common terminal of the DMM should be connected to ground as well.

6.4 • CIRCUITS FOR VOLTAGE AND CURRENT SOURCES

In this and previous chapters we have often made use of ideal current and voltage sources, which we assume provide the same value of current or voltage, respectively, regardless of how they are connected in a circuit. Our assumption of independence has its limits, of course, as mentioned in Sec. 5.2 when we discussed practical sources which included a "built-in" or inherent resistance. The effect of such a resistance was a reduction of the voltage output of a voltage source as more current was demanded, or a diminished current output as more voltage was required from a current source. As discussed in this section, it is possible to construct circuits with more reliable characteristics using op amps.

A Reliable Voltage Source

One of the most common means of providing a stable and consistent reference voltage is to make use of a nonlinear device known as a **Zener diode.** Its symbol is a triangle with a Z-like line across the top of the triangle, as shown for a 1N750 in the circuit of Fig. 6.18a. Diodes are characterized by

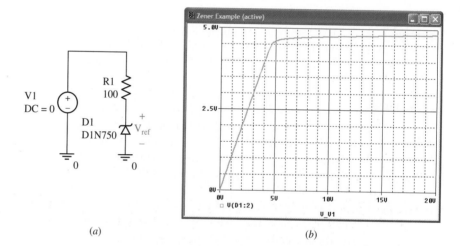

(a)   (b)

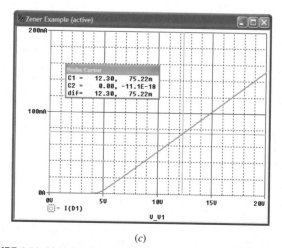

(c)

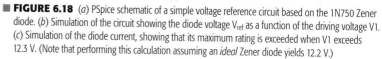

■ **FIGURE 6.18** (a) PSpice schematic of a simple voltage reference circuit based on the 1N750 Zener diode. (b) Simulation of the circuit showing the diode voltage V_{ref} as a function of the driving voltage V1. (c) Simulation of the diode current, showing that its maximum rating is exceeded when V1 exceeds 12.3 V. (Note that performing this calculation assuming an *ideal* Zener diode yields 12.2 V.)

a strongly asymmetric current-voltage relationship. For small voltages, they either conduct essentially zero current—or experience an exponentially increasing current—depending on the voltage polarity. In this way, they distinguish themselves from the simple resistor, where the magnitude of the current is the same for either voltage polarity and hence the resistor current-voltage relationship is symmetric. Consequently, the terminals of a diode are not interchangeable, and have unique names: the **anode** (the flat part of the triangle) and the **cathode** (the point of the triangle).

A *Zener* diode is a special type of diode designed to be used with a positive voltage at the cathode with respect to the anode; when connected this way, the diode is said to be *reverse biased*. For low voltages, the diode acts like a resistor with a small linear increase in current flow as the voltage is increased. Once a certain voltage (V_{BR}) is reached, however—known as the *reverse breakdown voltage* or **Zener voltage** of the diode—the voltage does not significantly increase further, but essentially any current can flow up to the maximum rating of the diode (75 mA for a 1N750, whose Zener voltage is 4.7 V).

Let's consider the simulation result presented in Fig. 6.18*b*, which shows the voltage V_{ref} across the diode as the voltage source V1 is swept from 0 to 20 V. Provided V1 remains above 5 V, *the voltage across our diode is essentially constant.* Thus, we could replace V1 with a 9 V battery, and not be too concerned with changes in our voltage reference as the battery voltage begins to drop as it discharges. The purpose of R1 in this circuit is simply to provide the necessary voltage drop between the battery and the diode; its value should be chosen to ensure that the diode is operating at its Zener voltage but below its maximum rated current. For example, Fig. 6.18*c* shows that the 75 mA rating is exceeded in our circuit if the source voltage V1 is much greater than 12 V. Thus, the value of resistor R1 should be sized corresponding to the source voltage available, as we explore in Example 6.4.

EXAMPLE 6.4

Design a circuit based on the 1N750 Zener diode that runs on a single 9 V battery and provides a reference voltage of 4.7 V.

The 1N750 has a maximum current rating of 75 mA, and a Zener voltage of 4.7 V. The voltage of a 9 V battery can vary slightly depending on its state of charge, but we neglect this for the present design.

A simple circuit such as the one shown in Fig. 6.19*a* is adequate for our purposes; the only issue is determining a suitable value for the resistor R_{ref}.

If 4.7 V is dropped across the diode, then $9 - 4.7 = 4.3$ V must be dropped across R_{ref}. Thus,

$$R_{ref} = \frac{9 - V_{ref}}{I_{ref}} = \frac{4.3}{I_{ref}}$$

We determine R_{ref} by specifying a current value. We know that I_{ref} should not be allowed to exceed 75 mA for this diode, and large currents will discharge the battery more quickly. However, as seen in Fig. 6.19*b*, we cannot simply select I_{ref} arbitrarily; very low currents do not allow

(Continued on next page)

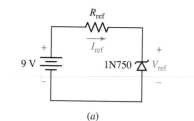

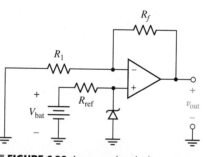

the diode to operate in the Zener breakdown region. In the absence of a detailed equation for the diode's current-voltage relationship (which is clearly nonlinear), we design for 50 percent of the maximum rated current as a rule of thumb. Thus,

$$R_{ref} = \frac{4.3}{0.0375} = 115\ \Omega$$

Detailed "tweaking" can be obtained by performing a PSpice simulation of the final circuit, although we see from Fig. 6.19c that our first pass is reasonably close (within 1 percent) to our target value.

The basic Zener diode voltage reference circuit of Fig. 6.18a works very well in many situations, but we are limited somewhat in the value of the voltage depending on which Zener diodes are available. Also, we often find that the circuit shown is not well suited to applications requiring more than a few milliamperes of current. In such instances, we may use the Zener reference circuit in conjunction with a simple amplifier stage, as shown in Fig. 6.20. The result is a stable voltage that can be controlled by adjusting the value of either R_1 or R_f, without having to switch to a different Zener diode.

■ **FIGURE 6.19** (*a*) A voltage reference circuit based on the 1N750 Zener diode. (*b*) Diode *I-V* relationship. (*c*) PSpice simulation of the final design.

■ **FIGURE 6.20** An op amp–based voltage source using on a Zener voltage reference.

PRACTICE

6.4 Design a circuit to provide a reference voltage of 6 V using a 1N750 Zener diode and a noninverting amplifier.

Ans: Using the circuit topology shown in Fig. 6.20, choose $V_{bat} = 9$ V, $R_{ref} = 115\ \Omega$, $R_1 = 1$ kΩ, and $R_f = 268\ \Omega$.

A Reliable Current Source

Consider the circuit shown in Fig. 6.21a, where V_{ref} is provided by a regulated voltage source such as the one shown in Fig. 6.19a. The reader may recognize this circuit as a simple inverting amplifier configuration, assuming we tap the output pin of the op amp. We can also use this circuit as a current source, however, where R_L represents a resistive load.

The input voltage V_{ref} appears across reference resistor R_{ref}, since the noninverting input of the op amp is connected to ground. With no current feedback is the unpleasant sensation we feel as our hand draws near a flame. The closer we move toward the flame, the larger the negative signal sent from our hand. Overdoing the proportion of negative feedback, however, might cause us to abhor heat, and eventually freeze to death. **Positive feedback** is the process where some fraction of the output signal is added back to the input. A common example is when a microphone is directed toward a speaker—a very soft sound is rapidly amplified over and over until the system "screams." Positive feedback generally leads to an unstable system.

All of the circuits considered in this chapter incorporate negative feedback through the presence of a resistor between the output pin and the inverting input. The resulting loop between the output and the input reduces the dependency of the output voltage on the actual value of the open-loop gain (as seen in Example 6.6). This obviates the need to measure the precise open-loop gain of each op amp we use, as small variations in A will not significantly impact the operation of the circuit. Negative feedback also provides increased stability in situations where A is sensitive to the op amp's surroundings. For example, if A suddenly increases in response to a change in the ambient temperature, a larger feedback voltage is added to the inverting input. This acts to reduce the differential input voltage v_d, and therefore the change in output voltage Av_d is smaller. We should note that the closed-loop circuit gain is always less than the open-loop device gain; this is the price we pay for stability and reduced sensitivity to parameter variations.

Saturation

So far, we have treated the op amp as a purely linear device, assuming that its characteristics are independent of the way in which it is connected in a circuit. In reality, it is necessary to supply power to an op amp in order to run the internal circuitry, as shown in Fig. 6.27. A positive supply, typically in the range of 5 to 24 V dc, is connected to the terminal marked V^+, and a negative supply of equal magnitude is connected to the terminal marked V^-. There are also a number of applications where a single voltage supply is acceptable, as well as situations where the two voltage magnitudes may be unequal. The op amp manufacturer will usually specify a maximum power supply voltage, beyond which damage to the internal transistors will occur.

The power supply voltages are a critical choice when designing an op amp circuit, because they represent the maximum possible output voltage of the op amp.[3] For example, consider the op amp circuit shown in Fig. 6.26, now connected as a noninverting amplifier having a gain of 10. As shown in the PSpice simulation in Fig. 6.28, we do in fact observe linear behavior from the op amp, but only in the range of ± 1.71 V for the input voltage. Outside of this range, the output voltage is no longer proportional to the input, reaching a peak magnitude of 17.6 V. This important nonlinear effect is known as **saturation,** which refers to the fact that further increases in the input voltage do not result in a change in the output voltage. This phenomenon refers to the fact that the output of a real op amp cannot exceed its

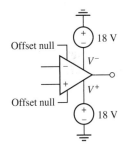

■ **FIGURE 6.27** Op amp with positive and negative voltage supplies connected. Two 18 V supplies are used as an example; note the polarity of each source.

---

(3) In practice, we find the maximum output voltage is slightly less than the supply voltage by as much as a volt or so.

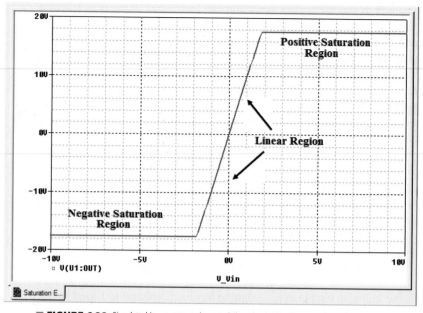

■ FIGURE 6.28 Simulated input-output characteristics of a μA741 connected as a noninverting amplifier with a gain of 10, and powered by $\pm$18 V supplies.

supply voltages. For example, if we choose to run the op amp with a +9 V supply and a −5 V supply, then our output voltage will be limited to the range of −5 to +9 V. The output of the op amp is a linear response bounded by the positive and negative saturation regions, and as a general rule, we try to design our op amp circuits so that we do not accidentally enter the saturation region. This requires us to select the operating voltage carefully based on the closed-loop gain and maximum expected input voltage.

Input Offset Voltage

As we are discovering, there are a number of practical considerations to keep in mind when working with op amps. One particular nonideality worth mentioning is the tendency for real op amps to have a nonzero output even when the two input terminals are shorted together. The value of the output under such conditions is known as the offset voltage, and the input voltage required to reduce the output to zero is referred to as the ***input offset voltage.*** Referring to Table 6.3, we see that typical values for the input offset voltage are on the order of a few millivolts or less.

Most op amps are provided with two pins marked either "offset null" or "balance." These terminals can be used to adjust the output voltage by connecting them to a variable resistor. A variable resistor is a three-terminal device commonly used for such applications as volume controls on radios. The device comes with a knob that can be rotated to select the actual value of resistance, and has three terminals. Measured between the two extreme terminals, its resistance is fixed regardless of the position of the knob. Using the middle terminal and one of the end terminals creates a resistor whose value depends on the knob position. Figure 6.29 shows a typical circuit used to adjust the output voltage of an op amp; the manufacturer's data sheet may suggest alternative circuitry for a particular device.

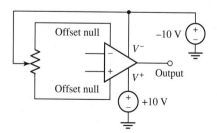

■ FIGURE 6.29 Suggested external circuitry for obtaining a zero output voltage. The $\pm$10 V supplies are shown as an example; the actual supply voltages used in the final circuit would be chosen in practice.

known as an ***instrumentation amplifier,*** which is actually three op amp devices in a single package.

An example of the common instrumentation amplifier configuration is shown in Fig. 6.38*a*, and its symbol is shown in Fig. 6.38*b*. Each input is fed directly into a voltage follower stage, and the output of both voltage followers is fed into a difference amplifier stage. It is particularly well suited to applications where the input voltage signal is very small (for example, on the order of millivolts), such as that produced by thermocouples or strain gauges, and where a significant common-mode noise signal of several volts may be present.

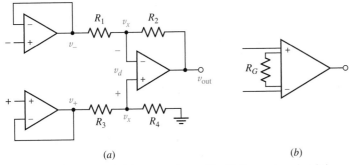

(a)   (b)

■ **FIGURE 6.38** (*a*) The basic instrumentation amplifier. (*b*) Commonly used symbol.

If components of the instrumentation amplifier are fabricated all on the same silicon "chip," then it is possible to obtain well-matched device characteristics and to achieve precise ratios for the two sets of resistors. In order to maximize the CMRR of the instrumentation amplifier, we expect $R_4/R_3 = R_2/R_1$, so that equal amplification of common-mode components of the input signals is obtained. To explore this further, we identify the voltage at the output of the top voltage follower as "v_-," and the voltage at the output of the bottom voltage follower as "v_+." Assuming all three op amps are ideal and naming the voltage at either input of the difference stage v_x, we may write the following nodal equations:

$$\frac{v_x - v_-}{R_1} + \frac{v_x - v_{out}}{R_2} = 0 \qquad [20]$$

and

$$\frac{v_x - v_+}{R_3} + \frac{v_x}{R_4} = 0 \qquad [21]$$

Solving Eq. [21] for v_x, we find that

$$v_x = \frac{v_+}{1 + R_3/R_4} \qquad [22]$$

and upon substituting into Eq. [20], obtain an expression for v_{out} in terms of the input:

$$v_{out} = \frac{R_4}{R_3}\left(\frac{1 + R_2/R_1}{1 + R_4/R_3}\right)v_+ - \frac{R_2}{R_1}v_- \qquad [23]$$

From Eq. [23] it is clear that the general case allows amplification of common-mode components to the two inputs. In the specific case where

$R_4/R_3 = R_2/R_1 = K$, however, Eq. [23] reduces to $K(v_+ - v_-) = Kv_d$, so that (asssuming ideal op amps) only the difference is amplified and the gain is set by the resistor ratio. Since these resistors are internal to the instrumentation amplifier and not accessible to the user, devices such as the AD622 allow the gain to be set anywhere in the range of 1 to 1000 by connecting an external resistor between two pins (shown as R_G in Fig. 6.38b).

SUMMARY AND REVIEW

In this chapter we introduced a new circuit element—a three-terminal device—called the operational amplifier (or more commonly, the *op amp*). In many circuit analysis situations it is approximated as an ideal device, which leads to two rules that are applied. We studied several op amp circuits in detail, including the *inverting amplifier* with gain R_f/R_1, the *noninverting amplifer* with gain $1 + R_f/R_1$, and the *summing amplifier*. We were also introduced to the *voltage follower* and the *difference amplifier*, although the analysis of these two circuits was left for the reader. The concept of cascaded stages was found to be particularly useful, as it allows a design to be broken down into distinct units, each of which has a specific function. We took a slight detour and introduced briefly a two-terminal nonlinear circuit element, the *Zener diode*, as it provides a practical and straightforward voltage reference. We then used this element to contruct practical voltage and current sources using op amps, removing some of the mystery as to their origins.

Modern op amps have nearly ideal characteristics, as we found when we opted for a more detailed model based on a dependent source. Still, nonidealities are encountered occasionally, so we considered the role *of negative feedback* in reducing the effect of temperature and manufacturing-related variations in various parameters, *common-mode rejection*, and *saturation*. One of the most interesting nonideal characteristics of any op amp is *slew rate*. By simulating three different cases, we were able to see how the output voltage can struggle to follow the form of the input voltage signal once its frequency becomes high enough. We concluded the chapter with two special cases: the *comparator*, which intentionally makes use of our ability to saturate a practical (nonideal) op amp, and the instrumentation amplifier, which is routinely used to amplify very small voltages.

This is a good point to pause, take a breath, and recap some of the key points. At the same time, we will highlight relevant examples as an aid to the reader.

❑ There are two fundamental rules that must be applied when analyzing *ideal* op amp circuits:

1. No current ever flows into either input terminal. (Example 6.1)

2. No voltage ever exists between the input terminals.

❑ Op amp circuits are usually analyzed for an output voltage in terms of some input quantity or quantities. (Examples 6.1, 6.2)

❑ Nodal analysis is typically the best choice in analyzing op amp circuits, and it is usually better to begin at the input, and work toward the output. (Examples 6.1, 6.2)

❑ The output current of an op amp cannot be assumed; it must be found after the output voltage has been determined independently. (Example 6.2)

❑ The gain of an inverting op amp circuit is given by the equation

$$v_{\text{out}} = -\frac{R_f}{R_1} v_{\text{in}}$$

❑ The gain of a noninverting op amp circuit is given by the equation

$$v_{\text{out}} = \left(1 + \frac{R_f}{R_1}\right) v_{\text{in}}$$

(Example 6.1)

❑ Cascaded stages may be analyzed one stage at a time to relate the output to the input. (Example 6.3)

❑ Zener diodes provide a convenient voltage reference. They are not symmetric, however, meaning the two terminals are not interchangeable. (Example 6.4)

❑ Op amps can be used to construct current sources which are independent of the load resistance over a specific current range. (Example 6.5)

❑ A resistor is almost always connected from the output pin of an op amp to its inverting input pin, which incorporates negative feedback into the circuit for increased stability.

❑ The ideal op amp model is based on the approximation of infinite open-loop gain A, infinite input resistance R_i, and zero output resistance R_o. (Example 6.6)

❑ In practice, the output voltage range of an op amp is limited by the supply voltages used to power the device. (Example 6.7)

❑ Comparators are op amps designed to be driven into saturation. These circuits operate in open loop, and hence have no external feedback resistor. (Example 6.8)

READING FURTHER

Two very readable books which deal with a variety of op amp applications are:

R. Mancini (ed.), *Op Amps Are For Everyone*, 2nd ed. Amsterdam: Newnes, 2003. Also available on the Texas Instruments website (www.ti.com).

W. G. Jung, *Op Amp Cookbook*, 3rd ed. Upper Saddle River, N.J.: Prentice-Hall, 1997.

Characteristics of Zener and other types of diodes are covered in Chapter 1 of

W. H. Hayt, Jr., and G. W. Neudeck, *Electronic Circuit Analysis and Design*, 2nd ed. New York: Wiley, 1995.

One of the first reports of the implementation of an "operational amplifier" can be found in

J. R. Ragazzini, R. M. Randall, and F. A. Russell, "Analysis of problems in dynamics by electronic circuits," *Proceedings of the IRE* **35**(5), 1947, pp. 444–452.

And an early applications guide for the op amp can be found on the Analog Devices, Inc. website (www.analog.com):

George A. Philbrick Researches, Inc., *Applications Manual for Computing Amplifiers for Modelling, Measuring, Manipulating & Much Else.* Norwood, Mass.: Analog Devices, 1998.

EXERCISES

6.2 The Ideal Op Amp

1. For the op amp circuit shown in Fig. 6.39, calculate v_{out} if (*a*) $R_1 = R_2 = 100 \ \Omega$ and $v_{in} = 5$ V; (*b*) $R_2 = 200R_1$ and $v_{in} = 1$ V; (*c*) $R_1 = 4.7$ kΩ, $R_2 = 47$ kΩ, and $v_{in} = 20 \sin 5t$ V.

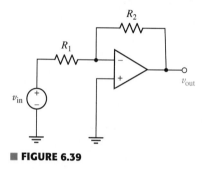

■ **FIGURE 6.39**

2. Determine the power dissipated by a 100 Ω resistor connected between ground and the output pin of the op amp of Fig. 6.39 if $v_{in} = 4$ V and (*a*) $R_1 = 2R_2$; (*b*) $R_1 = 1$ kΩ and $R_2 = 22$ kΩ; (*c*) $R_1 = 100 \ \Omega$ and $R_2 = 101 \ \Omega$.

3. Connect a 1 Ω resistor between ground and the output terminal of the op amp of Fig. 6.39, and sketch $v_{out}(t)$ if (*a*) $R_1 = R_2 = 10 \ \Omega$ and $v_{in} = 5 \sin 10t$ V; (*b*) $R_1 = 0.2R_2 = 1$ kΩ, and $v_{in} = 5 \cos 10t$ V; (*c*) $R_1 = 10 \ \Omega$, $R_2 = 200 \ \Omega$, and $v_{in} = 1.5 + 5e^{-t}$ V.

4. For the circuit of Fig. 6.40, calculate v_{out} if (*a*) $R_1 = R_2 = 100$ kΩ, $R_L = 100 \ \Omega$, and $v_{in} = 5$ V; (*b*) $R_1 = 0.1R_2$, $R_L = \infty$, and $v_{in} = 2$ V; (*c*) $R_1 = 1$ kΩ, $R_2 = 0$, $R_L = 1 \ \Omega$, and $v_{in} = 43.5$ V.

5. (*a*) Design a circuit which converts a voltage $v_1(t) = 9 \cos 5t$ V into $9 \sin 5t$ V. (*b*) Verify your design by analyzing the final circuit.

6. A certain load resistor requires a constant 5 V dc supply. Unfortunately, its resistance value changes with temperature. Design a circuit which supplies the requisite voltage if only 9 V batteries and standard 10% tolerance resistor values are available.

7. For the circuit of Fig. 6.40, $R_1 = R_L = 50 \ \Omega$. Calculate the value of R_2 required to deliver 5 W to R_L if V_{in} equals (*a*) 5 V; (*b*) 1.5 V. (*c*) Repeat parts (*a*) and (*b*) if R_L is reduced to 22 Ω.

8. Calculate v_{out} as labeled in the schematic of Fig. 6.41 if (*a*) $i_{in} = 1$ mA, $R_p = 2.2$ kΩ, and $R_3 = 1$ kΩ; (*b*) $i_{in} = 2$ A, $R_p = 1.1 \ \Omega$, and $R_3 = 8.5 \ \Omega$. (*c*) For each case, state whether the circuit is wired as a noninverting or an inverting amplifier. *Explain your reasoning.*

9. (*a*) Design a circuit using only a single op amp which adds two voltages v_1 and v_2 and provides an output voltage twice their sum (i.e., $v_{out} = 2v_1 + 2v_2$). (*b*) Verify your design by analyzing the final circuit.

10. (*a*) Design a circuit that provides a current i which is equal in magnitude to the sum of three input voltages v_1, v_2, and v_3. (Compare volts to amperes.) (*b*) Verify your design by analyzing the final circuit.

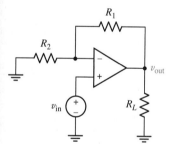

■ **FIGURE 6.40**

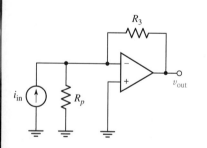

■ **FIGURE 6.41**

11. (*a*) Design a circuit that provides a voltage v_{out} which is equal to the difference between two voltages v_2 and v_1 (i.e., $v_{out} = v_2 - v_1$), if you have only the following resistors from which to choose: two 1.5 kΩ resistors, four 6 kΩ resistors, or three 500 Ω resistors. (*b*) Verify your design by analyzing the final circuit.

12. Analyze the circuit of Fig. 6.42 and determine a value for V_1, which is referenced to ground.

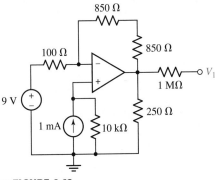

■ **FIGURE 6.42**

13. Derive an expression for v_{out} as a function of v_1 and v_2 for the circuit represented in Fig. 6.43.

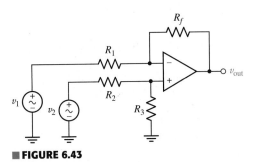

■ **FIGURE 6.43**

14. Explain what is wrong with each diagram in Fig. 6.44 if the two op amps are known to be *perfectly ideal*.

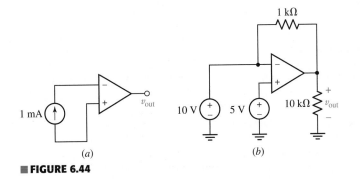

(*a*)   (*b*)

■ **FIGURE 6.44**

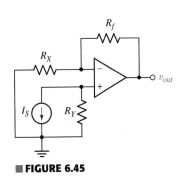

■ **FIGURE 6.45**

15. For the circuit depicted in Fig. 6.45, calculate v_{out} if $I_s = 2$ mA, $R_Y = 4.7$ kΩ, $R_X = 1$ kΩ, and $R_f = 500$ Ω.

16. Consider the amplifier circuit shown in Fig. 6.45. What value of R_f will yield $v_{out} = 2$ V when $I_s = 10$ mA and $R_Y = 2R_X = 500$ Ω?

17. With respect to the circuit shown in Fig. 6.46, calculate v_{out} if v_s equals (a) $2 \cos 100t$ mV; (b) $2 \sin(4t + 19°)$ V.

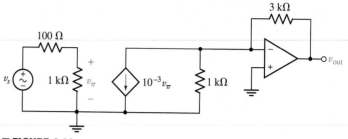

■ **FIGURE 6.46**

6.3 Cascaded Stages

18. Calculate v_{out} as labeled in the circuit of Fig. 6.47 if $R_x = 1$ kΩ.

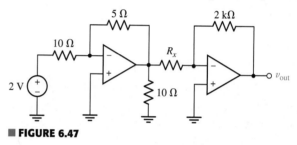

■ **FIGURE 6.47**

19. For the circuit of Fig. 6.47, determine the value of R_x that will result in a value of $v_{out} = 10$ V.

20. Referring to Fig. 6.48, sketch v_{out} as a function of (a) v_{in} over the range of $-2 \text{ V} \leq v_{in} \leq +2$ V, if $R_4 = 2$ kΩ; (b) R_4 over the range of $1 \text{ k}\Omega \leq R_4 \leq 10$ kΩ, if $v_{in} = 300$ mV.

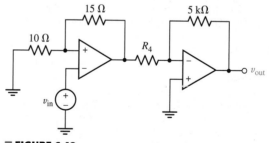

■ **FIGURE 6.48**

21. Obtain an expression for v_{out} as labeled in the circuit of Fig. 6.49 if v_1 equals (a) 0 V; (b) 1 V; (c) -5 V; (d) $2 \sin 100t$ V.

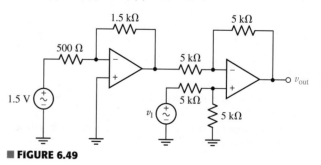

■ **FIGURE 6.49**

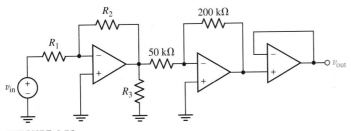

22. The 1.5 V source of Fig. 6.49 is disconnected, and the output of the circuit shown in Fig. 6.48 is connected to the left-hand terminal of the 500 Ω resistor instead. Calculate v_{out} if $R_4 = 2$ kΩ and (a) $v_{in} = 2$ V, $v_1 = 1$ V; (b) $v_{in} = 1$ V, $v_1 = 0$; (c) $v_{in} = 1$ V, $v_1 = -1$ V.

23. For the circuit shown in Fig. 6.50, compute v_{out} if (a) $v_1 = 2v_2 = 0.5v_3 = 2.2$ V and $R_1 = R_2 = R_3 = 50$ kΩ; (b) $v_1 = 0$, $v_2 = -8$ V, $v_3 = 9$ V, and $R_1 = 0.5R_2 = 0.4R_3 = 100$ kΩ.

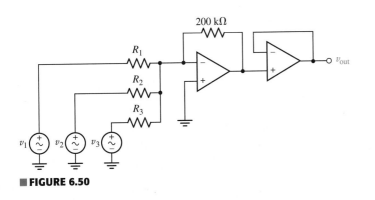

■ **FIGURE 6.50**

24. (a) Design a circuit which will add the voltages produced by three separate pressure sensors, each in the range of $0 \leq v_{sensor} \leq 5$ V, and produce a positive voltage v_{out} linearly correlated to the voltage sum such that $v_{out} = 0$ when all three voltages are zero, and $v_{out} = 2$ V when all three voltages are at their maximum. (b) Verify your design by analyzing the final circuit.

25. (a) Design a circuit which produces an output voltage v_{out} proportional to the difference of two positive voltages v_1 and v_2 such that $v_{out} = 0$ when both voltages are equal, and $v_{out} = 10$ V when $v_1 - v_2 = 1$ V. (b) Verify your design by analyzing the final circuit.

26. (a) Three pressure-sensitive sensors are used to double-check the weight readings obtained from the suspension systems of a long-range jet airplane. Each sensor is calibrated such that 10 μV corresponds to 1 kg. Design a circuit which adds the three voltage signals to produce an output voltage calibrated such that 10 V corresponds to 400,000 kg, the maximum takeoff weight of the aircraft. (b) Verify your design by analyzing the final circuit.

27. (a) The oxygen supply to a particular bathysphere consists of four separate tanks, each equipped with a pressure sensor capable of measuring between 0 (corresponding to 0 V output) and 500 bar (corresponding to 5 V output). Design a circuit which produces a voltage proportional to the total pressure in all tanks, such that 1.5 V corresponds to 0 bar and 3 V corresponds to 2000 bar. (b) Verify your design by analyzing the final circuit.

28. For the circuit shown in Fig. 6.51, let $v_{in} = 8$ V, and select values for R_1, R_2, and R_3 to ensure an output voltage $v_{out} = 4$ V.

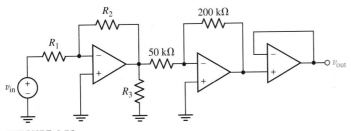

■ **FIGURE 6.51**

29. For the circuit of Fig. 6.52, derive an expression for v_{out} in terms of v_{in}.

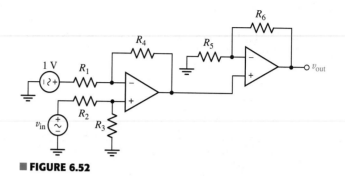

■ FIGURE 6.52

6.4 Circuits for Voltage and Current Sources

 30. Construct a circuit based on the 1N4740 diode which provides a reference voltage of 10 V if only 9 V batteries are available. Note that the breakdown voltage of this diode is equal to 10 V at a current of 25 mA.

 31. Employ a 1N4733 Zener diode to construct a circuit which provides a 4 V reference voltage to a 1 kΩ load, if only 9 V batteries are available as sources. Note that the Zener breakdown voltage of this diode is 5.1 V at a current of 76 mA.

 32. (a) Design a circuit which provides a 5 V dc reference voltage to an unknown (nonzero resistance) load, if only a 9 V battery is available as a supply. (b) Verify your design with an appropriate simulation. As part of this, determine the acceptable range for the load resistor.

 33. A particular passive network can be represented by a Thévenin equivalent resistance between 10 Ω and 125 Ω depending on the operating temperature. (a) Design a circuit which provides a constant 2.2 V to this network regardless of temperature. (b) Verify your design with an appropriate simulation (resistance can be varied from within a single simulation, as described in Chap. 8).

 34. Calculate the voltage V_1 as labeled in the circuit of Fig. 6.53 if the battery is rated at V_{batt} equal to (a) 9 V; (b) 12 V. (c) Verify your solutions with appropriate simulations, commenting on the possible origin of any discrepancies.

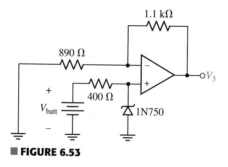

■ FIGURE 6.53

 35. (a) Design a current source based on the 1N750 diode which is capable of providing a dc current of 750 μA to a load R_L, such that 1 kΩ < R_L < 50 kΩ. (b) Verify your design with an appropriate simulation (note that resistance can be varied within a single simulation, as described in Chap. 8).

36. (a) Design a current source able to provide a dc current of 50 mA to an unspecified load. Use a 1N4733 diode (V_{br} = 5.1 V at 76 mA). (b) Use an appropriate simulation to determine the permissible range of load resistance for your design.

37. (*a*) Design a current source able to provide a dc current of 10 mA to an unspecified load. Use a 1N4747 diode ($V_{br} = 20$ V at 12.5 mA). (*b*) Use an appropriate simulation to determine the permissible range of load resistance for your design.

38. The circuit depicted in Fig. 6.54 is known as a Howland current source. Derive expressions for v_{out} and I_L, respectively as a function of V_1 and V_2.

39. For the circuit depicted in Fig. 6.54, known as a Howland current source, set $V_2 = 0$, $R_1 = R_3$, and $R_2 = R_4$; then solve for the current I_L when $R_1 = 2R_2 = 1$ kΩ and $R_L = 100$ Ω.

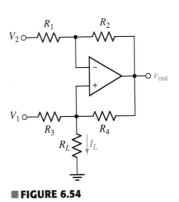

■ FIGURE 6.54

6.5 Practical Considerations

40. (*a*) Employ the parameters listed in Table 6.3 for the μA741 op amp to analyze the circuit of Fig. 6.55 and compute a value for v_{out}. (*b*) Compare your result to what is predicted using the ideal op amp model.

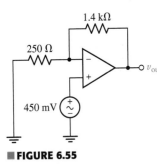

■ FIGURE 6.55

41. (*a*) Employ the parameters listed in Table 6.3 for the μA741 op amp to analyze the circuit of Fig. 6.10 if $R = 1.5$ kΩ, $v_1 = 2$ V, and $v_2 = 5$ V. (*b*) Compare your solution to what is predicted using the ideal op amp model.

42. Define the following terms, and explain when and how each can impact the performance of an op amp circuit: (*a*) common-mode rejection ratio; (*b*) slew rate; (*c*) saturation; (*d*) feedback.

43. For the circuit of Fig. 6.56, replace the 470 Ω resistor with a short circuit, and compute v_{out} using (*a*) the ideal op amp model; (*b*) the parameters listed in Table 6.3 for the μA741 op amp; (*c*) an appropriate PSpice simulation. (*d*) Compare the values obtained in parts (*a*) to (*c*) and comment on the possible origin of any discrepancies.

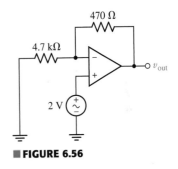

■ FIGURE 6.56

44. If the circuit of Fig. 6.55 is analyzed using the detailed model of an op amp (as opposed to the ideal op amp model), calculate the value of open-loop gain A required to achieve a closed-loop gain within 2% of its ideal value.

45. Replace the 2 V source in Fig. 6.56 with a sinusoidal voltage source having a magnitude of 3 V and radian frequency $\omega = 2\pi f$. (*a*) Which device, a μA741 op amp or an LF411 op amp, will track the source frequency better over the range 1 Hz $< f <$ 10 MHz? *Explain.* (*b*) Compare the frequency performance of the circuit over the range 1 Hz $< f <$ 10 MHz using appropriate PSpice simulations, and compare the results to your prediction in part (*a*).

46. (*a*) For the circuit of Fig. 6.56, if the op amp (assume LF411) is powered by matched 9 V supplies, estimate the maximum value to which the 470 Ω resistor can be increased before saturation effects become apparent. (*b*) Verify your prediction with an appropriate simulation.

47. For the circuit of Fig. 6.55, calculate the differential input voltage and the input bias current if the op amp is a(n) (*a*) μA741; (*b*) LF411; (*c*) AD549K; (*d*) OPA690.

48. Calculate the common-mode gain for each device listed in Table 6.3. Express your answer in units of V/V, not dB.

6.6 Comparators and the Instrumentation Amplifier

49. Human skin, especially when damp, is a reasonable conductor of electricity. If we assume a resistance of less than 10 MΩ for a fingertip pressed across two terminals, design a circuit which provides a +1 V output if this nonmechanical switch is "closed" and −1 V if it is "open."

50. Design a circuit which provides an output voltage v_{out} based on the behavior of another voltage v_{in}, such that

$$v_{out} = \begin{cases} +2.5\ \text{V} & v_{in} > 1\ \text{V} \\ 1.2\ \text{V} & \text{otherwise} \end{cases}$$

51. For the instrumentation amplifier shown in Fig. 6.38a, assume that the three internal op amps are ideal, and determine the CMRR of the circuit if (a) $R_1 = R_3$ and $R_2 = R_4$; (b) all four resistors have different values.

52. For the circuit depicted in Fig. 6.57, sketch the expected output voltage v_{out} as a function of v_{active} for $-5\ \text{V} \le v_{active} \le +5\ \text{V}$, if v_{ref} is equal to (a) −3 V; (b) +3 V.

53. For the circuit depicted in Fig. 6.58, (a) sketch the expected output voltage v_{out} as a function of v_1 for $-5\ \text{V} \le v_1 \le +5\ \text{V}$, if $v_2 = +2\ \text{V}$; (b) sketch the expected output voltage v_{out} as a function of v_2 for $-5\ \text{V} \le v_2 \le +5\ \text{V}$, if $v_1 = +2\ \text{V}$.

54. For the circuit depicted in Fig. 6.59, sketch the expected output voltage v_{out} as a function of v_{active}, if $-2\ \text{V} \le v_{active} \le +2\ \text{V}$. Verify your solution using a μA741 (although it is slow compared to op amps designed specifically for use as comparators, its PSpice model works well, and as this is a dc application speed is not an issue). Submit a properly labeled schematic with your results.

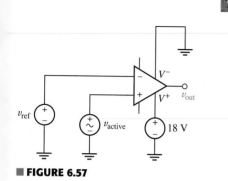

■ FIGURE 6.57

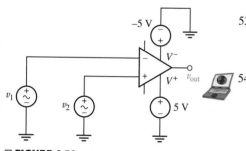

■ FIGURE 6.58

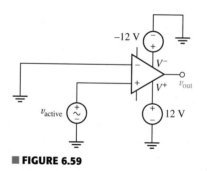

■ FIGURE 6.59

55. In digital logic applications, a +5 V signal represents a logic "1" state, and a 0 V signal represents a logic "0" state. In order to process real-world information using a digital computer, some type of interface is required, which typically includes an analog-to-digital (A/D) converter—a device that converts analog signals into digital signals. Design a circuit that acts as a simple 1-bit A/D, with any signal less than 1.5 V resulting in a logic "0" and any signal greater than 1.5 V resulting in a logic "1."

56. A common application for instrumentation amplifiers is to measure voltages in resistive strain gauge circuits. These strain sensors work by exploiting the changes in resistance that result from geometric distortions, as in Eq. [6] of Chap. 2. They are often part of a bridge circuit, as shown in Fig. 6.60a, where the strain gauge is identified as R_G. (a) Show that $V_{out} = V_{in}\left[\frac{R_2}{R_1+R_2} - \frac{R_3}{R_3+R_{Gauge}}\right]$. (b) Verify that $V_{out} = 0$ when the three fixed-value resistors R_1, R_2, and R_3 are all chosen to be equal to the unstrained gauge resistance R_{Gauge}. (c) For the intended application, the gauge selected has an unstrained resistance of 5 kΩ, and a maximum resistance increase of

50 mΩ is expected. Only ±12 V supplies are available. Using the instrumentation amplifier of Fig. 6.60b, design a circuit that will provide a voltage signal of +1 V when the strain gauge is at its maximum loading.

AD622 Specifications

Amplifier gain G can be varied from 2 to 1000 by connecting a resistor between pins 1 and 8 with a value calculated by $R = \dfrac{50.5}{G-1}$ kΩ.

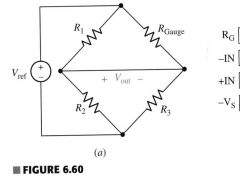

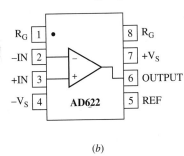

(a)

(b)

■ **FIGURE 6.60**

© *Analog Devices.*

Chapter-Integrating Exercises

57. (a) You're given an electronic switch which requires 5 V at 1 mA in order to close; it is open with no voltage present at its input. If the only microphone available produces a peak voltage of 250 mV, design a circuit which will energize the switch when someone speaks into the microphone. Note that the audio level of a general voice may not correspond to the peak voltage of the microphone. (b) Discuss any issues that may need to be addressed if your circuit were to be implemented.

58. You've formed a band, despite advice to the contrary. Actually, the band is pretty good except for the fact that the lead singer (who owns the drum set, the microphones, and the garage where you practice) is a bit tone-deaf. Design a circuit that takes the output from each of the five microphones your band uses, and adds the voltages to create a single voltage signal which is fed to the amplifer. Except not all voltages should be equally amplified. One microphone output should be attenuated such that its peak voltage is 10% of any other microphone's peak voltage.

59. Cadmium sulfide (CdS) is commonly used to fabricate resistors whose value depends on the intensity of light shining on the surface. In Fig. 6.61 a CdS "photocell" is used as the feedback resistor R_f. In total darkness, it has a resistance of 100 kΩ, and a resistance of 10 kΩ under a light intensity of 6 candela. R_L represents a circuit that is activated when a voltage of 1.5 V or less is applied to its terminals. Choose R_1 and V_s so that the circuit represented by R_L is activated by a light of 2 candela or brighter.

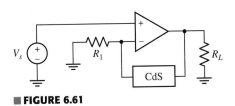

■ **FIGURE 6.61**

60. A fountain outside a certain office building is designed to reach a maximum height of 5 meters at a flow rate of 100 l/s. A variable position valve in line with the water supply to the fountain can be controlled electrically, such that 0 V applied results in the valve being fully open, and 5 V results in the valve being closed. In adverse wind conditions the maximum height of the fountain needs to be adjusted; if the wind velocity exceeds 50 km/h, the height cannot exceed 2 meters. A wind velocity sensor is available which provides a voltage calibrated such that 1 V corresponds to a wind velocity of 25 km/h. Design a circuit which uses the velocity sensor to control the fountain according to specifications.

61. For the circuit of Fig. 6.43, let all resistor values equal 5 kΩ. Sketch v_{out} as a function of time if (a) $v_1 = 5 \sin 5t$ V and $v_2 = 5 \cos 5t$ V; (b) $v_1 = 4e^{-t}$ V and $v_2 = 5e^{-2t}$ V; (c) $v_1 = 2$ V and $v_2 = e^{-t}$ V.

Capacitors and Inductors

INTRODUCTION

In this chapter we introduce two new passive circuit elements, the *capacitor* and the *inductor*, each of which has the ability to both store and deliver *finite* amounts of energy. They differ from ideal sources in this respect, since they cannot sustain a finite average power flow over an infinite time interval. Although they are classed as linear elements, the current-voltage relationships for these new elements are time-dependent, leading to many interesting circuits. The range of capacitance and inductance values we might encounter can be huge, so that at times they may dominate circuit behavior, and at other times be essentially insignificant. Such issues continue to be relevant in modern circuit applications, particularly as computer and communication systems move to increasingly higher operating frequencies and component densities.

7.1 THE CAPACITOR

Ideal Capacitor Model

Previously, we termed independent and dependent sources *active* elements, and the linear resistor a *passive* element, although our definitions of active and passive are still slightly fuzzy and need to be brought into sharper focus. We now define an ***active element*** as an element that is capable of furnishing an average power greater than zero to some external device, where the average is taken over an infinite time interval. Ideal sources are active elements, and the operational amplifier is also an active device. A ***passive element***, however, is defined as an element that cannot supply an average power that is greater than zero over an infinite time interval. The resistor falls into this category; the energy it receives is usually transformed into heat, and it never supplies energy.

We now introduce a new passive circuit element, the **capacitor.** We define capacitance C by the voltage-current relationship

$$i = C\frac{dv}{dt} \qquad [1]$$

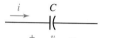

■ FIGURE 7.1 Electrical symbol and current-voltage conventions for a capacitor.

where v and i satisfy the conventions for a passive element, as shown in Fig. 7.1. We should bear in mind that v and i are functions of time; if needed, we can emphasize this fact by writing $v(t)$ and $i(t)$ instead. From Eq. [1], we may determine the unit of capacitance as an ampere-second per volt, or coulomb per volt. We will now define the *farad*[1] (F) as one coulomb per volt, and use this as our unit of capacitance.

The ideal capacitor defined by Eq. [1] is only a mathematical model of a real device. A capacitor consists of two conducting surfaces on which charge may be stored, separated by a thin insulating layer that has a very large resistance. If we assume that this resistance is sufficiently large that it may be considered infinite, then equal and opposite charges placed on the capacitor "plates" can never recombine, at least by any path within the element. The construction of the physical device is suggested by the circuit symbol shown in Fig. 7.1.

Let us visualize some external device connected to this capacitor and causing a positive current to flow into one plate of the capacitor and out of the other plate. Equal currents are entering and leaving the two terminals, and this is no more than we expect for any circuit element. Now let us examine the interior of the capacitor. The positive current entering one plate represents positive charge moving toward that plate through its terminal lead; this charge cannot pass through the interior of the capacitor, and it therefore accumulates on the plate. As a matter of fact, the current and the increasing charge are related by the familiar equation

$$i = \frac{dq}{dt}$$

Now let us consider this plate as an overgrown node and apply Kirchhoff's current law. It apparently does not hold; current is approaching the plate from the external circuit, but it is not flowing out of the plate into the "internal circuit." This dilemma bothered a famous Scottish scientist, James Clerk Maxwell, more than a century ago. The unified electromagnetic theory that he subsequently developed hypothesizes a "displacement current" that is present wherever an electric field or a voltage is varying with time. The displacement current flowing internally between the capacitor plates is exactly equal to the conduction current flowing in the capacitor leads; Kirchhoff's current law is therefore satisfied if we include both conduction and displacement currents. However, circuit analysis is not concerned with this internal displacement current, and since it is fortunately equal to the conduction current, we may consider Maxwell's hypothesis as relating the conduction current to the changing voltage across the capacitor.

A capacitor constructed of two parallel conducting plates of area A, separated by a distance d, has a capacitance $C = \varepsilon A/d$, where ε is the permittivity, a constant of the insulating material between the plates; this assumes

(1) Named in honor of Michael Faraday.

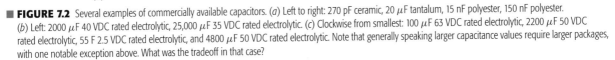

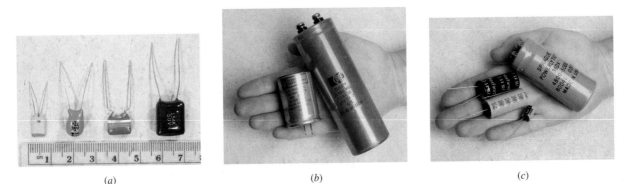

■ **FIGURE 7.2** Several examples of commercially available capacitors. (a) Left to right: 270 pF ceramic, 20 μF tantalum, 15 nF polyester, 150 nF polyester. (b) Left: 2000 μF 40 VDC rated electrolytic, 25,000 μF 35 VDC rated electrolytic. (c) Clockwise from smallest: 100 μF 63 VDC rated electrolytic, 2200 μF 50 VDC rated electrolytic, 55 F 2.5 VDC rated electrolytic, and 4800 μF 50 VDC rated electrolytic. Note that generally speaking larger capacitance values require larger packages, with one notable exception above. What was the tradeoff in that case?

the linear dimensions of the conducting plates are all very much greater than d. For air or vacuum, $\varepsilon = \varepsilon_0 = 8.854$ pF/m. Most capacitors employ a thin dielectric layer with a larger permittivity than air in order to minimize the device size. Examples of various types of commercially available capacitors are shown in Fig. 7.2, although we should remember that any two conducting surfaces not in direct contact with each other may be characterized by a nonzero (although probably small) capacitance. We should also note that a capacitance of several hundred *microfarads* (μF) is considered "large."

Several important characteristics of our new mathematical model can be discovered from the defining equation, Eq. [1]. A constant voltage across a capacitor results in zero current passing through it; a capacitor is thus an "*open circuit to dc.*" This fact is pictorially represented by the capacitor symbol. It is also apparent that a sudden jump in the voltage requires an infinite current. Since this is physically impossible, we will therefore prohibit the voltage across a capacitor to change in zero time.

EXAMPLE 7.1

Determine the current i flowing through the capacitor of Fig. 7.1 for the two voltage waveforms of Fig. 7.3 if $C = 2$ F.

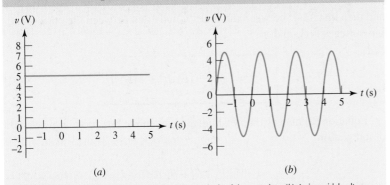

■ **FIGURE 7.3** (a) A dc voltage applied to the terminals of the capacitor. (b) A sinusoidal voltage waveform applied to the capacitor terminals.

(Continued on next page)

The current i is related to the voltage v across the capacitor by Eq. [1]:

$$i = C \frac{dv}{dt}$$

For the voltage waveform depicted in Fig. 7.3a, $dv/dt = 0$, so $i = 0$; the result is plotted in Fig. 7.4a. For the case of the sinusoidal waveform of Fig. 7.3b, we expect a cosine current waveform to flow in response, having the same frequency and twice the magnitude (since $C = 2$ F). The result is plotted in Fig. 7.4b.

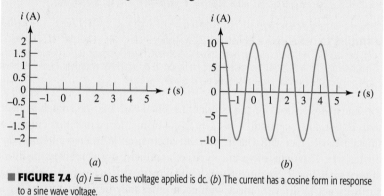

(a)   (b)

■ **FIGURE 7.4** (a) $i = 0$ as the voltage applied is dc. (b) The current has a cosine form in response to a sine wave voltage.

PRACTICE

7.1 Determine the current flowing through a 5 mF capacitor in response to a voltage v equal to: (a) -20 V; (b) $2e^{-5t}$ V.

Ans: 0 A; $-50e^{-5t}$ mA.

Integral Voltage-Current Relationships

The capacitor voltage may be expressed in terms of the current by integrating Eq. [1]. We first obtain

$$dv = \frac{1}{C} i(t)\, dt$$

and then integrate[2] between the times t_0 and t and between the corresponding voltages $v(t_0)$ and $v(t)$:

$$\boxed{v(t) = \frac{1}{C} \int_{t_0}^{t} i(t')\, dt' + v(t_0)} \qquad [2]$$

Equation [2] may also be written as an indefinite integral plus a constant of integration:

$$v(t) = \frac{1}{C} \int i\, dt + k$$

(2) Note that we are employing the mathematically correct procedure of defining a *dummy variable* t' in situations where the integration variable t is also a limit.

Finally, in many situations we will find that $v(t_0)$, the voltage initially across the capacitor, is not able to be discerned. In such instances it is mathematically convenient to set $t_0 = -\infty$ and $v(-\infty) = 0$, so that

$$v(t) = \frac{1}{C} \int_{-\infty}^{t} i \, dt'$$

Since the integral of the current over any time interval is the corresponding charge accumulated on the capacitor plate into which the current is flowing, we may also define capacitance as

$$q(t) = Cv(t)$$

where $q(t)$ and $v(t)$ represent instantaneous values of the charge on either plate and the voltage between the plates, respectively.

EXAMPLE 7.2

Find the capacitor voltage that is associated with the current shown graphically in Fig. 7.5a. The value of the capacitance is 5 μF.

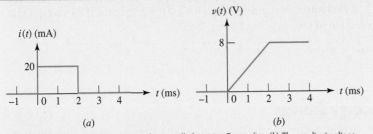

■ **FIGURE 7.5** (a) The current waveform applied to a 5 μF capacitor. (b) The resultant voltage waveform obtained by graphical integration.

Equation [2] is the appropriate expression here:

$$v(t) = \frac{1}{C} \int_{t_0}^{t} i(t') \, dt' + v(t_0)$$

but now it needs to be interpreted graphically. To do this, we note that the difference in voltage between times t and t_0 is proportional to the area under the current curve defined by the same two times. The constant of proportionality is $1/C$.

From Fig. 7.5a, we see three separate intervals: $t \le 0, 0 \le t \le 2$ ms, and $t \ge 2$ ms. Defining the first interval more specifically as between $-\infty$ and 0, so that $t_0 = -\infty$, we note two things, both a consequence of the fact that the current has always been zero up to $t = 0$: First,

$$v(t_0) = v(-\infty) = 0$$

Second, the integral of the current between $t_0 = -\infty$ and 0 is simply zero, since $i = 0$ in that interval. Thus,

$$v(t) = 0 + v(-\infty) \qquad -\infty \le t \le 0$$

or

$$v(t) = 0 \qquad t \le 0$$

(Continued on next page)

If we now consider the time interval represented by the rectangular pulse, we obtain

$$v(t) = \frac{1}{5 \times 10^{-6}} \int_0^t 20 \times 10^{-3} \, dt' + v(0)$$

Since $v(0) = 0$,

$$v(t) = 4000t \qquad 0 \le t \le 2 \text{ ms}$$

For the semi-infinite interval following the pulse, the integral of $i(t)$ is once again zero, so that

$$v(t) = 8 \qquad t \ge 2 \text{ ms}$$

The results are expressed much more simply in a sketch than by these analytical expressions, as shown in Fig. 7.5b.

PRACTICE

7.2 Determine the current through a 100 pF capacitor if its voltage as a function of time is given by Fig. 7.6.

Ans: $0 \text{ A}, -\infty \le t \le 1 \text{ ms}$; $200 \text{ nA}, 1 \text{ ms} \le t \le 2 \text{ ms}$; $0 \text{ A}, t \ge 2 \text{ ms}$.

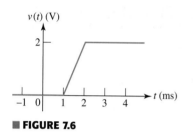

■ FIGURE 7.6

Energy Storage

To determine the energy stored in a capacitor, we begin with the power delivered to it:

$$p = vi = Cv \frac{dv}{dt}$$

The change in energy stored in its electric field is simply

$$\int_{t_0}^t p \, dt' = C \int_{t_0}^t v \frac{dv}{dt'} \, dt' = C \int_{v(t_0)}^{v(t)} v' \, dv' = \frac{1}{2} C \left\{ [v(t)]^2 - [v(t_0)]^2 \right\}$$

and thus

$$w_C(t) - w_C(t_0) = \tfrac{1}{2} C \left\{ [v(t)]^2 - [v(t_0)]^2 \right\} \qquad [3]$$

where the stored energy is $w_C(t_0)$ in joules (J) and the voltage at t_0 is $v(t_0)$. If we select a zero-energy reference at t_0, implying that the capacitor voltage is also zero at that instant, then

$$\boxed{w_C(t) = \tfrac{1}{2} Cv^2} \qquad [4]$$

Let us consider a simple numerical example. As sketched in Fig. 7.7, a sinusoidal voltage source is in parallel with a 1 MΩ resistor and a 20 μF capacitor. The parallel resistor may be assumed to represent the finite resistance of the dielectric between the plates of the physical capacitor (an *ideal* capacitor has infinite resistance).

EXAMPLE **7.3**

Find the maximum energy stored in the capacitor of Fig. 7.7 and the energy dissipated in the resistor over the interval $0 < t < 0.5$ s.

▶ *Identify the goal of the problem.*
The energy stored in the capacitor varies with time; we are asked for the *maximum* value over a specific time interval. We are also asked to find the *total* amount of energy dissipated by the resistor over this interval. These are actually two completely different questions.

▶ *Collect the known information.*
The only source of energy in the circuit is the independent voltage source, which has a value of $100 \sin 2\pi t$ V. We are only interested in the time interval of $0 < t < 0.5$ s. The circuit is properly labeled.

▶ *Devise a plan.*
Determine the energy in the capacitor by evaluating the voltage. To find the energy dissipated in the resistor during the same time interval, integrate the dissipated *power,* $p_R = i_R^2 \cdot R$.

▶ *Construct an appropriate set of equations.*
The energy stored in the capacitor is simply

$$w_C(t) = \tfrac{1}{2}Cv^2 = 0.1 \sin^2 2\pi t \quad \text{J}$$

We obtain an expression for the power dissipated by the resistor in terms of the current i_R:

$$i_R = \frac{v}{R} = 10^{-4} \sin 2\pi t \quad \text{A}$$

and so

$$p_R = i_R^2 R = (10^{-4})(10^6) \sin^2 2\pi t$$

so that the energy dissipated in the resistor between 0 and 0.5 s is

$$w_R = \int_0^{0.5} p_R \, dt = \int_0^{0.5} 10^{-2} \sin^2 2\pi t \, dt \quad \text{J}$$

▶ *Determine if additional information is required.*
We have an expression for the energy stored in the capacitor; a sketch is shown in Fig. 7.8. The expression derived for the energy dissipated by the resistor does not involve any unknown quantities, and so may also be readily evaluated.

▶ *Attempt a solution.*
From our sketch of the expression for the energy stored in the capacitor, we see that it increases from zero at $t = 0$ to a maximum of 100 mJ at $t = \frac{1}{4}$ s, and then decreases to zero in another $\frac{1}{4}$ s. Thus, $w_{C_{\max}} = 100$ mJ. Evaluating our integral expression for the energy dissipated in the resistor, we find that $w_R = 2.5$ mJ.

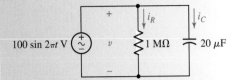

■ **FIGURE 7.7** A sinusoidal voltage source is applied to a parallel *RC* network. The 1 MΩ resistor might represent the finite resistance of the "real" capacitor's dielectric layer.

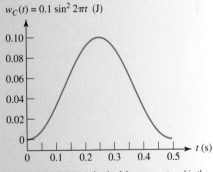

■ **FIGURE 7.8** A sketch of the energy stored in the capacitor as a function of time.

(Continued on next page)

▶ **Verify the solution. Is it reasonable or expected?**

We do not expect to calculate a *negative* stored energy, which is borne out in our sketch. Further, since the maximum value of $\sin 2\pi t$ is 1, the maximum energy expected anywhere would be $(1/2)(20 \times 10^{-6})(100)^2 = 100$ mJ.

The resistor dissipated 2.5 mJ in the period of 0 to 500 ms, although the capacitor stored a maximum of 100 mJ at one point during that interval. What happened to the "other" 97.5 mJ? To answer this, we compute the capacitor current

$$i_C = 20 \times 10^{-6} \frac{dv}{dt} = 0.004\pi \cos 2\pi t$$

and the current i_s defined as flowing *into* the voltage source

$$i_s = -i_C - i_R$$

both of which are plotted in Fig. 7.9. We observe that the current flowing through the resistor is a small fraction of the source current, not entirely surprising as 1 MΩ is a relatively large resistance value. As current flows from the source, a small amount is diverted to the resistor, with the rest flowing into the capacitor as it charges. After $t = 250$ ms, the source current is seen to change sign; current is now flowing from the capacitor back into the source. Most of the energy stored in the capacitor is being returned to the ideal voltage source, except for the small fraction dissipated in the resistor.

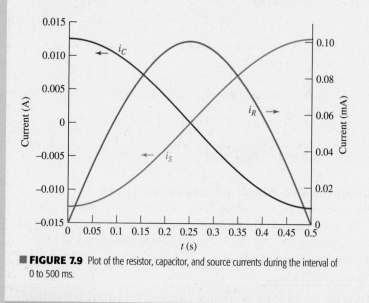

■ **FIGURE 7.9** Plot of the resistor, capacitor, and source currents during the interval of 0 to 500 ms.

PRACTICE

7.3 Calculate the energy stored in a 1000 μF capacitor at $t = 50\,\mu$s if the voltage across it is $1.5 \cos 10^5 t$ volts.

Ans: 90.52 μJ.

Important Characteristics of an Ideal Capacitor

1. There is no current through a capacitor if the voltage across it is not changing with time. A capacitor is therefore an *open circuit to dc*.

2. A finite amount of energy can be stored in a capacitor even if the current through the capacitor is zero, such as when the voltage across it is constant.

3. It is impossible to change the voltage across a capacitor by a finite amount in zero time, as this requires an infinite current through the capacitor. (A capacitor resists an abrupt change in the voltage across it in a manner analogous to the way a spring resists an abrupt change in its displacement.)

4. A capacitor never dissipates energy, but only stores it. Although this is true for the *mathematical model*, it is not true for a *physical* capacitor due to finite resistances associated with the dielectric as well as the packaging.

7.2 THE INDUCTOR

Ideal Inductor Model

In the early 1800s the Danish scientist Oersted showed that a current-carrying conductor produced a magnetic field (compass needles were affected in the presence of a wire when current was flowing). Shortly thereafter, Ampère made some careful measurements which demonstrated that this magnetic field was *linearly* related to the current which produced it. The next step occurred some 20 years later when the English experimentalist Michael Faraday and the American inventor Joseph Henry discovered almost simultaneously[3] that a changing magnetic field could induce a voltage in a neighboring circuit. They showed that this voltage was proportional to the time rate of change of the current producing the magnetic field. The constant of proportionality is what we now call the **inductance,** symbolized by L, and therefore

$$v = L \frac{di}{dt} \qquad [5]$$

where we must realize that v and i are both functions of time. When we wish to emphasize this, we may do so by using the symbols $v(t)$ and $i(t)$.

The circuit symbol for the inductor is shown in Fig. 7.10, and it should be noted that the passive sign convention is used, just as it was with the resistor and the capacitor. The unit in which inductance is measured is the **henry** (H), and the defining equation shows that the henry is just a shorter expression for a volt-second per ampere.

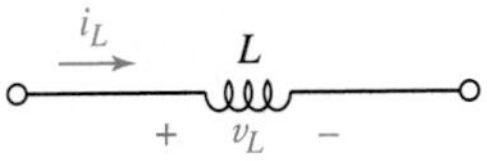

■ **FIGURE 7.10** Electrical symbol and current-voltage conventions for an inductor.

(3) Faraday won.

The inductor whose inductance is defined by Eq. [5] is a mathematical model; it is an *ideal* element which we may use to approximate the behavior of a *real* device. A physical inductor may be constructed by winding a length of wire into a coil. This serves effectively to increase the current that is causing the magnetic field and also to increase the "number" of neighboring circuits into which Faraday's voltage may be induced. The result of this twofold effect is that the inductance of a coil is approximately proportional to the square of the number of complete turns made by the conductor out of which it is formed. For example, an inductor or "coil" that has the form of a long helix of very small pitch is found to have an inductance of $\mu N^2 A/s$, where A is the cross-sectional area, s is the axial length of the helix, N is the number of complete turns of wire, and μ (mu) is a constant of the material inside the helix, called the permeability. For free space (and very closely for air), $\mu = \mu_0 = 4\pi \times 10^{-7}$ H/m $= 4\pi$ nH/cm. Several examples of commercially available inductors are shown in Fig. 7.11.

Let us now scrutinize Eq. [5] to determine some of the electrical characteristics of the mathematical model. This equation shows that the voltage across an inductor is proportional to the time rate of change of the current through it. In particular, it shows that there is no voltage across an inductor carrying a constant current, regardless of the magnitude of this current. Accordingly, we may view an inductor as a *short circuit to dc*.

Another fact that can be obtained from Eq. [5] is that a sudden or discontinuous change in the current must be associated with an infinite voltage across the inductor. In other words, if we wish to produce an abrupt change in an inductor current, we must apply an infinite voltage. Although an infinite-voltage forcing function might be amusing theoretically, it can never be a part of the phenomena displayed by a real physical device. As we

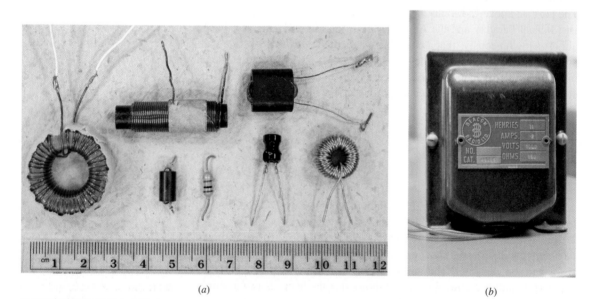

(a) (b)

■ **FIGURE 7.11** (*a*) Several different types of commercially available inductors, sometimes also referred to as "chokes." Clockwise, starting from far left: 287 μH ferrite core toroidal inductor, 266 μH ferrite core cylindrical inductor, 215 μH ferrite core inductor designed for VHF frequencies, 85 μH iron powder core toroidal inductor, 10 μH bobbin-style inductor, 100 μH axial lead inductor, and 7 μH lossy-core inductor used for RF suppression. (*b*) An 11 H inductor, measuring 10 cm (tall) × 8 cm (wide) × 8 cm (deep).

shall see shortly, an abrupt change in the inductor current also requires an abrupt change in the energy stored in the inductor, and this sudden change in energy requires infinite power at that instant; infinite power is again not a part of the real physical world. In order to avoid infinite voltage and infinite power, an inductor current must not be allowed to jump *instantaneously* from one value to another.

If an attempt is made to open-circuit a physical inductor through which a finite current is flowing, an arc may appear across the switch. This is useful in the ignition system of some automobiles, where the current through the spark coil is interrupted by the distributor and the arc appears across the spark plug. Although this does not occur instantaneously, it happens in a very short timespan, leading to the creation of a large voltage. The presence of a large voltage across a short distance equates to a very large electric field; the stored energy is dissipated in ionizing the air in the path of the arc.

Equation [5] may also be interpreted (and solved, if necessary) by graphical methods, as seen in Example 7.4.

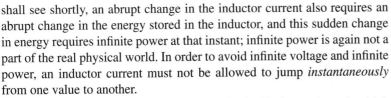

EXAMPLE **7.4**

Given the waveform of the current in a 3 H inductor as shown in Fig. 7.12*a*, determine the inductor voltage and sketch it.

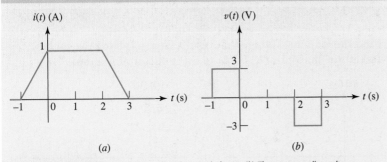

■ **FIGURE 7.12** (*a*) The current waveform in a 3 H inductor. (*b*) The corresponding voltage waveform, $v = 3\, di/dt$.

Defining the voltage v and the current i to satisfy the passive sign convention, we may obtain v from Fig. 7.12*a* using Eq. [5]:

$$v = 3\frac{di}{dt}$$

Since the current is zero for $t < -1$ s, the voltage is zero in this interval. The current then begins to increase at the linear rate of 1 A/s, and thus a constant voltage of $L\, di/dt = 3$ V is produced. During the following 2 s interval, the current is constant and the voltage is therefore zero. The final decrease of the current results in $di/dt = -1$ A/s, yielding $v = -3$ V. For $t > 3$ s, $i(t)$ is a constant (zero), so that $v(t) = 0$ for that interval. The complete voltage waveform is sketched in Fig. 7.12*b*.

PRACTICE

7.4 The current through a 200 mH inductor is shown in Fig. 7.13. Assume the passive sign convention, and find v_L at t equal to (a) 0; (b) 2 ms; (c) 6 ms.

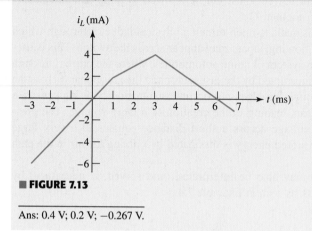

■ **FIGURE 7.13**

Ans: 0.4 V; 0.2 V; −0.267 V.

Let us now investigate the effect of a more rapid rise and decay of the current between the 0 and 1 A values.

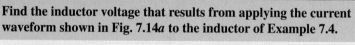

EXAMPLE **7.5**

Find the inductor voltage that results from applying the current waveform shown in Fig. 7.14a to the inductor of Example 7.4.

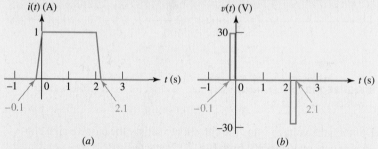

(a)   (b)

■ **FIGURE 7.14** (a) The time required for the current of Fig. 7.12a to change from 0 to 1 and from 1 to 0 is decreased by a factor of 10. (b) The resultant voltage waveform. The pulse widths are exaggerated for clarity.

Note that the intervals for the rise and fall have decreased to 0.1 s. Thus, the magnitude of each derivative will be 10 times larger; this condition is shown in the current and voltage sketches of Fig. 7.14a and b. In the voltage waveforms of Fig. 7.13b and 7.14b, it is interesting to note that the area under each voltage pulse is 3 V · s.

Just for curiosity's sake, let's continue in the same vein for a moment. A further decrease in the rise and fall times of the current waveform will produce a proportionally larger voltage magnitude, but only within the interval in which the current is increasing or decreasing. An abrupt change in the current will cause the infinite voltage "spikes" (each having an area of 3 V · s) that are suggested by the waveforms of Fig. 7.15a and b; or, from the equally valid but opposite point of view, these infinite voltage spikes are required to produce the abrupt changes in the current.

PRACTICE

7.5 The current waveform of Fig. 7.14a has equal rise and fall times of duration 0.1 s (100 ms). Calculate the maximum positive and negative voltages across the same inductor if the rise and fall times, respectively, are changed to (a) 1 ms, 1 ms; (b) 12 μs, 64 μs; (c) 1 s, 1 ns.

Ans: 3 kV, −3 kV; 250 kV, −46.88 kV; 3 V, −3 GV.

Integral Voltage-Current Relationships

We have defined inductance by a simple differential equation,

$$v = L \frac{di}{dt}$$

and we have been able to draw several conclusions about the characteristics of an inductor from this relationship. For example, we have found that we may consider an inductor to be a short circuit to direct current, and we have agreed that we cannot permit an inductor current to change abruptly from one value to another, because this would require that an infinite voltage and power be associated with the inductor. The simple defining equation for inductance contains still more information, however. Rewritten in a slightly different form,

$$di = \frac{1}{L} v \, dt$$

it invites integration. Let us first consider the limits to be placed on the two integrals. We desire the current i at time t, and this pair of quantities therefore provides the upper limits on the integrals appearing on the left and right sides of the equation, respectively; the lower limits may also be kept general by merely assuming that the current is $i(t_0)$ at time t_0. Thus,

$$\int_{i(t_0)}^{i(t)} di' = \frac{1}{L} \int_{t_0}^{t} v(t') \, dt'$$

which leads to the equation

$$i(t) - i(t_0) = \frac{1}{L} \int_{t_0}^{t} v \, dt'$$

or

$$i(t) = \frac{1}{L} \int_{t_0}^{t} v \, dt' + i(t_0) \qquad [6]$$

Equation [5] expresses the inductor voltage in terms of the current, whereas Eq. [6] gives the current in terms of the voltage. Other forms are

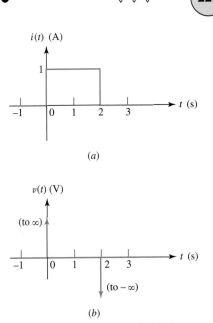

(a)

(b)

■ **FIGURE 7.15** (a) The time required for the current of Fig. 7.14a to change from 0 to 1 and from 1 to 0 is decreased to zero; the rise and fall are abrupt. (b) The resultant voltage across the 3 H inductor consists of a positive and a negative infinite spike.

also possible for the latter equation. We may write the integral as an indefinite integral and include a constant of integration k:

$$i(t) = \frac{1}{L} \int v \, dt + k \qquad [7]$$

We also may assume that we are solving a realistic problem in which the selection of t_0 as $-\infty$ ensures no current or energy in the inductor. Thus, if $i(t_0) = i(-\infty) = 0$, then

$$i(t) = \frac{1}{L} \int_{-\infty}^{t} v \, dt' \qquad [8]$$

Let us investigate the use of these several integrals by working a simple example where the voltage across an inductor is specified.

EXAMPLE 7.6

The voltage across a 2 H inductor is known to be 6 cos 5t V. Determine the resulting inductor current if $i(t = -\pi/2) = 1$ A.

From Eq. [6],

$$i(t) = \frac{1}{2} \int_{t_0}^{t} 6 \cos 5t' \, dt' + i(t_0)$$

or

$$i(t) = \frac{1}{2} \left(\frac{6}{5} \right) \sin 5t - \frac{1}{2} \left(\frac{6}{5} \right) \sin 5t_0 + i(t_0)$$

$$= 0.6 \sin 5t - 0.6 \sin 5t_0 + i(t_0)$$

The first term indicates that the inductor current varies sinusoidally; the second and third terms together represent a constant which becomes known when the current is numerically specified at some instant of time. Using the fact that the current is 1 A at $t = -\pi/2$ s, we identify t_0 as $-\pi/2$ with $i(t_0) = 1$, and find that

$$i(t) = 0.6 \sin 5t - 0.6 \sin(-2.5\pi) + 1$$

or

$$i(t) = 0.6 \sin 5t + 1.6$$

Alternatively, from Eq. [6],

$$i(t) = 0.6 \sin 5t + k$$

and we establish the numerical value of k by forcing the current to be 1 A at $t = -\pi/2$:

$$1 = 0.6 \sin(-2.5\pi) + k$$

or

$$k = 1 + 0.6 = 1.6$$

and so, as before,

$$i(t) = 0.6 \sin 5t + 1.6$$

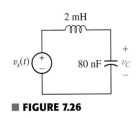

■ FIGURE 7.26

PRACTICE

7.9 If $v_C(t) = 4 \cos 10^5 t$ V in the circuit in Fig. 7.26, find $v_s(t)$.

Ans: $-2.4 \cos 10^5 t$ V.

We will not attempt the solution of integrodifferential equations here. It is worthwhile pointing out, however, that when the voltage forcing functions are sinusoidal functions of time, it will be possible to define a voltage-current ratio (called **impedance**) or a current-voltage ratio (called **admittance**) for each of the three passive elements. The factors operating on the two node voltages in the preceding equations will then become simple multiplying factors, and the equations will be linear algebraic equations once again. These we may solve by determinants or a simple elimination of variables as before.

We may also show that the benefits of linearity apply to *RLC* circuits as well. In accordance with our previous definition of a linear circuit, these circuits are also linear because the voltage-current relationships for the inductor and capacitor are linear relationships. For the inductor, we have

$$v = L \frac{di}{dt}$$

and multiplication of the current by some constant K leads to a voltage that is also greater by a factor K. In the integral formulation,

$$i(t) = \frac{1}{L} \int_{t_0}^{t} v \, dt' + i(t_0)$$

it can be seen that, if each term is to increase by a factor of K, then the initial value of the current must also increase by this same factor.

A corresponding investigation of the capacitor shows that it, too, is linear. Thus, a circuit composed of independent sources, linear dependent sources, and linear resistors, inductors, and capacitors is a linear circuit.

In this linear circuit the response is again proportional to the forcing function. The proof of this statement is accomplished by first writing a general system of integrodifferential equations. Let us place all the terms having the form of Ri, $L \, di/dt$, and $1/C \int i \, dt$ on the left side of each equation, and keep the independent source voltages on the right side. As a simple example, one of the equations might have the form

$$Ri + L \frac{di}{dt} + \frac{1}{C} \int_{t_0}^{t} i \, dt' + v_C(t_0) = v_s$$

If every independent source is now increased by a factor K, then the right side of each equation is greater by the factor K. Now each term on the left side is either a linear term involving some loop current or an initial capacitor voltage. In order to cause all the responses (loop currents) to increase by a factor K, it is apparent that we must also increase the initial capacitor voltages by a factor K. That is, we must treat the initial capacitor voltage as an independent source voltage and increase it also by a factor K. In a similar manner, initial inductor currents appear as independent source currents in nodal analysis.

The principle of proportionality between source and response can thus be extended to the general *RLC* circuit, and it follows that the principle of superposition also applies. It should be emphasized that initial inductor currents and capacitor voltages must be treated as independent sources in applying the superposition principle; each initial value must take its turn in being rendered inactive. In Chap. 5 we learned that the principle of superposition is a natural consequence of the linear nature of resistive circuits. The resistive circuits are linear because the voltage-current relationship for the resistor is linear and Kirchhoff's laws are linear.

Before we can apply the superposition principle to *RLC* circuits, however, it is first necessary to develop methods of solving the equations describing these circuits when only one independent source is present. At this time we should feel convinced that a linear circuit will possess a response whose amplitude is proportional to the amplitude of the source. We should be prepared to apply superposition later, considering an inductor current or capacitor voltage specified at $t = t_0$ as a source that must be deactivated when its turn comes.

Thévenin's and Norton's theorems are based on the linearity of the initial circuit, the applicability of Kirchhoff's laws, and the superposition principle. The general *RLC* circuit conforms perfectly to these requirements, and it follows, therefore, that all linear circuits that contain any combinations of independent voltage and current sources, linear dependent voltage and current sources, and linear resistors, inductors, and capacitors may be analyzed with the use of these two theorems, if we wish.

7.5 SIMPLE OP AMP CIRCUITS WITH CAPACITORS

In Chap. 6 we were introduced to several different types of amplifier circuits based on the ideal op amp. In almost every case, we found that the output was related to the input voltage by some combination of resistance ratios. If we replace one or more of these resistors with a capacitor, it is possible to obtain some interesting circuits in which the output is proportional to either the derivative or integral of the input voltage. Such circuits find widespread use in practice. For example, a velocity sensor can be connected to an op amp circuit that provides a signal proportional to the acceleration, or an output signal can be obtained that represents the total charge incident on a metal electrode during a specific period of time by simply integrating the measured current.

To create an integrator using an ideal op amp, we ground the noninverting input, install an ideal capacitor as a feedback element from the output back to the inverting input, and connect a signal source v_s to the inverting input through an ideal resistor as shown in Fig. 7.27.

Performing nodal analysis at the inverting input,

$$0 = \frac{v_a - v_s}{R_1} + i$$

We can relate the current i to the voltage across the capacitor,

$$i = C_f \frac{dv_{C_f}}{dt}$$

resulting in

$$0 = \frac{v_a - v_s}{R_1} + C_f \frac{dv_{C_f}}{dt}$$

Invoking ideal op amp rule 2, we know that $v_a = v_b = 0$, so

$$0 = \frac{-v_s}{R_1} + C_f \frac{dv_{C_f}}{dt}$$

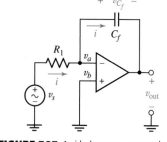

■ **FIGURE 7.27** An ideal op amp connected as an integrator.

Integrating and solving for v_{out}, we obtain

$$v_{C_f} = v_a - v_{\text{out}} = 0 - v_{\text{out}} = \frac{1}{R_1 C_f} \int_0^t v_s \, dt' + v_{C_f}(0)$$

or

$$v_{\text{out}} = -\frac{1}{R_1 C_f} \int_0^t v_s \, dt' - v_{C_f}(0) \qquad [17]$$

We therefore have combined a resistor, a capacitor, and an op amp to form an integrator. Note that the first term of the output is $1/RC$ times the negative of the integral of the input from $t' = 0$ to t, and the second term is the negative of the initial value of v_{C_f}. The value of $(RC)^{-1}$ can be made equal to unity, if we wish, by choosing $R = 1\ \text{M}\Omega$ and $C = 1\ \mu\text{F}$, for example; other selections may be made that will increase or decrease the output voltage.

Before we leave the integrator circuit, we might anticipate a question from an inquisitive reader, "*Could we use an inductor in place of the capacitor and obtain a differentiator?*" Indeed we could, but circuit designers usually avoid the use of inductors whenever possible because of their size, weight, cost, and associated resistance and capacitance. Instead, it is possible to interchange the positions of the resistor and capacitor in Fig. 7.27 and obtain a differentiator.

EXAMPLE 7.10

Derive an expression for the output voltage of the op amp circuit shown in Fig. 7.28.

We begin by writing a nodal equation at the inverting input pin, with $v_{C_1} \triangleq v_a - v_s$:

$$0 = C_1 \frac{dv_{C_1}}{dt} + \frac{v_a - v_{\text{out}}}{R_f}$$

Invoking ideal op amp rule 2, $v_a = v_b = 0$. Thus,

$$C_1 \frac{dv_{C_1}}{dt} = \frac{v_{\text{out}}}{R_f}$$

Solving for v_{out},

$$v_{\text{out}} = R_f C_1 \frac{dv_{C_1}}{dt}$$

Since $v_{C_1} = v_a - v_s = -v_s$,

$$v_{\text{out}} = -R_f C_1 \frac{dv_s}{dt}$$

So, simply by swapping the resistor and capacitor in the circuit of Fig. 7.27, we obtain a differentiator instead of an integrator.

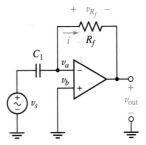

■ **FIGURE 7.28** An ideal op amp connected as a differentiator.

PRACTICE

7.10 Derive an expression for v_{out} in terms of v_s for the circuit shown in Fig. 7.29.

Ans: $v_{\text{out}} = -L_f/R_1 \, dv_s/dt$.

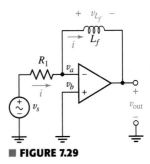

■ **FIGURE 7.29**

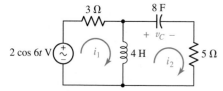

FIGURE 7.30 A given circuit to which the definition of duality may be applied to determine the dual circuit. Note that $v_c(0) = 10$ V.

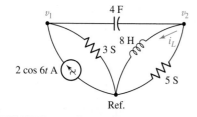

FIGURE 7.31 The exact dual of the circuit of Fig. 7.30.

7.6 · DUALITY

The concept of **duality** applies to many fundamental engineering concepts. In this section, we shall define duality in terms of the circuit equations. Two circuits are "duals" if the mesh equations that characterize one of them have the *same mathematical form* as the nodal equations that characterize the other. They are said to be exact duals if each mesh equation of one circuit is numerically identical with the corresponding nodal equation of the other; the current and voltage variables themselves cannot be identical, of course. Duality itself merely refers to any of the properties exhibited by dual circuits.

Let us use the definition to construct an exact dual circuit by writing the two mesh equations for the circuit shown in Fig. 7.30. Two mesh currents i_1 and i_2 are assigned, and the mesh equations are

$$3i_1 + 4\frac{di_1}{dt} - 4\frac{di_2}{dt} = 2\cos 6t \qquad [18]$$

$$-4\frac{di_1}{dt} + 4\frac{di_2}{dt} + \frac{1}{8}\int_0^t i_2\,dt' + 5i_2 = -10 \qquad [19]$$

We may now construct the two equations that describe the exact dual of our circuit. We wish these to be nodal equations, and thus begin by replacing the mesh currents i_1 and i_2 in Eqs. [18] and [19] by the two nodal voltages v_1 and v_2, respectively. We obtain

$$3v_1 + 4\frac{dv_1}{dt} - 4\frac{dv_2}{dt} = 2\cos 6t \qquad [20]$$

$$-4\frac{dv_1}{dt} + 4\frac{dv_2}{dt} + \frac{1}{8}\int_0^t v_2\,dt' + 5v_2 = -10 \qquad [21]$$

and we now seek the circuit represented by these two nodal equations.

Let us first draw a line to represent the reference node, and then we may establish two nodes at which the positive references for v_1 and v_2 are located. Equation [20] indicates that a current source of $2\cos 6t$ A is connected between node 1 and the reference node, oriented to provide a current entering node 1. This equation also shows that a 3 S conductance appears between node 1 and the reference node. Turning to Eq. [21], we first consider the nonmutual terms, i.e., those terms which do not appear in Eq. [20], and they instruct us to connect an 8 H inductor and a 5 S conductance (in parallel) between node 2 and the reference. The two similar terms in Eqs. [20] and [21] represent a 4 F capacitor present mutually at nodes 1 and 2; the circuit is completed by connecting this capacitor between the two nodes. The constant term on the right side of Eq. [21] is the value of the inductor current at $t = 0$; in other words, $i_L(0) = 10$ A. The dual circuit is shown in Fig. 7.31; since the two sets of equations are numerically identical, the circuits are exact duals.

Dual circuits may be obtained more readily than by this method, for the equations need not be written. In order to construct the dual of a given circuit, we think of the circuit in terms of its mesh equations. With each mesh we must associate a nonreference node, and, in addition, we must supply the reference node. On a diagram of the given circuit we therefore place a

❏ Series and parallel combinations of capacitors work the *opposite* way as they do for resistors. (Example 7.8)

❏ Since capacitors and inductors are linear elements, KVL, KCL, superposition, Thévenin's and Norton's theorems, and nodal and mesh analysis apply to their circuits as well. (Example 7.9)

❏ A capacitor as the feedback element in an inverting op amp leads to an output voltage proportional to the *integral* of the input voltage. Swapping the input resistor and the feedback capacitor leads to an output voltage proportional to the *derivative* of the input voltage. (Example 7.10)

❏ PSpice allows us to set the initial voltage across a capacitor, and the initial current through an inductor. A transient analysis provides details of the time-dependent response of circuits containing these types of elements. (Example 7.11)

READING FURTHER

A detailed guide to characteristics and selection of various capacitor and inductor types can be found in:

H. B. Drexler, *Passive Electronic Component Handbook,* 2nd ed., C. A. Harper, ed. New York: McGraw-Hill, 2003, pp. 69–203.

C. J. Kaiser, *The Inductor Handbook,* 2nd ed. Olathe, Kans.: C.J. Publishing, 1996.

Two books that describe capacitor-based op amp circuits are:

R. Mancini (ed.), *Op Amps Are For Everyone,* 2nd ed. Amsterdam: Newnes, 2003.

W. G. Jung, *Op Amp Cookbook,* 3rd ed. Upper Saddle River, N.J.: Prentice-Hall, 1997.

EXERCISES

7.1 The Capacitor

1. Making use of the passive sign convention, determine the current flowing through a 220 nF capacitor for $t \geq 0$ if its voltage $v_C(t)$ is given by (a) -3.35 V; (b) $16.2e^{-9t}$ V; (c) $8\cos 0.01t$ mV; (d) $5 + 9\sin 0.08t$ V.

2. Sketch the current flowing through a 13 pF capacitor for $t \geq 0$ as a result of the waveforms shown in Fig. 7.40. Assume the passive sign convention.

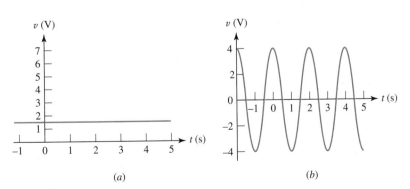

(a)   (b)

■ **FIGURE 7.40**

3. (a) If the voltage waveform depicted in Fig. 7.41 is applied across the terminals of a 1 μF electrolytic capacitor, graph the resulting current, assuming the passive sign convention. (b) Repeat part (a) if the capacitor is replaced with a 17.5 pF capacitor.

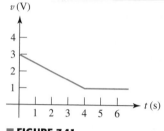

v (V)

■ FIGURE 7.41

4. A capacitor is constructed from two copper plates, each measuring 1 mm × 2.5 mm and 155 μm thick. The two plates are placed such that they face each other and are separated by a 1 μm gap. Calculate the resulting capacitance if (a) the intervening dielectric has a permittivity of $1.35\varepsilon_0$; (b) the intervening dielectric has a permittivity of $3.5\varepsilon_0$; (c) the plate separation is increased by 3.5 μm and the gap is filled with air; (d) the plate area is doubled and the 1 μm gap is filled with air.

5. Two pieces of gadolinium, each measuring 100 μm × 750 μm and 604 nm thick, are used to construct a capacitor. The two plates are arranged such that they face each other and are separated by a 100 nm gap. Calculate the resulting capacitance if (a) the intervening dielectric has a permittivity of $13.8\varepsilon_0$; (b) the intervening dielectric has a permittivity of $500\varepsilon_0$; (c) the plate separation is increased by 100 nm and the gap is filled with air; (d) the plate area is quadrupled and the 100 nm gap is filled with air.

6. Design a 100 nF capacitor constructed from 1 μm thick gold foil, and which fits entirely within a volume equal to that of a standard AAA battery, if the only dielectric available has a permittivity of $3.1\varepsilon_0$.

7. Design a capacitor whose capacitance can be varied mechanically with a simple vertical motion, between the values of 100 nF and 300 nF.

8. Design a capacitor whose capacitance can be varied mechanically over the range of 50 nF and 100 nF by rotating a knob 90°.

9. A silicon pn junction diode is characterized by a junction capacitance defined as

$$C_j = \frac{K_s \varepsilon_0 A}{W}$$

where $K_s = 11.8$ for silicon, ε_0 is the vacuum permittivity, A = the cross-sectional area of the junction, and W is known as the depletion width of the junction. Width W depends not only on how the diode is fabricated, but also on the voltage applied to its two terminals. It can be computed using

$$W = \sqrt{\frac{2K_s \varepsilon_0}{qN} (V_{bi} - V_A)}$$

Thus, diodes are frequently used in electronic circuits, since they can be thought of as voltage-controlled capacitors. Assuming parameter values of $N = 10^{18}$ cm^{-3}, $V_{bi} = 0.57$ V, and using $q = 1.6 \times 10^{-19}$ C, calculate the capacitance of a diode with cross-sectional area $A = 1$ μm × 1 μm at applied voltages of $V_A = -1, -5$, and -10 volts.

10. Assuming the passive sign convention, sketch the voltage which develops across the terminals of a 2.5 F capacitor in response to the current waveforms shown in Fig. 7.42.

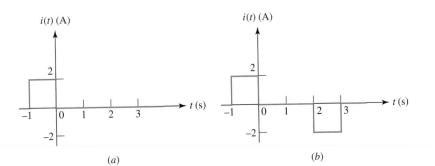

(a)

(b)

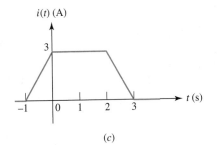

(c)

■ FIGURE 7.42

11. The current flowing through a 33 mF capacitor is shown graphically in Fig. 7.43. (a) Assuming the passive sign convention, sketch the resulting voltage waveform across the device. (b) Compute the voltage at 300 ms, 600 ms, and 1.1 s.

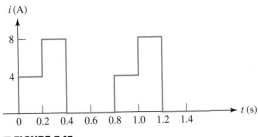

■ FIGURE 7.43

12. Calculate the energy stored in a capacitor at time $t = 1$ s if (a) $C = 1.4$ F and $v_C = 8$ V, $t > 0$; (b) $C = 23.5$ pF and $v_C = 0.8$ V, $t > 0$; (c) $C = 17$ nF, $v_C(1) = 12$ V, $v_C(0) = 2$ V, and $w_C(0) = 295$ nJ.

13. A 137 pF capacitor is connected to a voltage source such that $v_C(t) = 12e^{-2t}$V, $t \geq 0$ and $v_C(t) = 12$ V, $t < 0$. Calculate the energy stored in the capacitor at t equal to (a) 0; (b) 200 ms; (c) 500 ms; (d) 1 s.

14. Calculate the power dissipated in the 40 Ω resistor and the voltage labeled v_C in each of the circuits depicted in Fig. 7.44.

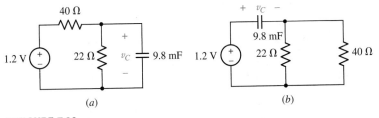

(a)

(b)

■ FIGURE 7.44

15. For each circuit shown in Fig. 7.45, calculate the voltage labeled v_C.

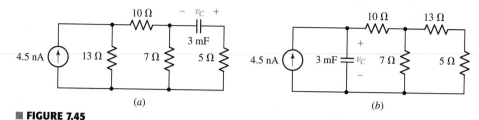

(a)   (b)

■ **FIGURE 7.45**

7.2 The Inductor

16. Design a 30 nH inductor using 29 AWG solid soft copper wire. Include a sketch of your design and label geometrical parameters as necessary for clarity. Assume the coil is filled with air only.

17. If the current flowing through a 75 mH inductor has the waveform shown in Fig. 7.46, (a) sketch the voltage which develops across the inductor terminals for $t \geq 0$, assuming the passive sign convention; and (b) calculate the voltage at $t = 1$ s, 2.9 s, and 3.1 s.

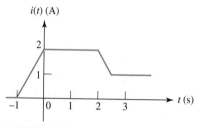

$i(t)$ (A)

■ **FIGURE 7.46**

$i(t)$ (A)

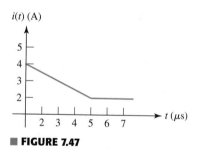

■ **FIGURE 7.47**

18. The current through a 17 nH aluminum inductor is shown in Fig. 7.47. Sketch the resulting voltage waveform for $t \geq 0$, assuming the passive sign convention.

19. Determine the voltage for $t \geq 0$ which develops across the terminals of a 4.2 mH inductor, if the current (defined consistent with the passive sign convention) is (a) -10 mA; (b) $3 \sin 6t$ A; (c) $11 + 115\sqrt{2} \cos(100\pi t - 9°)$ A; (d) $13e^{-t}$ nA; (e) $3 + te^{-14t}$ A.

20. Determine the voltage for $t \geq 0$ which develops across the terminals of an 8 pH inductor, if the current (defined consistent with the passive sign convention) is (a) 8 mA; (b) 800 mA; (c) 8 A; (d) $4e^{-t}$ A; (e) $-3 + te^{-t}$ A.

21. Calculate v_L and i_L for each of the circuits depicted in Fig. 7.48, if $i_s = 1$ mA and $v_s = 2.1$ V.

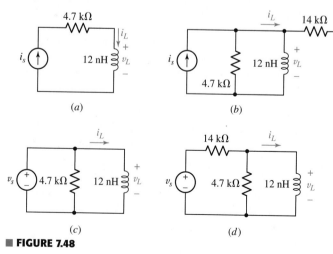

(a)   (b)

(c)   (d)

■ **FIGURE 7.48**

22. The current waveform shown in Fig. 7.14 has a rise time of 0.1 (100 ms) and a fall time of the same duration. If the current is applied to the "+" voltage reference terminal of a 200 nH inductor, sketch the expected voltage waveform if the rise and fall times are changed, respectively, to (a) 200 ms, 200 ms; (b) 10 ms, 50 ms; (c) 10 ns, 20 ns.

23. Determine the inductor voltage which results from the current waveform shown in Fig. 7.49 (assuming the passive sign convention) at t equal to (a) −1 s; (b) 0 s; (c) 1.5 s; (d) 2.5 s; (e) 4 s; (f) 5 s.

FIGURE 7.49

24. Determine the current flowing through a 6 mH inductor if the voltage (defined such that it is consistent with the passive sign convention) is given by (a) 5 V; (b) 100 sin 120πt, $t \geq 0$ and 0, $t < 0$.

25. The voltage across a 2 H inductor is given by $v_L = 4.3t$, $0 \leq t \leq 50$ ms. With the knowledge that $i_L(-0.1) = 100$ μA, calculate the current (assuming it is defined consistent with the passive sign convention) at t equal to (a) 0; (b) 1.5 ms; (c) 45 ms.

26. Calculate the energy stored in a 1 nH inductor if the current flowing through it is (a) 0 mA; (b) 1 mA; (c) 20 A; (d) 5 sin 6t mA, $t > 0$.

27. Determine the amount of energy stored in a 33 mH inductor at $t = 1$ ms as a result of a current i_L given by (a) 7 A; (b) $3 - 9e^{-10^3t}$ mA.

28. Making the assumption that the circuits in Fig. 7.50 have been connected for a very long time, determine the value for each current labeled i_x.

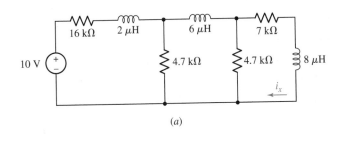

(a)

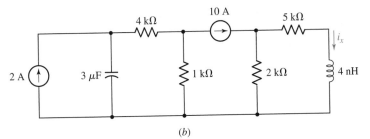

(b)

FIGURE 7.50

29. Calculate the voltage labeled v_x in Fig. 7.51, assuming the circuit has been running a very long time, if (a) a 10 Ω resistor is connected between terminals x and y; (b) a 1 H inductor is connected between terminals x and y; (c) a 1 F capacitor is connected between terminals x and y; (d) a 4 H inductor in parallel with a 1 Ω resistor is connected between terminals x and y.

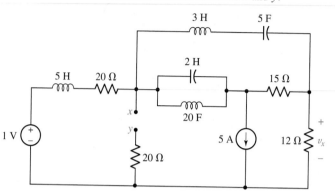

■ **FIGURE 7.51**

30. For the circuit shown in Fig. 7.52, (a) compute the Thévenin equivalent seen by the inductor; (b) determine the power being dissipated by both resistors; (c) calculate the energy stored in the inductor.

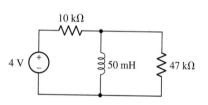

■ **FIGURE 7.52**

7.3 Inductance and Capacitance Combinations

31. If each capacitor has a value of 1 F, determine the equivalent capacitance of the network shown in Fig. 7.53.

32. Determine an equivalent inductance for the network shown in Fig. 7.54 if each inductor has value L.

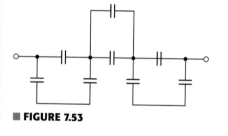

■ **FIGURE 7.53**

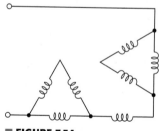

■ **FIGURE 7.54**

33. Using as many 1 nH inductors as you like, design two networks, each of which has an equivalent inductance of 1.25 nH.

34. Compute the equivalent capacitance C_{eq} as labeled in Fig. 7.55.

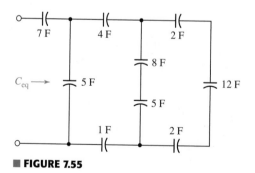

■ **FIGURE 7.55**

35. Determine the equivalent capacitance C_{eq} of the network shown in Fig. 7.56.

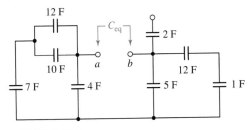

■ **FIGURE 7.56**

36. Apply combinatorial techniques as appropriate to obtain a value for the equivalent inductance L_{eq} as labeled on the network of Fig. 7.57.

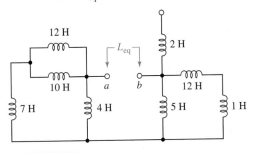

■ **FIGURE 7.57**

37. Reduce the circuit depicted in Fig. 7.58 to as few components as possible.

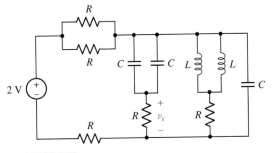

■ **FIGURE 7.58**

38. Refer to the network shown in Fig. 7.59 and find (a) R_{eq} if each element is a 10 Ω resistor; (b) L_{eq} if each element is a 10 H inductor; and (c) C_{eq} if each element is a 10 F capacitor.

39. Determine the equivalent inductance seen looking into the terminals marked a and b of the network represented in Fig. 7.60.

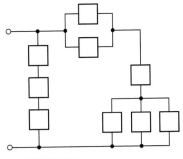

■ **FIGURE 7.59**

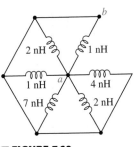

■ **FIGURE 7.60**

40. Reduce the circuit represented in Fig. 7.61 to the smallest possible number of components.

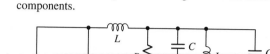

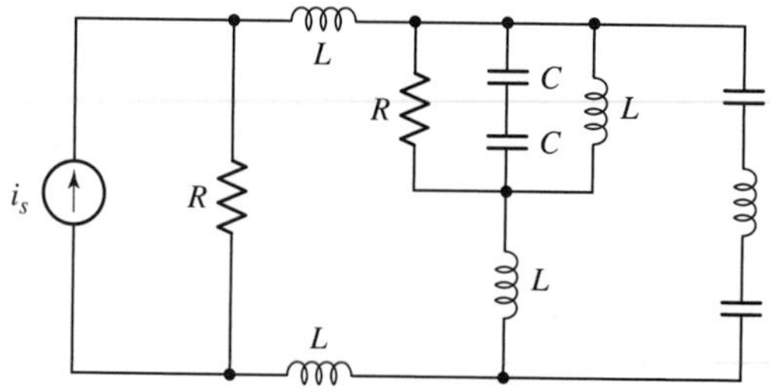

■ **FIGURE 7.61**

41. Reduce the network of Fig. 7.62 to the smallest possible number of components if each inductor is 1 nH and each capacitor is 1 mF.

■ **FIGURE 7.62**

42. For the network of Fig. 7.63, $L_1 = 1$ H, $L_2 = L_3 = 2$ H, $L_4 = L_5 = L_6 = 3$ H. (a) Find the equivalent inductance. (b) Derive an expression for a general network of this type having N stages, assuming stage N is composed of N inductors, each having inductance N henrys.

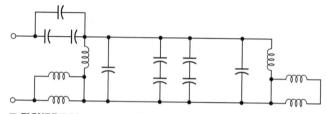

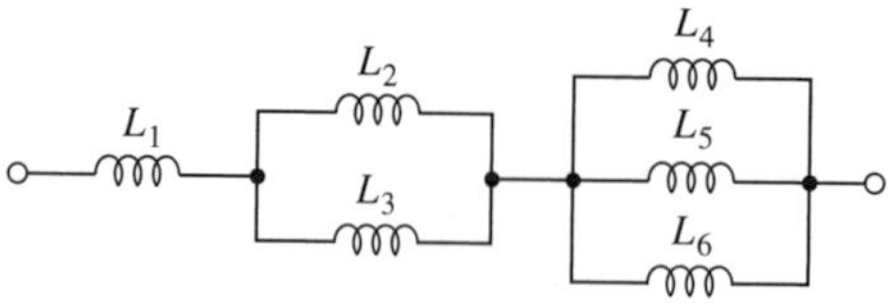

■ **FIGURE 7.63**

43. Extend the concept of Δ-Y transformations to simplify the network of Fig. 7.64 if each element is a 2 pF capacitor.

44. Extend the concept of Δ-Y transformations to simplify the network of Fig. 7.64 if each element is a 1 nH inductor.

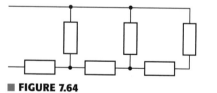

■ **FIGURE 7.64**

7.4 Consequences of Linearity

45. With regard to the circuit represented in Fig. 7.65, (a) write a complete set of nodal equations and (b) write a complete set of mesh equations.

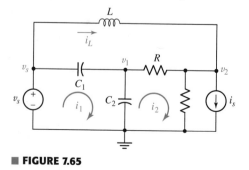

■ **FIGURE 7.65**

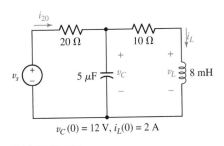

46. (*a*) Write nodal equations for the circuit of Fig. 7.66. (*b*) Write mesh equations for the same circuit.

47. In the circuit shown in Fig. 7.67, let $i_s = 60e^{-200t}$ mA with $i_1(0) = 20$ mA. (*a*) Find $v(t)$ for all t. (*b*) Find $i_1(t)$ for $t \geq 0$. (*c*) Find $i_2(t)$ for $t \geq 0$.

48. Let $v_s = 100e^{-80t}$ V and $v_1(0) = 20$ V in the circuit of Fig. 7.68. (*a*) Find $i(t)$ for all t. (*b*) Find $v_1(t)$ for $t \geq 0$. (*c*) Find $v_2(t)$ for $t \geq 0$.

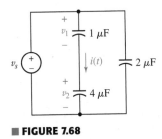

■ FIGURE 7.68

49. If it is assumed that all the sources in the circuit of Fig. 7.69 have been connected and operating for a very long time, use the superposition principle to find $v_C(t)$ and $v_L(t)$.

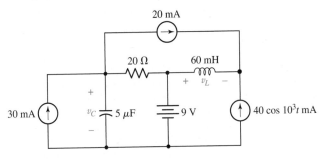

■ FIGURE 7.69

50. For the circuit of Fig. 7.70, assume no energy is stored at $t = 0$, and write a complete set of nodal equations.

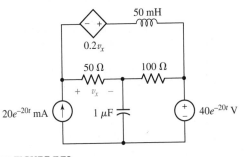

■ FIGURE 7.70

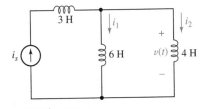

$v_C(0) = 12$ V, $i_L(0) = 2$ A

■ FIGURE 7.66

■ FIGURE 7.67

7.5 Simple Op Amp Circuits with Capacitors

51. Interchange the location of R_1 and C_f in the circuit of Fig. 7.27, and assume that $R_i = \infty$, $R_o = 0$, and $A = \infty$ for the op amp. (*a*) Find $v_{out}(t)$ as a function of $v_s(t)$. (*b*) Obtain an equation relating $v_o(t)$ and $v_s(t)$ if A is not assumed to be infinite.

52. For the integrating amplifier circuit of Fig. 7.27, $R_1 = 100$ kΩ, $C_f = 500$ μF, and $v_s = 20\sin 540t$ mV. Calculate v_{out} if (*a*) $A = \infty$, $R_i = \infty$, and $R_o = 0$; (*b*) $A = 5000$, $R_i = 1$ MΩ, and $R_o = 3$ Ω.

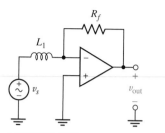

■ **FIGURE 7.71**

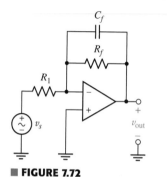

■ **FIGURE 7.72**

53. Derive an expression for v_{out} in terms of v_s for the amplifier circuit shown in Fig. 7.71.

54. In practice, circuits such as those depicted in Fig. 7.27 may not function correctly unless there is a conducting pathway between the output and input terminals of the op amp. (*a*) Analyze the modified integrating amplifier circuit shown in Fig. 7.72 to obtain an expression for v_{out} in terms of v_s, and (*b*) compare this expression to Eq. [17].

DP 55. A new piece of equipment designed to make crystals from molten constituents is experiencing too many failures (cracked products). The production manager wants to monitor the cooling rate to see if this is related to the problem. The system has two output terminals available, where the voltage across them is linearly proportional to the crucible temperature such that 30 mV corresponds to 30°C and 1 V corresponds to 1000°C. Design a circuit whose voltage output represents the cooling rate, calibrated such that 1 V = 1°C/s.

DP 56. A confectionary company has decided to increase the production rate of its milk chocolate bars to compensate for a recent increase in the cost of raw materials. However, the wrapping unit cannot accept more than 1 bar per second, or it drops bars. A 200 mV peak-to-peak sinusoidal voltage signal is available from the bar-making system which feeds into the wrapping unit, such that its frequency matches the bar production frequency (i.e., 1 Hz = 1 bar/s). Design a circuit that provides a voltage output sufficient to power a 12 V audible alarm when the production rate exceeds the capacity of the wrapping unit.

DP 57. One problem satellites face is exposure to high-energy particles, which can cause damage to sensitive electronics as well as solar arrays used to provide power. A new communications satellite is equipped with a high-energy proton detector measuring 1 cm × 1 cm. It provides a current directly equal to the number of protons impinging the surface per second. Design a circuit whose output voltage provides a running total of the number of proton hits, calibrated such that 1 V = 1 million hits.

DP 58. The output of a velocity sensor attached to a sensitive piece of mobile equipment is calibrated to provide a signal such that 10 mV corresponds to linear motion at 1 m/s. If the equipment is subjected to sudden shock, it can be damaged. Since force = mass × acceleration, monitoring of the rate of change of velocity can be used to determine if the equipment is transported improperly. (*a*) Design a circuit to provide a voltage proportional to the linear acceleration such that 10 mV = 1 m/s². (*b*) How many sensor-circuit combinations does this application require?

DP 59. A floating sensor in a certain fuel tank is connected to a variable resistor (often called a potentiometer) such that a full tank (100 liters) corresponds to 1 Ω and an empty tank corresponds to 10 Ω. (*a*) Design a circuit that provides an output voltage which indicates the amount of fuel remaining, so that 1 V = empty and 5 V = full. (*b*) Design a circuit to indicate the rate of fuel consumption by providing a voltage output calibrated to yield 1 V = 1 l/s.

7.6 Duality

60. (*a*) Draw the exact dual of the circuit depicted in Fig. 7.73. (*b*) Label the new (dual) variables. (*c*) Write nodal equations for both circuits.

61. (*a*) Draw the exact dual of the simple circuit shown in Fig. 7.74. (*b*) Label the new (dual) variables. (*c*) Write mesh equations for both circuits.

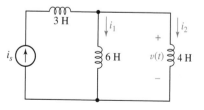

■ **FIGURE 7.73**

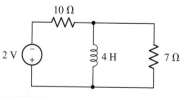

■ **FIGURE 7.74**

62. (*a*) Draw the exact dual of the simple circuit shown in Fig. 7.75. (*b*) Label the new (dual) variables. (*c*) Write mesh equations for both circuits.

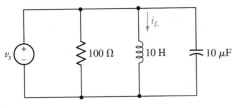

■ **FIGURE 7.75**

63. (*a*) Draw the exact dual of the simple circuit shown in Fig. 7.76. (*b*) Label the new (dual) variables. (*c*) Write nodal and mesh equations for both circuits.

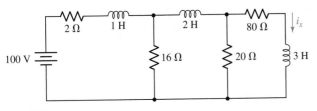

■ **FIGURE 7.76**

64. Draw the exact dual of the circuit shown in Fig. 7.77. Keep it neat!

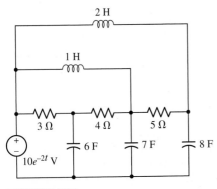

■ **FIGURE 7.77**

7.7 Modeling Capacitors and Inductors with PSpice

65. Taking the bottom node in the circuit of Fig. 7.78 as the reference terminal, calculate (*a*) the current through the inductor and (*b*) the power dissipated by the 7 Ω resistor. (*c*) Verify your answers with an appropriate PSpice simulation.

66. For the four-element circuit shown in Fig. 7.79, (*a*) calculate the power absorbed in each resistor; (*b*) determine the voltage across the capacitor; (*c*) compute the energy stored in the capacitor; and (*d*) verify your answers with an appropriate PSpice simulation. (Recall that calculations can be performed in Probe.)

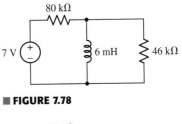

■ **FIGURE 7.78**

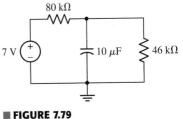

■ **FIGURE 7.79**

67. (a) Compute i_L and v_x as indicated in the circuit of Fig. 7.80. (b) Determine the energy stored in the inductor and in the capacitor. (c) Verify your answers with an appropriate PSpice simulation.

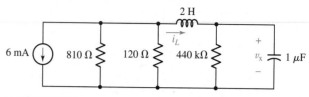

■ **FIGURE 7.80**

68. For the circuit depicted in Fig. 7.81, the value of $i_L(0) = 1$ mA. (a) Compute the energy stored in the element at $t = 0$. (b) Perform a transient simulation of the circuit over the range of $0 \leq t \leq 500$ ns. Determine the value of i_L at $t = 0$, 130 ns, 260 ns, and 500 ns. (c) What fraction of the initial energy remains in the inductor at $t = 130$ ns? At $t = 500$ ns?

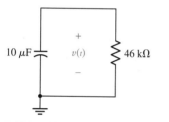

■ **FIGURE 7.81**

69. Assume an initial voltage of 9 V across the 10 μF capacitor shown in Fig. 7.82 (i.e., $v(0) = 9$ V). (a) Compute the initial energy stored in the capacitor. (b) For $t > 0$, do you expect the energy to remain in the capacitor? *Explain*. (c) Perform a transient simulation of the circuit over the range of $0 \leq t \leq 2.5$ s and determine $v(t)$ at $t = 460$ ms, 920 ms, and 2.3 s. (c) What fraction of the initial energy remains stored in the capacitor at $t = 460$ ms? At $t = 2.3$ s?

70. Referring to the circuit of Fig. 7.83, (a) calculate the energy stored in each energy storage element; (b) verify your answers with an appropriate PSpice simulation.

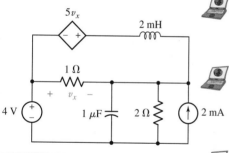

■ **FIGURE 7.82**

Chapter-Integrating Exercises

71. For the circuit of Fig. 7.28, (a) sketch v_{out} over the range of $0 \leq t \leq 5$ ms if $R_f = 1$ kΩ, $C_1 = 100$ mF, and v_s is a 1 kHz sinusoidal source having a peak voltage of 2 V. (b) Verify your answer with an appropriate transient simulation, plotting both v_s and v_{out} in Probe. (*Hint:* Between plotting traces, add a second y axis using **Plot, Add Y Axis.** This allows both traces to be seen clearly.)

72. (a) Sketch the output function v_{out} of the amplifier circuit in Fig. 7.29 over the range of $0 \leq t \leq 100$ ms if v_s is a 60 Hz sinusoidal source having a peak voltage of 400 mV, R_1 is 1 kΩ, and L_f is 80 nH. (b) Verify your answer with an appropriate transient simulation, plotting both v_s and v_{out} in Probe. (*Hint:* Between plotting traces, add a second y axis using **Plot, Add Y Axis.** This allows both traces to be seen clearly.)

73. For the circuit of Fig. 7.71, (a) sketch v_{out} over the range of $0 \leq t \leq 2.5$ ms if $R_f = 100$ kΩ, $L_1 = 100$ mH, and v_s is a 2 kHz sinusoidal source having a peak voltage of 5 V. (b) Verify your answer with an appropriate transient simulation, plotting both v_s and v_{out} in Probe. (*Hint:* Between plotting traces, add a second y axis using **Plot, Add Y Axis.** This allows both traces to be seen clearly.)

■ **FIGURE 7.83**

74. Consider the modified integrator depicted in Fig. 7.72. Take $R_1 = 100$ Ω, $R_f = 10$ MΩ, and $C_1 = 10$ mF. The source v_s provides a 10 Hz sinusoidal voltage having a peak amplitude of 0.5 V. (a) Sketch v_{out} over the range of $0 \leq t \leq 500$ ms. (b) Verify your answer with an appropriate transient simulation, plotting both v_s and v_{out} in Probe. (*Hint:* Between plotting traces, add a second y axis using **Plot, Add Y Axis.** This allows both traces to be seen clearly.)

Basic *RL* and *RC* Circuits

INTRODUCTION

In Chap. 7 we wrote equations for the response of several circuits containing both inductance and capacitance, but we did not solve any of them. Now we are ready to proceed with the solution of the simpler circuits, namely, those which contain only resistors and inductors, or only resistors and capacitors.

Although the circuits we are about to consider have a very elementary appearance, they are also of practical importance. Networks of this form find use in electronic amplifiers, automatic control systems, operational amplifiers, communications equipment, and many other applications. Familiarity with these simple circuits will enable us to predict the accuracy with which the output of an amplifier can follow an input that is changing rapidly with time, or to predict how quickly the speed of a motor will change in response to a change in its field current. Our understanding of simple *RL* and *RC* circuits will also enable us to suggest modifications to the amplifier or motor in order to obtain a more desirable response.

8.1 THE SOURCE-FREE *RL* CIRCUIT

The analysis of circuits containing inductors and/or capacitors is dependent upon the formulation and solution of the integrodifferential equations that characterize the circuits. We will call the special type of equation we obtain a *homogeneous linear differential equation*, which is simply a differential equation in which every term is of the first degree in the dependent variable or one of its derivatives. A solution is obtained when we have found an expression for the dependent variable that satisfies both the differential equation and also the prescribed energy distribution in the inductors or capacitors at a prescribed instant of time, usually $t = 0$.

The solution of the differential equation represents a response of the circuit, and it is known by many names. Since this response depends upon the general "nature" of the circuit (the types of elements, their sizes, the interconnection of the elements), it is often called a **natural response.** However, any real circuit we construct cannot store energy forever; the resistances intrinsically associated with inductors and capacitors will eventually convert all stored energy into heat. The response must eventually die out, and for this reason it is frequently referred to as the **transient response.** Finally, we should also be familiar with the mathematicians' contribution to the nomenclature; they call the solution of a homogeneous linear differential equation a *complementary function.*

When we consider independent sources acting on a circuit, part of the response will resemble the nature of the particular source (or *forcing function*) used; this part of the response, called the *particular solution,* the *steady-state response,* or the **forced response,** will be "complemented" by the complementary response produced in the source-free circuit. The complete response of the circuit will then be given by the sum of the complementary function and the particular solution. In other words, the complete response is the sum of the natural response and the forced response. The source-free response may be called the *natural* response, the *transient* response, the *free* response, or the *complementary function,* but because of its more descriptive nature, we will most often call it the *natural* response.

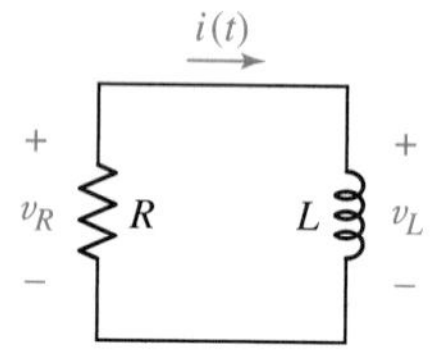

We will consider several different methods of solving these differential equations. The mathematical manipulation, however, is not circuit analysis. Our greatest interest lies in the solutions themselves, their meaning, and their interpretation, and we will try to become sufficiently familiar with the form of the response that we are able to write down answers for new circuits by just plain thinking. Although complicated analytical methods are needed when simpler methods fail, a well-developed intuition is an invaluable resource in such situations.

We begin our study of transient analysis by considering the simple series *RL* circuit shown in Fig. 8.1. Let us designate the time-varying current as $i(t)$; we will represent the value of $i(t)$ at $t = 0$ as I_0; in other words, $i(0) = I_0$. We therefore have

$$Ri + v_L = Ri + L\frac{di}{dt} = 0$$

or

$$\frac{di}{dt} + \frac{R}{L}i = 0 \qquad [1]$$

Our goal is an expression for $i(t)$ which satisfies this equation *and* also has the value I_0 at $t = 0$. The solution may be obtained by several different methods.

A Direct Approach

One very direct method of solving a differential equation consists of writing the equation in such a way that the variables are separated, and then

FIGURE 8.1 A series *RL* circuit for which $i(t)$ is to be determined, subject to the initial condition that $i(0) = I_0$.

It may seem pretty strange to discuss a time-varying current flowing in a circuit with no sources! Keep in mind that we only know the current at the time specified as $t = 0$; we don't know the current prior to that time. In the same vein, we don't know what the circuit looked like prior to $t = 0$, either. In order for a current to be flowing, a source had to have been present at some point, but we are not privy to this information. Fortunately, it is not required in order to analyze the circuit we are given.

integrating each side of the equation. The variables in Eq. [1] are i and t, and it is apparent that the equation may be multiplied by dt, divided by i, and arranged with the variables separated:

$$\frac{di}{i} = -\frac{R}{L}\,dt \qquad [2]$$

Since the current is I_0 at $t = 0$ and $i(t)$ at time t, we may equate the two definite integrals which are obtained by integrating each side between the corresponding limits:

$$\int_{I_0}^{i(t)} \frac{di'}{i'} = \int_0^t -\frac{R}{L}\,dt'$$

Performing the indicated integration,

$$\ln i'\Big|_{I_0}^{i} = -\frac{R}{L}t'\Big|_0^t$$

which results in

$$\ln i - \ln I_0 = -\frac{R}{L}(t - 0)$$

After a little manipulation, we find that the current $i(t)$ is given by

$$i(t) = I_0 e^{-Rt/L} \qquad [3]$$

We check our solution by first showing that substitution of Eq. [3] in Eq. [1] yields the identity $0 = 0$, and then showing that substitution of $t = 0$ in Eq. [3] produces $i(0) = I_0$. Both steps are necessary; the solution must satisfy the differential equation which characterizes the circuit, and it must also satisfy the initial condition.

EXAMPLE 8.1

If the inductor of Fig. 8.2 has a current $i_L = 2$ A at $t = 0$, find an expression for $i_L(t)$ valid for $t > 0$, and its value at $t = 200\ \mu s$.

This is the identical type of circuit just considered, so we expect an inductor current of the form

$$i_L = I_0 e^{-Rt/L}$$

where $R = 200\ \Omega$, $L = 50$ mH and I_0 is the initial current flowing through the inductor at $t = 0$. Thus,

$$i_L(t) = 2e^{-4000t}$$

Substituting $t = 200 \times 10^{-6}$ s, we find that $i_L(t) = 898.7$ mA, less than half the initial value.

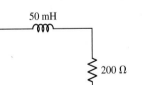

■ **FIGURE 8.2** A simple *RL* circuit in which energy is stored in the inductor at $t = 0$.

PRACTICE

8.1 Determine the current i_R through the resistor of Fig. 8.3 at $t = 1$ ns if $i_R(0) = 6$ A.

Ans: 812 mA.

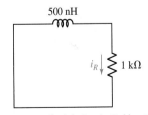

■ **FIGURE 8.3** Circuit for Practice Problem 8.1.

An Alternative Approach

The solution may also be obtained by a slight variation of the method we just described. After separating the variables, we now also include a constant of integration. Thus,

$$\int \frac{di}{i} = -\int \frac{R}{L}\,dt + K$$

and integration gives us

$$\ln i = -\frac{R}{L}t + K \qquad [4]$$

The constant K cannot be evaluated by substitution of Eq. [4] in the original differential equation [1]; the identity $0 = 0$ will result, because Eq. [4] is a solution of Eq. [1] for *any* value of K (try it out on your own). The constant of integration must be selected to satisfy the initial condition $i(0) = I_0$. Thus, at $t = 0$, Eq. [4] becomes

$$\ln I_0 = K$$

and we use this value for K in Eq. [4] to obtain the desired response

$$\ln i = -\frac{R}{L}t + \ln I_0$$

or

$$i(t) = I_0 e^{-Rt/L}$$

as before.

A More General Solution Approach

Either of these methods can be used when the variables are separable, but this is not always the situation. In the remaining cases we will rely on a very powerful method, the success of which will depend upon our intuition or experience. We simply guess or assume a form for the solution and then test our assumptions, first by substitution in the differential equation, and then by applying the given initial conditions. Since we cannot be expected to guess the exact numerical expression for the solution, we will assume a solution containing several unknown constants and select the values for these constants in order to satisfy the differential equation and the initial conditions. Many of the differential equations encountered in circuit analysis have a solution which may be represented by the exponential function or by the sum of several exponential functions. Let us assume a solution of Eq. [1] in exponential form,

$$i(t) = Ae^{s_1 t} \qquad [5]$$

where A and s_1 are constants to be determined. After substituting this assumed solution in Eq. [1], we have

$$As_1 e^{s_1 t} + A\frac{R}{L}e^{s_1 t} = 0$$

or

$$\left(s_1 + \frac{R}{L}\right) A e^{s_1 t} = 0 \qquad [6]$$

In order to satisfy this equation for all values of time, it is necessary that $A = 0$, or $s_1 = -\infty$, or $s_1 = -R/L$. But if $A = 0$ or $s_1 = -\infty$, then every response is zero; neither can be a solution to our problem. Therefore, we must choose

$$s_1 = -\frac{R}{L} \qquad [7]$$

and our assumed solution takes on the form

$$i(t) = A e^{-Rt/L}$$

The remaining constant must be evaluated by applying the initial condition $i(0) = I_0$. Thus, $A = I_0$, and the final form of the assumed solution is (again)

$$i(t) = I_0 e^{-Rt/L}$$

A summary of the basic approach is outlined in Fig. 8.4.

A Direct Route: The Characteristic Equation

In fact, there is a more direct route that we can take. In obtaining Eq. [7], we solved

$$s_1 + \frac{R}{L} = 0 \qquad [8]$$

which is known as the **characteristic equation**. We can obtain the characteristic equation directly from the differential equation, without the need for substitution of our trial solution. Consider the general first-order differential equation

$$a \frac{df}{dt} + bf = 0$$

where a and b are constants. We substitute s^1 for df/dt and s^0 for f, resulting in

$$a \frac{df}{dt} + bf = (as + b)f = 0$$

From this we may directly obtain the characteristic equation

$$as + b = 0$$

which has the single root $s = -b/a$. The solution to our differential equation is then

$$f = A e^{-bt/a}$$

This basic procedure is easily extended to second-order differential equations, as we will explore in Chap. 9.

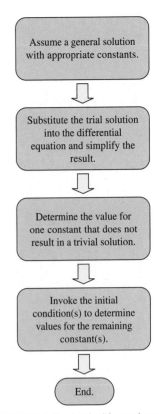

Assume a general solution with appropriate constants.

Substitute the trial solution into the differential equation and simplify the result.

Determine the value for one constant that does not result in a trivial solution.

Invoke the initial condition(s) to determine values for the remaining constant(s).

End.

■ **FIGURE 8.4** Flowchart for the general approach to solution of first-order differential equations where, based on experience, we can guess the form of the solution.

EXAMPLE **8.2**

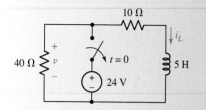

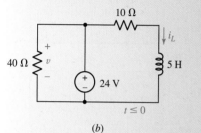

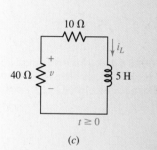

■ FIGURE 8.5 (*a*) A simple *RL* circuit with a switch thrown at time *t* = 0. (*b*) The circuit as it exists prior to *t* = 0. (*c*) The circuit after the switch is thrown, and the 24 V source is removed.

For the circuit of Fig. 8.5a, find the voltage labeled v at t = 200 ms.

▶ **Identify the goal of the problem.**
The schematic of Fig. 8.5a actually represents *two different* circuits: one with the switch closed (Fig. 8.5b) and one with the switch open (Fig. 8.5c). We are asked to find v(0.2) for the circuit shown in Fig. 8.5c.

▶ **Collect the known information.**
Both new circuits are drawn and labeled correctly. We next make the assumption that the circuit in Fig. 8.5b has been connected for a long time, so that any transients have dissipated. We may make such an assumption as a general rule unless instructed otherwise. This circuit determines $i_L(0)$.

▶ **Devise a plan.**
The circuit of Fig. 8.5c may be analyzed by writing a KVL equation. Ultimately we want a differential equation with only v and t as variables; we will then solve the differential equation for v(t).

▶ **Construct an appropriate set of equations.**
Referring to Fig. 8.5c, we write

$$-v + 10i_L + 5\frac{di_L}{dt} = 0$$

Substituting $i_L = -v/40$, we find that

$$\frac{5}{40}\frac{dv}{dt} + \left(\frac{10}{40} + 1\right)v = 0$$

or, more simply,

$$\frac{dv}{dt} + 10v = 0 \qquad [9]$$

▶ **Determine if additional information is required.**
From previous experience, we know that a complete expression for v will require knowledge of v at a specific instant of time, with t = 0 being the most convenient. We might be tempted to look at Fig. 8.5b and write v(0) = 24 V, but this is only true *just before the switch opens*. The resistor voltage can change to any value in the instant that the switch is thrown; only the inductor current must remain unchanged.
In the circuit of Fig. 8.5b, i_L = 24/10 = 2.4 A since the inductor acts like a short circuit to a dc current. Therefore, $i_L(0)$ = 2.4 A in the circuit of Fig. 8.5c, as well—a key point in analyzing this type of circuit. Therefore, in the circuit of Fig. 8.5c, v(0) = (40)(−2.4) = −96 V.

▶ **Attempt a solution.**
Any of the three basic solution techniques can be brought to bear; let's start by writing the characteristic equation corresponding to Eq. [9]:

$$s + 10 = 0$$

Solving, we find that $s = -10$, so

$$v(t) = Ae^{-10t} \qquad [10]$$

(which, upon substitution into the left-hand side of Eq. [9], results in

$$-10Ae^{-10t} + 10Ae^{-10t} = 0$$

as expected.)

We find A by setting $t = 0$ in Eq. [10] and employing the fact that $v(0) = -96$ V. Thus,

$$v(t) = -96e^{-10t} \qquad [11]$$

and so $v(0.2) = -12.99$ V, down from a maximum of -96 V.

▶ *Verify the solution. Is it reasonable or expected?*

Instead of writing a differential equation in v, we could have written our differential equation in terms of i_L:

$$40i_L + 10i_L + 5\frac{di_L}{dt} = 0$$

or

$$\frac{di_L}{dt} + 10i_L = 0$$

which has the solution $i_L = Be^{-10t}$. With $i_L(0) = 2.4$, we find that $i_L(t) = 2.4e^{-10t}$. Since $v = -40i_L$, we once again obtain Eq. [11]. We should note: it is **no coincidence** that the inductor current and the resistor voltage have the same exponential dependence!

PRACTICE

8.2 Determine the inductor voltage v in the circuit of Fig. 8.6 for $t > 0$.

Ans: $-25e^{-2t}$ V.

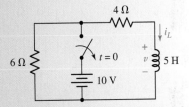

■ **FIGURE 8.6** Circuit for Practice Problem 8.2.

Accounting for the Energy

Before we turn our attention to the interpretation of the response, let us return to the circuit of Fig. 8.1, and check the power and energy relationships. The power being dissipated in the resistor is

$$p_R = i^2 R = I_0^2 R e^{-2Rt/L}$$

and the total energy turned into heat in the resistor is found by integrating the instantaneous power from zero time to infinite time:

$$w_R = \int_0^\infty p_R \, dt = I_0^2 R \int_0^\infty e^{-2Rt/L} \, dt$$

$$= I_0^2 R \left(\frac{-L}{2R}\right) e^{-2Rt/L} \Big|_0^\infty = \frac{1}{2} L I_0^2$$

This is the result we expect, because the total energy stored initially in the inductor is $\frac{1}{2}L I_0^2$, and there is no longer any energy stored in the inductor at infinite time since its current eventually drops to zero. All the initial energy therefore is accounted for by dissipation in the resistor.

8.2 • PROPERTIES OF THE EXPONENTIAL RESPONSE

Let us now consider the nature of the response in the series *RL* circuit. We have found that the inductor current is represented by

$$i(t) = I_0 e^{-Rt/L}$$

At $t = 0$, the current has value I_0, but as time increases, the current decreases and approaches zero. The shape of this decaying exponential is seen by the plot of $i(t)/I_0$ versus t shown in Fig. 8.7. Since the function we are plotting is $e^{-Rt/L}$, the curve will not change if R/L remains unchanged. Thus, the same curve must be obtained for every series *RL* circuit having the same L/R ratio. Let us see how this ratio affects the shape of the curve.

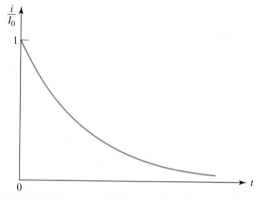

■ **FIGURE 8.7** A plot of $e^{-Rt/L}$ versus t.

If we double the ratio of L to R, the exponent will be unchanged only if t is also doubled. In other words, the original response will occur at a later time, and the new curve is obtained by moving each point on the original curve twice as far to the right. With this larger L/R ratio, the current takes longer to decay to any given fraction of its original value. We might have a tendency to say that the "width" of the curve is doubled, or that the width is proportional to L/R. However, we find it difficult to define our term *width*, because each curve extends from $t = 0$ to ∞! Instead, let us consider the time that would be required for the current to drop to zero *if it continued to drop at its initial rate*.

The initial rate of decay is found by evaluating the derivative at zero time:

$$\frac{d}{dt}\frac{i}{I_0}\bigg|_{t=0} = -\frac{R}{L}e^{-Rt/L}\bigg|_{t=0} = -\frac{R}{L}$$

We designate the value of time it takes for i/I_0 to drop from unity to zero, assuming a constant rate of decay, by the Greek letter τ (tau). Thus,

$$\left(\frac{R}{L}\right)\tau = 1$$

or

$$\boxed{\tau = \frac{L}{R}}$$

[12]

The ratio L/R has the units of seconds, since the exponent $-Rt/L$ must be dimensionless. This value of time τ is called the **time constant** and is shown pictorially in Fig. 8.8. The time constant of a series RL circuit may be found graphically from the response curve; it is necessary only to draw the tangent to the curve at $t = 0$ and determine the intercept of this tangent line with the time axis. This is often a convenient way of approximating the time constant from the display on an oscilloscope.

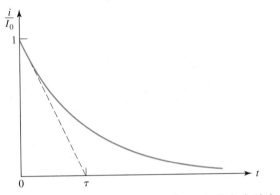

■ **FIGURE 8.8** The time constant τ is L/R for a series RL circuit. It is the time required for the response curve to drop to zero if it decays at a constant rate equal to its initial rate of decay.

An equally important interpretation of the time constant τ is obtained by determining the value of $i(t)/I_0$ at $t = \tau$. We have

$$\frac{i(\tau)}{I_0} = e^{-1} = 0.3679 \qquad \text{or} \qquad i(\tau) = 0.3679 I_0$$

Thus, in one time constant the response has dropped to 36.8 percent of its initial value; the value of τ may also be determined graphically from this fact, as indicated by Fig. 8.9. It is convenient to measure the decay of the current at intervals of one time constant, and recourse to a hand calculator shows that $i(t)/I_0$ is 0.3679 at $t = \tau$, 0.1353 at $t = 2\tau$, 0.04979 at $t = 3\tau$, 0.01832 at $t = 4\tau$, and 0.006738 at $t = 5\tau$. At some point three to five time constants after zero time, most of us would agree that the current is a

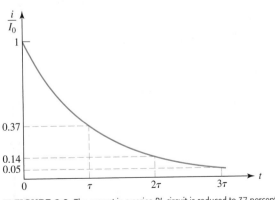

■ **FIGURE 8.9** The current in a series RL circuit is reduced to 37 percent of its initial value at $t = \tau$, 14 percent at $t = 2\tau$, and 5 percent at $t = 3\tau$.

negligible fraction of its former self. Thus, if we are asked, "*How long does it take for the current to decay to zero?*" our answer might be, "*About five time constants.*" At that point, the current is less than 1 percent of its original value!

PRACTICE

8.3 In a source-free series *RL* circuit, find the numerical value of the ratio: (*a*) $i(2\tau)/i(\tau)$; (*b*) $i(0.5\tau)/i(0)$; (*c*) t/τ if $i(t)/i(0) = 0.2$; (*d*) t/τ if $i(0) - i(t) = i(0) \ln 2$.

Ans: 0.368; 0.607; 1.609; 1.181.

COMPUTER-AIDED ANALYSIS

The transient analysis capability of PSpice is very useful when considering the response of source-free circuits. In this example, we make use of a special feature that allows us to vary a component parameter, similar to the way we varied the dc voltage in other simulations. We do this by adding the component **PARAM** to our schematic; it may be placed anywhere, as we will not wire it into the circuit. Our complete *RL* circuit is shown in Fig. 8.10, which includes an initial inductor current of 1 mA.

In order to relate our resistor value to the proposed parameter sweep, we must perform three tasks. First, we provide a name for our parameter, which we choose to call Resistance for the sake of simplicity. This is accomplished by double-clicking on the **PARAMETERS:** label in the schematic, which opens the Property Editor for this pseudo-component. Clicking on **New Column** results in the dialog box shown in Fig. 8.11*a*, in which we enter *Resistance* under **Name** and a placeholder value of 1 under **Value.** Our second task consists of linking the

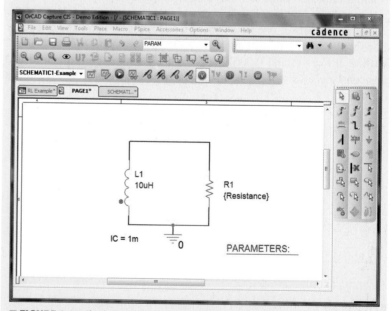

■ **FIGURE 8.10** Simple *RL* circuit drawn using the schematic capture tool.

value of R1 to our parameter sweep, which we accomplish by double-clicking on the default value of R1 on the schematic, resulting in the dialog box of Fig. 8.11*b*. Under **Value,** we simply enter *{Resistance}.* (Note the curly brackets are required.)

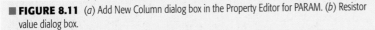

(*a*)   (*b*)

■ **FIGURE 8.11** (*a*) Add New Column dialog box in the Property Editor for PARAM. (*b*) Resistor value dialog box.

Our third task consists of setting up the simulation, which includes setting transient analysis parameters as well as the values we desire for R1. Under **PSpice** we select **New Simulation Profile** (Fig. 8.12*a*), in which we select **Time Domain (Transient)** for **Analysis type,** 300 ns for **Run to time,** and tick the **Parametric Sweep box** under **Options.** This last action results in the dialog box shown in Fig. 8.12*b*, in which we select **Global parameter** for **Sweep variable** and enter *Resistance* for **Parameter name.** The final setup step required is to select **Logarithmic** under **Sweep type,** a **Start value** of 10, an **End value** of 1000, and 1 **Points/Decade;** alternatively we could list the desired resistor values using **Value list.**

After running the simulation, the notification box shown in Fig. 8.13 appears, listing the available data sets for plotting (**Resistance** = 10, 100, and 1000 in this case). A particular data set is selected by highlighting it; we select all three for this example. Upon selecting the

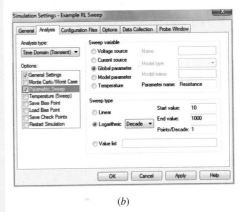

(*a*)

(*b*)

■ **FIGURE 8.12** (*a*) Simulation dialog box. (*b*) Parameter sweep dialog box.

■ **FIGURE 8.13** Available data sections dialog box.

(Continued on next page)

inductor current from our **Trace** variable choices in Probe, we obtain three graphs at once, as shown (after labeling by hand) in Fig. 8.14.

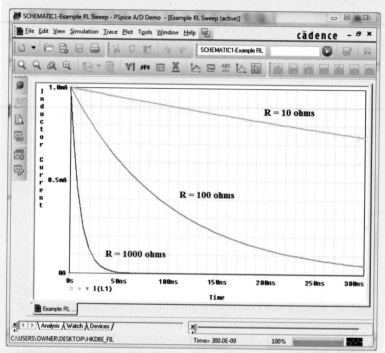

■ **FIGURE 8.14** Probe output for the three resistances.

Why does a larger value of the time constant L/R produce a response curve that decays more slowly? Let us consider the effect of each element.

In terms of the time constant τ, the response of the series RL circuit may be written simply as

$$i(t) = I_0 e^{-t/\tau}$$

An increase in L allows a greater energy storage for the same initial current, and this larger energy requires a longer time to be dissipated in the resistor. We may also increase L/R by reducing R. In this case, the power flowing into the resistor is less for the same initial current; again, a greater time is required to dissipate the stored energy. This effect is seen clearly in our simulation result of Fig. 8.14.

8.3 • THE SOURCE-FREE *RC* CIRCUIT

Circuits based on resistor-capacitor combinations are more common than their resistor-inductor analogs. The principal reasons for this are the smaller losses present in a physical capacitor, lower cost, better agreement between the simple mathematical model and the actual device behavior, and also smaller size and lighter weight, both of which are particularly important for integrated-circuit applications.

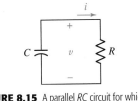

Let us see how closely the analysis of the parallel (or is it series?) *RC* circuit shown in Fig. 8.15 corresponds to that of the *RL* circuit. We will assume an initial stored energy in the capacitor by selecting

$$v(0) = V_0$$

The total current leaving the node at the top of the circuit diagram must be zero, so we may write

$$C\frac{dv}{dt} + \frac{v}{R} = 0$$

Division by *C* gives us

$$\frac{dv}{dt} + \frac{v}{RC} = 0 \qquad [13]$$

Equation [13] has a familiar form; comparison with Eq. [1]

$$\frac{di}{dt} + \frac{R}{L}i = 0 \qquad [1]$$

shows that the replacement of *i* by *v* and *L/R* by *RC* produces the identical equation we considered previously. It should, for the *RC* circuit we are now analyzing is the dual of the *RL* circuit we considered first. This *duality* forces *v(t)* for the *RC* circuit and *i(t)* for the *RL* circuit to have identical expressions if the resistance of one circuit is equal to the reciprocal of the resistance of the other circuit, and if *L* is numerically equal to *C*. Thus, the response of the *RL* circuit

$$i(t) = i(0)e^{-Rt/L} = I_0 e^{-Rt/L}$$

enables us to immediately write

$$v(t) = v(0)e^{-t/RC} = V_0 e^{-t/RC} \qquad [14]$$

for the *RC* circuit.

Suppose instead that we had selected the current *i* as our variable in the *RC* circuit, rather than the voltage *v*. Applying Kirchhoff's voltage law,

$$\frac{1}{C}\int_{t_0}^{t} i\, dt' - v_0(t_0) + Ri = 0$$

we obtain an integral equation as opposed to a differential equation. However, taking the time derivative of both sides of this equation,

$$\frac{i}{C} + R\frac{di}{dt} = 0 \qquad [15]$$

and replacing *i* with *v/R*, we obtain Eq. [13] again:

$$\frac{v}{RC} + \frac{dv}{dt} = 0$$

Equation [15] could have been used as our starting point, but the application of duality principles would not have been as natural.

Let us discuss the physical nature of the voltage response of the *RC* circuit as expressed by Eq. [14]. At *t* = 0 we obtain the correct initial condition, and as *t* becomes infinite, the voltage approaches zero. This latter result agrees with our thinking that if there were any voltage remaining across the capacitor, then energy would continue to flow into the resistor and be dissipated as heat. *Thus, a final voltage of zero is necessary.* The time constant of the *RC*

FIGURE 8.15 A parallel *RC* circuit for which *v(t)* is to be determined, subject to the initial condition that $v(0) = V_0$.

circuit may be found by using the duality relationships on the expression for the time constant of the *RL* circuit, or it may be found by simply noting the time at which the response has dropped to 37 percent of its initial value:

$$\frac{\tau}{RC} = 1$$

so that

$$\boxed{\tau = RC} \tag{16}$$

Our familiarity with the negative exponential and the significance of the time constant τ enables us to sketch the response curve readily (Fig. 8.16). Larger values of R or C provide larger time constants and slower dissipation of the stored energy. A larger resistance will dissipate a smaller power with a given voltage across it, thus requiring a greater time to convert the stored energy into heat; a larger capacitance stores a larger energy with a given voltage across it, again requiring a greater time to lose this initial energy.

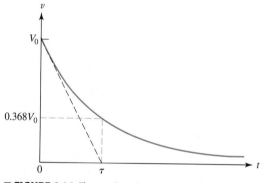

■ **FIGURE 8.16** The capacitor voltage $v(t)$ in the parallel *RC* circuit is plotted as a function of time. The initial value of $v(t)$ is V_0.

EXAMPLE 8.3

For the circuit of Fig. 8.17a, find the voltage labeled v at $t = 200$ μs.

To find the requested voltage, we will need to draw and analyze two separate circuits: one corresponding to before the switch is thrown (Fig. 8.17b), and one corresponding to after the switch is thrown (Fig. 8.17c).

The sole purpose of analyzing the circuit of Fig. 8.17b is to obtain an initial capacitor voltage; we assume any transients in that circuit died out long ago, leaving a purely dc circuit. With no current through either the capacitor or the 4 Ω resistor, then,

$$v(0) = 9 \text{ V} \tag{17}$$

We next turn our attention to the circuit of Fig. 8.17c, recognizing that

$$\tau = RC = (2+4)(10 \times 10^{-6}) = 60 \times 10^{-6} \text{ s}$$

Thus, from Eq. [14],

$$v(t) = v(0)e^{-t/RC} = v(0)e^{-t/60 \times 10^{-6}} \tag{18}$$

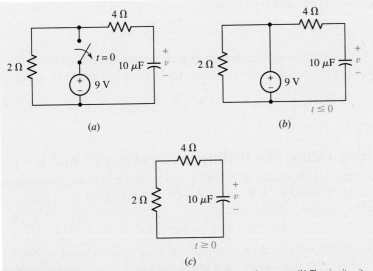

(a)

(b)

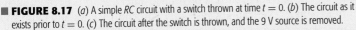

(c)

■ **FIGURE 8.17** (a) A simple RC circuit with a switch thrown at time $t = 0$. (b) The circuit as it exists prior to $t = 0$. (c) The circuit after the switch is thrown, and the 9 V source is removed.

The capacitor voltage must be the same in both circuits at $t = 0$; no such restriction is placed on any other voltage or current. Substituting Eq. [17] into Eq. [18],

$$v(t) = 9e^{-t/60 \times 10^{-6}} \text{ V}$$

so that $v(200 \times 10^{-6}) = 321.1$ mV (less than 4 percent of its maximum value).

PRACTICE

8.4 Noting carefully how the circuit changes once the switch in the circuit of Fig. 8.18 is thrown, determine $v(t)$ at $t = 0$ and at $t = 160 \ \mu$s.

Ans: 50 V, 18.39 V.

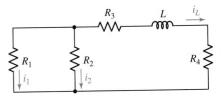

■ **FIGURE 8.18**

8.4 A MORE GENERAL PERSPECTIVE

As seen indirectly from Examples 8.2 and 8.3, regardless of how many resistors we have in the circuit, we obtain a single time constant (either $\tau = L/R$ or $\tau = RC$) when only one energy storage element is present. We can formalize this by realizing that the value needed for R is in fact the Thévenin equivalent resistance seen by our energy storage element. (*Strange as it may seem, it is even possible to compute a time constant for a circuit containing dependent sources!*)

General RL Circuits

As an example, consider the circuit shown in Fig. 8.19. The equivalent resistance the inductor faces is

$$R_{eq} = R_3 + R_4 + \frac{R_1 R_2}{R_1 + R_2}$$

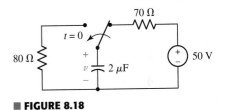

■ **FIGURE 8.19** A source-free circuit containing one inductor and several resistors is analyzed by determining the time constant $\tau = L/R_{eq}$.

and the time constant is therefore

$$\tau = \frac{L}{R_{\text{eq}}} \qquad [19]$$

If several inductors are present in a circuit and can be combined using series and/or parallel combination, then Eq. [19] can be further generalized to

$$\tau = \frac{L_{\text{eq}}}{R_{\text{eq}}} \qquad [20]$$

where L_{eq} represents the equivalent inductance.

> We could also write
> $$\tau = \frac{L}{R_{TH}},$$
> where R_{TH} is the Thévenin equivalent resistance "seen" by the inductor L.

Slicing Thinly: The Distinction Between 0^+ and 0^-

Let's return to the circuit of Fig. 8.19, and assume that some finite amount of energy is stored in the inductor at $t = 0$, so that $i_L(0) \neq 0$.

The inductor current i_L is

$$i_L = i_L(0)e^{-t/\tau}$$

and this represents what we might call the basic solution to the problem. It is quite possible that some current or voltage other than i_L is needed, such as the current i_2 in R_2. We can always apply Kirchhoff's laws and Ohm's law to the resistive portion of the circuit without any difficulty, but current division provides the quickest answer in this circuit:

$$i_2 = -\frac{R_1}{R_1 + R_2}[i_L(0)e^{-t/\tau}]$$

It may also happen that we know the initial value of some current *other* than the inductor current. Since *the current in a resistor may change instantaneously,* we will indicate the instant after any change that might have occurred at $t = 0$ by the use of the symbol 0^+; in more mathematical language, $i_1(0^+)$ is the limit from the right of $i_1(t)$ as t approaches zero.[1] Thus, if we are given the initial value of i_1 as $i_1(0^+)$, then the initial value of i_2 is

$$i_2(0^+) = i_1(0^+)\frac{R_1}{R_2}$$

> Note that $i_L(0^+)$ is *always* equal to $i_L(0^-)$. This is not necessarily true for the inductor voltage or any resistor voltage or current, since they may change in zero time.

From these values, we obtain the necessary initial value of i_L:

$$i_L(0^+) = -[i_1(0^+) + i_2(0^+)] = -\frac{R_1 + R_2}{R_2}i_1(0^+)$$

and the expression for i_2 becomes

$$i_2 = i_1(0^+)\frac{R_1}{R_2}e^{-t/\tau}$$

Let us see if we can obtain this last expression more directly. Since the inductor current decays exponentially as $e^{-t/\tau}$, *every current throughout the circuit must follow the same functional behavior.* This is made clear by considering the inductor current as a source current that is being applied to a resistive network. Every current and voltage in the resistive network must have the same time dependence. Using these ideas, we therefore express i_2 as

$$i_2 = Ae^{-t/\tau}$$

where

$$\tau = \frac{L}{R_{\text{eq}}}$$

---

(1) Note that this is a notational convenience *only*. When faced with $t = 0^+$ or its companion $t = 0^-$ in an equation, we simply use the value zero. This notation allows us to clearly differentiate between the time before and after an event, such as a switch opening or closing, or a power supply being turned on or off.

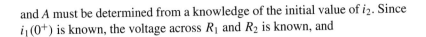

and A must be determined from a knowledge of the initial value of i_2. Since $i_1(0^+)$ is known, the voltage across R_1 and R_2 is known, and

$$R_2 i_2(0^+) = R_1 i_1(0^+)$$

leads to

$$i_2(0^+) = i_1(0^+) \frac{R_1}{R_2}$$

Therefore,

$$i_2(t) = i_1(0^+) \frac{R_1}{R_2} e^{-t/\tau}$$

A similar sequence of steps will provide a rapid solution to a large number of problems. We first recognize the time dependence of the response as an exponential decay, determine the appropriate time constant by combining resistances, write the solution with an unknown amplitude, and then determine the amplitude from a given initial condition.

This same technique can be applied to any circuit with one inductor and any number of resistors, as well as to those special circuits containing two or more inductors and also two or more resistors that may be simplified by resistance or inductance combination to one inductor and one resistor.

EXAMPLE **8.4**

Determine both i_1 and i_L in the circuit shown in Fig. 8.20a for $t > 0$.

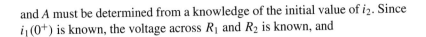

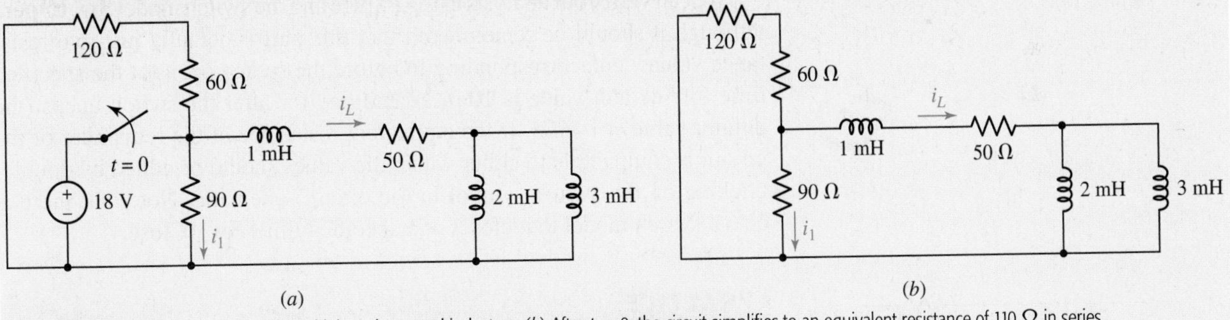

(a)   (b)

■ **FIGURE 8.20** (a) A circuit with multiple resistors and inductors. (b) After $t = 0$, the circuit simplifies to an equivalent resistance of 110 Ω in series with $L_{eq} = 2.2$ mH.

After $t = 0$, when the voltage source is disconnected as shown in Fig. 8.20b, we easily calculate an equivalent inductance,

$$L_{eq} = \frac{2 \times 3}{2 + 3} + 1 = 2.2 \text{ mH}$$

an equivalent resistance, in series with the equivalent inductance,

$$R_{eq} = \frac{90(60 + 120)}{90 + 180} + 50 = 110 \text{ }\Omega$$

(Continued on next page)

and the time constant,

$$\tau = \frac{L_{\text{eq}}}{R_{\text{eq}}} = \frac{2.2 \times 10^{-3}}{110} = 20 \ \mu\text{s}$$

Thus, the form of the natural response is $Ke^{-50,000t}$, where K is an unknown constant. Considering the circuit just prior to the switch opening ($t = 0^-$), $i_L = 18/50$ A. Since $i_L(0^+) = i_L(0^-)$, we know that $i_L = 18/50$ A or 360 mA at $t = 0^+$ and so

$$i_L = \begin{cases} 360 \text{ mA} & t < 0 \\ 360e^{-50,000t} \text{ mA} & t \geq 0 \end{cases}$$

There is no restriction on i_1 changing instantaneously at $t = 0$, so its value at $t = 0^-$ (18/90 A or 200 mA) is not relevant to finding i_1 for $t > 0$. Instead, we must find $i_1(0^+)$ through our knowledge of $i_L(0^+)$. Using current division,

$$i_1(0^+) = -i_L(0^+)\frac{120 + 60}{120 + 60 + 90} = -240 \text{ mA}$$

Hence,

$$i_1 = \begin{cases} 200 \text{ mA} & t < 0 \\ -240e^{-50,000t} \text{ mA} & t \geq 0 \end{cases}$$

We can verify our analysis using PSpice and the switch model **Sw_tOpen,** although it should be remembered that this part is actually just two resistance values: one corresponding to before the switch opens at the specified time (the default value is 10 mΩ), and one for after the switch opens (the default value is 1 MΩ). If the equivalent resistance of the remainder of the circuit is comparable to either value, the values should be edited by double-clicking on the switch symbol in the circuit schematic. Note that there is also a switch model that closes at a specified time: **Sw_tClose.**

PRACTICE

8.5 At $t = 0.15$ s in the circuit of Fig. 8.21, find the value of (a) i_L; (b) i_1; (c) i_2.

Ans: 0.756 A; 0; 1.244 A.

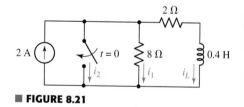

■ **FIGURE 8.21**

We have now considered the task of finding the natural response of any circuit which can be represented by an equivalent inductor in series with an equivalent resistor. *A circuit containing several resistors and several inductors does not always possess a form which allows either the resistors or the inductors to be combined into single equivalent elements.* In such instances, there is no single negative exponential term or single time constant associated with the circuit. Rather, there will, in general, be several negative exponential terms, the number of terms being equal to the number of inductors that remain after all possible inductor combinations have been made.

General *RC* Circuits

Many of the *RC* circuits for which we would like to find the natural response contain more than a single resistor and capacitor. Just as we did for the *RL* circuits, we first consider those cases in which the given circuit may be reduced to an equivalent circuit consisting of only one resistor and one capacitor.

Let us suppose first that we are faced with a circuit containing a single capacitor, but any number of resistors. It is possible to replace the two-terminal resistive network which is across the capacitor terminals with an equivalent resistor, and we may then write down the expression for the capacitor voltage immediately. In such instances, the circuit has an effective time constant given by

$$\tau = R_{\text{eq}}C$$

where R_{eq} is the equivalent resistance of the network. An alternative perspective is that R_{eq} is in fact the Thévenin equivalent resistance "seen" by the capacitor.

If the circuit has more than one capacitor, but they may be replaced somehow using series and/or parallel combinations with an equivalent capacitance C_{eq}, then the circuit has an effective time constant given by

$$\tau = RC_{\text{eq}}$$

with the general case expressed as

$$\tau = R_{\text{eq}}C_{\text{eq}}$$

It is worth noting, however, that parallel capacitors replaced by an equivalent capacitance would have to have identical initial conditions.

EXAMPLE **8.5**

Find $v(0^+)$ and $i_1(0^+)$ for the circuit shown in Fig. 8.22a if $v(0^-) = V_0$.

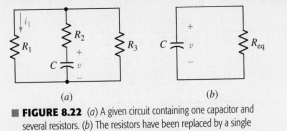

(*a*) (*b*)

■ **FIGURE 8.22** (*a*) A given circuit containing one capacitor and several resistors. (*b*) The resistors have been replaced by a single equivalent resistor; the time constant is simply $\tau = R_{\text{eq}}C$.

We first simplify the circuit of Fig. 8.22*a* to that of Fig. 8.22*b*, enabling us to write

$$v = V_0 e^{-t/R_{\text{eq}}C}$$

(Continued on next page)

where

$$v(0^+) = v(0^-) = V_0 \qquad \text{and} \qquad R_{\text{eq}} = R_2 + \frac{R_1 R_3}{R_1 + R_3}$$

Every current and voltage in the resistive portion of the network must have the form $Ae^{-t/R_{\text{eq}}C}$, where A is the initial value of that current or voltage. Thus, the current in R_1, for example, may be expressed as

$$i_1 = i_1(0^+)e^{-t/\tau}$$

where

$$\tau = \left(R_2 + \frac{R_1 R_3}{R_1 + R_3} \right) C$$

and $i_1(0^+)$ remains to be determined from the initial condition. Any current flowing in the circuit at $t = 0^+$ must come from the capacitor. Therefore, since v cannot change instantaneously, $v(0^+) = v(0^-) = V_0$ and

$$i_1(0^+) = \frac{V_0}{R_2 + R_1 R_3/(R_1 + R_3)} \frac{R_3}{R_1 + R_3}$$

PRACTICE

8.6 Find values of v_C and v_o in the circuit of Fig. 8.23 at t equal to (*a*) 0^-; (*b*) 0^+; (*c*) 1.3 ms.

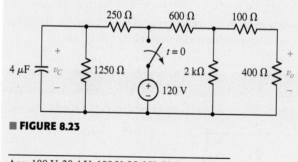

■ **FIGURE 8.23**

Ans: 100 V, 38.4 V; 100 V, 25.6 V; 59.5 V, 15.22 V.

Our method can be applied to circuits with one energy storage element and one or more dependent sources as well. In such instances, we may write an appropriate KCL or KVL equation along with any necessary supporting equations, distill this down into a single differential equation, and extract the characteristic equation to find the time constant. Alternatively, we may begin by finding the Thévenin equivalent resistance of the network connected to the capacitor or inductor, and use this in calculating the appropriate *RL* or *RC* time constant—unless the dependent source is controlled by a voltage or current associated with the energy storage element, in which case the Thévenin approach cannot be used.

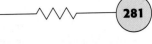

EXAMPLE **8.6**

For the circuit of Fig. 8.24a, find the voltage labeled v_C for $t > 0$ if $v_C(0^-) = 2$ V.

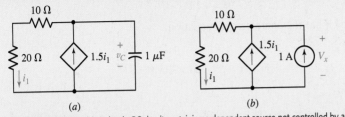

(a)   *(b)*

■ **FIGURE 8.24** *(a)* A simple *RC* circuit containing a dependent source not controlled by a capacitor voltage or current. *(b)* Circuit for finding the Thévenin equivalent of the network connected to the capacitor.

The dependent source is not controlled by a capacitor voltage or current, so we can start by finding the Thévenin equivalent of the network to the left of the capacitor. Connecting a 1 A test source as in Fig. 8.24b,

$$V_x = (1 + 1.5i_1)(30)$$

where

$$i_1 = \left(\frac{1}{20}\right)\frac{20}{10+20}V_x = \frac{V_x}{30}$$

Performing a little algebra, we find that $V_x = -60$ V, so the network has a Thévenin equivalent resistance of -60 Ω (unusual, but not impossible when dealing with a dependent source). Our circuit therefore has a *negative* time constant

$$\tau = -60(1 \times 10^{-6}) = -60\ \mu s$$

The capacitor voltage is therefore

$$v_C(t) = Ae^{t/60\times10^{-6}} \quad \text{V}$$

where $A = v_C(0^+) = v_C(0^-) = 2$ V. Thus,

$$v_C(t) = 2e^{t/60\times10^{-6}} \quad \text{V} \tag{21}$$

which, interestingly enough is unstable: it grows exponentially with time. This cannot continue indefinitely; one or more elements in the circuit will eventually fail.

Alternatively, we could write a simple KCL equation for the top node of Fig. 8.24a

$$v_C = 30\left(1.5i_1 - 10^{-6}\frac{dv_C}{dt}\right) \tag{22}$$

where

$$i_1 = \frac{v_C}{30} \tag{23}$$

(Continued on next page)

Substituting Eq. [23] into Eq. [22] and performing some algebra, we obtain

$$\frac{dv_C}{dt} - \frac{1}{60 \times 10^{-6}} v_C = 0$$

which has the characteristic equation

$$s - \frac{1}{60 \times 10^{-6}} = 0$$

Thus,

$$s = \frac{1}{60 \times 10^{-6}}$$

and so

$$v_C(t) = Ae^{t/60 \times 10^{-6}} \qquad \text{V}$$

as we found before. Substitution of $A = v_C(0^+) = 2$ results in Eq. [21], our expression for the capacitor voltage for $t > 0$.

PRACTICE

8.7 (*a*) Regarding the circuit of Fig. 8.25, determine the voltage $v_C(t)$ for $t > 0$ if $v_C(0^-) = 11$ V. (*b*) Is the circuit "stable"?

Ans: (*a*) $v_C(t) = 11e^{-2 \times 10^3 t/3}$ V, $t > 0$. (*b*) Yes; it decays (exponentially) rather than grows with time.

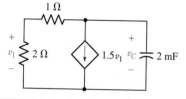

■ **FIGURE 8.25** Circuit for Practice Problem 8.7.

Some circuits containing a number of both resistors and capacitors may be replaced by an equivalent circuit containing only one resistor and one capacitor; it is necessary that the original circuit be one which can be broken into two parts, one containing all resistors and the other containing all capacitors, such that the two parts are connected by only two ideal conductors. Otherwise, multiple time constants and multiple exponential terms will be required to describe the behavior of the circuit (one time constant for each energy storage element remaining in the circuit after it is reduced as much as possible).

As a parting comment, we should be wary of certain situations involving only ideal elements which are suddenly connected together. For example, we may imagine connecting two ideal capacitors in series having unequal voltages prior to $t = 0$. This poses a problem using our mathematical model of an ideal capacitor; however, real capacitors have resistances associated with them through which energy can be dissipated.

8.5 THE UNIT-STEP FUNCTION

We have been studying the response of *RL* and *RC* circuits when no sources or forcing functions were present. We termed this response the *natural response,* because its form depends only on the nature of the circuit. The reason that any response at all is obtained arises from the presence of initial energy storage within the inductive or capacitive elements in the circuit. In some cases we were confronted with circuits containing sources and switches; we were informed that certain switching operations were performed at $t = 0$ in order to remove all the sources from the circuit, while leaving known amounts of energy stored here and there. In other words, we have been solving problems in which energy sources are suddenly *removed* from the circuit; now we must consider that type of response which results when energy sources are suddenly *applied* to a circuit.

We will focus on the response which occurs when the energy sources suddenly applied are dc sources. Since every electrical device is intended to be energized at least once, and since most devices are turned on and off many times in the course of their lifetimes, our study applies to many practical cases. Even though we are now restricting ourselves to dc sources, there are still many cases in which these simpler examples correspond to the operation of physical devices. For example, the first circuit we will analyze could represent the buildup of the current when a dc motor is started. The generation and use of the rectangular voltage pulses needed to represent a number or a command in a microprocessor provide many examples in the field of electronic or transistor circuitry. Similar circuits are found in the synchronization and sweep circuits of television receivers, in communication systems using pulse modulation, and in radar systems, to name but a few examples.

We have been speaking of the "sudden application" of an energy source, and by this phrase we imply its application in zero time.[2] The operation of a switch in series with a battery is thus equivalent to a forcing function which is zero up to the instant that the switch is closed and is equal to the battery voltage thereafter. The forcing function has a break, or discontinuity, at the instant the switch is closed. Certain special forcing functions which are discontinuous or have discontinuous derivatives are called **singularity functions,** the two most important of these singularity functions being the **unit-step function** and the **unit-impulse function.**

We define the unit-step forcing function as a function of time which is zero for all values of its argument less than zero and which is unity for all positive values of its argument. If we let $(t - t_0)$ be the argument and represent the unit-step function by u, then $u(t - t_0)$ must be zero for all values of t less than t_0, and it must be unity for all values of t greater than t_0. At $t = t_0$, $u(t - t_0)$ changes *abruptly* from 0 to 1. Its value at $t = t_0$ is not defined, but its value is known for all instants of time that are arbitrarily close to $t = t_0$. We often indicate this by writing $u(t_0^-) = 0$ and $u(t_0^+) = 1$. The concise mathematical definition of the unit-step forcing function is

$$u(t - t_0) = \begin{cases} 0 & t < t_0 \\ 1 & t > t_0 \end{cases}$$

and the function is shown graphically in Fig. 8.26. Note that a vertical line of unit length is shown at $t = t_0$. Although this "riser" is not strictly a part of the definition of the unit step, it is usually shown in each drawing.

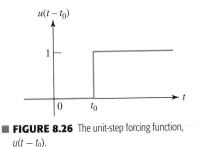

■ FIGURE 8.26 The unit-step forcing function, $u(t - t_0)$.

(2) Of course, this is not physically possible. However, if the time scale over which such an event occurs is very short compared to all other relevant time scales that describe the operation of a circuit, this is approximately true, and mathematically convenient.

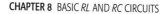

We also note that the unit step need not be a time function. For example, $u(x - x_0)$ could be used to denote a unit-step function where x might be a distance in meters, for example, or a frequency.

Very often in circuit analysis a discontinuity or a switching action takes place at an instant that is defined as $t = 0$. In that case $t_0 = 0$, and we then represent the corresponding unit-step forcing function by $u(t - 0)$, or more simply $u(t)$. This is shown in Fig. 8.27. Thus

$$u(t) = \begin{cases} 0 & t < 0 \\ 1 & t > 0 \end{cases}$$

The unit-step forcing function is in itself dimensionless. If we wish it to represent a voltage, it is necessary to multiply $u(t - t_0)$ by some constant voltage, such as 5 V. Thus, $v(t) = 5u(t - 0.2)$ V is an ideal voltage source which is zero before $t = 0.2$ s and a constant 5 V after $t = 0.2$ s. This forcing function is shown connected to a general network in Fig. 8.28a.

Physical Sources and the Unit-Step Function

Perhaps we should ask what physical source is the equivalent of this discontinuous forcing function. By equivalent, we mean simply that the voltage-current characteristics of the two networks are identical. For the step-voltage source of Fig. 8.28a, the voltage-current characteristic is simple: the voltage is zero prior to $t = 0.2$ s, it is 5 V after $t = 0.2$ s, and the current may be any (finite) value in either time interval. Our first thoughts might produce the attempt at an equivalent shown in Fig. 8.28b, a 5 V dc source in series with a switch which closes at $t = 0.2$ s. This network is not equivalent for $t < 0.2$ s, however, because the voltage across the battery and switch is completely unspecified in this time interval. The "equivalent" source is an open circuit, and the voltage across it *may be anything*. After $t = 0.2$ s, the networks are equivalent, and if this is the only time interval in which we are interested, and if the initial currents which flow from the two networks are identical at $t = 0.2$ s, then Fig. 8.28b becomes a useful equivalent of Fig. 8.28a.

In order to obtain an exact equivalent for the voltage-step forcing function, we may provide a single-pole double-throw switch. Before $t = 0.2$ s, the switch serves to ensure zero voltage across the input terminals of the general network. After $t = 0.2$ s, the switch is thrown to provide a constant input voltage of 5 V. At $t = 0.2$ s, the voltage is indeterminate (as is the step forcing function), and the battery is momentarily short-circuited (it is

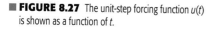

FIGURE 8.27 The unit-step forcing function $u(t)$ is shown as a function of t.

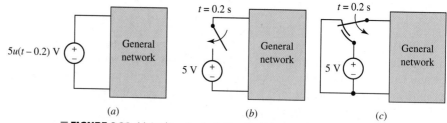

FIGURE 8.28 (a) A voltage-step forcing function is shown as the source driving a general network. (b) A simple circuit which, although not the exact equivalent of part (a), may be used as its equivalent in many cases. (c) An exact equivalent of part (a).

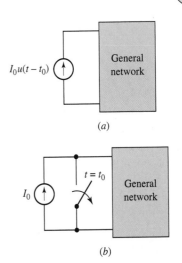

fortunate that we are dealing with mathematical models!). This exact equivalent of Fig. 8.28a is shown in Fig. 8.28c.

Figure 8.29a shows a current-step forcing function driving a general network. If we attempt to replace this circuit by a dc source in parallel with a switch (which opens at $t = t_0$), we must realize that the circuits are equivalent after $t = t_0$ but that the responses after $t = t_0$ are alike only if the initial conditions are the same. The circuit in Fig. 8.29b implies no voltage exists across the current source terminals for $t < t_0$. This is not the case for the circuit of Fig. 8.29a. However, we may often use the circuits of Fig. 8.29a and b interchangeably. The exact equivalent of Fig. 8.29a is the dual of the circuit of Fig. 8.28c; the exact equivalent of Fig. 8.29b cannot be constructed with current- and voltage-step forcing functions alone.[3]

FIGURE 8.29 (a) A current-step forcing function is applied to a general network. (b) A simple circuit which, although not the exact equivalent of part (a), may be used as its equivalent in many cases.

The Rectangular Pulse Function

Some very useful forcing functions may be obtained by manipulating the unit-step forcing function. Let us define a rectangular voltage pulse by the following conditions:

$$v(t) = \begin{cases} 0 & t < t_0 \\ V_0 & t_0 < t < t_1 \\ 0 & t > t_1 \end{cases}$$

The pulse is drawn in Fig. 8.30. Can this pulse be represented in terms of the unit-step forcing function? Let us consider the difference of the two unit steps, $u(t - t_0) - u(t - t_1)$. The two step functions are shown in Fig. 8.31a, and their difference is a rectangular pulse. The source $V_0 u(t - t_0) - V_0 u(t - t_1)$ which provides us with the desired voltage is indicated in Fig. 8.31b.

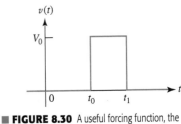

FIGURE 8.30 A useful forcing function, the rectangular voltage pulse.

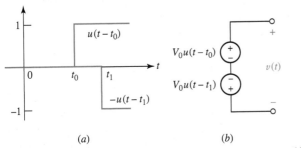

(a) (b)

FIGURE 8.31 (a) The unit steps $u(t - t_0)$ and $-u(t - t_1)$. (b) A source which yields the rectangular voltage pulse of Fig. 8.30.

If we have a sinusoidal voltage source $V_m \sin \omega t$ which is suddenly connected to a network at $t = t_0$, then an appropriate voltage forcing function would be $v(t) = V_m u(t - t_0) \sin \omega t$. If we wish to represent one burst of energy from the transmitter for a radio-controlled car operating at 47 MHz (295 Mrad/s), we may turn the sinusoidal source off 70 ns later by a second unit-step forcing function.[4] The voltage pulse is thus

$$v(t) = V_m [u(t - t_0) - u(t - t_0 - 7 \times 10^{-8})] \sin(295 \times 10^6 t)$$

This forcing function is sketched in Fig. 8.32.

(3) The equivalent can be drawn if the current through the switch prior to $t = t_0$ is known.
(4) Apparently, we're pretty good at the controls of this car. A reaction time of 70 ns?

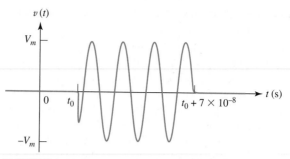

■ **FIGURE 8.32** A 47 MHz radio-frequency pulse, described by
$v(t) = V_m[u(t - t_0) - u(t - t_0 - 7 \times 10^{-8})] \sin(259 \times 10^6 t)$.

PRACTICE

8.8 Evaluate each of the following at $t = 0.8$: (a) $3u(t) - 2u(-t) + 0.8u(1 - t)$; (b) $[4u(t)]u(-t)$; (c) $2u(t) \sin \pi t$.

Ans: 3.8; 0; 1.176.

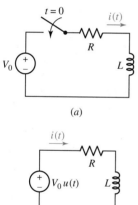

(a)

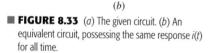

(b)

■ **FIGURE 8.33** (a) The given circuit. (b) An equivalent circuit, possessing the same response $i(t)$ for all time.

8.6 DRIVEN *RL* CIRCUITS

We are now ready to subject a simple network to the sudden application of a dc source. The circuit consists of a battery whose voltage is V_0 in series with a switch, a resistor R, and an inductor L. The switch is closed at $t = 0$, as indicated on the circuit diagram of Fig. 8.33a. It is evident that the current $i(t)$ is zero before $t = 0$, and we are therefore able to replace the battery and switch by a voltage-step forcing function $V_0 u(t)$, which also produces no response prior to $t = 0$. After $t = 0$, the two circuits are clearly identical. Hence, we seek the current $i(t)$ either in the given circuit of Fig. 8.33a or in the equivalent circuit of Fig. 8.33b.

We will find $i(t)$ at this time by writing the appropriate circuit equation and then solving it by separation of the variables and integration. After we obtain the answer and investigate the two parts of which it is composed, we will see that there is physical significance to each of these two terms. With a more intuitive understanding of how each term originates, we will be able to produce more rapid and more meaningful solutions to every problem involving the sudden application of any source.

Applying Kirchhoff's voltage law to the circuit of Fig. 8.33b, we have

$$Ri + L \frac{di}{dt} = V_0 u(t)$$

Since the unit-step forcing function is discontinuous at $t = 0$, we will first consider the solution for $t < 0$ and then for $t > 0$. The application of zero voltage since $t = -\infty$ forces a zero response, so that

$$i(t) = 0 \qquad t < 0$$

For positive time, however, $u(t)$ is unity and we must solve the equation

$$Ri + L \frac{di}{dt} = V_0 \qquad t > 0$$

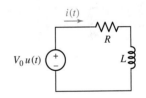

depend on (but *not* always be equal to) the initial value of the complete response and thus on the initial value of the forcing function also.

The Forced Response

We next observe that the first term of Eq. [30] depends on the functional form of $Q(t)$, the forcing function. Whenever we have a circuit in which the natural response dies out as t becomes infinite, this first term must completely describe the form of the response after the natural response has disappeared. This term is typically called the forced response; it is also called the *steady-state response*, the *particular solution*, or the *particular integral*.

For the present, we have elected to consider only those problems involving the sudden application of dc sources, and $Q(t)$ will therefore be a constant for all values of time. If we wish, we can now evaluate the integral in Eq. [30], obtaining the forced response

$$i_f = \frac{Q}{P} \qquad [32]$$

and the complete response

$$i(t) = \frac{Q}{P} + Ae^{-Pt} \qquad [33]$$

For the *RL* series circuit, Q/P is the constant current V_0/R and $1/P$ is the time constant τ. We should see that the forced response might have been obtained without evaluating the integral, because it must be the complete response at infinite time; it is merely the source voltage divided by the series resistance. The forced response is thus obtained by inspection of the final circuit.

Determination of the Complete Response

Let us use the simple *RL* series circuit to illustrate how to determine the complete response by the addition of the natural and forced responses. The circuit shown in Fig. 8.35 was analyzed earlier, but by a longer method. The desired response is the current $i(t)$, and we first express this current as the sum of the natural and the forced current,

$$i = i_n + i_f$$

The functional form of the natural response must be the same as that obtained without any sources. We therefore replace the step-voltage source by a short circuit and recognize the old *RL* series loop. Thus,

$$i_n = Ae^{-Rt/L}$$

where the amplitude A is yet to be determined; since the initial condition applies to the *complete* response, we cannot simply assume $A = i(0)$.

We next consider the forced response. In this particular problem the forced response must be constant, because the source is a constant V_0 for all positive values of time. After the natural response has died out, there can be no voltage across the inductor; hence, a voltage V_0 appears across R, and the forced response is simply

$$i_f = \frac{V_0}{R}$$

■ **FIGURE 8.35** A series *RL* circuit that is used to illustrate the method by which the complete response is obtained as the sum of the natural and forced responses.

Note that the forced response is determined completely; there is no unknown amplitude. We next combine the two responses to obtain

$$i = Ae^{-Rt/L} + \frac{V_0}{R}$$

and apply the initial condition to evaluate A. The current is zero prior to $t = 0$, and it cannot change value instantaneously since it is the current flowing through an inductor. Thus, the current is zero immediately after $t = 0$, and

$$0 = A + \frac{V_0}{R}$$

so

$$i = \frac{V_0}{R}(1 - e^{-Rt/L}) \qquad [34]$$

Note carefully that A is not the initial value of i, since $A = -V_0/R$, while $i(0) = 0$. In considering source-free circuits, we found that A was the initial value of the response. When forcing functions are present, however, we must first find the initial value of the response and then substitute this in the equation for the complete response to find A.

This response is plotted in Fig. 8.36, and we can see the manner in which the current builds up from its initial value of zero to its final value of V_0/R. The transition is effectively accomplished in a time 3τ. If our circuit represents the field coil of a large dc motor, we might have $L = 10\,\text{H}$ and $R = 20\,\Omega$, obtaining $\tau = 0.5\,\text{s}$. The field current is thus established in about 1.5 s. In one time constant, the current has attained 63.2 percent of its final value.

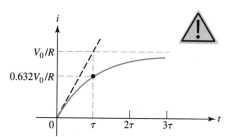

■ FIGURE 8.36 The current flowing through the inductor of Fig. 8.35 is shown graphically. A line extending the initial slope meets the constant forced response at $t = \tau$.

EXAMPLE 8.8

Determine $i(t)$ for all values of time in the circuit of Fig. 8.37.

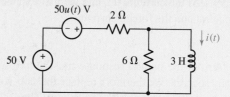

■ FIGURE 8.37 The circuit of Example 8.8.

The circuit contains a dc voltage source as well as a step-voltage source. We might choose to replace everything to the left of the inductor by the Thévenin equivalent, but instead let us merely recognize the form of that equivalent as a resistor in series with some voltage source. The circuit contains only one energy storage element, the inductor. We first note that

$$\tau = \frac{L}{R_{\text{eq}}} = \frac{3}{1.5} = 2\,\text{s}$$

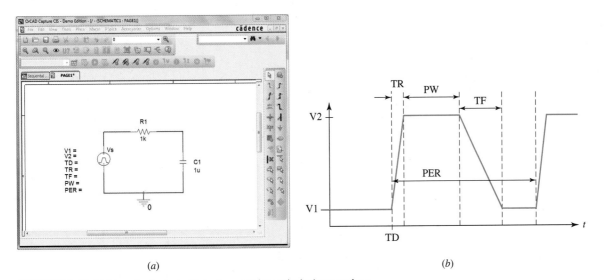

(a)　　　　　　　　　　　　　　　　　　　(b)

■ **FIGURE 8.47** (a) Schematic of a simple *RC* circuit connected to a pulsed voltage waveform. (b) Diagram of the SPICE **VPULSE** parameter definitions.

begins, which can be useful in allowing initial transient responses to decay for some circuit configurations.

For the purposes of this discussion, we set a zero time delay, V1 = 0, and V2 = 9 V. The circuit time constant is $\tau = RC = 1$ ms, so we set the rise and fall times to be 1 ns. Although SPICE will not allow a voltage to change in zero time since it solves the differential equations using discrete time intervals, compared to our circuit time constant 1 ns is a reasonable approximation to "instantaneous."

We will consider four basic cases, summarized in Table 8.1. In the first two cases, the pulse width W_p is much longer than the circuit time constant τ, so we expect the transients resulting from the beginning of the pulse to die out before the pulse is over. In the latter two cases, the opposite is true: the pulse width is so short that the capacitor does not have time to fully charge before the pulse ends. A similar issue arises when we consider the response of the circuit when the time between pulses $(T - W_p)$ is either short (Case II) or long (Case III) compared to the circuit time constant.

TABLE 8.1 Four Separate Cases of Pulse Width and Period Relative to the Circuit Time Constant of 1 ms

Case	Pulse Width W_p	Period T
I	10 ms ($\tau \ll W_p$)	20 ms ($\tau \ll T - W_p$)
II	10 ms ($\tau \ll W_p$)	10.1 ms ($\tau \gg T - W_p$)
III	0.1 ms ($\tau \gg W_p$)	10.1 ms ($\tau \ll T - W_p$)
IV	0.1 ms ($\tau \gg W_p$)	0.2 ms ($\tau \gg T - W_p$)

We qualitatively sketch the circuit response for each of the four cases in Fig. 8.48, arbitrarily selecting the capacitor voltage as the quantity of interest as any voltage or current is expected to have the same time dependence.

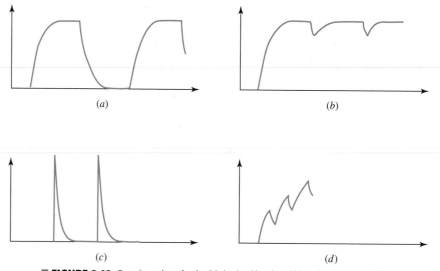

■ **FIGURE 8.48** Capacitor voltage for the *RC* circuit, with pulse width and period as in (*a*) Case I; (*b*) Case II; (*c*) Case III; and (*d*) Case IV.

In Case I, the capacitor has time to both fully charge and fully discharge (Fig. 8.48*a*), whereas in Case II (Fig. 8.48*b*), when the time between pulses is reduced, it no longer has time to fully discharge. In contrast, the capacitor does not have time to fully charge in either Case III (Fig. 8.48*c*) or Case IV (Fig. 8.48*d*).

Case I: Time Enough to Fully Charge and Fully Discharge

We can obtain exact values for the response in each case, of course, by performing a series of analyses. We consider Case I first. Since the capacitor has time to fully charge, the forced response will correspond to the 9 V dc driving voltage. The complete response to the first pulse is therefore

$$v_C(t) = 9 + Ae^{-1000t} \quad \text{V}$$

With $v_C(0) = 0$, $A = -9$ V and so

$$v_C(t) = 9(1 - e^{-1000t}) \qquad \text{V} \qquad [39]$$

in the interval of $0 < t < 10$ ms. At $t = 10$ ms, the source drops suddenly to 0 V, and the capacitor begins to discharge through the resistor. In this time interval we are faced with a simple "source-free" *RC* circuit, and we can write the response as

$$v_C(t) = Be^{-1000(t-0.01)} \qquad 10 < t < 20 \text{ ms} \qquad [40]$$

where $B = 8.99959$ V is found by substituting $t = 10$ ms in Eq. [39]; we will be pragmatic here and round this to 9 V, noting that the value calculated is consistent with our assumption that the initial transient dissipates before the pulse ends.

At $t = 20$ ms, the voltage source jumps immediately back to 9 V. The capacitor voltage just prior to this event is given by substituting $t = 20$ ms in Eq. [40], leading to $v_C(20\,\text{ms}) = 408.6\,\mu\text{V}$, essentially zero compared to the peak value of 9 V.

If we keep to our convention of rounding to four significant digits, the capacitor voltage at the beginning of the second pulse is zero, which is the same as our starting point. Thus, Eqs. [39] and [40] form the basis of the response for all subsequent pulses, and we may write

$$v_C(t) = \begin{cases} 9(1 - e^{-1000t}) \text{ V} & 0 \le t \le 10 \text{ ms} \\ 9e^{-1000(t-0.01)} \text{ V} & 10 < t \le 20 \text{ ms} \\ 9(1 - e^{-1000(t-0.02)}) \text{ V} & 20 < t \le 30 \text{ ms} \\ 9e^{-1000(t-0.03)} \text{ V} & 30 < t \le 40 \text{ ms} \end{cases}$$

and so on.

Case II: Time Enough to Fully Charge But Not Fully Discharge

Next we consider what happens if the capacitor is not allowed to completely discharge (Case II). Equation [39] still describes the situation in the interval of $0 < t < 10$ ms, and Eq. [40] describes the capacitor voltage in the interval between pulses, which has been reduced to $10 < t < 10.1$ ms.

Just prior to the onset of the second pulse at $t = 10.1$ ms, v_C is now 8.144 V; the capacitor has only had 0.1 ms to discharge, and therefore still retains 82 percent of its maximum energy when the next pulse begins. Thus, in the next interval,

$$v_C(t) = 9 + Ce^{-1000(t-10.1 \times 10^{-3})} \qquad \text{V} \qquad 10.1 < t < 20.1 \text{ ms}$$

where $v_C(10.1 \text{ ms}) = 9 + C = 8.144$ V, so $C = -0.856$ V and

$$v_C(t) = 9 - 0.856e^{-1000(t-10.1 \times 10^{-3})} \qquad \text{V} \qquad 10.1 < t < 20.1 \text{ ms}$$

which reaches the peak value of 9 V much more quickly than for the previous pulse.

Case III: No Time to Fully Charge But Time to Fully Discharge

What if it isn't clear that the transient will dissipate before the end of the voltage pulse? In fact, this situation arises in Case III. Just as we wrote for Case I,

$$v_C(t) = 9 + Ae^{-1000t} \qquad \text{V} \qquad\qquad\qquad [41]$$

still applies to this situation, but now only in the interval $0 < t < 0.1$ ms. Our initial condition has not changed, so $A = -9$ V as before. Now, however, just before this first pulse ends at $t = 0.1$ ms, we find that $v_C = 0.8565$ V. This is a far cry from the maximum of 9 V possible if we allow the capacitor time to fully charge, and is a direct result of the pulse lasting only one-tenth of the circuit time constant.

The capacitor now begins to discharge, so that

$$v_C(t) = Be^{-1000(t-1 \times 10^{-4})} \qquad \text{V} \qquad 0.1 < t < 10.1 \text{ ms} \qquad [42]$$

We have already determined that $v_C(0.1^- \text{ ms}) = 0.8565$ V, so $v_C(0.1^+ \text{ ms}) = 0.8565$ V and substitution into Eq. [42] yields $B = 0.8565$ V. Just prior to the onset of the second pulse at $t = 10.1$ ms, the capacitor voltage has decayed to essentially 0 V; this is the initial condition at the start of the second pulse and so Eq. [41] can be rewritten as

$$v_C(t) = 9 - 9e^{-1000(t-10.1 \times 10^{-3})} \qquad \text{V} \qquad 10.1 < t < 10.2 \text{ ms} \quad [43]$$

to describe the corresponding response.

Case IV: No Time to Fully Charge or Even Fully Discharge

In the last case, we consider the situation where the pulse width and period are so short that the capacitor can neither fully charge nor fully discharge in any one period. Based on experience, we can write

$$v_C(t) = 9 - 9e^{-1000t} \qquad \text{V} \qquad 0 < t < 0.1 \text{ ms} \qquad [44]$$

$$v_C(t) = 0.8565e^{-1000(t-1\times10^{-4})} \qquad \text{V} \qquad 0.1 < t < 0.2 \text{ ms} \qquad [45]$$

$$v_C(t) = 9 + Ce^{-1000(t-2\times10^{-4})} \qquad \text{V} \qquad 0.2 < t < 0.3 \text{ ms} \qquad [46]$$

$$v_C(t) = De^{-1000(t-3\times10^{-4})} \qquad \text{V} \qquad 0.3 < t < 0.4 \text{ ms} \qquad [47]$$

Just prior to the onset of the second pulse at $t = 0.2$ ms, the capacitor voltage has decayed to $v_C = 0.7750$ V; with insufficient time to fully discharge, it retains a large fraction of the little energy it had time to store initially. For the interval of $0.2 < t < 0.3$ ms, substitution of $v_C(0.2^+) = v_C(0.2^-) = 0.7750$ V into Eq. [46] yields $C = -8.225$ V. Continuing, we evaluate Eq. [46] at $t = 0.3$ ms and calculate $v_C = 1.558$ V just prior to the end of the second pulse. Thus, $D = 1.558$ V and our

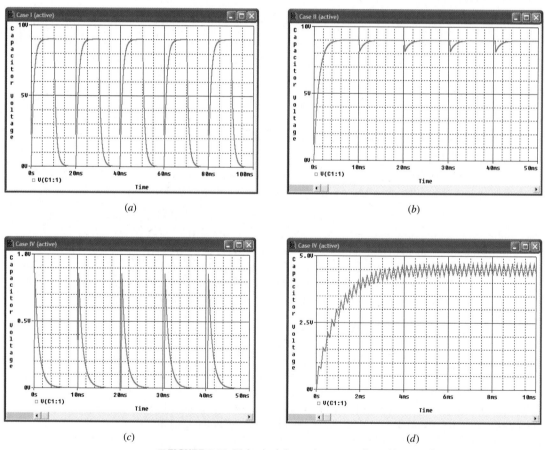

FIGURE 8.49 PSpice simulation results corresponding to (*a*) Case I; (*b*) Case II; (*c*) Case III; (*d*) Case IV.

capacitor is slowly charging to ever increase voltage levels over several pulses. At this stage it might be useful if we plot the detailed responses, so we show the PSpice simulation results of Cases I through IV in Fig. 8.49. Note in particular that in Fig. 8.49*d*, the small charge/discharge transient response similar in shape to that shown in Fig. 8.49*a–c* is superimposed on a charging-type response of the form $(1 - e^{-t/\tau})$. Thus, it takes about 3 to 5 circuit time constants for the capacitor to charge to its maximum value in situations where a single period does not allow it to fully charge or discharge!

What we have not yet done is predict the behavior of the response for $t \gg 5\tau$, although we would be interested in doing so, especially if it was not necessary to consider a very long sequence of pulses one at a time. We note that the response of Fig. 8.49*d* has an *average* value of 4.50 V from about 4 ms onward. This is exactly half the value we would expect if the voltage source pulse width allowed the capacitor to fully charge. In fact, this long-term average value can be computed by multiplying the dc capacitor voltage by the ratio of the pulse width to the period.

PRACTICE

8.14 With regard to Fig. 8.50*a*, sketch $i_L(t)$ in the range of $0 < t < 6$ s for (*a*) $v_S(t) = 3u(t) - 3u(t-2) + 3u(t-4) - 3u(t-6) + \cdots$; (*b*) $v_S(t) = 3u(t) - 3u(t-2) + 3u(t-2.1) - 3u(t-4.1) + \cdots$.

Ans: See Fig. 8.50*b*; see Fig. 8.50*c*.

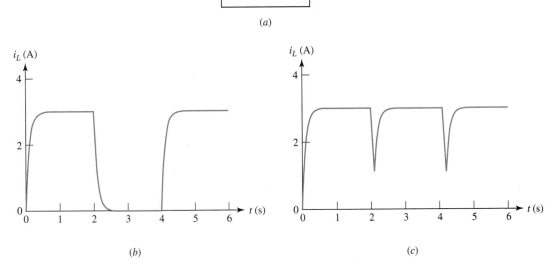

■ FIGURE 8.50 (*a*) Circuit for Practice Problem 8.14; (*b*) solution to part (*a*); (*c*) solution to part (*b*).

Frequency Limits in Digital Integrated Circuits

Modern digital integrated circuits such as programmable array logic (PALs) and microprocessors (Fig. 8.51) are composed of interconnected transistor circuits known as *gates*.

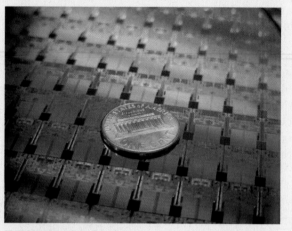

■ **FIGURE 8.51** A silicon wafer with multiple, identical integrated circuit dies. Each die is smaller than a US 1 cent coin. *Reprinted with permission of Intel Corporation.*

Digital signals are represented symbolically by combinations of ones and zeros, and can be either data or instructions (such as "add" or "subtract"). Electrically, we represent a logic "1" by a "high" voltage, and a logic "0" by a "low" voltage. In practice, there is a range of voltages that correspond to each; for example, in the 7400 series of TTL logic integrated circuits, any voltage between 2 and 5 V will be interpreted as a logic "1," and any voltage between 0 and 0.8 V will be interpreted as a logic "0." Voltages between 0.8 and 2 V do not correspond to either logic state, as shown in Fig. 8.52.

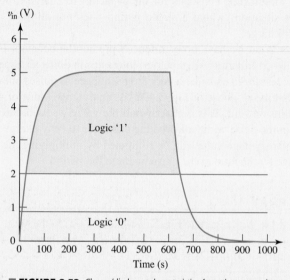

■ **FIGURE 8.52** Charge/discharge characteristic of a pathway capacitance identifying the TTL voltage ranges for logic "1" and logic "0," respectively.

A key parameter in digital circuits is the speed at which we can effectively use them. In this sense, "speed" refers to how quickly we can switch a gate from one logic state to another (either logic "0" to logic "1" or vice

SUMMARY AND REVIEW

In this chapter we learned that circuits containing a single energy storage element (either an inductor or a capacitor) can be described by a characteristic time scale, namely, the *circuit time constant* ($\tau = L/R$, or $\tau = RC$, respectively). If we attempt to change the amount of energy stored in the element (either charging or discharging), *every* voltage and current in the circuit will include an exponential term of the form $e^{-t/\tau}$. After approximately *5 time constants* from the moment we attempted to alter the amount of stored energy, the *transient* response has *essentially disappeared* and we are left simply with a *forced* response which arises from the independent sources driving the circuit at times $t > 0$. When determining the forced response in a purely dc circuit, we may treat inductors as short circuits and capacitors as open circuits.

versa), and the time delay required to convey the output of one gate to the input of the next gate. Although transistors contain "built-in" capacitances that affect their switching speed, it is the *interconnect pathways* that presently limit the speed of the fastest digital integrated circuits. We can model the interconnect pathway between two logic gates using a simple *RC* circuit (although as feature sizes continue to decrease in modern designs, more detailed models are required to accurately predict circuit performance). For example, consider a 2000 μm long pathway 2 μm wide. We can model this pathway in a typical silicon-based integrated circuit as having a capacitance of 0.5 pF and a resistance of 100 Ω, shown schematically in Fig. 8.53.

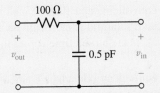

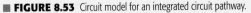

■ **FIGURE 8.53** Circuit model for an integrated circuit pathway.

Let's assume the voltage v_{out} represents the output voltage of a gate that is changing from a logic "0" state to a logic "1" state. The voltage v_{in} appears across the input of a second gate, and we are interested in how long it takes v_{in} to reach the same value as v_{out}.

Assuming the 0.5 pF capacitance that characterizes the interconnect pathway is initially discharged [i.e., $v_{in}(0) = 0$], calculating the *RC* time constant for our

pathway as $\tau = RC = 50$ ps, and defining $t = 0$ as when v_{out} changes, we obtain the expression

$$v_{in}(t) = Ae^{-t/\tau} + v_{out}(0)$$

Setting $v_{in}(0) = 0$, we find that $A = -v_{out}(0)$ so that

$$v_{in}(t) = v_{out}(0)[1 - e^{-t/\tau}]$$

Upon examining this equation, we see that v_{in} will reach the value $v_{out}(0)$ after $\sim 5\tau$ or 250 ps. If the voltage v_{out} changes again before this transient time period is over, then the capacitance does not have sufficient time to fully charge. In such situations, v_{in} will be less than $v_{out}(0)$. Assuming that $v_{out}(0)$ equals the minimum logic "1" voltage, for example, this means that v_{in} will not correspond to a logic "1." If v_{out} now suddenly changes to 0 V (logic "0"), the capacitance will begin to discharge so that v_{in} decreases further. Thus, by switching our logic states too quickly, we are unable to transfer the information from one gate to another.

The fastest speed at which we can change logic states is therefore $(5\tau)^{-1}$. This can be expressed in terms of the maximum operating frequency:

$$f_{max} = \frac{1}{2(5\tau)} = 2 \text{ GHz}$$

where the factor of 2 represents a charge/discharge period. If we desire to operate our integrated circuit at a higher frequency so that calculations can be performed faster, we need to reduce the interconnect capacitance and/or the interconnect resistance.

We started our analysis with so-called source-free circuits to introduce the idea of time constants without unnecessary distractions; such circuits have zero forced response and a transient response derived entirely from the energy stored at $t = 0$. We reasoned that a capacitor cannot change its voltage in zero time (or an infinite current results), and indicated this by introducing the notation $v_C(0^+) = v_C(0^-)$. Similarly, the current through an inductor cannot change in zero time, or $i_L(0^+) = i_L(0^-)$. The *complete* response is always the sum of the transient response and the forced response. Applying the initial condition to the complete response allows us to determine the unknown constant which multiplies the transient term.

We spent a little time discussing modeling switches, both analytically and within the context of PSpice. A common mathematical representation makes use of the unit-step function $u(t - t_0)$, which has zero value for $t < t_0$, unity value for $t > t_0$, and is indeterminate for $t = t_0$. Unit-step functions can "activate" a circuit (connecting sources so current can flow) for values of t preceding a specific time as well as after. Combinations of step functions can be used to create pulses and more complex waveforms. In the case of sequentially switched circuits, where sources are connected and disconnected repeatedly, we found the behavior of the circuits to depend strongly on both period and pulse width as they compare to the circuit time constant.

This is a good time to highlight some key points worth reviewing, along with relevant example(s).

❑ The response of a circuit having sources suddenly switched in or out of a circuit containing capacitors and inductors will always be composed of two parts: a *natural* response and a *forced* response.

❑ The form of the natural response (also referred to as the *transient response*) depends only on the component values and the way they are wired together. (Examples 8.1, 8.2)

❑ A circuit reduced to a single equivalent capacitance C and a single equivalent resistance R will have a natural response given by $v(t) = V_0 e^{-t/\tau}$, where $\tau = RC$ is the circuit time constant. (Examples 8.3, 8.5)

❑ A circuit reduced to a single equivalent inductance L and a single equivalent resistance R will have a natural response given by $i(t) = I_0 e^{-t/\tau}$, where $\tau = L/R$ is the circuit time constant. (Example 8.4)

❑ Circuits with dependent sources can be represented by a resistance using Thévenin procedures.

❑ The unit-step function is a useful way to model the closing or opening of a switch, provided we are careful to keep an eye on the initial conditions. (Examples 8.7, 8.9)

❑ The form of the forced response mirrors the form of the forcing function. Therefore, a dc forcing function always leads to a constant forced response. (Examples 8.7, 8.8)

❑ The *complete* response of an *RL* or *RC* circuit excited by a dc source will have the form $f(0^+) = f(\infty) + A$ and $f(t) = f(\infty) + [f(0^+) - f(\infty)]e^{-t/\tau}$, or total response = final value + (initial value − final value)$e^{-t/\tau}$. (Examples 8.9, 8.10, 8.11)

❑ The complete response for an *RL* or *RC* circuit may also be determined by writing a single differential equation for the quantity of interest and solving. (Examples 8.2, 8.11)

❑ When dealing with sequentially switched circuits, or circuits connected to pulsed waveforms, the relevant issue is whether the energy storage element has sufficient time to fully charge and to fully discharge, as measured relative to the circuit time constant.

READING FURTHER

A guide to solution techniques for differential equations can be found in:

W. E. Boyce and R. C. DiPrima, *Elementary Differential Equations and Boundary Value Problems,* 7th ed. New York: Wiley, 2002.

A detailed description of transients in electric circuits is given in:

E. Weber, *Linear Transient Analysis Volume I.* New York: Wiley, 1954. (Out of print, but in many university libraries.)

EXERCISES

8.1 The Source-Free *RL* Circuit

1. Setting $R = 1$ kΩ and $L = 1$ nH for the circuit represented in Fig. 8.1, and with the knowledge that $i(0) = -3$ mA, (a) write an expression for $i(t)$ valid for all $t \geq 0$; (b) compute $i(t)$ at $t = 0$, $t = 1$ ps, 2 ps, and 5 ps; and (c) calculate the energy stored in the inductor at $t = 0$, $t = 1$ ps, and $t = 5$ ps.

2. If $i(0) = 1$ A and $R = 100$ Ω for the circuit of Fig. 8.1, (a) select L such that $i(50$ ms$) = 368$ mA; (b) compute the energy stored in the inductor at $t = 0$, 50 ms, 100 ms, and 150 ms.

3. Referring to the circuit shown in Fig. 8.1, select values for both elements such that $L/R = 1$ and (a) calculate $v_R(t)$ at $t = 0, 1, 2, 3, 4$, and 5 s; (b) compute the power dissipated in the resistor at $t = 0$, 1 s, and 5 s. (c) At $t = 5$ s, what is the percentage of the initial energy still stored in the inductor?

4. The circuit depicted in Fig. 8.1 is constructed from components whose value is unknown. If a current $i(0)$ of 6 μA initially flows through the inductor, and it is determined that $i(1$ ms$) = 2.207$ μA, calculate the ratio of R to L.

5. Determine the characteristic equation of each of the following differential equations:

(a) $5v + 14\dfrac{dv}{dt} = 0$; (b) $-9\dfrac{di}{dt} - 18i = 0$;

(c) $\dfrac{di}{dt} + 18i + \dfrac{R}{B}i = 0$; (d) $\dfrac{d^2f}{dt^2} + 8\dfrac{df}{dt} + 2f = 0$.

6. For the following characteristic equations, write corresponding differential equations and find all roots, whether real, imaginary, or complex:
(a) $4s + 9 = 0$; (b) $2s - 4 = 0$; (c) $s^2 + 7s + 1 = 0$; (d) $5s^2 + 8s + 18 = 0$.

7. With the assumption that the switch in the circuit of Fig. 8.54 has been closed a long, long, long time, calculate $i_L(t)$ at (a) the instant just before the switch opens; (b) the instant just after the switch opens; (c) $t = 15.8$ μs; (d) $t = 31.5$ μs; (e) $t = 78.8$ μs.

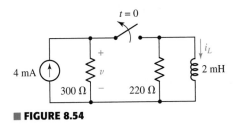

■ **FIGURE 8.54**

8. The switch in Fig. 8.54 has been closed since Catfish Hunter last pitched for the New York Yankees. Calculate the voltage labeled v as well as the energy stored in the inductor at (a) the instant just prior to the switch being thrown open; (b) the instant just after the switch is opened; (c) $t = 8$ μs; (d) $t = 80$ μs.

9. The switch in the circuit of Fig. 8.55 has been closed a ridiculously long time before suddenly being thrown open at $t = 0$. (a) Obtain expressions for i_L and v in the circuit of Fig. 8.55 which are valid for all $t \geq 0$. (b) Calculate $i_L(t)$ and

$v(t)$ at the instant just prior to the switch opening, at the instant just after the switch opening, and at $t = 470\ \mu s$.

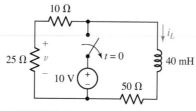

■ **FIGURE 8.55**

10. Assuming the switch initially has been open for a really, really long time, (*a*) obtain an expression for i_W in the circuit of Fig. 8.56 which is valid for all $t \geq 0$; (*b*) calculate i_W at $t = 0$ and $t = 1.3$ ns.

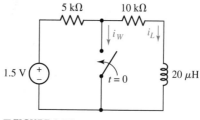

■ **FIGURE 8.56**

8.2 Properties of the Exponential Response

11. (*a*) Graph the function $f(t) = 10e^{-2t}$ over the range of $0 \leq t \leq 2.5$ s using linear scales for both *y* and *x* axes. (*b*) Replot with a logarithmic scale for the *y* axis. [*Hint:* the function semilogy() can be helpful here.] (*c*) What are the units of the 2 in the argument of the exponential? (*d*) At what time does the function reach a value of 9? 8? 1?

12. The current $i(t)$ flowing through a 1 Ω resistor is given by $i(t) = 5e^{-10t}$ mA, $t \geq 0$. (*a*) Determine the values of *t* for which the resistor voltage magnitude is equal to 5 V, 2.5 V, 0.5 V, and 5 mV. (*b*) Graph the function over the range of $0 \leq t \leq 1$ s using linear scales for both axes. (*c*) Draw a tangent to your curve at $t = 100$ ms, and determine where the tangent intersects the time axis.

13. The thickness of a solar cell must be chosen carefully to ensure photons are properly absorbed; even metals can be partly transparent when rolled out into very thin foils. If the incident light flux (number of photons per unit area per unit time) at the solar cell surface ($x = 0$) is given by Φ_0, and the intensity of light a distance *x* inside the solar cell is given by $\Phi(x)$, the behavior of $\Phi(x)$ is described by the equation $d\Phi/dx + \alpha\Phi = 0$. Here, α, known as the absorption coefficient, is a constant specific to a given semiconductor material. (*a*) What is the SI unit for α? (*b*) Obtain an expression for $\Phi(x)$ in terms of Φ_0, α, and *x*. (*c*) How thick should the solar cell be made in order to absorb at least 38% of the incident light? Express your answer in terms of α. (*d*) What happens to the light which enters the solar cell at $x = 0$ but is not absorbed?

14. For the circuit of Fig. 8.5, compute the time constant if the 10 Ω resistor is replaced with (*a*) a short circuit; (*b*) a 1 Ω resistor; (*c*) a series connection of two 5 Ω resistors; (*d*) a 100 Ω resistor. (*e*) Verify your answers with a suitable parameter sweep simulation. (*Hint:* the cursor tool might come in handy, and the answer does not depend on the initial current you choose for the inductor.)

15. Design a circuit which will produce a voltage of 1 V at some initial time, and a voltage of 368 mV at a time 5 seconds later. You may specify an initial inductor current without showing how it arises.

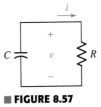

8.3 The Source-Free *RC* Circuit

16. The resistor in the circuit of Fig. 8.57 has been included to model the dielectric layer separating the plates of the 3.1 nF capacitor, and has a value of 55 MΩ. The capacitor is storing 200 mJ of energy just prior to $t = 0$. (*a*) Write an expression for $v(t)$ valid for $t \geq 0$. (*b*) Compute the energy remaining in the capacitor at $t = 170$ ms. (*c*) Graph $v(t)$ over the range of $0 < t < 850$ ms, and identify the value of $v(t)$ when $t = 2\tau$.

17. The resistor in the circuit of Fig. 8.57 has a value of 1 Ω and is connected to a 22 mF capacitor. The capacitor dielectric has infinite resistance, and the device is storing 891 mJ of energy just prior to $t = 0$. (*a*) Write an expression for $v(t)$ valid for $t \geq 0$. (*b*) Compute the energy remaining in the capacitor at $t = 11$ ms and 33 ms. (*c*) If it is determined that the capacitor dielectric is much leakier than expected, having a resistance as low as 100 kΩ, repeat parts (*a*) and (*b*).

18. Calculate the time constant of the circuit depicted in Fig. 8.57 if $C = 10$ mF and R is equal to (*a*) 1 Ω; (*b*) 10 Ω; (*c*) 100 Ω. (*d*) Verify your answers with an appropriate parameter sweep simulation. (*Hint:* the cursor tool might come in handy, and the time constant does not depend on the initial voltage across the capacitor.)

19. Design a capacitor-based circuit that will provide (*a*) a voltage of 9 V at some time $t = 0$, and a voltage of 1.2 V at a time 4 ms later; (*b*) a current of 1 mA at some time $t = 0$, and a reduced current of 50 μA at a time 100 ns later. (You can choose to design two separate circuits if desired, and do not need to show how the initial capacitor voltage is set.)

20. It is safe to assume that the switch drawn in the circuit of Fig. 8.58 has been closed such a long time that any transients which might have arisen from first connecting the voltage source have disappeared. (*a*) Determine the circuit time constant. (*b*) Calculate the voltage $v(t)$ at $t = \tau$, 2τ, and 5τ.

■ FIGURE 8.57

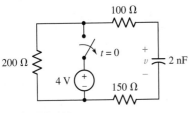

■ FIGURE 8.58

21. We can safely assume the switch in the circuit of Fig. 8.59 was closed a very long time prior to being thrown open at $t = 0$. (*a*) Determine the circuit time constant. (*b*) Obtain an expression for $i_1(t)$ which is valid for $t > 0$. (*c*) Determine the power dissipated by the 12 Ω resistor at $t = 500$ ms.

22. The switch above the 12 V source in the circuit of Fig. 8.60 has been closed since just after the wheel was invented. It is finally thrown open at $t = 0$. (*a*) Compute the circuit time constant. (*b*) Obtain an expression for $v(t)$ valid for $t > 0$. (*c*) Calculate the energy stored in the capacitor 170 ms after the switch is opened.

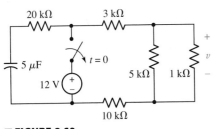

■ FIGURE 8.60

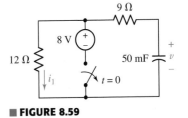

■ FIGURE 8.59

23. For the circuit represented schematically in Fig. 8.61, (*a*) calculate $v(t)$ at $t = 0$, $t = 984$ s, and $t = 1236$ s; (*b*) determine the energy still stored in the capacitor at $t = 100$ s.

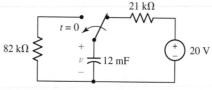

■ **FIGURE 8.61**

24. For the circuit depicted in Fig. 8.62, (*a*) compute the circuit time constant; (*b*) determine v in the instant just before the switch is closed; (*c*) obtain an expression for $v(t)$ valid for $t > 0$; (*d*) calculate $v(3\text{ ms})$.

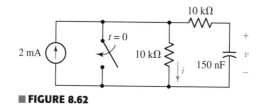

■ **FIGURE 8.62**

25. The switch drawn in Fig. 8.62 has been open a ponderously long time. (*a*) Determine the value of the current labeled i just prior to the switch being closed. (*b*) Obtain the value of i just after the switch is closed. (*c*) Compute the power dissipated in each resistor over the range of $0 < t < 15$ ms. (*d*) Graph your answer to part (*c*).

8.4 A More General Perspective

26. (*a*) Obtain an expression for $v(t)$, the voltage which appears across resistor R_3 in the circuit of Fig. 8.63, which is valid for $t > 0$. (*b*) If $R_1 = 2R_2 = 3R_3 = 4R_4 = 1.2$ kΩ, $L = 1$ mH, and $i_L(0^-) = 3$ mA, calculate $v(t = 500\text{ ns})$.

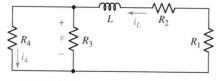

■ **FIGURE 8.63**

27. For the circuit of Fig. 8.64, determine i_x, i_L, and v_L at t equal to (*a*) 0^-; (*b*) 0^+.

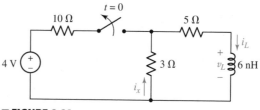

■ **FIGURE 8.64**

28. The switch shown in Fig. 8.65 has been closed for 6 years prior to being flipped open at $t = 0$. Determine i_L, v_L, and v_R at t equal to (a) 0^-; (b) 0^+; (c) 1 μs; (d) 10 μs.

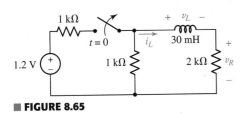

■ **FIGURE 8.65**

29. Obtain expressions for both $i_1(t)$ and $i_L(t)$ as labeled in Fig. 8.66, which are valid for $t > 0$.

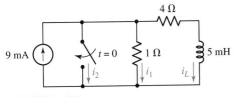

■ **FIGURE 8.66**

30. The voltage across the resistor in a simple source-free *RL* circuit is given by $5e^{-90t}$ V, $t > 0$. The inductor value is not known. (a) At what time will the inductor voltage be exactly one-half of its maximum value? (b) At what time will the inductor current reach 10% of its maximum value?

31. Referring to Fig. 8.67, calculate the currents i_1 and i_2 at t equal to (a) 1 ms; (b) 3 ms.

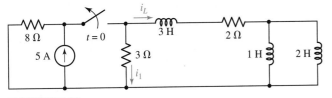

■ **FIGURE 8.67**

32. (a) Obtain an expression for v_x as labeled in the circuit of Fig. 8.68. (b) Evaluate v_x at $t = 5$ ms. (c) Verify your answer with an appropriate PSpice simulation. (*Hint:* employ the part named **Sw_tClose**.)

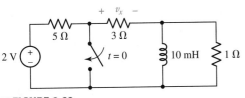

■ **FIGURE 8.68**

33. Design a complete circuit which provides a voltage v_{ab} across two terminals labeled a and b, respectively, such that $v_{ab} = 5$ V at $t = 0^-$, 2 V at $t = 1$ s, and less than 60 mV at $t = 5$. Verify the operation of your circuit using an appropriate PSpice simulation. (*Hint:* employ the part named **Sw_tOpen** or **Sw_tClose** as appropriate.)

34. For the part **Sw_tOpen**, PSpice actually employs a sequence of simulations where the part is first replaced with a resistor having value 1 MΩ, and then replaced with a resistor having value 10 mΩ corresponding to when the switch opens. (*a*) Evaluate the reliability of these default values by simulating the circuit of Fig. 8.55, and evaluating i_L at $t = 1$ ns. (*b*) Repeat part (*a*) with **RCLOSED** changed to 1 Ω. Did this change your answer? (*c*) Repeat part (*a*) with **ROPEN** changed to 100 kΩ and **RCLOSED** reset to its default value. Did this change your answer? (*Hint:* double-click on the part to access its attributes.)

35. Select values for the resistors R_0 and R_1 in the circuit of Fig. 8.69 such that $v_C(0.65) = 5.22$ V and $v_C(2.21) = 1$ V.

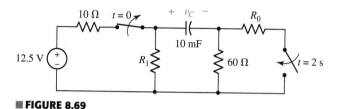

■ **FIGURE 8.69**

36. A quick measurement determines that the capacitor voltage v_C in the circuit of Fig. 8.70 is 2.5 V at $t = 0^-$. (*a*) Determine $v_C(0^+)$, $i_1(0^+)$, and $v(0^+)$. (*b*) Select a value of C so that the circuit time constant is equal to 14 s.

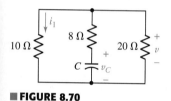

■ **FIGURE 8.70**

37. Determine $v_C(t)$ and $v_o(t)$ as labeled in the circuit represented by Fig. 8.71 for t equal to (*a*) 0^-; (*b*) 0^+; (*c*) 10 ms; (*d*) 12 ms.

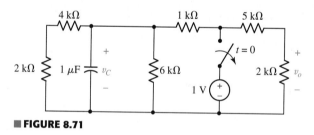

■ **FIGURE 8.71**

38. For the circuit shown in Fig. 8.72, determine (*a*) $v_C(0^-)$; (*b*) $v_C(0^+)$; (*c*) the circuit time constant; (*d*) $v_C(3$ ms$)$.

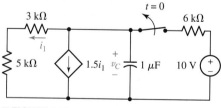

■ **FIGURE 8.72**

39. The switch in Fig. 8.73 is moved from *A* to *B* at $t = 0$ after being at *A* for a long time. This places the two capacitors in series, thus allowing equal and opposite dc voltages to be trapped on the capacitors. (*a*) Determine $v_1(0^-)$, $v_2(0^-)$, and $v_R(0^-)$. (*b*) Find $v_1(0^+)$, $v_2(0^+)$, and $v_R(0^+)$. (*c*) Determine the time constant of $v_R(t)$. (*d*) Find $v_R(t)$, $t > 0$. (*e*) Find $i(t)$. (*f*) Find $v_1(t)$ and $v_2(t)$ from $i(t)$ and the initial values. (*g*) Show that the stored energy at $t = \infty$ plus the total energy dissipated in the 20 kΩ resistor is equal to the energy stored in the capacitors at $t = 0$.

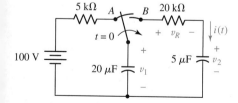

■ **FIGURE 8.73**

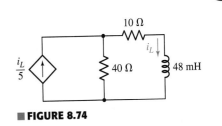

40. The inductor in Fig. 8.74 is storing 54 nJ at $t = 0^-$. Compute the energy remaining at t equal to (a) 0^+; (b) 1 ms; (c) 5 ms.

FIGURE 8.74

8.5 The Unit-Step Function

41. Evaluate the following functions at $t = -2$, 0, and $+2$: (a) $f(t) = 3u(t)$; (b) $g(t) = 5u(-t) + 3$; (c) $h(t) = 5u(t - 3)$; (d) $z(t) = 7u(1 - t) + 4u(t + 3)$.

42. Evaluate the following functions at $t = -1$, 0, and $+3$: (a) $f(t) = tu(1 - t)$; (b) $g(t) = 8 + 2u(2 - t)$; (c) $h(t) = u(t + 1) - u(t - 1) + u(t + 2) - u(t - 4)$; (d) $z(t) = 1 + u(3 - t) + u(t - 2)$.

43. Sketch the following functions over the range $-3 \leq t \leq 3$: (a) $v(t) = 3 - u(2 - t) - 2u(t)$ V; (b) $i(t) = u(t) - u(t - 0.5) + u(t - 1) - u(t - 1.5) + u(t - 2) - u(t - 2.5)$ A; (c) $q(t) = 8u(-t)$ C.

44. Use step functions to construct an equation that describes the waveform sketched in Fig. 8.75.

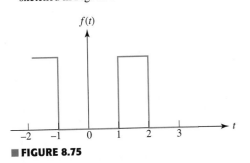

FIGURE 8.75

45. Employing step functions as appropriate, describe the voltage waveform graphed in Fig. 8.76.

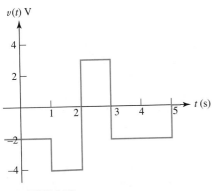

FIGURE 8.76

8.6 Driven *RL* Circuits

46. With reference to the simple circuit depicted in Fig. 8.77, compute $i(t)$ for (a) $t = 0^-$; (b) $t = 0^+$; (c) $t = 1^-$; (d) $t = 1^+$; (e) $t = 2$ ms.

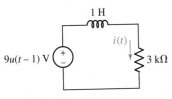

FIGURE 8.77

47. For the circuit given in Fig. 8.78, (*a*) determine $v_L(0^-)$, $v_L(0^+)$, $i_L(0^-)$, and $i_L(0^+)$; (*b*) calculate $i_L(150 \text{ ns})$. (*c*) Verify your answer to part (*b*) with an appropriate PSpice simulation.

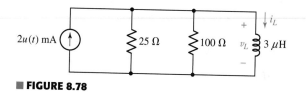

■ **FIGURE 8.78**

48. The circuit depicted in Fig. 8.79 contains two independent sources, one of which is only active for $t > 0$. (*a*) Obtain an expression for $i_L(t)$ valid for all t; (*b*) calculate $i_L(t)$ at $t = 10 \ \mu s$, $20 \ \mu s$, and $50 \ \mu s$.

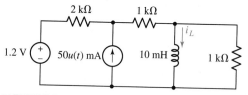

■ **FIGURE 8.79**

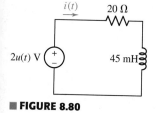

■ **FIGURE 8.80**

49. The circuit shown in Fig. 8.80 is powered by a source which is inactive for $t < 0$. (*a*) Obtain an expression for $i(t)$ valid for all t. (*b*) Graph your answer over the range of $-1 \text{ ms} \le t \le 10 \text{ ms}$.

50. For the circuit shown in Fig. 8.81, (*a*) obtain an expression for $i(t)$ valid for all time; (*b*) obtain an expression for $v_R(t)$ valid for all time; and (*c*) graph both $i(t)$ and $v_R(t)$ over the range of $-1 \text{ s} \le t \le 6 \text{ s}$.

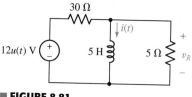

■ **FIGURE 8.81**

8.7 Natural and Forced Response

51. For the two-source circuit of Fig. 8.82, note that one source is always on. (*a*) Obtain an expression for $i(t)$ valid for all t; (*b*) determine at what time the energy stored in the inductor reaches 99% of its maximum value.

52. (*a*) Obtain an expression for i_L as labeled in Fig. 8.83 which is valid for all values of t. (*b*) Sketch your result over the range $-1 \text{ ms} \le t \le 3 \text{ ms}$.

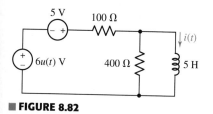

■ **FIGURE 8.82**

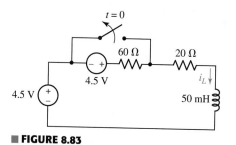

■ **FIGURE 8.83**

53. Obtain an expression for $i(t)$ as labeled in the circuit diagram of Fig. 8.84, and determine the power being dissipated in the 40 Ω resistor at $t = 2.5$ ms.

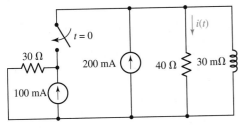

■ **FIGURE 8.84**

54. Obtain an expression for i_1 as indicated in Fig. 8.85 that is valid for all values of t.

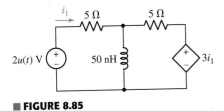

■ **FIGURE 8.85**

55. Plot the current $i(t)$ in Fig. 8.86 if (a) $R = 10$ Ω; (b) $R = 1$ Ω. In which case does the inductor (temporarily) store the most energy? *Explain.*

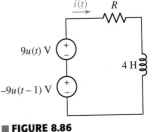

■ **FIGURE 8.86**

8.8 Driven *RC* Circuits

56. (a) Obtain an expression for v_C in the circuit of Fig. 8.87 valid for all values of t. (b) Sketch $v_C(t)$ over the range $0 \le t \le 4$ μs.

57. Obtain an equation which describes the behavior of i_A as labeled in Fig. 8.88 over the range of -1 ms $\le t \le 5$ ms.

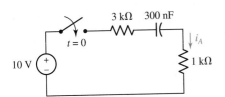

■ **FIGURE 8.88**

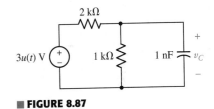

■ **FIGURE 8.87**

58. The switch in the circuit of Fig. 8.89 has been closed an incredibly long time, before being thrown open at $t = 0$. (a) Evaluate the current labeled i_x at $t = 70$ ms. (b) Verify your answer with an appropriate PSpice simulation.

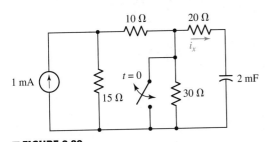

■ **FIGURE 8.89**

59. The switch in the circuit of Fig. 8.89 has been open a really, really incredibly long time, before being closed without further fanfare at $t = 0$. (*a*) Evaluate the current labeled i_x at $t = 70$ ms. (*b*) Verify your answer with an appropriate PSpice simulation.

60. The "make-before-break" switch shown in Fig. 8.90 has been in position *a* since the first episode of "Jonny Quest" aired on television. It is moved to position *b*, finally, at time $t = 0$. (*a*) Obtain expressions for $i(t)$ and $v_C(t)$ valid for all values of *t*. (*b*) Determine the energy remaining in the capacitor at $t = 33$ μs.

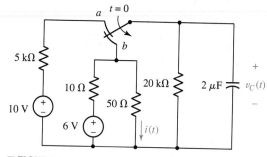

■ **FIGURE 8.90**

61. The switch in the circuit of Fig. 8.91, often called a *make-before-break* switch (since during switching it briefly makes contact to both parts of the circuit to ensure a smooth electrical transition), moves to position *b* at $t = 0$ only after being in position *a* long enough to ensure all initial transients arising from turning on the sources have long since decayed. (*a*) Determine the power dissipated by the 5 Ω resistor at $t = 0^-$. (*b*) Determine the power dissipated in the 3 Ω resistor at $t = 2$ ms.

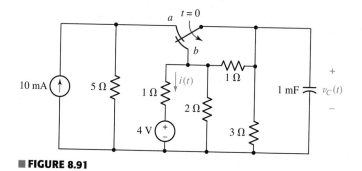

■ **FIGURE 8.91**

62. Referring to the circuit represented in Fig. 8.92, (*a*) obtain an equation which describes v_C valid for all values of *t*; (*b*) determine the energy remaining in the capacitor at $t = 0^+$, $t = 25$ μs, and $t = 150$ μs.

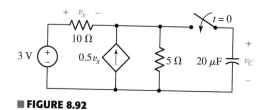

■ **FIGURE 8.92**

63. The dependent source shown in Fig. 8.92 is unfortunately installed upside down during manufacturing, so that the terminal corresponding to the arrowhead is actually wired to the negative reference terminal of the voltage source. This is not detected by the quality assurance team so the unit ships out wired improperly. The capacitor is initially discharged. If the 5 Ω resistor is only rated to 2 W, after what time t is the circuit likely to fail?

64. For the circuit represented in Fig. 8.93, (a) obtain an expression for v which is valid for all values of t; (b) sketch your result for $0 \leq t \leq 3$ s.

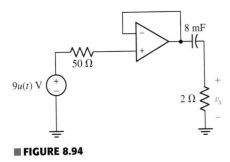

FIGURE 8.93

65. Obtain an expression for the voltage v_x as labeled in the op amp circuit of Fig. 8.94.

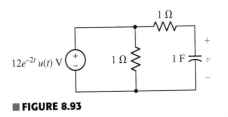

FIGURE 8.94

8.9 Predicting the Response of Sequentially Switched Circuits

66. Sketch the current i_L of the circuit in Fig. 8.50a if the 100 mH inductor is replaced by a 1 nH inductor, and is subjected to the waveform $v_s(t)$ equal to
(a) $5u(t) - 5u(t - 10^{-9}) + 5u(t - 2 \times 10^{-9})$ V, $0 \leq t \leq 4$ ns;
(b) $9u(t) - 5u(t - 10^{-8}) + 5u(t - 2 \times 10^{-8})$ V, $0 \leq t \leq 40$ ns.

67. The 100 mH inductor in the circuit of Fig. 8.50a is replaced with a 1 H inductor. Sketch the inductor current i_L if the source $v_s(t)$ is equal to
(a) $5u(t) - 5u(t - 0.01) + 5u(t - 0.02)$ V, $0 \leq t \leq 40$ ms;
(b) $5u(t) - 5u(t - 10) + 5u(t - 10.1)$ V, $0 \leq t \leq 11$ s.

68. Sketch the voltage v_C across the capacitor of Fig. 8.95 for at least 3 periods if $R = 1$ Ω, $C = 1$ F, and $v_s(t)$ is a pulsed waveform having (a) minimum of 0 V, maximum of 2 V, rise and fall times of 1 ms, pulse width of 10 s, and period of 10 s; (b) minimum of 0 V, maximum of 2 V, rise and fall times of 1 ms, pulse width of 10 ms, and period of 10 ms. (c) Verify your answers with appropriate PSpice simulations.

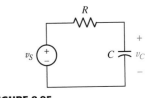

FIGURE 8.95

69. Sketch the voltage v_C across the capacitor of Fig. 8.95 for at least 3 periods if $R = 1$ Ω, $C = 1$ F, and $v_s(t)$ is a pulsed waveform having (a) minimum of 0 V, maximum of 2 V, rise and fall times of 1 ms, pulse width of 10 s, and period of 10 ms; (b) minimum of 0 V, maximum of 2 V, rise and fall times of 1 ms, pulse width of 10 ms, and period of 10 s. (c) Verify your answers with appropriate PSpice simulations.

Chapter-Integrating Exercises

70. The circuit in Fig. 8.96 contains two switches that always move in perfect synchronization. However, when switch A opens, switch B closes, and vice versa. Switch A is initially open, while switch B is initially closed; they change positions every 40 ms. Using the bottom node as the reference node, determine the voltage across the capacitor at t equal to (a) 0^-; (b) 0^+; (c) 40^- ms; (d) 40^+ ms; (e) 50 ms.

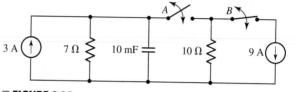

■ **FIGURE 8.96**

71. In the circuit of Fig. 8.96, when switch A opens, switch B closes, and vice versa. Switch A is initially open, while switch B is initially closed; they change positions every 400 ms. Determine the energy in the capacitor at t equal to (a) 0^-; (b) 0^+; (c) 200 ms; (d) 400^- ms; (e) 400^+ ms; (f) 700 ms.

72. Refer to the circuit of Fig. 8.97, which contains a voltage-controlled dependent voltage source in addition to two resistors. (a) Compute the circuit time constant. (b) Obtain an expression for v_x valid for all t. (c) Plot the power dissipated in the resistor over the range of 6 time constants. (d) Repeat parts (a) to (c) if the dependent source is installed in the circuit upside down. (e) Are both circuit configurations "stable"? *Explain*.

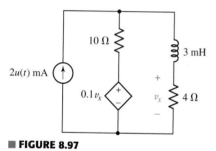

■ **FIGURE 8.97**

73. In the circuit of Fig. 8.97, a 3 mF capacitor is accidentally installed instead of the inductor. Unfortunately, that's not the end of the problems, as it's later determined that the real capacitor is not really well modeled by an ideal capacitor, and the dielectric has a resistance of 10 kΩ (which should be viewed as connected in parallel to the ideal capacitor). (a) Compute the circuit time constant with and without taking the dielectric resistance into account. By how much does the dielectric change your answer? (b) Calculate v_x at $t = 200$ ms. Does the dielectric resistance affect your answer significantly? *Explain*.

74. For the circuit of Fig. 8.98, assuming an ideal op amp, derive an expression for $v_o(t)$ if v_s is equal to (a) $4u(t)$ V; (b) $4e^{-130,000t}u(t)$ V.

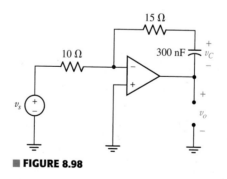

■ **FIGURE 8.98**

9

The *RLC* Circuit

INTRODUCTION

In Chap. 8 we studied circuits which contained only **one** energy storage element, combined with a passive network which partly determined how long it took either the capacitor or the inductor to charge/discharge. The differential equations which resulted from analysis were always first-order. In this chapter, we consider more complex circuits which contain **both** an inductor **and** a capacitor. The result is a **second-order** differential equation for any voltage or current of interest. What we learned in Chap. 8 is easily extended to the study of these so-called *RLC* circuits, although now we need **two** initial conditions to solve each differential equation. Such circuits occur routinely in a wide variety of applications, including oscillators and frequency filters. They are also very useful in modeling a number of practical situations, such as automobile suspension systems, temperature controllers, and even the response of an airplane to changes in elevator and aileron positions.

9.1 • THE SOURCE-FREE PARALLEL CIRCUIT

There are two basic types of *RLC* circuits: *parallel connected*, and *series connected*. We could start with either, but somewhat arbitrarily choose to begin by considering parallel *RLC* circuits. This particular combination of ideal elements is a reasonable model for portions of many communication networks. It represents, for example, an important part of certain electronic amplifiers found in radios, and enables the amplifiers to produce a large voltage amplification over a narrow band of signal frequencies (with almost zero amplification outside this band).

Just as we did with *RL* and *RC* circuits, we first consider the natural response of a parallel *RLC* circuit, where one or both of the energy storage elements have some nonzero initial energy (the origin of which for now is unimportant). This is represented by the inductor current and the capacitor voltage, both specified at $t = 0^+$. Once we're comfortable with this part of *RLC* circuit analysis, we can easily include dc sources, switches, or step sources in the circuit. Then we find the total response, which will be the sum of the natural response and the forced response.

Frequency selectivity of this kind enables us to listen to the transmission of one station while rejecting the transmission of any other station. Other applications include the use of parallel *RLC* circuits in frequency multiplexing and harmonic-suppression filters. However, even a simple discussion of these principles requires an understanding of such terms as *resonance, frequency response,* and *impedance,* which we have not yet discussed. Let it suffice to say, therefore, that an understanding of the natural behavior of the parallel *RLC* circuit is fundamentally important to future studies of communications networks and filter design, as well as many other applications.

When a *physical* capacitor is connected in parallel with an inductor and the capacitor has associated with it a finite resistance, the resulting network can be shown to have an equivalent circuit model like that shown in Fig. 9.1. The presence of this resistance can be used to model energy loss in the capacitor; over time, all real capacitors will eventually discharge, even if disconnected from a circuit. Energy losses in the physical inductor can also be taken into account by adding an ideal resistor (in series with the ideal inductor). For simplicity, however, we restrict our discussion to the case of an essentially ideal inductor in parallel with a "leaky" capacitor.

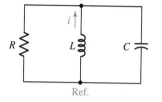

■ **FIGURE 9.1** The source-free parallel *RLC* circuit.

Obtaining the Differential Equation for a Parallel *RLC* Circuit

In the following analysis we will assume that energy may be stored initially in both the inductor and the capacitor; in other words, nonzero initial values of both inductor current and capacitor voltage may be present. With reference to the circuit of Fig. 9.1, we may then write the single nodal equation

$$\frac{v}{R} + \frac{1}{L} \int_{t_0}^{t} v \, dt' - i(t_0) + C \frac{dv}{dt} = 0 \qquad [1]$$

Note that the minus sign is a consequence of the assumed direction for i. We must solve Eq. [1] subject to the initial conditions

$$i(0^+) = I_0 \qquad [2]$$

and

$$v(0^+) = V_0 \qquad [3]$$

When both sides of Eq. [1] are differentiated once with respect to time, the result is the linear second-order homogeneous differential equation

$$C \frac{d^2v}{dt^2} + \frac{1}{R} \frac{dv}{dt} + \frac{1}{L} v = 0 \qquad [4]$$

whose solution $v(t)$ is the desired natural response.

Solution of the Differential Equation

There are a number of interesting ways to solve Eq. [4]. Most of these methods we will leave to a course in differential equations, selecting only the quickest and simplest method to use now. We will assume a solution, relying upon our intuition and modest experience to select one of the several possible forms that are suitable. Our experience with first-order equations might suggest that we at least try the exponential form once more. Thus, we *assume*

$$v = Ae^{st} \qquad [5]$$

being as general as possible by allowing A and s to be complex numbers if necessary. Substituting Eq. [5] in Eq. [4], we obtain

$$CAs^2 e^{st} + \frac{1}{R} As e^{st} + \frac{1}{L} Ae^{st} = 0$$

or

$$Ae^{st} \left(Cs^2 + \frac{1}{R}s + \frac{1}{L} \right) = 0$$

In order for this equation to be satisfied for all time, at least one of the three factors must be zero. If either of the first two factors is set equal to zero, then $v(t) = 0$. This is a trivial solution of the differential equation which cannot satisfy our given initial conditions. We therefore equate the remaining factor to zero:

$$Cs^2 + \frac{1}{R}s + \frac{1}{L} = 0 \qquad [6]$$

This equation is usually called the *auxiliary equation* or the **characteristic equation,** as we discussed in Sec. 8.1. If it can be satisfied, then our assumed solution is correct. Since Eq. [6] is a quadratic equation, there are two solutions, identified as s_1 and s_2:

$$s_1 = -\frac{1}{2RC} + \sqrt{\left(\frac{1}{2RC}\right)^2 - \frac{1}{LC}} \qquad [7]$$

and

$$s_2 = -\frac{1}{2RC} - \sqrt{\left(\frac{1}{2RC}\right)^2 - \frac{1}{LC}} \qquad [8]$$

If *either* of these two values is used for s in the assumed solution, then that solution satisfies the given differential equation; it thus becomes a valid solution of the differential equation.

Let us assume that we replace s by s_1 in Eq. [5], obtaining

$$v_1 = A_1 e^{s_1 t}$$

and, similarly,

$$v_2 = A_2 e^{s_2 t}$$

The former satisfies the differential equation

$$C \frac{d^2 v_1}{dt^2} + \frac{1}{R} \frac{dv_1}{dt} + \frac{1}{L} v_1 = 0$$

and the latter satisfies

$$C \frac{d^2 v_2}{dt^2} + \frac{1}{R} \frac{dv_2}{dt} + \frac{1}{L} v_2 = 0$$

Adding these two differential equations and combining similar terms, we have

$$C \frac{d^2(v_1 + v_2)}{dt^2} + \frac{1}{R} \frac{d(v_1 + v_2)}{dt} + \frac{1}{L}(v_1 + v_2) = 0$$

Linearity triumphs, and it is seen that the *sum* of the two solutions is also a solution. We thus have the general form of the natural response

$$v(t) = A_1 e^{s_1 t} + A_2 e^{s_2 t} \qquad [9]$$

where s_1 and s_2 are given by Eqs. [7] and [8]; A_1 and A_2 are two arbitrary constants which are to be selected to satisfy the two specified initial conditions.

Definition of Frequency Terms

The form of the natural response as given in Eq. [9] offers little insight into the nature of the curve we might obtain if $v(t)$ were plotted as a function of time. The relative amplitudes of A_1 and A_2, for example, will certainly be important in determining the shape of the response curve. Furthermore, the constants s_1 and s_2 can be real numbers or conjugate complex numbers, depending upon the values of R, L, and C in the given network. These two cases will produce fundamentally different response forms. Therefore, it will be helpful to make some simplifying substitutions in Eq. [9].

Since the exponents $s_1 t$ and $s_2 t$ must be dimensionless, s_1 and s_2 must have the unit of some dimensionless quantity "per second." From Eqs. [7] and [8] we therefore see that the units of $1/2RC$ and $1/\sqrt{LC}$ must also be s^{-1} (i.e., seconds^{-1}). Units of this type are called *frequencies.*

Let us define a new term, ω_0 (omega-sub-zero, or just omega-zero):

$$\omega_0 = \frac{1}{\sqrt{LC}} \qquad [10]$$

and reserve the term *resonant frequency* for it. On the other hand, we will call $1/2RC$ the *neper frequency,* or the *exponential damping coefficient,* and represent it by the symbol α (alpha):

$$\alpha = \frac{1}{2RC} \qquad [11]$$

This latter descriptive expression is used because α is a measure of how rapidly the natural response decays or damps out to its steady, final value (usually zero). Finally, s, s_1, and s_2, which are quantities that will form the basis for some of our later work, are called *complex frequencies.*

We should note that s_1, s_2, α, and ω_0 are merely symbols used to simplify the discussion of *RLC* circuits; they are not mysterious new properties of any kind. It is easier, for example, to say "*alpha*" than it is to say "*the reciprocal of 2RC*."

Let us collect these results. The natural response of the parallel *RLC* circuit is

$$v(t) = A_1 e^{s_1 t} + A_2 e^{s_2 t} \qquad [9]$$

where

$$s_1 = -\alpha + \sqrt{\alpha^2 - \omega_0^2} \qquad [12]$$

$$s_2 = -\alpha - \sqrt{\alpha^2 - \omega_0^2} \qquad [13]$$

$$\alpha = \frac{1}{2RC} \qquad [11]$$

$$\omega_0 = \frac{1}{\sqrt{LC}} \qquad [10]$$

and A_1 and A_2 must be found by applying the given initial conditions.

We note two basic scenarios possible with Eqs. [12] and [13] depending on the relative sizes of α and ω_0 (dictated by the values of R, L, and C). If $\alpha > \omega_0$, s_1 and s_2 will both be real numbers, leading to what is referred to as an **overdamped response.** In the opposite case, where $\alpha < \omega_0$, both s_1 and s_2 will have nonzero imaginary components, leading to what is known as an **underdamped response.** Both of these situations are considered separately in the following sections, along with the special case of $\alpha = \omega_0$, which leads to what is called a **critically damped response.** We should also note that the general response comprised by Eqs. [9] through [13] describes not only the voltage but all three branch currents in the parallel RLC circuit; the constants A_1 and A_2 will be different for each, of course.

The ratio of α to ω_0 is called the *damping ratio* by control system engineers and is designated by ζ (zeta).

Overdamped: $\alpha > \omega_0$
Critically damped: $\alpha = \omega_0$
Underdamped: $\alpha < \omega_0$

EXAMPLE **9.1**

Consider a parallel RLC circuit having an inductance of 10 mH and a capacitance of 100 μF. Determine the resistor values that would lead to overdamped and underdamped responses.

We first calculate the resonant frequency of the circuit:

$$\omega_0 = \sqrt{\frac{1}{LC}} = \sqrt{\frac{1}{(10 \times 10^{-3})(100 \times 10^{-6})}} = 10^3 \text{ rad/s}$$

An *overdamped* response will result if $\alpha > \omega_0$; an *underdamped* response will result if $\alpha < \omega_0$. Thus,

$$\frac{1}{2RC} > 10^3$$

and so

$$R < \frac{1}{(2000)(100 \times 10^{-6})}$$

or

$$R < 5 \ \Omega$$

leads to an overdamped response; $R > 5 \ \Omega$ leads to an underdamped response.

PRACTICE

9.1 A parallel RLC circuit contains a 100 Ω resistor and has the parameter values $\alpha = 1000 \text{ s}^{-1}$ and $\omega_0 = 800$ rad/s. Find (*a*) C; (*b*) L; (*c*) s_1; (*d*) s_2.

Ans: 5 μF; 312.5 mH; -400 s^{-1}; -1600 s^{-1}.

9.2 THE OVERDAMPED PARALLEL *RLC* CIRCUIT

A comparison of Eqs. [10] and [11] shows that α will be greater than ω_0 if $LC > 4R^2C^2$. In this case the radical used in calculating s_1 and s_2 will be real, and both s_1 and s_2 will be real. Moreover, the following inequalities

$$\sqrt{\alpha^2 - \omega_0^2} < \alpha$$

$$\left(-\alpha - \sqrt{\alpha^2 - \omega_0^2}\right) < \left(-\alpha + \sqrt{\alpha^2 - \omega_0^2}\right) < 0$$

may be applied to Eqs. [12] and [13] to show that both s_1 and s_2 are *negative* real numbers. Thus, the response $v(t)$ can be expressed as the (algebraic) sum of two decreasing exponential terms, both of which approach zero as time increases. In fact, since the absolute value of s_2 is larger than that of s_1, the term containing s_2 has the more rapid rate of decrease, and, for large values of time, we may write the limiting expression

$$v(t) \to A_1 e^{s_1 t} \to 0 \qquad \text{as } t \to \infty$$

The next step is to determine the arbitrary constants A_1 and A_2 in conformance with the initial conditions. We select a parallel *RLC* circuit with $R = 6\ \Omega$, $L = 7$ H, and, for ease of computation, $C = \frac{1}{42}$ F. The initial energy storage is specified by choosing an initial voltage across the circuit $v(0) = 0$ and an initial inductor current $i(0) = 10$ A, where v and i are defined in Fig. 9.2. We may easily determine the values of the several parameters

$$\alpha = 3.5 \qquad \omega_0 = \sqrt{6} \qquad \text{(all } s^{-1})$$
$$s_1 = -1 \qquad s_2 = -6$$

and immediately write the general form of the natural response

$$v(t) = A_1 e^{-t} + A_2 e^{-6t} \qquad [14]$$

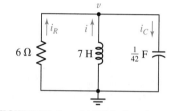

■ **FIGURE 9.2** A parallel *RLC* circuit used as a numerical example. The circuit is overdamped.

Finding Values for A_1 and A_2

Only the evaluation of the two constants A_1 and A_2 remains. If we knew the response $v(t)$ at two different values of time, these two values could be substituted in Eq. [14] and A_1 and A_2 easily found. However, we know only one instantaneous value of $v(t)$,

$$v(0) = 0$$

and, therefore,

$$0 = A_1 + A_2 \qquad [15]$$

We can obtain a second equation relating A_1 and A_2 by taking the derivative of $v(t)$ with respect to time in Eq. [14], determining the initial value of this derivative through the use of the remaining initial condition $i(0) = 10$, and equating the results. So, taking the derivative of both sides of Eq. [14],

$$\frac{dv}{dt} = -A_1 e^{-t} - 6A_2 e^{-6t}$$

and evaluating the derivative at $t = 0$,

$$\left.\frac{dv}{dt}\right|_{t=0} = -A_1 - 6A_2$$

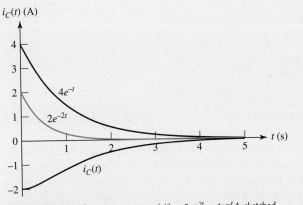

■ FIGURE 9.9 The current response $i_C(t) = 2e^{-2t} - 4e^{-t}$ A, sketched alongside its two components.

This equation can be solved using an iterative solver routine on a scientific calculator, which returns the solution $t_s = 5.296$ s. If such an option is not available, however, we can approximate Eq. [21] for $t \geq t_s$ as

$$-4e^{-t_s} = -0.02 \qquad [22]$$

Solving,

$$t_s = -\ln\left(\frac{0.02}{4}\right) = 5.298 \text{ s} \qquad [23]$$

which is reasonably close (better than 0.1% accuracy) to the exact solution.

PRACTICE

9.4 (*a*) Sketch the voltage $v_R(t) = 2e^{-t} - 4e^{-3t}$ V in the range $0 < t < 5$ s. (*b*) Estimate the settling time. (*c*) Calculate the maximum positive value and the time at which it occurs.

Ans: See Fig. 9.10; 5.9 s; 544 mV, 896 ms.

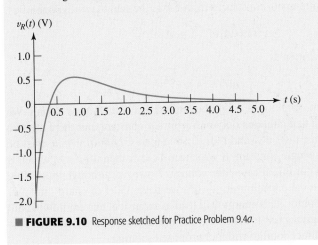

■ FIGURE 9.10 Response sketched for Practice Problem 9.4*a*.

9.3 • CRITICAL DAMPING

The overdamped case is characterized by

$$\alpha > \omega_0$$

or

$$LC > 4R^2C^2$$

and leads to negative real values for s_1 and s_2 and to a response expressed as the algebraic sum of two negative exponentials.

Now let us adjust the element values until α and ω_0 are equal. This is a very special case which is termed ***critical damping.*** If we were to attempt to construct a parallel *RLC* circuit that is critically damped, we would be attempting an essentially impossible task, for we could never make α exactly equal to ω_0. For completeness, however, we will discuss the critically damped circuit here, because it shows an interesting transition between overdamping and underdamping.

Critical damping is achieved when

or
$$\left.\begin{aligned} \alpha &= \omega_0 \\ LC &= 4R^2C^2 \\ L &= 4R^2C \end{aligned}\right\} \text{ critical damping}$$

We can produce critical damping by changing the value of any of the three elements in the numerical example discussed at the end of Sec. 9.1. We will select R, increasing its value until critical damping is obtained, and thus leave ω_0 unchanged. The necessary value of R is $7\sqrt{6}/2$ Ω; L is still 7 H, and C remains $\frac{1}{42}$ F. We thus find

$$\alpha = \omega_0 = \sqrt{6} \text{ s}^{-1}$$
$$s_1 = s_2 = -\sqrt{6} \text{ s}^{-1}$$

and recall the initial conditions that were specified, $v(0) = 0$ and $i(0) = 10$ A.

Form of a Critically Damped Response

We proceed to attempt to construct a response as the sum of two exponentials,

$$v(t) \stackrel{?}{=} A_1 e^{-\sqrt{6}t} + A_2 e^{-\sqrt{6}t}$$

which may be written as

$$v(t) \stackrel{?}{=} A_3 e^{-\sqrt{6}t}$$

At this point, some of us might be feeling that something's wrong. We have a response that contains only one arbitrary constant, but there are two initial conditions, $v(0) = 0$ and $i(0) = 10$ amperes, *both of which* must be satisfied by this single constant. If we select $A_3 = 0$, then $v(t) = 0$, which is consistent with our initial capacitor voltage. However, although there is no energy stored in the capacitor at $t = 0^+$, we have 350 J of energy initially stored in the inductor. This energy will lead to a transient current flowing out of the inductor, giving rise to a nonzero voltage across all three elements. This seems to be in direct conflict with our proposed solution.

"Impossible" is a pretty strong term. We make this statement because in practice it is unusual to obtain components that are closer than 1 percent of their specified values. Thus, obtaining *L precisely* equal to $4R^2C$ is theoretically possible, but not very likely, even if we're willing to measure a drawer full of components until we find the right ones.

If a mistake has not led to our difficulties, we must have begun with an incorrect assumption, and only one assumption has been made. We originally hypothesized that the differential equation could be solved by assuming an exponential solution, and this turns out to be incorrect for this single special case of critical damping. When $\alpha = \omega_0$, the differential equation, Eq. [4], becomes

$$\frac{d^2v}{dt^2} + 2\alpha \frac{dv}{dt} + \alpha^2 v = 0$$

The solution of this equation is not a tremendously difficult process, but we will avoid developing it here, since the equation is a standard type found in the usual differential-equation texts. The solution is

$$v = e^{-\alpha t}(A_1 t + A_2) \tag{24}$$

It should be noted that the solution is still expressed as the sum of two terms, where one term is the familiar negative exponential and the second is t times a negative exponential. We should also note that the solution contains the *two* expected arbitrary constants.

Finding Values for A_1 and A_2

Let us now complete our numerical example. After we substitute the known value of α in Eq. [24], obtaining

$$v = A_1 t e^{-\sqrt{6}t} + A_2 e^{-\sqrt{6}t}$$

we establish the values of A_1 and A_2 by first imposing the initial condition on $v(t)$ itself, $v(0) = 0$. Thus, $A_2 = 0$. This simple result occurs because the initial value of the response $v(t)$ was selected as zero; the more general case will require the solution of two equations simultaneously. The second initial condition must be applied to the derivative dv/dt just as in the overdamped case. We therefore differentiate, remembering that $A_2 = 0$:

$$\frac{dv}{dt} = A_1 t(-\sqrt{6})e^{-\sqrt{6}t} + A_1 e^{-\sqrt{6}t}$$

evaluate at $t = 0$:

$$\left.\frac{dv}{dt}\right|_{t=0} = A_1$$

and express the derivative in terms of the initial capacitor current:

$$\left.\frac{dv}{dt}\right|_{t=0} = \frac{i_C(0)}{C} = \frac{i_R(0)}{C} + \frac{i(0)}{C}$$

where reference directions for i_C, i_R, and i are defined in Fig. 9.2. Thus,

$$A_1 = 420 \text{ V}$$

The response is, therefore,

$$v(t) = 420t e^{-2.45t} \qquad \text{V} \tag{25}$$

Graphical Representation of the Critically Damped Response

Before plotting this response in detail, let us again try to anticipate its form by qualitative reasoning. The specified initial value is zero, and Eq. [25] concurs. It is not immediately apparent that the response also approaches zero as t becomes infinitely large, because $te^{-2.45t}$ is an indeterminate form. However, this obstacle is easily overcome by use of L'Hôspital's rule, which yields

$$\lim_{t\to\infty} v(t) = 420 \lim_{t\to\infty} \frac{t}{e^{2.45t}} = 420 \lim_{t\to\infty} \frac{1}{2.45e^{2.45t}} = 0$$

and once again we have a response that begins and ends at zero and has positive values at all other times. A maximum value v_m again occurs at time t_m; for our example,

$$t_m = 0.408 \text{ s} \qquad \text{and} \qquad v_m = 63.1 \text{ V}$$

This maximum is larger than that obtained in the overdamped case, and is a result of the smaller losses that occur in the larger resistor; the time of the maximum response is slightly later than it was with overdamping. The settling time may also be determined by solving

$$\frac{v_m}{100} = 420t_s e^{-2.45t_s}$$

for t_s (by trial-and-error methods or a calculator's SOLVE routine):

$$t_s = 3.12 \text{ s}$$

which is a considerably smaller value than that which arose in the overdamped case (5.15 s). As a matter of fact, it can be shown that, for given values of L and C, the selection of that value of R which provides critical damping will always give a shorter settling time than any choice of R that produces an overdamped response. However, a slight improvement (reduction) in settling time may be obtained by a further slight increase in resistance; a slightly underdamped response that will undershoot the zero axis before it dies out will yield the shortest settling time.

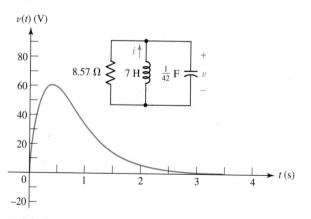

■ **FIGURE 9.11** The response $v(t) = 420te^{-2.45t}$ of the network shown in Fig. 9.2 with R changed to provide critical damping.

The response curve for critical damping is drawn in Fig. 9.11; it may be compared with the overdamped (and underdamped) case by reference to Fig. 9.16.

EXAMPLE **9.5**

Select a value for R_1 such that the circuit of Fig. 9.12 will be characterized by a critically damped response for $t > 0$, and a value for R_2 such that $v(0) = 2$ V.

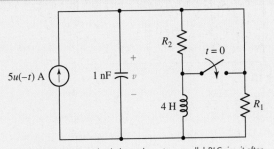

■ **FIGURE 9.12** A circuit that reduces to a parallel *RLC* circuit after the switch is thrown.

We note that at $t = 0^-$, the current source is on, and the inductor can be treated as a short circuit. Thus, $v(0^-)$ appears across R_2, and is given by

$$v(0^-) = 5R_2$$

and a value of 400 mΩ should be selected for R_2 to obtain $v(0) = 2$ V.

After the switch is thrown, the current source has turned itself off and R_2 is shorted. We are left with a parallel *RLC* circuit comprised of R_1, a 4 H inductor, and a 1 nF capacitor.

We may now calculate (for $t > 0$)

$$\alpha = \frac{1}{2RC}$$
$$= \frac{1}{2 \times 10^{-9}R_1}$$

and

$$\omega_0 = \frac{1}{\sqrt{LC}}$$
$$= \frac{1}{\sqrt{4 \times 10^{-9}}}$$
$$= 15{,}810 \text{ rad/s}$$

Therefore, to establish a critically damped response in the circuit for $t > 0$, we need to set $R_1 = 31.63$ kΩ. (*Note: since we have rounded to four significant figures, the pedantic can rightly argue that this is still not **exactly** a critically damped response—a difficult situation to create.*)

PRACTICE

9.5 (*a*) Choose R_1 in the circuit of Fig. 9.13 so that the response after $t = 0$ will be critically damped. (*b*) Now select R_2 to obtain $v(0) = 100$ V. (*c*) Find $v(t)$ at $t = 1$ ms.

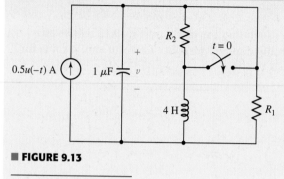

■ **FIGURE 9.13**

Ans: 1 kΩ; 250 Ω; -212 V.

9.4 THE UNDERDAMPED PARALLEL *RLC* CIRCUIT

Let us continue the process begun in Sec. 9.3 by increasing R once more to obtain what we will refer to as an ***underdamped*** response. Thus, the damping coefficient α decreases while ω_0 remains constant, α^2 becomes smaller than ω_0^2, and the radicand appearing in the expressions for s_1 and s_2 becomes negative. This causes the response to take on a much different character, but it is fortunately not necessary to return to the basic differential equation again. By using complex numbers, the exponential response turns into a *damped sinusoidal response;* this response is composed entirely of real quantities, the complex quantities being necessary only for the derivation.[1]

The Form of the Underdamped Response

We begin with the exponential form

$$v(t) = A_1 e^{s_1 t} + A_2 e^{s_2 t}$$

where

$$s_{1,2} = -\alpha \pm \sqrt{\alpha^2 - \omega_0^2}$$

and then let

$$\sqrt{\alpha^2 - \omega_0^2} = \sqrt{-1}\sqrt{\omega_0^2 - \alpha^2} = j\sqrt{\omega_0^2 - \alpha^2}$$

Electrical engineers use "*j*" instead of "*i*" to represent $\sqrt{-1}$ to avoid confusion with currents.

where $j \equiv \sqrt{-1}$.

We now take the new radical, which is real for the underdamped case, and call it ω_d, the ***natural resonant frequency:***

$$\omega_d = \sqrt{\omega_0^2 - \alpha^2}$$

The response may now be written as

$$v(t) = e^{-\alpha t}(A_1 e^{j\omega_d t} + A_2 e^{-j\omega_d t}) \qquad [26]$$

(1) A review of complex numbers is presented in Appendix 5.

or, in the longer but equivalent form,

$$v(t) = e^{-\alpha t} \left\{ (A_1 + A_2) \left[\frac{e^{j\omega_d t} + e^{-j\omega_d t}}{2} \right] + j(A_1 - A_2) \left[\frac{e^{j\omega_d t} - e^{-j\omega_d t}}{j2} \right] \right\}$$

Applying identities described in Appendix 5, the term in the first square brackets in the preceding equation is identically equal to $\cos \omega_d t$, and the second is identically $\sin \omega_d t$. Hence,

$$v(t) = e^{-\alpha t} [(A_1 + A_2) \cos \omega_d t + j(A_1 - A_2) \sin \omega_d t]$$

and the multiplying factors may be assigned new symbols:

$$v(t) = e^{-\alpha t} (B_1 \cos \omega_d t + B_2 \sin \omega_d t) \qquad [27]$$

where Eqs. [26] and [27] are identical.

It may seem a little odd that our expression originally appeared to have a complex component, and now is purely real. However, we should remember that we originally allowed for A_1 and A_2 to be complex as well as s_1 and s_2. In any event, if we are dealing with the underdamped case, we have now left complex numbers behind. This must be true since α, ω_d, and t are real quantities, so that $v(t)$ itself must be a real quantity (which might be presented on an oscilloscope, a voltmeter, or a sheet of graph paper). Equation [27] is the desired functional form for the underdamped response, and its validity may be checked by direct substitution in the original differential equation; this exercise is left to the doubters. The two real constants B_1 and B_2 are again selected to fit the given initial conditions.

We return to our simple parallel *RLC* circuit of Fig. 9.2 with $R = 6\ \Omega$, $C = 1/42$ F, and $L = 7$ H, but now increase the resistance further to $10.5\ \Omega$. Thus,

$$\alpha = \frac{1}{2RC} = 2\ \text{s}^{-1}$$

$$\omega_0 = \frac{1}{\sqrt{LC}} = \sqrt{6}\ \text{s}^{-1}$$

and

$$\omega_d = \sqrt{\omega_0^2 - \alpha^2} = \sqrt{2}\ \text{rad/s}$$

Except for the evaluation of the arbitrary constants, the response is now known:

$$v(t) = e^{-2t} (B_1 \cos \sqrt{2}t + B_2 \sin \sqrt{2}t)$$

Finding Values for B_1 and B_2

The determination of the two constants proceeds as before. If we still assume that $v(0) = 0$ and $i(0) = 10$, then B_1 must be zero. Hence

$$v(t) = B_2 e^{-2t} \sin \sqrt{2}t$$

The derivative is

$$\frac{dv}{dt} = \sqrt{2}B_2 e^{-2t} \cos \sqrt{2}t - 2B_2 e^{-2t} \sin \sqrt{2}t$$

and at $t = 0$ it becomes

$$\frac{dv}{dt}\bigg|_{t=0} = \sqrt{2}B_2 = \frac{i_C(0)}{C} = 420$$

where i_C is defined in Fig. 9.2. Therefore,

$$v(t) = 210\sqrt{2}e^{-2t}\sin\sqrt{2}t$$

Graphical Representation of the Underdamped Response

Notice that, as before, this response function has an initial value of zero because of the initial voltage condition we imposed, and a final value of zero because the exponential term vanishes for large values of t. As t increases from zero through small positive values, $v(t)$ increases as $210\sqrt{2}\sin\sqrt{2}t$, because the exponential term remains essentially equal to unity. But, at some time t_m, the exponential function begins to decrease more rapidly than $\sin\sqrt{2}t$ is increasing; thus $v(t)$ reaches a maximum v_m and begins to decrease. We should note that t_m is not the value of t for which $\sin\sqrt{2}t$ is a maximum, but must occur somewhat before $\sin\sqrt{2}t$ reaches its maximum.

When $t = \pi/\sqrt{2}$, $v(t)$ is zero. Thus, in the interval $\pi/\sqrt{2} < t < \sqrt{2}\pi$, the response is negative, becoming zero again at $t = \sqrt{2}\pi$. Hence, $v(t)$ is an *oscillatory* function of time and crosses the time axis an infinite number of times at $t = n\pi/\sqrt{2}$, where n is any positive integer. In our example, however, the response is only slightly underdamped, and the exponential term causes the function to die out so rapidly that most of the zero crossings will not be evident in a sketch.

The oscillatory nature of the response becomes more noticeable as α decreases. If α is zero, which corresponds to an infinitely large resistance, then $v(t)$ is an undamped sinusoid that oscillates with constant amplitude. There is never a time at which $v(t)$ drops and stays below 1 percent of its maximum value; the settling time is therefore infinite. This is not perpetual motion; we have merely assumed an initial energy in the circuit and have not provided any means to dissipate this energy. It is transferred from its initial location in the inductor to the capacitor, then returns to the inductor, and so on, forever.

The Role of Finite Resistance

A finite R in the parallel *RLC* circuit acts as a kind of electrical transfer agent. Every time energy is transferred from L to C or from C to L, the agent exacts a commission. Before long, the agent has taken all the energy, wantonly dissipating every last joule. The L and C are left without a joule of their own, without voltage and without current. Actual parallel *RLC* circuits can be made to have effective values of R so large that a natural undamped sinusoidal response can be maintained for years without supplying any additional energy.

Returning to our specific numerical problem, differentiation locates the first maximum of $v(t)$,

$$v_{m_1} = 71.8 \text{ V} \qquad \text{at} \qquad t_{m_1} = 0.435 \text{ s}$$

the succeeding minimum,

$$v_{m_2} = -0.845 \text{ V} \qquad \text{at} \qquad t_{m_2} = 2.66 \text{ s}$$

PRACTICE

9.9 Let $i_s = 10u(-t) - 20u(t)$ A in Fig. 9.30. Find (a) $i_L(0^-)$;
(b) $v_C(0^+)$; (c) $v_R(0^+)$; (d) $i_L(\infty)$; (e) $i_L(0.1 \text{ ms})$.

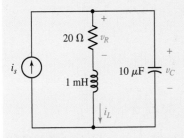

■ **FIGURE 9.30**

Ans: 10 A; 200 V; 200 V; −20 A; 2.07 A.

EXAMPLE 9.10

Complete the determination of the initial conditions in the circuit of Fig. 9.28, repeated in Fig. 9.31, by finding values at $t = 0^+$ for the first derivatives of the three voltage and three current variables defined on the circuit diagram.

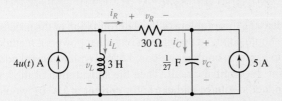

■ **FIGURE 9.31** Circuit of Fig. 9.28, repeated for Example 9.10.

We begin with the two energy storage elements. For the inductor,

$$v_L = L \frac{di_L}{dt}$$

and, specifically,

$$v_L(0^+) = L \frac{di_L}{dt}\bigg|_{t=0^+}$$

Thus,

$$\frac{di_L}{dt}\bigg|_{t=0^+} = \frac{v_L(0^+)}{L} = \frac{120}{3} = 40 \text{ A/s}$$

Similarly,

$$\frac{dv_C}{dt}\bigg|_{t=0^+} = \frac{i_C(0^+)}{C} = \frac{4}{1/27} = 108 \text{ V/s}$$

(Continued on next page)

The other four derivatives may be determined by realizing that KCL and KVL are both satisfied by the derivatives also. For example, at the left-hand node in Fig. 9.31,

$$4 - i_L - i_R = 0 \qquad t > 0$$

and thus,

$$0 - \frac{di_L}{dt} - \frac{di_R}{dt} = 0 \qquad t > 0$$

and therefore,

$$\left. \frac{di_R}{dt} \right|_{t=0^+} = -40 \text{ A/s}$$

The three remaining initial values of the derivatives are found to be

$$\left. \frac{dv_R}{dt} \right|_{t=0^+} = -1200 \text{ V/s}$$

$$\left. \frac{dv_L}{dt} \right|_{t=0^+} = -1092 \text{ V/s}$$

and

$$\left. \frac{di_C}{dt} \right|_{t=0^+} = -40 \text{ A/s}$$

Before leaving this problem of the determination of the necessary initial values, we should point out that at least one other powerful method of determining them has been omitted: we could have written general nodal or loop equations for the original circuit. Then the substitution of the known zero values of inductor voltage and capacitor current at $t = 0^-$ would uncover several other response values at $t = 0^-$ and enable the remainder to be found easily. A similar analysis at $t = 0^+$ must then be made. This is an important method, and it becomes a necessary one in more complicated circuits which cannot be analyzed by our simpler step-by-step procedures.

Now let us briefly complete the determination of the response $v_C(t)$ for the original circuit of Fig. 9.31. With both sources dead, the circuit appears as a series *RLC* circuit and s_1 and s_2 are easily found to be -1 and -9, respectively. The forced response may be found by inspection or, if necessary, by drawing the dc equivalent, which is similar to Fig. 9.29a, with the addition of a 4 A current source. The forced response is 150 V. Thus,

$$v_C(t) = 150 + Ae^{-t} + Be^{-9t}$$

and

$$v_C(0^+) = 150 = 150 + A + B$$

or

$$A + B = 0$$

Then,

$$\frac{dv_C}{dt} = -Ae^{-t} - 9Be^{-9t}$$

and

$$\left.\frac{dv_C}{dt}\right|_{t=0^+} = 108 = -A - 9B$$

Finally,

$$A = 13.5 \qquad B = -13.5$$

and

$$v_C(t) = 150 + 13.5(e^{-t} - e^{-9t}) \qquad \text{V}$$

A Quick Summary of the Solution Process

In summary, then, whenever we wish to determine the transient behavior of a simple three-element *RLC* circuit, we must first decide whether we are confronted with a series or a parallel circuit, so that we may use the correct relationship for α. The two equations are

$$\alpha = \frac{1}{2RC} \qquad \text{(parallel } RLC\text{)}$$

$$\alpha = \frac{R}{2L} \qquad \text{(series } RLC\text{)}$$

Our second decision is made after comparing α with ω_0, which is given for either circuit by

$$\omega_0 = \frac{1}{\sqrt{LC}}$$

If $\alpha > \omega_0$, the circuit is *overdamped*, and the natural response has the form

$$f_n(t) = A_1 e^{s_1 t} + A_2 e^{s_2 t}$$

where

$$s_{1,2} = -\alpha \pm \sqrt{\alpha^2 - \omega_0^2}$$

If $\alpha = \omega_0$, then the circuit is *critically damped* and

$$f_n(t) = e^{-\alpha t}(A_1 t + A_2)$$

And finally, if $\alpha < \omega_0$, then we are faced with the *underdamped* response,

$$f_n(t) = e^{-\alpha t}(A_1 \cos \omega_d t + A_2 \sin \omega_d t)$$

where

$$\omega_d = \sqrt{\omega_0^2 - \alpha^2}$$

Our last decision depends on the independent sources. If there are none acting in the circuit after the switching or discontinuity is completed, then the circuit is source-free and the natural response accounts for the complete response. If independent sources are still present, then the circuit is driven and a forced response must be determined. The complete response is then the sum

$$f(t) = f_f(t) + f_n(t)$$

This is applicable to any current or voltage in the circuit. Our final step is to solve for unknown constants given the initial conditions.

Modeling Automotive Suspension Systems

Earlier, we alluded to the fact that the concepts investigated in this chapter actually extend beyond the analysis of electric circuits. In fact, the general form of the differential equations we have been working with appears in many fields—we need only learn how to "translate" new parameter terminology. For example, consider a simple automotive suspension, as shown in Fig. 9.32. The piston is not attached to the cylinder, but *is* attached to both the spring and the wheel. The moving parts therefore are the spring, the piston, and the wheel.

We will model this physical system by first determining the forces in play. Defining a position function $p(t)$ which describes where the piston lies within the cylinder, we may write F_S, the force on the spring, as

$$F_S = Kp(t)$$

where K is known as the spring constant and has units of lb/ft. The force on the wheel F_W is equal to the mass of the wheel times its acceleration, or

$$F_W = m \frac{d^2 p(t)}{dt^2}$$

where m is measured in lb · s²/ft. Last but not least is the force of friction F_f acting on the piston

$$F_f = \mu_f \frac{dp(t)}{dt}$$

FIGURE 9.32 Typical automotive suspension system.
© Transtock Inc./Alamy.

where μ_f is the coefficient of friction, with units of lb · s/ft.

From our basic physics courses we know that all forces acting in our system must sum to zero, so that

$$m \frac{d^2 p(t)}{dt^2} + \mu_f \frac{dp(t)}{dt} + Kp(t) = 0 \qquad [34]$$

This equation most likely had the potential to give us nightmares at one point in our academic career, but no longer. We compare Eq. [32] to Eqs. [30] and [31] and immediately see a distinct resemblance, at least in the general form. Choosing Eq. [30], the differential equation describing the inductor current of a series-connected *RLC* circuit, we observe the following correspondences:

Mass	m	$\rightarrow$ inductance	L
Coefficient of friction	μ_f	$\rightarrow$ resistance	R
Spring constant	K	$\rightarrow$ inverse of the capacitance	C^{-1}
Position variable	$p(t)$	$\rightarrow$ current variable	$i(t)$

So, if we are willing to talk about feet instead of amperes, lb · s²/ft instead of H, ft/lb instead of F, and lb · s/ft instead of Ω, we can apply our newly found skills at modeling *RLC* circuits to the task of evaluating automotive shock absorbers.

Take a typical car wheel of 70 lb. The mass is found by dividing the weight by the earth's gravitational acceleration (32.17 ft/s²), resulting in $m = 2.176$ lb · s²/ft. The curb weight of our car is 1985 lb, and the static displacement of the spring is 4 inches (no passengers). The spring constant is obtained by dividing the weight on each shock absorber by the static displacement, so that we have $K = (\frac{1}{4})(1985)(3 \text{ ft}^{-1}) = 1489$ lb/ft. We are also told that the coefficient of friction for our piston-cylinder assembly is 65 lb · s/ft. Thus, we can simulate our shock absorber by modeling it with a series *RLC* circuit having $R = 65 \ \Omega$, $L = 2.176$ H, and $C = K^{-1} = 671.6 \ \mu$F.

The resonant frequency of our shock absorber is $\omega_0 = (LC)^{-1/2} = 26.16$ rad/s, and the damping coefficient is $\alpha = R/2L = 14.94$ s⁻¹. Since $\alpha < \omega_0$, our shock absorber represents an underdamped system; this means that we expect a bounce or two after we run over a pothole. A stiffer shock (larger coefficient of friction, or a larger resistance in our circuit model) is typically desirable when curves are taken at high speeds—at some point this corresponds to an overdamped response. However, if most of our driving is over unpaved roads, a slightly underdamped response is preferable.

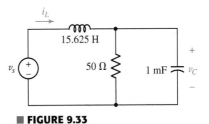

9.10 Let $v_s = 10 + 20u(t)$ V in the circuit of Fig. 9.33. Find (a) $i_L(0)$; (b) $v_C(0)$; (c) $i_{L,f}$; (d) $i_L(0.1 \text{ s})$.

Ans: 0.2 A; 10 V; 0.6 A; 0.319 A.

■ **FIGURE 9.33**

9.7 THE LOSSLESS *LC* CIRCUIT

When we considered the source-free *RLC* circuit, it became apparent that the resistor served to dissipate any initial energy stored in the circuit. At some point it might occur to us to ask what would happen if we could remove the resistor. If the value of the resistance in a parallel *RLC* circuit becomes infinite, or zero in the case of a series *RLC* circuit, we have a simple *LC* loop in which an oscillatory response can be maintained forever. Let us look briefly at an example of such a circuit, and then discuss another means of obtaining an identical response without the need of supplying any inductance.

Consider the source-free circuit of Fig. 9.34, in which the large values $L = 4$ H and $C = \frac{1}{36}$ F are used so that the calculations will be simple. We let $i(0) = -\frac{1}{6}$ A and $v(0) = 0$. We find that $\alpha = 0$ and $\omega_0^2 = 9$ s^{-2}, so that $\omega_d = 3$ rad/s. In the absence of exponential damping, the voltage v is simply

$$v = A \cos 3t + B \sin 3t$$

Since $v(0) = 0$, we see that $A = 0$. Next,

$$\left. \frac{dv}{dt} \right|_{t=0} = 3B = -\frac{i(0)}{1/36}$$

But $i(0) = -\frac{1}{6}$ ampere, and therefore $dv/dt = 6$ V/s at $t = 0$. We must have $B = 2$ V and so

$$v = 2 \sin 3t \qquad \text{V}$$

which is an undamped sinusoidal response; in other words, our voltage response does not decay.

Now let us see how we might obtain this voltage without using an *LC* circuit. Our intentions are to write the differential equation that v satisfies and then to develop a configuration of op amps that will yield the solution of the equation. Although we are working with a specific example, the technique is a general one that can be used to solve any linear homogeneous differential equation.

For the *LC* circuit of Fig. 9.34, we select v as our variable and set the sum of the downward inductor and capacitor currents equal to zero:

$$\frac{1}{4} \int_{t_0}^{t} v \, dt' - \frac{1}{6} + \frac{1}{36} \frac{dv}{dt} = 0$$

Differentiating once, we have

$$\frac{1}{4} v + \frac{1}{36} \frac{d^2 v}{dt^2} = 0$$

or

$$\frac{d^2 v}{dt^2} = -9v$$

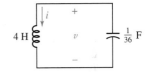

■ **FIGURE 9.34** This circuit is lossless, and it provides the undamped response $v = 2 \sin 3t$ V, if $v(0) = 0$ and $i(0) = -\frac{1}{6}$ A.

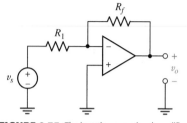

■ **FIGURE 9.35** The inverting operational amplifier provides a gain $v_o/v_s = -R_f/R_1$, assuming an ideal op amp.

In order to solve this equation, we plan to make use of the operational amplifier as an integrator. We assume that the highest-order derivative appearing in the differential equation here, d^2v/dt^2, is available in our configuration of op amps at an arbitrary point A. We now make use of the integrator, with $RC = 1$, as discussed in Sec. 7.5. The input is d^2v/dt^2, and the output must be $-dv/dt$, where the sign change results from using an inverting op amp configuration for the integrator. The initial value of dv/dt is 6 V/s, as we showed when we first analyzed the circuit, and thus an initial value of -6 V must be set in the integrator. The negative of the first derivative now forms the input to a second integrator. Its output is therefore $v(t)$, and the initial value is $v(0) = 0$. Now it only remains to multiply v by -9 to obtain the second derivative we assumed at point A. This is amplification by 9 with a sign change, and it is easily accomplished by using the op amp as an inverting amplifier.

Figure 9.35 shows the circuit of an inverting amplifier. For an ideal op amp, both the input current and the input voltage are zero. Thus, the current going "east" through R_1 is v_s/R_1, while that traveling west through R_f is v_o/R_f. Since their sum is zero, we have

$$\frac{v_o}{v_s} = -\frac{R_f}{R_1}$$

Thus, we can design for a gain of -9 by setting $R_f = 90\,\text{k}\Omega$ and $R_1 = 10\,\text{k}\Omega$, for example.

If we let R be 1 MΩ and C be 1 μF in each of the integrators, then

$$v_o = -\int_0^t v_s\, dt' + v_o(0)$$

in each case. The output of the inverting amplifier now forms the assumed input at point A, leading to the configuration of op amps shown in Fig. 9.36. If the left switch is closed at $t = 0$ while the two initial-condition switches are opened at the same time, the output of the second integrator will be the undamped sine wave $v = 2\sin 3t$ V.

Note that both the LC circuit of Fig. 9.34 and the op amp circuit of Fig. 9.36 have the same output, but the op amp circuit does not contain a

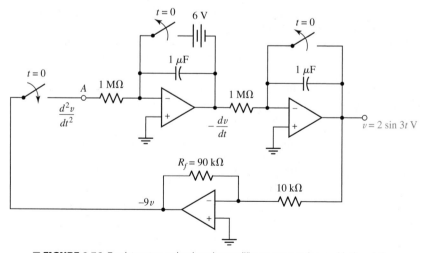

■ **FIGURE 9.36** Two integrators and an inverting amplifier are connected to provide the solution of the differential equation $d^2v/dt^2 = -9v$.

single inductor. It simply *acts* as though it contained an inductor, providing the appropriate sinusoidal voltage between its output terminal and ground. This can be a considerable practical or economic advantage in circuit design, as inductors are typically bulky, more costly than capacitors, and have more losses associated with them (and therefore are not as well approximated by the "ideal" model).

PRACTICE

9.11 Give new values for R_f and the two initial voltages in the circuit of Fig. 9.36 if the output represents the voltage $v(t)$ in the circuit of Fig. 9.37.

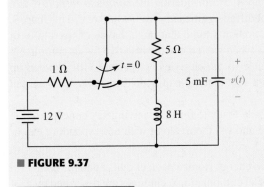

■ **FIGURE 9.37**

Ans: 250 kΩ; 400 V; 10 V.

SUMMARY AND REVIEW

The simple *RL* and *RC* circuits examined in Chap. 8 essentially did one of two things as the result of throwing a switch: *charge* or *discharge*. Which one happened was determined by the initial charge state of the energy storage element. In this chapter, we considered circuits that had two energy storage elements (a capacitor and an inductor), and found that things could get pretty interesting. There are two basic configurations of such *RLC* circuits: *parallel* connected and *series* connected. Analysis of such a circuit yields a *second-order* partial differential equation, consistent with the number of distinct energy storage elements (if we construct a circuit using only resistors and capacitors such that the capacitors cannot be combined using series/parallel techniques, we also obtain—eventually—a second-order partial differential equation).

Depending on the value of the resistance connected to our energy storage elements, we found the transient response of an *RLC* circuit could be either *overdamped* (decaying exponentially) or *underdamped* (decaying, but oscillatory), with a "special case" of *critically damped* which is difficult to achieve in practice. Oscillations can be useful (for example, in transmitting information over a wireless network) and not so useful (for example, in accidental feedback situations between an amplifier and a microphone at a concert). Although the oscillations are not sustained in the circuits we examined, we have at least seen one way to create them at will, and design for a specific frequency of operation if so desired. We didn't end up spending a great deal of time with the series connected *RLC* circuit because with the exception of α, the equations are the same; only a minor adjustment in how we employ initial conditions to find the two unknown constants characterizing the transient response is needed. Along those lines, there were two "tricks," if you will, that we encountered. One is that to employ the second initial condition, we need to take the derivative of our response equation. The second is that whether we're employing KCL or KVL to make use of that initial condition, we're doing so at the instant that $t = 0$; appreciating this fact can simplify equations dramatically by setting $t = 0$ early.

We wrapped up the chapter by considering the *complete response*, and our approach to this did not differ significantly from what we did in Chap. 8. We closed with a brief section on a topic that might have occurred to us at some point—what happens when we remove the resistive losses completely (by setting parallel resistance to ∞, or series resistance to 0)? We end up with an *LC* circuit, and we saw that we can approximate such an animal with an op amp circuit.

By now the reader is likely ready to finish reviewing key concepts of the chapter, so we'll stop here and list them, along with corresponding examples in the text.

❏ Circuits that contain two energy storage devices that cannot be combined using series-parallel combination techniques are described by a second-order differential equation.

❏ Series and parallel *RLC* circuits fall into one of three categories, depending on the relative values of R, L, and C:

Overdamped	$\alpha > \omega_0$
Critically damped	$\alpha = \omega_0$
Underdamped	$\alpha < \omega_0$

(Example 9.1)

❏ For series *RLC* circuits, $\alpha = R/2L$ and $\omega_0 = 1/\sqrt{LC}$. (Example 9.7)
❏ For parallel *RLC* circuits, $\alpha = 1/2RC$ and $\omega_0 = 1/\sqrt{LC}$. (Example 9.1)
❏ The typical form of an overdamped response is the sum of two exponential terms, one of which decays more quickly than the other: e.g., $A_1 e^{-t} + A_2 e^{-6t}$. (Examples 9.2, 9.3, 9.4)

❏ The typical form of a critically damped response is $e^{-\alpha t}(A_1 t + A_2)$. (Example 9.5)

❏ The typical form of an underdamped response is an exponentially damped sinusoid: $e^{-\alpha t}(B_1 \cos \omega_d t + B_2 \sin \omega_d t)$.
(Examples 9.6, 9.7, 9.8)

❏ During the transient response of an *RLC* circuit, energy is transferred between energy storage elements to the extent allowed by the resistive component of the circuit, which acts to dissipate the energy initially stored. (See Computer-Aided Analysis section.)

❏ The complete response is the sum of the forced and natural responses. In this case the total response must be determined before solving for the constants. (Examples 9.9, 9.10)

READING FURTHER

An excellent discussion of employing PSpice in the modeling of automotive suspension systems can be found in

R.W. Goody, *MicroSim PSpice for Windows,* vol. I, 2nd ed. Englewood Cliffs, N.J.: Prentice-Hall, 1998.

Many detailed descriptions of analogous networks can be found in Chap. 3 of

E. Weber, *Linear Transient Analysis Volume I.* New York: Wiley, 1954. (Out of print, but in many university libraries.)

EXERCISES

9.1 The Source-Free Parallel Circuit

1. For a certain source-free parallel *RLC* circuit, $R = 1$ kΩ, $C = 3$ μF, and L is such that the circuit response is overdamped. (*a*) Determine the value of L. (*b*) Write the equation for the voltage v across the resistor if it is known that $v(0^-) = 9$ V and $dv/dt|_{t=0^+} = 2$ V/s.

2. Element values of 10 mF and 2 nH are employed in the construction of a simple source-free parallel *RLC* circuit. (*a*) Select R so that the circuit is just barely overdamped. (*b*) Write the equation for the resistor current if its initial value is $i_R(0^+) = 13$ pA and $di_E/dt|_{t=0^+} = 1$ nA/s.

3. If a parallel *RLC* circuit is constructed from component values $C = 16$ mF and $L = 1$ mH, choose R such that the circuit is (*a*) just barely overdamped; (*b*) just barely underdamped; (*c*) critically damped. (*d*) Does your answer for part (*a*) change if the resistor tolerance is 1%? 10%? (*e*) Increase the exponential damping coefficient for part (*c*) by 20%. Is the circuit now underdamped, overdamped, or still critically damped? *Explain.*

4. Calculate α, ω_0, s_1, and s_2 for a source-free parallel *RLC* circuit if (*a*) $R = 4$ Ω, $L = 2.22$ H, and $C = 12.5$ mF; (*b*) $L = 1$ nH, $C = 1$ pF, and R is 1% of the value required to make the circuit underdamped. (*c*) Calculate the damping ratio for the circuits of parts (*a*) and (*b*).

5. You go to construct the circuit in Exercise 1, only to find no 1 kΩ resistors. In fact, all you are able to locate in addition to the capacitor and inductor is a 1 meter long piece of 24 AWG soft solid copper wire. Connecting it in parallel to the two components you did find, compute the value of α, ω_0, s_1, and s_2, and verify that the circuit is still overdamped.

6. Consider a source-free parallel *RLC* circuit having $\alpha = 10^8$ s^{-1}, $\omega_0 = 10^3$ rad/s, and $\omega_0 L = 5$ Ω. (*a*) Show that the stated units of $\omega_0 L$ are correct. (*b*) Compute s_1 and s_2. (*c*) Write the general form of the natural response for the capacitor voltage. (*d*) By appropriate substitution, verify that your answer to part (*c*) is indeed a solution to Eq. [1] if the inductor and capacitor each initially store 1 mJ of energy, respectively.

7. A parallel *RLC* circuit is constructed with $R = 500$ Ω, $C = 10$ μF, and L such that it is critically damped. (*a*) Determine L. Is this value large or small for a printed-circuit board mounted component? (*b*) Add a resistor in parallel to the existing components such that the damping ratio is equal to 10. (*c*) Does increasing the damping ratio further lead to an overdamped, critically damped, or underdamped circuit? *Explain.*

9.2 The Overdamped Parallel *RLC* Circuit

8. The circuit of Fig. 9.2 is modified substantially, with the resistor being replaced with a 1 kΩ resistor, the inductor swapped out for a smaller 7 mH version, the capacitor replaced with a 1 nF alternative, and now the inductor is initially discharged while the capacitor is storing 7.2 mJ. (*a*) Compute α, ω_0, s_1, and s_2, and verify that the circuit is still overdamped. (*b*) Obtain an expression

for the current flowing through the resistor which is valid for $t > 0$. (*c*) Calculate the magnitude of the resistor current at $t = 10$ μs.

9. The voltage across a capacitor is found to be given by $v_C(t) = 10e^{-10t} - 5e^{-4t}$ V. (*a*) Sketch each of the two components over the range of $0 \le t \le 1.5$ s. (*b*) Graph the capacitor voltage over the same time range.

10. The current flowing through a certain inductor is found to be given by $i_L(t) = 0.20e^{-2t} - 0.6e^{-3t}$ V. (*a*) Sketch each of the two components over the range of $0 \le t \le 1.5$ s. (*b*) Graph the inductor current over the same time range. (*c*) Graph the energy remaining in the inductor over $0 \le t \le 1.5$ s.

11. The current flowing through a 5 Ω resistor in a source-free parallel *RLC* circuit is determined to be $i_R(t) = 2e^{-t} - 3e^{-8t}$ V, $t > 0$. Determine (*a*) the maximum current and the time at which it occurs; (*b*) the settling time; (*c*) the time t corresponding to the resistor absorbing 2.5 W of power.

12. For the circuit of Fig. 9.38, obtain an expression for $v_C(t)$ valid for all $t > 0$.

13. Consider the circuit depicted in Fig. 9.38. (*a*) Obtain an expression for $i_L(t)$ valid for all $t > 0$. (*b*) Obtain an expression for $i_R(t)$ valid for all $t > 0$. (*c*) Determine the settling time for both i_L and i_R.

14. With regard to the circuit represented in Fig. 9.39, determine (*a*) $i_C(0^-)$; (*b*) $i_L(0^-)$; (*c*) $i_R(0^-)$; (*d*) $v_C(0^-)$; (*e*) $i_C(0^+)$; (*f*) $i_L(0^+)$; (*g*) $i_R(0^+)$; (*h*) $v_C(0^+)$.

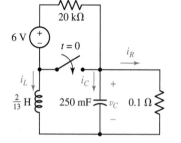

■ **FIGURE 9.38**

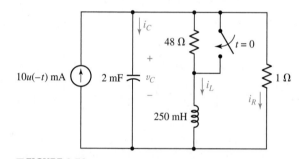

■ **FIGURE 9.39**

15. (*a*) Assuming the passive sign convention, obtain an expression for the voltage across the 1 Ω resistor in the circuit of Fig. 9.39 which is valid for all $t > 0$. (*b*) Determine the settling time of the resistor voltage.

16. With regard to the circuit presented in Fig. 9.40, (*a*) obtain an expression for $v(t)$ which is valid for all $t > 0$; (*b*) calculate the maximum inductor current and identify the time at which it occurs; (*c*) determine the settling time.

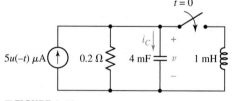

■ **FIGURE 9.40**

17. Obtain expressions for the current $i(t)$ and voltage $v(t)$ as labeled in the circuit of Fig. 9.41 which are valid for all $t > 0$.

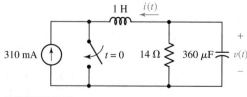

■ **FIGURE 9.41**

18. Replace the 14 Ω resistor in the circuit of Fig. 9.41 with a 1 Ω resistor. (a) Obtain an expression for the energy stored in the capacitor as a function of time, valid for $t > 0$. (b) Determine the time at which the energy in the capacitor has been reduced to one-half its maximum value. (c) Verify your answer with an appropriate PSpice simulation.

19. Design a complete source-free parallel RLC circuit which exhibits an overdamped response, has a settling time of 1 s, and has a damping ratio of 15.

20. For the circuit represented by Fig. 9.42, the two resistor values are $R_1 = 0.752$ Ω and $R_2 = 1.268$ Ω, respectively. (a) Obtain an expression for the energy stored in the capacitor, valid for all $t > 0$; (b) determine the settling time of the current labeled i_A.

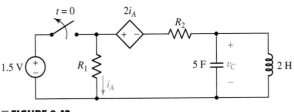

■ **FIGURE 9.42**

9.3 Critical Damping

21. A motor coil having an inductance of 8 H is in parallel with a 2 μF capacitor and a resistor of unknown value. The response of the parallel combination is determined to be critically damped. (a) Determine the value of the resistor. (b) Compute α. (c) Write the equation for the current flowing into the resistor if the top node is labeled v, the bottom node is grounded, and $v = Ri_r$. (d) Verify that your equation is a solution to the circuit differential equation,

$$\frac{di_r}{dt} + 2\alpha \frac{di_r}{dt} + \alpha^2 i_r = 0$$

22. The condition for critical damping in an RLC circuit is that the resonant frequency ω_0 and the exponential damping factor α are equal. This leads to the relationship $L = 4R^2C$, which implies that 1 H = 1 Ω² · F. Verify this equivalence by breaking down each of the three units to fundamental SI units (see Chap. 2).

23. A critically damped parallel RLC circuit is constructed from component values 40 Ω, 8 nF, and 51.2 μH, respectively. (a) Verify that the circuit is indeed critically damped. (b) Explain why, in practice, the circuit once fabricated is unlikely to be truly critically damped. (c) The inductor initially stores 1 mJ of energy while the capacitor is initially discharged. Determine the magnitude of the capacitor voltage at $t = 500$ ns, the maximum absolute capacitor voltage, and the settling time.

24. Design a complete (i.e., with all necessary switches or step function sources) parallel RLC circuit which has a critically damped response such that the capacitor voltage at $t = 1$ s is equal to 9 V and the circuit is source-free for all $t > 0$.

25. A critically damped parallel RLC circuit is constructed from component values 40 Ω and 2 pF. (a) Determine the value of L, taking care not to overround. (b) Explain why, in practice, the circuit once fabricated is unlikely to be truly critically damped. (c) The inductor initially stores no energy while the capacitor is initially storing 10 pJ. Determine the power absorbed by the resistor at $t = 2$ ns, the maximum absolute inductor current $|i_L|$, and the settling time.

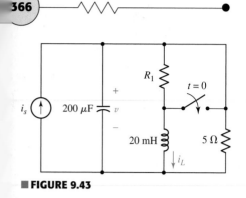

■ **FIGURE 9.43**

26. For the circuit of Fig. 9.43, $i_s(t) = 30u(-t)$ mA. (*a*) Select R_1 so that $v(0^+) = 6$ V. (*b*) Compute $v(2$ ms). (*c*) Determine the settling time of the capacitor voltage. (*d*) Is the inductor current settling time the same as your answer to part (*c*)?

27. The current source in Fig. 9.43 is $i_s(t) = 10u(1 - t)$ μA. (*a*) Select R_1 such that $i_L(0^+) = 2$ μA. Compute i_L at $t = 500$ ms and $t = 1.002$ ms.

28. The inductor in the circuit of Fig. 9.41 is changed such that the circuit response is now critically damped. (*a*) Determine the new inductor value. (*b*) Calculate the energy stored in both the inductor and the capacitor at $t = 10$ ms.

29. The circuit of Fig. 9.42 is rebuilt such that the quantity controlling the dependent source is now $100i_A$, a 2 μF capacitor is used instead, and $R_1 = R_2 = 10$ Ω. (*a*) Calculate the inductor value required to obtain a critically damped response. (*b*) Determine the power being absorbed by R_2 at $t = 300$ μs.

9.4 The Underdamped Parallel *RLC* Circuit

30. (*a*) With respect to the parallel *RLC* circuit, derive an expression for R in terms of C and L to ensure the response is underdamped. (*b*) If $C = 1$ nF and $L = 10$ mH, select R such that an underdamped response is (just barely) achieved. (*c*) If the damping ratio is increased, does the circuit become more or less underdamped? *Explain*. (*d*) Compute α and ω_d for the value of R you selected in part (*b*).

31. The circuit of Fig. 9.1 is constructed using component values 10 kΩ, 72 μH, and 18 pF. (*a*) Compute α, ω_d, and ω_0. Is the circuit overdamped, critically damped, or underdamped? (*b*) Write the form of the natural capacitor voltage response $v(t)$. (*c*) If the capacitor initially stores 1 nJ of energy, compute v at $t = 300$ ns.

32. The source-free circuit depicted in Fig. 9.1 is constructed using a 10 mH inductor, a 1 mF capacitor, and a 1.5 kΩ resistor. (*a*) Calculate α, ω_d, and ω_0. (*b*) Write the equation which describes the current i for $t > 0$. (*c*) Determine the maximum value of i, and the time at which it occurs, if the inductor initially stores no energy and $v(0^-) = 9$ V.

33. (*a*) Graph the current i for the circuit described in Exercise 32 for resistor values 1.5 kΩ, 15 kΩ, and 150 kΩ. Make three separate graphs and be sure to extend the corresponding time axis to $6\pi/\omega_d$ in each case. (*b*) Determine the corresponding settling times.

34. Analyze the circuit described in Exercise 32 to find $v(t)$, $t > 0$, if R is equal to (*a*) 2 kΩ; (*b*) 2 Ω. (*c*) Graph both responses over the range of $0 \le t \le 60$ ms. (*d*) Verify your answers with appropriate PSpice simulations.

35. For the circuit of Fig. 9.44, determine (*a*) $i_C(0^-)$; (*b*) $i_L(0^-)$; (*c*) $i_R(0^-)$; (*d*) $v_C(0^-)$; (*e*) $i_C(0^+)$; (*f*) $i_L(0^+)$; (*g*) $i_R(0^+)$; (*h*) $v_C(0^+)$.

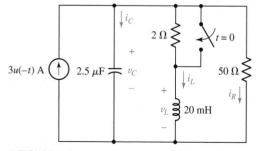

■ **FIGURE 9.44**

36. Obtain an expression for $v_L(t)$, $t > 0$, for the circuit shown in Fig. 9.44. Plot the waveform for at least two periods of oscillation.

37. For the circuit of Fig. 9.45, determine (*a*) the first time $t > 0$ when $v(t) = 0$; (*b*) the settling time.

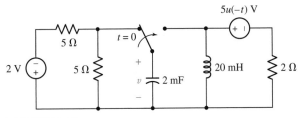

■ **FIGURE 9.45**

38. (*a*) Design a parallel *RLC* circuit that provides a capacitor voltage which oscillates with a frequency of 100 rad/s, with a maximum value of 10 V occurring at $t = 0$, and the second and third maxima both in excess of 6 V. (*b*) Verify your design with an appropriate PSpice simulation.

39. The circuit depicted in Fig. 9.46 is just barely underdamped. (*a*) Compute α and ω_d. (*b*) Obtain an expression for $i_L(t)$ valid for $t > 0$. (*c*) Determine how much energy is stored in the capacitor, and in the inductor, at $t = 200$ ms.

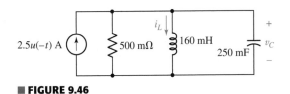

■ **FIGURE 9.46**

40. When constructing the circuit of Fig. 9.46, you inadvertently install a 500 MΩ resistor by mistake. (*a*) Compute α and ω_d. (*b*) Obtain an expression for $i_L(t)$ valid for $t > 0$. (*c*) Determine how long it takes for the energy stored in the inductor to reach 10% of its maximum value.

9.5 The Source-Free Series *RLC* Circuit

41. The circuit of Fig. 9.21*a* is constructed with a 160 mF capacitor and a 250 mH inductor. Determine the value of R needed to obtain (*a*) a critically damped response; (*b*) a "just barely" underdamped response. (*c*) Compare your answers to parts (*a*) and (*b*) if the circuit was a parallel *RLC* circuit.

42. Component values of $R = 2\ \Omega$, $C = 1$ mF, and $L = 2$ mH are used to construct the circuit represented in Fig. 9.21*a*. If $v_C(0^-) = 1$ V and no current initially flows through the inductor, calculate $i(t)$ at $t = 1$ ms, 2 ms, and 3 ms.

43. The series *RLC* circuit described in Exercise 42 is modified slightly by adding a 2 Ω resistor in parallel to the existing resistor. The initial capacitor voltage remains 1 V, and there is still no current flowing in the inductor prior to $t = 0$. (*a*) Calculate $v_C(t)$ at 4 ms. (*b*) Sketch $v_C(t)$ over the interval $0 \leq t \leq 10$ s.

44. The simple three-element series *RLC* circuit of Exercise 42 is constructed with the same component values, but the initial capacitor voltage $v_C(0^-) = 2$ V and the initial inductor current $i(0^-) = 1$ mA. (*a*) Obtain an expression for $i(t)$ valid for all $t > 0$. (*b*) Verify your solution with an appropriate simulation.

45. The series *RLC* circuit of Fig. 9.22 is constructed using $R = 1$ kΩ, $C = 2$ mF, and $L = 1$ mH. The initial capacitor voltage v_C is -4 V at $t = 0^-$. There is no current initially flowing through the inductor. (*a*) Obtain an expression for $v_C(t)$ valid for $t > 0$. (*b*) Sketch over $0 \leq t \leq 6\ \mu$s.

46. With reference to the circuit depicted in Fig. 9.47, calculate (*a*) α; (*b*) ω_0; (*c*) $i(0^+)$; (*d*) $di/dt|_{0^+}$; (*e*) $i(t)$ at $t = 6$ s.

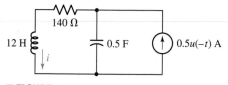

■ FIGURE 9.47

47. Obtain an equation for v_C as labeled in the circuit of Fig. 9.48 valid for all $t > 0$.

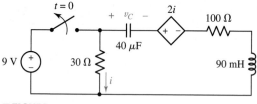

■ FIGURE 9.48

 48. With reference to the series *RLC* circuit of Fig. 9.48, (*a*) obtain an expression for *i*, valid for $t > 0$; (*b*) calculate $i(0.8$ ms$)$ and $i(4$ ms$)$; (*c*) verify your answers to part (*b*) with an appropriate PSpice simulation.

49. Obtain an expression for i_1 as labeled in Fig. 9.49 which is valid for all $t > 0$.

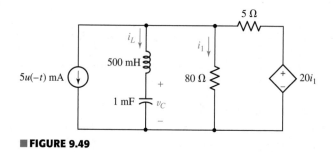

■ FIGURE 9.49

9.6 The Complete Response of the *RLC* Circuit

50. In the series circuit of Fig. 9.50, set $R = 1$ Ω. (*a*) Compute α and ω_0. (*b*) If $i_s = 3u(-t) + 2u(t)$ mA, determine $v_R(0^-)$, $i_L(0^-)$, $v_C(0^-)$, $v_R(0^+)$, $i_L(0^+)$, $v_C(0^+)$, $i_L(\infty)$, and $v_C(\infty)$.

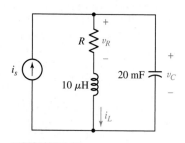

■ FIGURE 9.50

The real power of the phasor-based analysis technique lies in the fact that it is possible to define *algebraic* relationships between the voltage and current for inductors and capacitors, just as we have always been able to do in the case of resistors. Now that we are able to transform into and out of the frequency domain, we can proceed to our simplification of sinusoidal steady-state analysis by establishing the relationship between the phasor voltage and phasor current for each of the three passive elements.

The Resistor

The resistor provides the simplest case. In the time domain, as indicated by Fig. 10.13a, the defining equation is

$$v(t) = Ri(t)$$

Now let us apply the complex voltage

$$v(t) = V_m e^{j(\omega t + \theta)} = V_m \cos(\omega t + \theta) + jV_m \sin(\omega t + \theta) \qquad [16]$$

and assume the complex current response

$$i(t) = I_m e^{j(\omega t + \phi)} = I_m \cos(\omega t + \phi) + jI_m \sin(\omega t + \phi) \qquad [17]$$

so that

$$V_m e^{j(\omega t + \theta)} = Ri(t) = RI_m e^{j(\omega t + \phi)}$$

Dividing throughout by $e^{j\omega t}$, we find

$$V_m e^{j\theta} = RI_m e^{j\phi}$$

or, in polar form,

$$V_m \underline{/\theta} = RI_m \underline{/\phi}$$

But $V_m \underline{/\theta}$ and $I_m \underline{/\phi}$ merely represent the general voltage and current phasors **V** and **I**. Thus,

$$\mathbf{V} = R\mathbf{I} \qquad [18]$$

The voltage-current relationship in phasor form for a resistor has the same form as the relationship between the time-domain voltage and current. The defining equation in phasor form is illustrated in Fig. 10.13b. The angles θ and ϕ are equal, so that the current and voltage are always in phase.

As an example of the use of both the time-domain and frequency-domain relationships, let us assume that a voltage of $8\cos(100t - 50°)$ V is across a 4 Ω resistor. Working in the time domain, we find that the current must be

$$i(t) = \frac{v(t)}{R} = 2\cos(100t - 50°) \qquad \text{A}$$

The phasor form of the same voltage is $8\underline{/-50°}$ V, and therefore

$$\mathbf{I} = \frac{\mathbf{V}}{R} = 2\underline{/-50°} \qquad \text{A}$$

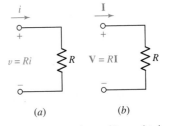

■ **FIGURE 10.13** A resistor and its associated voltage and current in (a) the time domain, $v = Ri$; and (b) the frequency domain, **V** = R**I**.

Ohm's law holds true both in the time domain and in the frequency domain. In other words, the voltage across a resistor is always given by the resistance times the current flowing through the element.

If we transform this answer back to the time domain, it is evident that the same expression for the current is obtained. We conclude that there is no saving in time or effort when a *resistive* circuit is analyzed in the frequency domain.

The Inductor

Let us now turn to the inductor. The time-domain representation is shown in Fig. 10.14a, and the defining equation, a time-domain expression, is

$$v(t) = L \frac{di(t)}{dt} \tag{19}$$

After substituting the complex voltage equation [16] and complex current equation [17] in Eq. [19], we have

$$V_m e^{j(\omega t + \theta)} = L \frac{d}{dt} I_m e^{j(\omega t + \phi)}$$

Taking the indicated derivative:

$$V_m e^{j(\omega t + \theta)} = j\omega L I_m e^{j(\omega t + \phi)}$$

and dividing through by $e^{j\omega t}$:

$$V_m e^{j\theta} = j\omega L I_m e^{j\phi}$$

we obtain the desired phasor relationship

$$\boxed{\mathbf{V} = j\omega L \mathbf{I}} \tag{20}$$

The time-domain differential equation [19] has become the algebraic equation [20] in the frequency domain. The phasor relationship is indicated in Fig. 10.14b. Note that the angle of the factor $j\omega L$ is exactly $+90°$ and that **I** must therefore lag **V** by 90° in an inductor.

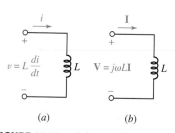

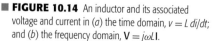

■ **FIGURE 10.14** An inductor and its associated voltage and current in (a) the time domain, $v = L\, di/dt$; and (b) the frequency domain, $\mathbf{V} = j\omega L \mathbf{I}$.

EXAMPLE 10.4

Apply the voltage $8\underline{/-50°}$ V at a frequency $\omega = 100$ rad/s to a 4 H inductor, and determine the phasor current and the time-domain current.

We make use of the expression we just obtained for the inductor,

$$\mathbf{I} = \frac{\mathbf{V}}{j\omega L} = \frac{8\underline{/-50°}}{j\,100(4)} = -j0.02\underline{/-50°} = (1\underline{/-90°})(0.02\underline{/-50°})$$

or

$$\mathbf{I} = 0.02\underline{/-140°} \text{ A}$$

If we express this current in the time domain, it becomes

$$i(t) = 0.02\cos(100t - 140°) \text{ A} = 20\cos(100t - 140°) \text{ mA}$$

The Capacitor

The final element to consider is the capacitor. The time-domain current-voltage relationship is

$$i(t) = C\frac{dv(t)}{dt}$$

The equivalent expression in the frequency domain is obtained once more by letting $v(t)$ and $i(t)$ be the complex quantities of Eqs. [16] and [17], taking the indicated derivative, suppressing $e^{j\omega t}$, and recognizing the phasors **V** and **I**. Doing this, we find

$$\mathbf{I} = j\omega C\mathbf{V} \qquad\qquad [21]$$

Thus, **I** leads **V** by 90° in a capacitor. This, of course, does not mean that a current response is present one-quarter of a period earlier than the voltage that caused it! We are studying steady-state response, and we find that the current maximum is caused by the increasing voltage that occurs 90° earlier than the voltage maximum.

The time-domain and frequency-domain representations are compared in Fig. 10.15*a* and *b*. We have now obtained the **V-I** relationships for the three passive elements. These results are summarized in Table 10.1, where the time-domain *v-i* expressions and the frequency-domain **V-I** relationships are shown in adjacent columns for the three circuit elements. All the phasor equations are algebraic. Each is also linear, and the equations relating to inductance and capacitance bear a great similarity to Ohm's law. In fact, we will indeed *use* them as we use Ohm's law.

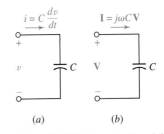

■ **FIGURE 10.15** (*a*) The time-domain and (*b*) the frequency-domain relationships between capacitor current and voltage.

TABLE **10.1** Comparison of Time-Domain and Frequency-Domain Voltage-Current Expressions

Time Domain		Frequency Domain	
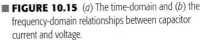	$v = Ri$	$\mathbf{V} = R\mathbf{I}$	
	$v = L\dfrac{di}{dt}$	$\mathbf{V} = j\omega L\mathbf{I}$	
	$v = \dfrac{1}{C}\displaystyle\int i\,dt$	$\mathbf{V} = \dfrac{1}{j\omega C}\mathbf{I}$	

Kirchhoff's Laws Using Phasors

Kirchhoff's voltage law in the time domain is

$$v_1(t) + v_2(t) + \cdots + v_N(t) = 0$$

We now use Euler's identity to replace each real voltage v_i by a complex voltage having the same real part, suppress $e^{j\omega t}$ throughout, and obtain

$$\mathbf{V}_1 + \mathbf{V}_2 + \cdots + \mathbf{V}_N = 0$$

Thus, we see that Kirchhoff's voltage law applies to phasor voltages just as it did in the time domain. Kirchhoff's current law can be shown to hold for phasor currents by a similar argument.

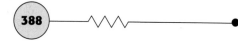

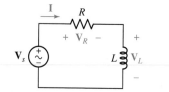

■ **FIGURE 10.16** The series RL circuit with a phasor voltage applied.

Now let us look briefly at the series RL circuit that we have considered several times before. The circuit is shown in Fig. 10.16, and a phasor current and several phasor voltages are indicated. We may obtain the desired response, a time-domain current, by first finding the phasor current. From Kirchhoff's voltage law,

$$\mathbf{V}_R + \mathbf{V}_L = \mathbf{V}_s$$

and using the recently obtained **V-I** relationships for the elements, we have

$$R\mathbf{I} + j\omega L\mathbf{I} = \mathbf{V}_s$$

The phasor current is then found in terms of the source voltage $\mathbf{V}_s$:

$$\mathbf{I} = \frac{\mathbf{V}_s}{R + j\omega L}$$

Let us select a source-voltage amplitude of V_m and phase angle of $0°$. Thus,

$$\mathbf{I} = \frac{V_m\underline{/0°}}{R + j\omega L}$$

The current may be transformed to the time domain by first writing it in polar form:

$$\mathbf{I} = \frac{V_m}{\sqrt{R^2 + \omega^2 L^2}}\underline{/[-\tan^{-1}(\omega L/R)]}$$

and then following the familiar sequence of steps to obtain in a very simple manner the same result we obtained the "hard way" earlier in this chapter.

EXAMPLE 10.5

For the *RLC* circuit of Fig. 10.17, determine $\mathbf{I}_s$ and $i_s(t)$ if both sources operate at $\omega = 2$ rad/s, and $\mathbf{I}_C = 2\underline{/28°}$ A.

The fact that we are given $\mathbf{I}_C$ and asked for $\mathbf{I}_s$ is all the prompting we need to consider applying KCL. If we label the capacitor voltage $\mathbf{V}_C$ consistent with the passive sign convention, then

$$\mathbf{V}_C = \frac{1}{j\omega C}\mathbf{I}_C = \frac{-j}{2}\mathbf{I}_C = \frac{-j}{2}(2\underline{/28°}) = (0.5\underline{/-90°})(2\underline{/28°}) = 1\underline{/-62°} \text{ V}$$

This voltage also appears across the 2 Ω resistor, so that the current $\mathbf{I}_{R_2}$ flowing downward through that branch is

$$\mathbf{I}_{R_2} = \frac{1}{2}\mathbf{V}_C = \frac{1}{2}\underline{/-62°} \text{ A}$$

KCL then yields $\mathbf{I}_s = \mathbf{I}_{R_2} + \mathbf{I}_C = 1\underline{/-62°} + \frac{1}{2}\underline{/-62°} = (3/2)\underline{/-62°}$ A. (We should note the addition of these polar quantities was trivial since the resistor and capacitor currents have the same angle, i.e., are in phase.)

Thus $\mathbf{I}_s$ and knowledge of ω permit us to write $i_s(t)$ directly:

$$i_s(t) = 1.5 \cos(2t - 62°) \text{ A}$$

PRACTICE

10.8 In the circuit of Fig. 10.17, both sources operate at $\omega = 1$ rad/s. If $\mathbf{I}_C = 2\underline{/28°}$ A and $\mathbf{I}_L = 3\underline{/53°}$ A, calculate (*a*) $\mathbf{I}_s$; (*b*) $\mathbf{V}_s$; (*c*) $i_{R_1}(t)$.

Ans: $3\underline{/-62°}$ A; (*b*) $3.71\underline{/-4.5°}$ V; (*c*) $3.22 \cos(t - 4.5°)$ A.

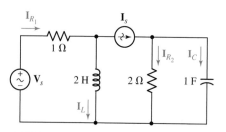

■ **FIGURE 10.17** A three-mesh circuit. Each source operates at the same frequency ω.

10.5 IMPEDANCE AND ADMITTANCE

The current-voltage relationships for the three passive elements in the frequency domain are (assuming that the passive sign convention is satisfied)

$$\mathbf{V} = R\mathbf{I} \qquad \mathbf{V} = j\omega L\mathbf{I} \qquad \mathbf{V} = \frac{\mathbf{I}}{j\omega C}$$

If these equations are written as phasor voltage/phasor current ratios

$$\frac{\mathbf{V}}{\mathbf{I}} = R \qquad \frac{\mathbf{V}}{\mathbf{I}} = j\omega L \qquad \frac{\mathbf{V}}{\mathbf{I}} = \frac{1}{j\omega C}$$

we find that these ratios are simple quantities that depend on element values (and frequency also, in the case of inductance and capacitance). We treat these ratios in the same manner that we treat resistances, provided we remember that they are complex quantities.

Let us define the ratio of the phasor voltage to the phasor current as *impedance,* symbolized by the letter **Z**. The impedance is a complex quantity having the dimensions of ohms. Impedance is not a phasor and cannot be transformed to the time domain by multiplying by $e^{j\omega t}$ and taking the real part. Instead, we think of an inductor as being represented in the time domain by its inductance L and in the frequency domain by its impedance $j\omega L$. A capacitor in the time domain has a capacitance C; in the frequency domain, it has an impedance $1/j\omega C$. Impedance is a part of the frequency domain and not a concept that is a part of the time domain.

$$Z_R = R$$
$$Z_L = j\omega L$$
$$Z_C = \frac{1}{j\omega C}$$

Series Impedance Combinations

The validity of Kirchhoff's two laws in the frequency domain leads to the fact that impedances may be combined in series and parallel by the same rules we established for resistances. For example, at $\omega = 10 \times 10^3$ rad/s, a 5 mH inductor in series with a 100 μF capacitor may be replaced by the sum of the individual impedances. The impedance of the inductor is

$$\mathbf{Z}_L = j\omega L = j50 \ \Omega$$

and the impedance of the capacitor is

$$\mathbf{Z}_C = \frac{1}{j\omega C} = \frac{-j}{\omega C} = -j1 \ \Omega$$

Note that $\frac{1}{j} = -j$.

The impedance of the series combination is therefore

$$\mathbf{Z}_{eq} = \mathbf{Z}_L + \mathbf{Z}_C = j50 - j1 = j49 \ \Omega$$

The impedance of inductors and capacitors is a function of frequency, and this equivalent impedance is thus applicable only at the single frequency at which it was calculated, $\omega = 10,000$ rad/s. If we change the frequency to $\omega = 5000$ rad/s, for example, $\mathbf{Z}_{eq} = j23 \ \Omega$.

Parallel Impedance Combinations

The *parallel* combination of the 5 mH inductor and the 100 μF capacitor at $\omega = 10,000$ rad/s is calculated in exactly the same fashion in which we calculated parallel resistances:

$$\mathbf{Z}_{eq} = \frac{(j50)(-j1)}{j50 - j1} = \frac{50}{j49} = -j1.020 \ \Omega$$

At $\omega = 5000$ rad/s, the parallel equivalent is $-j2.17 \ \Omega$.

Reactance

Of course, we may choose to express impedance in either *rectangular* ($\mathbf{Z} = R + jX$) or *polar* ($\mathbf{Z} = |\mathbf{Z}|\underline{/\theta}$) form. In rectangular form, we can see clearly the real part which arises only from real resistances, and an imaginary component, termed the **reactance,** which arises from the energy storage elements. Both resistance and reactance have units of ohms, but reactance will always depend upon frequency. An ideal resistor has zero reactance; an ideal inductor or capacitor is purely reactive (i.e., characterized by zero resistance). Can a series or parallel combination include *both* a capacitor and an inductor, and yet have *zero reactance?* Sure! Consider the series connection of a 1 Ω resistor, a 1 F capacitor, and a 1 H inductor driven at $\omega = 1$ rad/s. $\mathbf{Z}_{eq} = 1 - j(1)(1) + j(1)(1) = 1 \ \Omega$. At that particular frequency, the equivalent is a simple 1 Ω resistor. However, even small deviations from $\omega = 1$ rad/s lead to nonzero reactance.

EXAMPLE 10.6

Determine the equivalent impedance of the network shown in Fig. 10.18a, given an operating frequency of 5 rad/s.

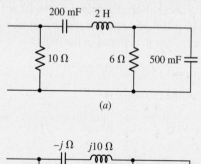

(a)

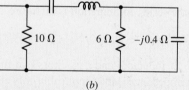

(b)

■ **FIGURE 10.18** (*a*) A network that is to be replaced by a single equivalent impedance. (*b*) The elements are replaced by their impedances at $\omega = 5$ rad/s.

We begin by converting the resistors, capacitors, and inductor into the corresponding impedances as shown in Fig. 10.18*b*.

Upon examining the resulting network, we observe that the 6 Ω impedance is in parallel with $-j0.4$ Ω. This combination is equivalent to

$$\frac{(6)(-j0.4)}{6-j0.4} = 0.02655 - j0.3982 \text{ Ω}$$

which is in series with both the $-j$ Ω and $j10$ Ω impedances, so that we have

$$0.0265 - j0.3982 - j + j10 = 0.02655 + j8.602 \text{ Ω}$$

This new impedance is in parallel with 10 Ω, so that the equivalent impedance of the network is

$$10 \parallel (0.02655 + j8.602) = \frac{10(0.02655 + j8.602)}{10 + 0.02655 + j8.602}$$
$$= 4.255 + j4.929 \text{ Ω}$$

Alternatively, we can express the impedance in polar form as $6.511\underline{/49.20°}$ Ω.

PRACTICE

10.9 With reference to the network shown in Fig. 10.19, find the input impedance $\mathbf{Z}_{in}$ that would be measured between terminals: (a) a and g; (b) b and g; (c) a and b.

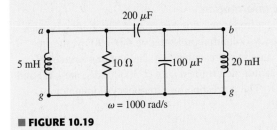

$\omega = 1000$ rad/s

■ FIGURE 10.19

Ans: $2.81 + j4.49$ Ω; $1.798 - j1.124$ Ω; $0.1124 - j3.82$ Ω.

It is important to note that the resistive component of the impedance is not necessarily equal to the resistance of the resistor that is present in the network. For example, a 10 Ω resistor and a 5 H inductor in series at $\omega = 4$ rad/s have an equivalent impedance $\mathbf{Z} = 10 + j20$ Ω, or, in polar form, $22.4\underline{/63.4°}$ Ω. In this case, the resistive component of the impedance is equal to the resistance of the series resistor because the network is a simple series network. However, if these same two elements are placed in parallel, the equivalent impedance is $10(j20)/(10 + j20)$ Ω, or $8 + j4$ Ω. The resistive component of the impedance is now 8 Ω.

EXAMPLE 10.7

Find the current $i(t)$ in the circuit shown in Fig. 10.20a.

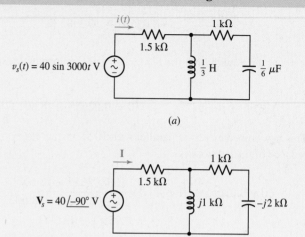

(a)

(b)

■ **FIGURE 10.20** (a) An *RLC* circuit for which the sinusoidal forced response $i(t)$ is desired. (b) The frequency-domain equivalent of the given circuit at $\omega = 3000$ rad/s.

▸ **Identify the goal of the problem.**

We need to find the sinusoidal steady-state current flowing through the 1.5 kΩ resistor due to the 3000 rad/s voltage source.

▸ **Collect the known information.**

We begin by drawing a frequency-domain circuit. The source is transformed to the frequency-domain representation $40\underline{/-90^\circ}$ V, the frequency domain response is represented as **I**, and the impedances of the inductor and capacitor, determined at $\omega = 3000$ rad/s, are j kΩ and $-j2$ kΩ, respectively. The corresponding frequency-domain circuit is shown in Fig. 10.20b.

▸ **Devise a plan.**

We will analyze the circuit of Fig. 10.20b to obtain **I**; combining impedances and invoking Ohm's law is one possible approach. We will then make use of the fact that we know $\omega = 3000$ rad/s to convert **I** into a time-domain expression.

▸ **Construct an appropriate set of equations.**

$$\mathbf{Z}_{\text{eq}} = 1.5 + \frac{(j)(1-2j)}{j+1-2j} = 1.5 + \frac{2+j}{1-j}$$

$$= 1.5 + \frac{2+j}{1-j}\frac{1+j}{1+j} = 1.5 + \frac{1+j3}{2}$$

$$= 2 + j1.5 = 2.5\underline{/36.87^\circ} \text{ k}\Omega$$

The phasor current is then simply

$$I = \frac{V_s}{Z_{eq}}$$

▶ **Determine if additional information is required.**

Substituting known values, we find that

$$I = \frac{40\underline{/-90°}}{2.5\underline{/36.87°}} \text{ mA}$$

which, along with the knowledge that $\omega = 3000$ rad/s, is sufficient to solve for $i(t)$.

▶ **Attempt a solution.**

This complex expression is easily simplified to a single complex number in polar form:

$$I = \frac{40}{2.5}\underline{/-90° - 36.87°} \text{ mA} = 16.00\underline{/-126.9°} \text{ mA}$$

Upon transforming the current to the time domain, the desired response is obtained:

$$i(t) = 16\cos(3000t - 126.9°) \qquad \text{mA}$$

▶ **Verify the solution. Is it reasonable or expected?**

The effective impedance connected to the source has an angle of $+36.87°$, indicating that it has a net inductive character, or that the current will lag the voltage. Since the voltage source has a phase angle of $-90°$ (once converted to a cosine source), we see that our answer is consistent.

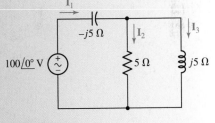

PRACTICE

10.10 In the frequency-domain circuit of Fig. 10.21, find (*a*) I_1; (*b*) I_2; (*c*) I_3.

Ans: $28.3\underline{/45°}$ A; $20\underline{/90°}$ A; $20\underline{/0°}$ A.

■ **FIGURE 10.21**

Before we begin to write great numbers of equations in the time domain or in the frequency domain, it is very important that we shun the construction of equations that are partly in the time domain, partly in the frequency domain, and wholly incorrect. One clue that a faux pas of this type has been committed is the sight of both a complex number and a t in the same equation, except in the factor $e^{j\omega t}$. And, since $e^{j\omega t}$ plays a much bigger role in derivations than in applications, it is pretty safe to say that students who find they have just created an equation containing j and t, or $\underline{/}$ and t, have created a monster that the world would be better off without.

For example, a few equations back we saw

$$\mathbf{I} = \frac{\mathbf{V}_s}{\mathbf{Z}_{eq}} = \frac{40\underline{/-90°}}{2.5\underline{/36.9°}} = 16\underline{/-126.9°} \text{ mA}$$

Please do not try anything like the following:

$$i(t) \not= \frac{40 \sin 3000t}{2.5\underline{/36.9°}} \qquad \text{or} \qquad i(t) \not= \frac{40 \sin 3000t}{2 + j1.5}$$

Admittance

Although the concept of impedance is very useful, and familiar in a way based on our experience with resistors, the reciprocal is often just as valuable. We define this quantity as the ***admittance*** $\mathbf{Y}$ of a circuit element or passive network, and it is simply the ratio of current to voltage:

The real part of the admittance is the ***conductance*** G, and the imaginary part is the ***susceptance*** B. All three quantities ($\mathbf{Y}$, G, and B) are measured in siemens.

The real part of the admittance is the ***conductance*** G, and the imaginary part of the admittance is the ***susceptance*** B. Thus,

$$\mathbf{Y} = G + jB = \frac{1}{\mathbf{Z}} = \frac{1}{R + jX} \qquad [22]$$

Equation [22] should be scrutinized carefully; it does *not* state that the real part of the admittance is equal to the reciprocal of the real part of the impedance, or that the imaginary part of the admittance is equal to the reciprocal of the imaginary part of the impedance!

$$Y_R = \frac{1}{R}$$

$$Y_L = \frac{1}{j\omega L}$$

$$Y_C = j\omega C$$

There is a general (unitless) term for both impedance and admittance–***immitance***–which is sometimes used, but not very often.

PRACTICE

10.11 Determine the admittance (in rectangular form) of (*a*) an impedance $\mathbf{Z} = 1000 + j400 \ \Omega$; (*b*) a network consisting of the parallel combination of an 800 Ω resistor, a 1 mH inductor, and a 2 nF capacitor, if $\omega = 1$ Mrad/s; (*c*) a network consisting of the series combination of an 800 Ω resistor, a 1 mH inductor, and a 2 nF capacitor, if $\omega = 1$ Mrad/s.

Ans: $0.862 - j0.345$ mS; $1.25 + j1$ mS; $0.899 - j0.562$ mS.

10.6 • NODAL AND MESH ANALYSIS

We previously achieved a great deal with nodal and mesh analysis techniques, and it's reasonable to ask if a similar procedure might be valid in terms of phasors and impedances for the sinusoidal steady state. We already know that both of Kirchhoff's laws are valid for phasors; also, we have an Ohm-like law for the passive elements $\mathbf{V} = \mathbf{ZI}$. We may therefore analyze circuits by nodal techniques in the sinusoidal steady state. Using similar arguments, we can establish that mesh analysis methods are valid (and often useful) as well.

EXAMPLE **10.8**

Find the time-domain node voltages $v_1(t)$ and $v_2(t)$ in the circuit shown in Fig. 10.22.

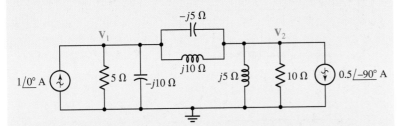

■ **FIGURE 10.22** A frequency-domain circuit for which node voltages V_1 and V_2 are identified.

Two current sources are given as phasors, and phasor node voltages V_1 and V_2 are indicated. At the left node we apply KCL, yielding:

$$\frac{V_1}{5} + \frac{V_1}{-j10} + \frac{V_1 - V_2}{-j5} + \frac{V_1 - V_2}{j10} = 1\underline{/0°} = 1 + j0$$

At the right node,

$$\frac{V_2 - V_1}{-j5} + \frac{V_2 - V_1}{j10} + \frac{V_2}{j5} + \frac{V_2}{10} = -(0.5\underline{/-90°}) = j0.5$$

Combining terms, we have

$$(0.2 + j0.2)V_1 - j0.1V_2 = 1$$

and

$$-j0.1V_1 + (0.1 - j0.1)V_2 = j0.5$$

These equations are easily solved on most scientific calculators, resulting in $V_1 = 1 - j2$ V and $V_2 = -2 + j4$ V.

The time-domain solutions are obtained by expressing V_1 and V_2 in polar form:

$$V_1 = 2.24\underline{/-63.4°}$$
$$V_2 = 4.47\underline{/116.6°}$$

and passing to the time domain:

$$v_1(t) = 2.24\cos(\omega t - 63.4°) \quad \text{V}$$
$$v_2(t) = 4.47\cos(\omega t + 116.6°) \quad \text{V}$$

Note that the value of ω would have to be known in order to compute the impedance values given on the circuit diagram. Also, *both sources must be operating at the same frequency.*

(Continued on next page)

PRACTICE

10.12 Use nodal analysis on the circuit of Fig. 10.23 to find $\mathbf{V}_1$ and $\mathbf{V}_2$.

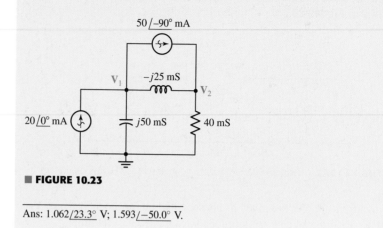

■ **FIGURE 10.23**

Ans: $1.062\underline{/23.3°}$ V; $1.593\underline{/-50.0°}$ V.

Now let us look at an example of mesh analysis, keeping in mind again that all sources must be operating at the same frequency. Otherwise, it is impossible to define a numerical value for any reactance in the circuit. As we see in the next section, the only way out of such a dilemma is to apply superposition.

EXAMPLE 10.9

Obtain expressions for the time-domain currents i_1 and i_2 in the circuit given as Fig. 10.24a.

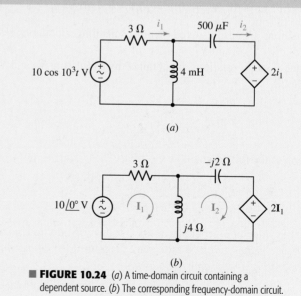

(a)

(b)

■ **FIGURE 10.24** (a) A time-domain circuit containing a dependent source. (b) The corresponding frequency-domain circuit.

PRACTICE

10.14 If superposition is used on the circuit of Fig. 10.30, find $\mathbf{V}_1$ with (*a*) only the $20\underline{/0°}$ mA source operating; (*b*) only the $50\underline{/-90°}$ mA source operating.

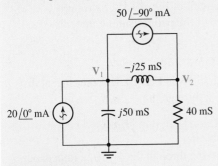

■ FIGURE 10.30

Ans: $0.1951 - j0.556$ V; $0.780 + j0.976$ V.

EXAMPLE 10.11

Determine the Thévenin equivalent seen by the $-j10\ \Omega$ impedance of Fig. 10.31*a*, and use this to compute $\mathbf{V}_1$.

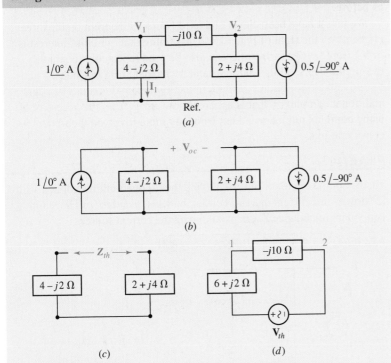

■ FIGURE 10.31 (*a*) Circuit of Fig. 10.29*b*. The Thévenin equivalent seen by the $-j10\ \Omega$ impedance is desired. (*b*) $\mathbf{V}_{oc}$ is defined. (*c*) $\mathbf{Z}_{th}$ is defined. (*d*) The circuit is redrawn using the Thévenin equivalent.

(Continued on next page)

The open-circuit voltage, defined in Fig. 10.31b, is

$$\mathbf{V}_{oc} = (1\underline{/0^\circ})(4 - j2) - (-0.5\underline{/-90^\circ})(2 + j4)$$
$$= 4 - j2 + 2 - j1 = 6 - j3 \quad \text{V}$$

The impedance of the inactive circuit of Fig. 10.31c *as viewed from the load terminals* is simply the sum of the two remaining impedances. Hence,

$$\mathbf{Z}_{th} = 6 + j2 \ \Omega$$

Thus, when we reconnect the circuit as in Fig. 10.31d, the current directed from node 1 toward node 2 through the $-j10\ \Omega$ load is

$$\mathbf{I}_{12} = \frac{6 - j3}{6 + j2 - j10} = 0.6 + j0.3 \ \text{A}$$

We now know the current flowing through the $-j10\ \Omega$ impedance of Fig. 10.31a. *Note that we are unable to compute* $\mathbf{V}_1$ *using the circuit of Fig. 10.31d as the reference node no longer exists.* Returning to the original circuit, then, and subtracting the $0.6 + j0.3$ A current from the left source current, the downward current through the $(4 - j2)\ \Omega$ branch is found:

$$\mathbf{I}_1 = 1 - 0.6 - j0.3 = 0.4 - j0.3 \quad \text{A}$$

and, thus,

$$\mathbf{V}_1 = (0.4 - j0.3)(4 - j2) = 1 - j2 \quad \text{V}$$

as before.

We might have been clever and used Norton's theorem on the three elements on the right of Fig. 10.31a, assuming that our chief interest is in $\mathbf{V}_1$. Source transformations can also be used repeatedly to simplify the circuit. Thus, all the shortcuts and tricks that arose in Chaps. 4 and 5 are available for circuit analysis in the frequency domain. The slight additional complexity that is apparent now arises from the necessity of using complex numbers and not from any more involved theoretical considerations.

PRACTICE

10.15 For the circuit of Fig. 10.32, find the (a) open-circuit voltage $\mathbf{V}_{ab}$; (b) downward current in a short circuit between a and b; (c) Thévenin equivalent impedance $\mathbf{Z}_{ab}$ in parallel with the current source.

Ans: $16.77\underline{/-33.4^\circ}$ V; $2.60 + j1.500$ A; $2.5 - j5\ \Omega$.

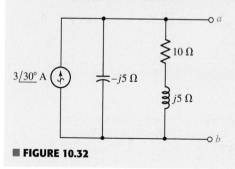

■ **FIGURE 10.32**

EXAMPLE **10.12**

Determine the power dissipated by the 10 Ω resistor in the circuit
of Fig. 10.33*a*.

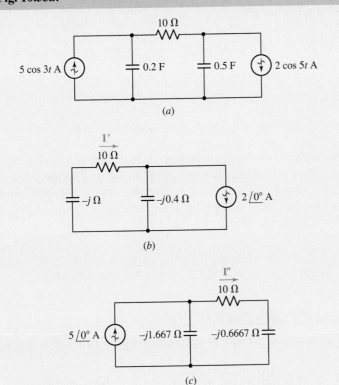

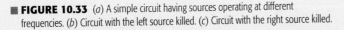

■ **FIGURE 10.33** (*a*) A simple circuit having sources operating at different
frequencies. (*b*) Circuit with the left source killed. (*c*) Circuit with the right source killed.

After glancing at the circuit, we might be tempted to write two quick
nodal equations, or perhaps perform two sets of source transformations
and launch immediately into finding the voltage across the 10 Ω resistor.

Unfortunately, this is impossible, since we have *two* sources operating at *different* frequencies. In such a situation, there is no way to compute the impedance of any capacitor or inductor in the circuit—which
ω would we use?

The only way out of this dilemma is to employ superposition, grouping all sources with the same frequency in the same subcircuit, as shown
in Fig. 10.33*b* and *c*.

In the subcircuit of Fig. 10.33*b*, we quickly compute the current **I**′
using current division:

$$\mathbf{I}' = 2\underline{/0^\circ}\left[\frac{-j0.4}{10 - j - j0.4}\right]$$

$$= 79.23\underline{/-82.03^\circ} \text{ mA}$$

(Continued on next page)

In future studies of signal processing, we will also be
introduced to the method of Jean-Baptiste Joseph
Fourier, a French mathematician who developed a
technique for representing almost any arbitrary function
by a combination of sinusoids. When working with linear circuits, once we know the response of a particular
circuit to a general sinusoidal forcing function, we can
easily predict the response of the circuit to an arbitrary
waveform represented by a Fourier series function,
simply by using superposition.

so that

$$i' = 79.23 \cos(5t - 82.03°) \qquad \text{mA}$$

Likewise, we find that

$$\mathbf{I}'' = 5\underline{/0°}\left[\frac{-j1.667}{10 - j0.6667 - j1.667}\right]$$
$$= 811.7\underline{/-76.86°} \text{ mA}$$

so that

$$i'' = 811.7 \cos(3t - 76.86°) \qquad \text{mA}$$

It should be noted at this point that no matter how tempted we might be to add the two phasor currents $\mathbf{I}'$ and $\mathbf{I}''$, in Fig. 10.33b and c, this *would be incorrect*. Our next step is to add the two time-domain currents, square the result, and multiply by 10 to obtain the power absorbed by the 10 Ω resistor in Fig. 10.33a:

$$p_{10} = (i' + i'')^2 \times 10$$
$$= 10[79.23 \cos(5t - 82.03°) + 811.7 \cos(3t - 76.86°)]^2 \qquad \mu\text{W}$$

PRACTICE

10.16 Determine the current i through the 4 Ω resistor of Fig. 10.34.

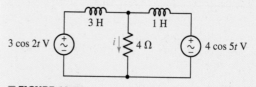

■ **FIGURE 10.34**

Ans: $i = 175.6 \cos(2t - 20.55°) + 547.1 \cos(5t - 43.16°)$ mA.

COMPUTER-AIDED ANALYSIS

We have several options in PSpice for the analysis of circuits in the sinusoidal steady state. Perhaps the most straightforward approach is to make use of two specially designed sources: VAC and IAC. The magnitude and phase of either source is selected by double-clicking on the part.

Let's simulate the circuit of Fig. 10.20a, shown redrawn in Fig. 10.35.

The frequency of the source is not selected through the Property Editor, but rather through the ac sweep analysis dialog box. This is accomplished by choosing **AC Sweep/Noise** for **Analysis** when presented with the Simulation Settings window. We select a **Linear** sweep

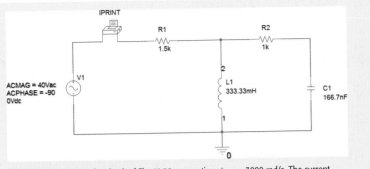

■ **FIGURE 10.35** The circuit of Fig. 10.20a, operating at $\omega = 3000$ rad/s. The current through the 1.5 kΩ resistor is desired.

and set **Total Points** to 1. Since we are only interested in the frequency of 3000 rad/s (477.5 Hz), we set both **Start Frequency** and **End Frequency** to 477.5 as shown in Fig. 10.36.

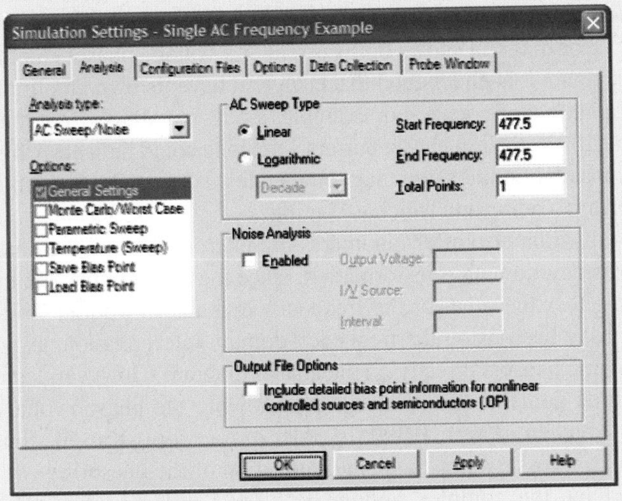

■ **FIGURE 10.36** Dialog box for setting source frequency.

Note that an additional "component" appears in the schematic. This component is called IPRINT, and allows a variety of current parameters to be printed. In this simulation, we are interested in the **AC, MAG,** and **PHASE** attributes. In order for PSpice to print these quantities, double-click on the IPRINT symbol in the schematic, and enter *yes* in each of the appropriate fields.

The simulation results are obtained by choosing **Vie̲w Output File** under **PSpice** in the Capture CIS window.

```
FREQ         IM(V_PRINT1)  IP(V_PRINT1)

4.775E+02    1.600E-02     -1.269E+02
```

Thus, the current magnitude is 16 mA, and the phase angle is $-126.9°$, so that the current through the 1.5 kΩ resistor is

$$i = 16\cos(3000t - 126.9°) \qquad \text{mA}$$
$$= 16\sin(3000t - 36.9°) \qquad \text{mA}$$

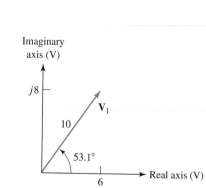

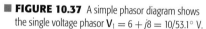

FIGURE 10.37 A simple phasor diagram shows the single voltage phasor $\mathbf{V}_1 = 6 + j8 = 10\underline{/53.1°}$ V.

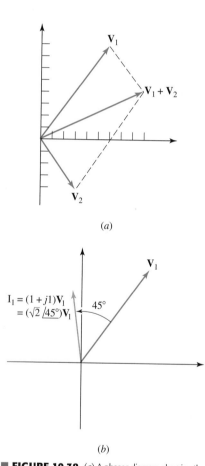

(a)

(b)

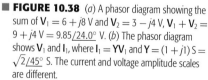

FIGURE 10.38 (a) A phasor diagram showing the sum of $\mathbf{V}_1 = 6 + j8$ V and $\mathbf{V}_2 = 3 - j4$ V, $\mathbf{V}_1 + \mathbf{V}_2 = 9 + j4$ V $= 9.85\underline{/24.0°}$ V. (b) The phasor diagram shows $\mathbf{V}_1$ and $\mathbf{I}_1$, where $\mathbf{I}_1 = \mathbf{Y}\mathbf{V}_1$ and $\mathbf{Y} = (1 + j1)$ S $= \sqrt{2}\underline{/45°}$ S. The current and voltage amplitude scales are different.

10.8 PHASOR DIAGRAMS

The *phasor diagram* is a name given to a sketch in the complex plane showing the relationships of the phasor voltages and phasor currents throughout a specific circuit. We are already familiar with the use of the complex plane in the graphical identification of complex numbers and in their addition and subtraction. Since phasor voltages and currents are complex numbers, they may also be identified as points in a complex plane. For example, the phasor voltage $\mathbf{V}_1 = 6 + j8 = 10\underline{/53.1°}$ V is identified on the complex voltage plane shown in Fig. 10.37. The x axis is the real voltage axis, and the y axis is the imaginary voltage axis; the voltage $\mathbf{V}_1$ is located by an arrow drawn from the origin. Since addition and subtraction are particularly easy to perform and display on a complex plane, phasors may be easily added and subtracted in a phasor diagram. Multiplication and division result in the addition and subtraction of angles and a change of amplitude. Figure 10.38a shows the sum of $\mathbf{V}_1$ and a second phasor voltage $\mathbf{V}_2 = 3 - j4 = 5\underline{/-53.1°}$ V, and Fig. 10.38b shows the current $\mathbf{I}_1$, which is the product of $\mathbf{V}_1$ and the admittance $\mathbf{Y} = 1 + j1$ S.

This last phasor diagram shows both current and voltage phasors on the same complex plane; it is understood that each will have its own amplitude scale, but a common angle scale. For example, a phasor voltage 1 cm long might represent 100 V, while a phasor current 1 cm long could indicate 3 mA. Plotting both phasors on the same diagram enables us to easily determine which waveform is leading and which is lagging.

The phasor diagram also offers an interesting interpretation of the time-domain to frequency-domain transformation, since the diagram may be interpreted from either the time- or the frequency-domain viewpoint. Up to this point, we have been using the frequency-domain interpretation, as we have been showing phasors directly on the phasor diagram. However, let us proceed to a time-domain viewpoint by first showing the phasor voltage $\mathbf{V} = V_m\underline{/\alpha}$ as sketched in Fig. 10.39a. In order to transform $\mathbf{V}$ to the time domain, the next necessary step is the multiplication of the phasor by $e^{j\omega t}$; thus we now have the complex voltage $V_m e^{j\alpha} e^{j\omega t} = V_m\underline{/\omega t + \alpha}$. This voltage may also be interpreted as a phasor, one which possesses a phase angle that increases linearly with time. On a phasor diagram it therefore represents a rotating line segment, the instantaneous position being ωt radians ahead (counterclockwise) of $V_m\underline{/\alpha}$. Both $V_m\underline{/\alpha}$ and $V_m\underline{/\omega t + \alpha}$ are shown on the phasor diagram of Fig. 10.39b. The passage to the time

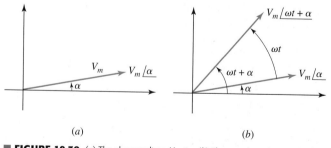

(a)   (b)

FIGURE 10.39 (a) The phasor voltage $V_m\underline{/\alpha}$. (b) The complex voltage $V_m\underline{/\omega t + \alpha}$ is shown as a phasor at a particular instant of time. This phasor leads $V_m\underline{/\alpha}$ by ωt radians.

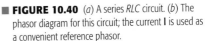

domain is now completed by taking the real part of $V_m \underline{/\omega t + \alpha}$. The real part of this complex quantity is the projection of $V_m \underline{/\omega t + \alpha}$ on the real axis: $V_m \cos(\omega t + \alpha)$.

In summary, then, the frequency-domain phasor appears on the phasor diagram, and the transformation to the time domain is accomplished by allowing the phasor to rotate in a counterclockwise direction at an angular velocity of ω rad/s and then visualizing the projection on the real axis. It is helpful to think of the arrow representing the phasor **V** on the phasor diagram as the photographic snapshot, taken at $\omega t = 0$, of a rotating arrow whose projection on the real axis is the instantaneous voltage $v(t)$.

Let us now construct the phasor diagrams for several simple circuits. The series *RLC* circuit shown in Fig. 10.40*a* has several different voltages associated with it, but only a single current. The phasor diagram is constructed most easily by employing the single current as the reference phasor. Let us arbitrarily select $\mathbf{I} = I_m \underline{/0°}$ and place it along the real axis of the phasor diagram, Fig. 10.40*b*. The resistor, capacitor, and inductor voltages may then be calculated and placed on the diagram, where the 90° phase relationships stand out clearly. The sum of these three voltages is the source voltage, and for this circuit, which is in what we will define in a subsequent chapter as the "resonant condition" since $\mathbf{Z}_C = -\mathbf{Z}_L$, the source voltage and resistor voltage are equal. The total voltage across the resistor and inductor or resistor and capacitor is obtained from the diagram by adding the appropriate phasors as shown.

Figure 10.41*a* is a simple parallel circuit in which it is logical to use the single voltage between the two nodes as a reference phasor. Suppose that $\mathbf{V} = 1\underline{/0°}$ V. The resistor current, $\mathbf{I}_R = 0.2\underline{/0°}$ A, is in phase with this voltage, and the capacitor current, $\mathbf{I}_C = j0.1$ A, leads the reference voltage by 90°. After these two currents are added to the phasor diagram, shown as Fig. 10.41*b*, they may be summed to obtain the source current. The result is $\mathbf{I}_s = 0.2 + j0.1$ A.

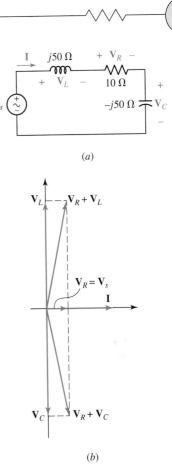

FIGURE 10.40 (*a*) A series *RLC* circuit. (*b*) The phasor diagram for this circuit; the current **I** is used as a convenient reference phasor.

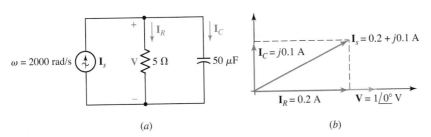

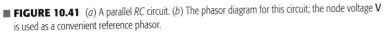

FIGURE 10.41 (*a*) A parallel *RC* circuit. (*b*) The phasor diagram for this circuit; the node voltage **V** is used as a convenient reference phasor.

If the source current is specified initially as the convenient value of $1\underline{/0°}$ A and the node voltage is not initially known, it is still convenient to begin construction of the phasor diagram by assuming a node voltage (for example, $\mathbf{V} = 1\underline{/0°}$ V once again) and using it as the reference phasor. The diagram is then completed as before, and the source current that flows as a result of the assumed node voltage is again found to be $0.2 + j0.1$ A. The true source current is $1\underline{/0°}$ A, however, and thus the true node voltage is obtained by multiplying the assumed node voltage by $1\underline{/0°}/(0.2 + j0.1)$;

the true node voltage is therefore $4 - j2$ V $= \sqrt{20}\underline{/-26.6°}$ V. The assumed voltage leads to a phasor diagram which differs from the true phasor diagram by a change of scale (the assumed diagram is smaller by a factor of $1/\sqrt{20}$) and an angular rotation (the assumed diagram is rotated counterclockwise through 26.6°).

EXAMPLE 10.13

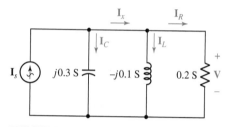

■ **FIGURE 10.42** A simple circuit for which several currents are required.

Construct a phasor diagram showing I_R, I_L, and I_C for the circuit of Fig. 10.42. Combining these currents, determine the angle by which I_s leads I_R, I_C, and I_x.

We begin by choosing a suitable reference phasor. Upon examining the circuit and the variables to be determined, we see that once $\mathbf{V}$ is known, I_R, I_L, and I_C can be computed by simple application of Ohm's law. Thus, we select $\mathbf{V} = 1\underline{/0°}$ V for simplicity's sake, and subsequently compute

$$I_R = (0.2)1\underline{/0°} \quad = 0.2\underline{/0°} \text{ A}$$
$$I_L = (-j0.1)1\underline{/0°} = 0.1\underline{/-90°} \text{ A}$$
$$I_C = (j0.3)1\underline{/0°} \quad = 0.3\underline{/90°} \text{ A}$$

The corresponding phasor diagram is shown in Fig. 10.43a. We also need to find the phasor currents I_s and I_x. Figure 10.43b shows the determination of $I_x = I_L + I_R = 0.2 - j0.1 = 0.224\underline{/-26.6°}$ A, and Fig. 10.43c shows the determination of $I_s = I_C + I_x = 0.283\underline{/45°}$ A. From Fig. 10.43c, we ascertain that I_s leads I_R by 45°, I_C by −45°, and I_x by $45° + 26.6° = 71.6°$. These angles are only relative, however; the exact numerical values will depend on I_s, upon which the actual value of $\mathbf{V}$ (assumed here to be $1\underline{/0°}$ V for convenience) also depends.

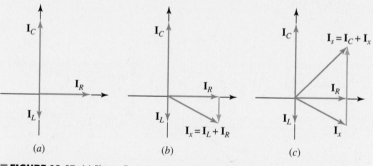

(a)   (b)   (c)

■ **FIGURE 10.43** (a) Phasor diagram constructed using a reference value of $\mathbf{V} = 1/0°$. (b) Graphical determination of $I_x = I_L + I_R$. (c) Graphical determination of $I_s = I_C + I_x$.

PRACTICE

10.17 Select some convenient reference value for I_C in the circuit of Fig. 10.44; draw a phasor diagram showing $\mathbf{V}_R$, $\mathbf{V}_2$, $\mathbf{V}_1$, and $\mathbf{V}_s$; and measure the ratio of the lengths of (a) $\mathbf{V}_s$ to $\mathbf{V}_1$; (b) $\mathbf{V}_1$ to $\mathbf{V}_2$; (c) $\mathbf{V}_s$ to $\mathbf{V}_R$.

Ans: 1.90; 1.00; 2.12

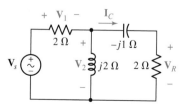

■ **FIGURE 10.44**

SUMMARY AND REVIEW

This chapter dealt with the steady-state response of circuits to sinusoidal excitation. This is a limited analysis of a circuit in some respects, as the transient behavior is completely ignored. In many situations, such an approach is more than adequate, and reducing the amount of information we seek about a circuit speeds up the analysis considerably. The fundamental idea behind what we did was that an *imaginary* source was added to every *real* sinusoidal source; then Euler's identity converted the source to a complex exponential. Since the derivative of an exponential is simply another exponential, what would otherwise be integrodifferential equations arising from mesh or nodal analysis become *algebraic equations*.

A few new terms were introduced: *lagging, leading, impedance, admittance,* and a particularly important one *phasor*. Phasor relationships between current and voltage gave rise to the concept of impedance, where resistors are represented by a real number (resistance, as before), and inductors are represented by $\mathbf{Z} = j\omega L$ while capacitors are represented by $-j/\omega C$ (ω being the operating frequency of our sources). From that point forward, all the circuit analysis techniques learned in Chaps. 3 to 5 apply.

It might seem odd to have an imaginary number as part of our solution, but we found that recovering the time-domain solution to our analysis is straightforward once the voltage or current is expressed in polar form. The magnitude of our quantity of interest is the magnitude of the cosine function, the phase angle is the phase of the cosine term, and the frequency is obtained from the original circuit (it disappears from view during the analysis, but the circuits we are analyzing do not change it in any way). We concluded the chapter with an introduction to the concept of phasor diagrams. Prior to inexpensive scientific calculators such tools were invaluable in analyzing many sinusoidal circuits. They still find use in analysis of ac power systems, as we see in subsequent chapters.

A concise list of key concepts of the chapter is presented below for the convenience of the reader, along with the corresponding example numbers.

- ❑ If two sine waves (or two cosine waves) both have positive magnitudes and the same frequency, it is possible to determine which waveform is leading and which is lagging by comparing their phase angles.

- ❑ The forced response of a linear circuit to a sinusoidal voltage or current source can always be written as a single sinusoid having the same frequency as the sinusoidal source. (Example 10.1)

- ❑ A phasor has both a magnitude and a phase angle; the frequency is understood to be that of the sinusoidal source driving the circuit. (Example 10.2)

- ❑ A phasor transform may be performed on any sinusoidal function, and vice versa: $\mathbf{V}_m \cos(\omega t + \phi) \leftrightarrow \mathbf{V}_m \underline{/\phi}$. (Example 10.3)

- ❑ When transforming a time-domain circuit into the corresponding frequency-domain circuit, resistors, capacitors, and inductors are replaced by impedances (or, occasionally, by admittances). (Examples 10.4, 10.6)

  - • The impedance of a resistor is simply its resistance.

- The impedance of a capacitor is $1/j\omega C$ Ω.
- The impedance of an inductor is $j\omega L$ Ω.

❑ Impedances combine both in series and in parallel combinations in the same manner as resistors. (Example 10.6)

❑ All analysis techniques previously used on resistive circuits apply to circuits with capacitors and/or inductors once all elements are replaced by their frequency-domain equivalents. (Examples 10.5, 10.7, 10.8, 10.9, 10.10, 10.11)

❑ Phasor analysis can only be performed on single-frequency circuits. Otherwise, superposition must be invoked, and the *time-domain* partial responses added to obtain the complete response. (Example 10.12)

❑ The power behind phasor diagrams is evident when a convenient forcing function is used initially, and the final result scaled appropriately. (Example 10.13)

READING FURTHER

A good reference to phasor-based analysis techniques can be found in:

R. A. DeCarlo and P. M. Lin, *Linear Circuit Analysis,* 2nd ed. New York: Oxford University Press, 2001.

Frequency-dependent transistor models are discussed from a phasor perspective in Chap. 7 of:

W. H. Hayt, Jr., and G. W. Neudeck, *Electronic Circuit Analysis and Design,* 2nd ed. New York: Wiley, 1995.

EXERCISES

10.1 Characteristics of Sinusoids

1. Evaluate the following: (*a*) $5 \sin (5t - 9°)$ at $t = 0, 0.01$, and 0.1 s; (*b*) $4 \cos 2t$ and $4 \sin (2t + 90°)$ at $t = 0, 1$, and 1.5 s; (*c*) $3.2 \cos (6t + 15°)$ and $3.2 \sin (6t + 105°)$ at $t = 0, 0.01$, and 0.1 s.

2. (*a*) Express each of the following as a single *cosine* function: $5 \sin 300t$, $1.95 \sin (\pi t - 92°)$, $2.7 \sin (50t + 5°) - 10 \cos 50t$. (*b*) Express each of the following as a single *sine* function: $66 \cos (9t - 10°)$, $4.15 \cos 10t$, $10 \cos (100t - 9°) + 10 \sin (100t + 19°)$.

3. Determine the angle by which v_1 *leads* i_1 if $v_1 = 10 \cos (10t - 45°)$ and i_1 is equal to (*a*) $5 \cos 10t$; (*b*) $5 \cos (10t - 80°)$; (*c*) $5 \cos (10t - 40°)$; (*d*) $5 \cos (10t + 40°)$; (*e*) $5 \sin (10t - 19°)$.

4. Determine the angle by which v_1 *lags* i_1 if $v_1 = 34 \cos (10t + 125°)$ and i_1 is equal to (*a*) $5 \cos 10t$; (*b*) $5 \cos (10t - 80°)$; (*c*) $5 \cos (10t - 40°)$; (*d*) $5 \cos (10t + 40°)$; (*e*) $5 \sin (10t - 19°)$.

5. Determine which waveform in each of the following pairs is lagging: (*a*) $\cos 4t$, $\sin 4t$; (*b*) $\cos (4t - 80°)$, $\cos (4t)$; (*c*) $\cos (4t + 80°)$, $\cos 4t$; (*d*) $-\sin 5t$, $\cos (5t + 2°)$; (*e*) $\sin 5t + \cos 5t$, $\cos (5t - 45°)$.

6. Calculate the first three instants in time ($t > 0$) for which the following functions are zero, by first converting to a single sinusoid: (*a*) $\cos 3t - 7 \sin 3t$; (*b*) $\cos (10t + 45°)$; (*c*) $\cos 5t - \sin 5t$; (*d*) $\cos 2t + \sin 2t - \cos 5t + \sin 5t$.

7. (*a*) Determine the first two instants in time ($t > 0$) for which each of the functions in Exercise 6 are equal to unity, by first converting to a single sinusoid. (*b*) Verify your answers by plotting each waveform using an appropriate software package.

22. For the circuit of Fig. 10.51, if $i_s = 5 \cos 10t$ A, use a suitable complex source replacement to obtain a steady-state expression for $i_L(t)$.

23. In the circuit depicted in Fig. 10.51, i_s is modified such that the 2 Ω resistor is replaced by a 20 Ω resistor. If $i_L(t) = 62.5\ \underline{/31.3^\circ}$ mA, determine i_s.

24. Employ a suitable complex source to determine the steady-state current i_L in the circuit of Fig. 10.52.

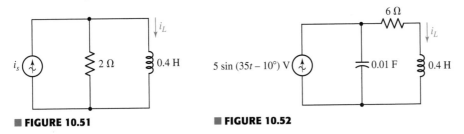

■ **FIGURE 10.51**   ■ **FIGURE 10.52**

10.4 The Phasor

25. Transform each of the following into phasor form: (a) 75.928 cos (110.1t); (b) 5 cos (55t − 42°); (c) − sin (8000t + 14°); (d) 3 cos 10t − 8 cos (10t + 80°).

26. Transform each of the following into phasor form: (a) 11 sin 100t; (b) 11 cos 100t; (c) 11 cos(100t − 90°); (d) 3 cos 100t − 3 sin 100t.

27. Assuming an operating frequency of 1 kHz, transform the following phasor expressions into a single cosine function in the time domain: (a) $9\underline{/65^\circ}$ V; (b) $\dfrac{2\underline{/31^\circ}}{4\underline{/25^\circ}}$ A; (c) $22\underline{/14^\circ} - 8\underline{/33^\circ}$ V.

28. The following complex voltages are written in a combination of rectangular and polar form. Rewrite each, using conventional phasor notation (i.e., a magnitude and angle): (a) $\dfrac{2 - j}{5\underline{/45^\circ}}$ V; (b) $\dfrac{6\underline{/20^\circ}}{1000} - jV$; (c) $(j)(52.5\underline{/-90^\circ})$ V.

29. Assuming an operating frequency of 50 Hz, compute the instantaneous voltage at $t = 10$ ms and $t = 25$ ms for each of the quantities represented in Exercise 26.

30. Assuming an operating frequency of 50 Hz, compute the instantaneous voltage at $t = 10$ ms and $t = 25$ ms for each of the quantities represented in Exercise 27.

31. Assuming the passive sign convention and an operating frequency of 5 rad/s, calcualte the phasor voltage which develops across the following when driven by the phasor current $\mathbf{I} = 2\underline{/0^\circ}$ mA: (a) a 1 kΩ resistor; (b) a 1 mF capacitor; (c) a 1 nH inductor.

32. (a) A series connection is formed between a 1 Ω resistor, a 1 F capacitor, and a 1 H inductor, in that order. Assuming operation at $\omega = 1$ rad/s, what are the magnitude and phase angle of the phasor current which yields a voltage of $1\underline{/30^\circ}$ V across the resistor (assume the passive sign convention)? (b) Compute the ratio of the phasor voltage across the resistor to the phasor voltage which appears across the capacitor-inductor combination. (c) The frequency is doubled. Calculate the new ratio of the phasor voltage across the resistor to the phasor voltage across the capacitor-inductor combination.

33. Assuming the passive sign convention and an operating frequency of 314 rad/s, calculate the phasor voltage $\mathbf{V}$ which appears across each of the following when driven by the phasor current $\mathbf{I} = 10\underline{/0^\circ}$ mA: (a) a 2 Ω resistor; (b) a 1 F capacitor; (c) a 1 H inductor; (d) a 2 Ω resistor in series with a 1 F capacitor; (e) a 2 Ω resistor in series with a 1 H inductor. (f) Calculate the instantaneous value of each voltage determined in parts (a) to (e) at $t = 0$.

34. In the circuit of Fig. 10.53, which is shown in the phasor (frequency) domain, $\mathbf{I}_{10}$ is determined to be $2\underline{/42^\circ}$ mA. If $\mathbf{V} = 40\underline{/132^\circ}$ mV: (a) what is the likely type of element connected to the right of the 10 Ω resistor and (b) what is its value, assuming the voltage source operates at a frequency of 1000 rad/s?

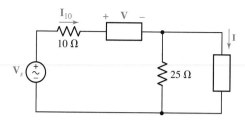

■ FIGURE 10.53

35. The circuit of Fig. 10.53 is shown represented in the phasor (frequency) domain. If $\mathbf{I}_{10} = 4\underline{/35°}$ A, $\mathbf{V} = 10\underline{/35°}$, and $\mathbf{I} = 2\underline{/35°}$ A, (a) across what type of element does $\mathbf{V}$ appear, and what is its value? (b) Determine the value of $\mathbf{V}_s$.

10.5 Impedance and Admittance

36. (a) Obtain an expression for the equivalent impedance $\mathbf{Z}_{eq}$ of a 1 Ω resistor in series with a 10 mH inductor as a function of ω. (b) Plot the magnitude of $\mathbf{Z}_{eq}$ as a function of ω over the range $1 < \omega < 100$ krad/s (use a logarithmic scale for the frequency axis). (c) Plot the angle (in degrees) of $\mathbf{Z}_{eq}$ as a function of ω over the range $1 < \omega < 100$ krad/s (use a logarithmic scale for the frequency axis). [Hint: semilogx() in MATLAB is a useful plotting function.]

37. Determine the equivalent impedance of the following, assuming an operating frequency of 20 rad/s: (a) 1 kΩ in series with 1 mF; (b) 1 kΩ in parallel with 1 mH; (c) 1 kΩ in parallel with the series combination of 1 F and 1 H.

38. (a) Obtain an expression for the equivalent impedance $\mathbf{Z}_{eq}$ of a 1 Ω resistor in series with a 10 mF capacitor as a function of ω. (b) Plot the magnitude of $\mathbf{Z}_{eq}$ as a function of ω over the range $1 < \omega < 100$ krad/s (use a logarithmic scale for the frequency axis). (c) Plot the angle (in degrees) of $\mathbf{Z}_{eq}$ as a function of ω over the range $1 < \omega < 100$ krad/s (use a logarithmic scale for the frequency axis). [Hint: semilogx() in MATLAB is a useful plotting function.]

39. Determine the equivalent admittance of the following, assuming an operating frequency of 1000 rad/s: (a) 25 Ω in series with 20 mH; (b) 25 Ω in parallel with 20 mH; (c) 25 Ω in parallel with 20 mH in parallel with 20 mF; (d) 1 Ω in series with 1 F in series with 1 H; (e) 1 Ω in parallel with 1 F in parallel with 1 H.

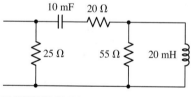

■ FIGURE 10.54

40. Consider the network depicted in Fig. 10.54, and determine the equivalent impedance seen looking into the open terminals if (a) $\omega = 1$ rad/s; (b) $\omega = 10$ rad/s; (c) $\omega = 100$ rad/s.

41. Exchange the capacitor and inductor in the network shown in Fig. 10.54, and calculate the equivalent impedance looking into the open terminals if $\omega = 25$ rad/s.

42. Find $\mathbf{V}$ in Fig. 10.55 if the box contains (a) 3 Ω in series with 2 mH; (b) 3 Ω in series with 125 μF; (c) 3 Ω, 2 mH, and 125 μF in series; (d) 3 Ω, 2 mH, and 125 μF in series, but $\omega = 4$ krad/s.

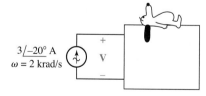

■ FIGURE 10.55

43. Calculate the equivalent impedance seen at the open terminals of the network shown in Fig. 10.56 if f is equal to (a) 1 Hz; (b) 1 kHz; (c) 1 MHz; (d) 1 GHz; (e) 1 THz.

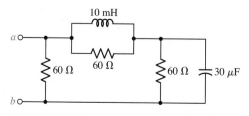

■ FIGURE 10.56

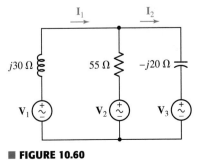

44. Employ phasor-based analysis to obtain an expression for $i(t)$ in the circuit of Fig. 10.57.

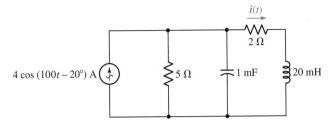

$4 \cos (100t - 20°)$ A

$5\,\Omega$ 1 mF $2\,\Omega$ $i(t)$ 20 mH

■ **FIGURE 10.57**

45. Design a suitable combination of resistors, capacitors, and/or inductors which has an equivalent impedance at $\omega = 100$ rad/s of (a) $1\,\Omega$ using at least one inductor; (b) $7\underline{/10°}\,\Omega$; (c) $3 - j4\,\Omega$.

46. Design a suitable combination of resistors, capacitors, and/or inductors which has an equivalent admittance at $\omega = 10$ rad/s of (a) 1 S using at least one capacitor; (b) $12\underline{/-18°}$ S ; (c) $2 + j$ mS.

10.6 Nodal and Mesh Analysis

47. For the circuit depicted in Fig. 10.58, (a) redraw with appropriate phasors and impedances labeled; (b) employ nodal analysis to determine the two nodal voltages $v_1(t)$ and $v_2(t)$.

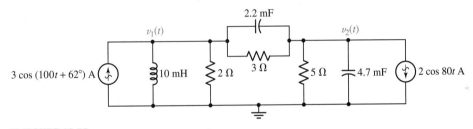

$3 \cos (100t + 62°)$ A $v_1(t)$ 10 mH $2\,\Omega$ $3\,\Omega$ 2.2 mF $v_2(t)$ $5\,\Omega$ 4.7 mF $2 \cos 80t$ A

■ **FIGURE 10.58**

48. For the circuit illustrated in Fig. 10.59, (a) redraw, labeling appropriate phasor and impedance quantities; (b) determine expressions for the three time-domain mesh currents.

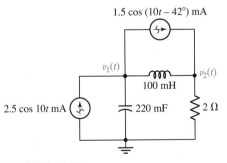

$1.5 \cos (10t - 42°)$ mA

$2.5 \cos 10t$ mA $v_1(t)$ 100 mH $v_2(t)$ 220 mF $2\,\Omega$

■ **FIGURE 10.59**

49. Referring to the circuit of Fig. 10.59, employ phasor-based analysis techniques to determine the two nodal voltages.

50. In the phasor-domain circuit represented by Fig. 10.60, let $\mathbf{V}_1 = 10\underline{/-80°}$ V, $\mathbf{V}_2 = 4\underline{/-0°}$ V, and $\mathbf{V}_3 = 2\underline{/-23°}$ V. Calculate $\mathbf{I}_1$ and $\mathbf{I}_2$.

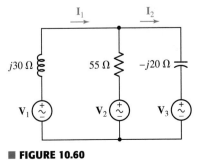

■ **FIGURE 10.60**

51. With regard to the two-mesh phasor-domain circuit depicted in Fig. 10.60, calculate the ratio of $\mathbf{I}_1$ to $\mathbf{I}_2$ if $\mathbf{V}_1 = 3\underline{/0°}$ V, $\mathbf{V}_2 = 5.5\underline{/-130°}$ V, and $\mathbf{V}_3 = 1.5\underline{/17°}$ V.

52. Employ phasor analysis techniques to obtain expressions for the two mesh currents i_1 and i_2 as shown in Fig. 10.61.

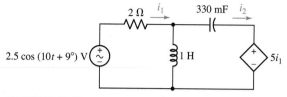

■ FIGURE 10.61

53. Determine $\mathbf{I}_B$ in the circuit of Fig. 10.62 if $\mathbf{I}_1 = 5\underline{/-18°}$ A and $\mathbf{I}_2 = 2\underline{/5°}$ A.

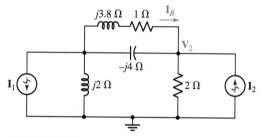

■ FIGURE 10.62

54. Determine $\mathbf{V}_2$ in the circuit of Fig. 10.62 if $\mathbf{I}_1 = 15\underline{/0°}$ A and $\mathbf{I}_2 = 25\underline{/131°}$ A.

55. Employ phasor analysis to obtain an expression for v_x as labeled in the circuit of Fig. 10.63.

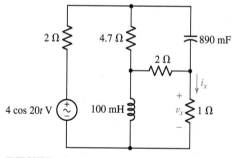

■ FIGURE 10.63

56. Determine the current i_x in the circuit of Fig. 10.63.

57. Obtain an expression for each of the four (clockwise) mesh currents for the circuit of Fig. 10.64 if $v_1 = 133 \cos(14t + 77°)$ V and $v_2 = 55 \cos(14t + 22°)$ V.

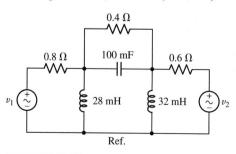

■ FIGURE 10.64

58. Determine the nodal voltages for the circuit of Fig. 10.64, using the bottom node as the reference node, if $v_1 = 0.009 \cos(500t + 0.5°)$ V and $v_2 = 0.004 \cos(500t + 1.5°)$ V.

59. The op amp shown in Fig. 10.65 has an infinite input impedance, zero output impedance, and a large but finite (positive, real) gain, $A = -\mathbf{V}_o/\mathbf{V}_i$. (a) Construct a basic differentiator by letting $\mathbf{Z}_f = R_f$, find $\mathbf{V}_o/\mathbf{V}_s$, and then show that $\mathbf{V}_o/\mathbf{V}_s \rightarrow -j\omega C_1 R_f$ as $A \rightarrow \infty$. (b) Let $\mathbf{Z}_f$ represent C_f and R_f in parallel, find $\mathbf{V}_o/\mathbf{V}_s$, and then show that $\mathbf{V}_o/\mathbf{V}_s \rightarrow -j\omega C_1 R_f/(1 + j\omega C_f R_f)$ as $A \rightarrow \infty$.

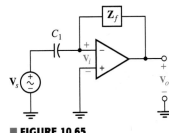

■ FIGURE 10.65

60. Obtain an expression for each of the four mesh currents labeled in the circuit of Fig. 10.66.

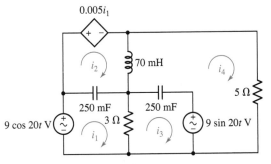

■ FIGURE 10.66

10.7 Superposition, Source Transformations, and Thévenin's Theorem

61. Determine the individual contribution each current source makes to the two nodal voltages $\mathbf{V}_1$ and $\mathbf{V}_2$ as represented in Fig. 10.67.

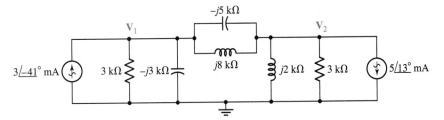

■ FIGURE 10.67

62. Determine $\mathbf{V}_1$ and $\mathbf{V}_2$ in Fig. 10.68 if $\mathbf{I}_1 = 33\underline{/3°}$ mA and $\mathbf{I}_2 = 51\underline{/-91°}$ mA.

63. The phasor domain circuit of Fig. 10.68 was drawn assuming an operating frequency of 2.5 rad/s. Unfortunately, the manufacturing unit installed the wrong sources, each operating at a different frequency. If $i_1(t) = 4 \cos 40t$ mA and $i_2(t) = 4 \sin 30t$ mA, calculate $v_1(t)$ and $v_2(t)$.

64. Obtain the Thévenin equivalent seen by the $(2 - j)$ Ω impedance of Fig. 10.69, and employ it to determine the current $\mathbf{I}_1$.

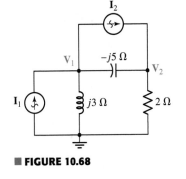

■ FIGURE 10.68

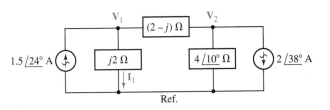

■ FIGURE 10.69

65. The $(2 - j)\,\Omega$ impedance in the circuit of Fig. 10.69 is replaced with a $(1 + j)$ Ω impedance. Perform a source transformation on each source, simplify the resulting circuit as much as possible, and calculate the current flowing through the $(1 + j)\,\Omega$ impedance.

66. With regard to the circuit depicted in Fig. 10.70, (a) calculate the Thévenin equivalent seen looking into the terminals marked a and b; (b) determine the Norton equivalent seen looking into the terminals marked a and b; (c) compute the current flowing from a to b if a $(7 - j2)\,\Omega$ impedance is connected across them.

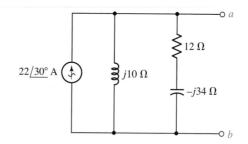

■ **FIGURE 10.70**

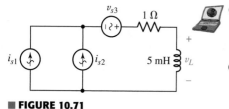

■ **FIGURE 10.71**

67. In the circuit of Fig. 10.71, $i_{s1} = 8\cos(4t - 9°)$ mA, $i_{s2} = 5\cos 4t$ and $v_{s3} = 2\sin 4t$. (a) Redraw the circuit in the phasor domain; (b) reduce the circuit to a single current source with the assistance of a source transformation; (c) calculate $v_L(t)$. (d) Verify your solution with an appropriate PSpice simulation.

68. Determine the individual contribution of each source in Fig. 10.72 to the voltage $v_1(t)$.

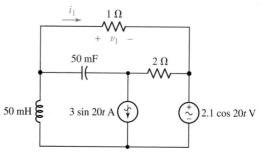

■ **FIGURE 10.72**

69. Determine the power dissipated by the $1\,\Omega$ resistor in the circuit of Fig. 10.73. Verify your solution with an appropriate PSpice simulation.

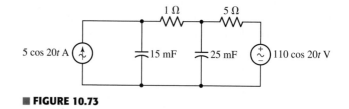

■ **FIGURE 10.73**

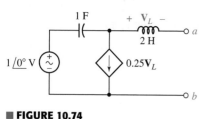

■ **FIGURE 10.74**

70. Use $\omega = 1$ rad/s, and find the Norton equivalent of the network shown in Fig. 10.74. Construct the Norton equivalent as a current source $\mathbf{I}_N$ in parallel with a resistance R_N and either an inductance L_N or a capacitance C_N.

10.8 Phasor Diagrams

71. The source $\mathbf{I}_s$ in the circuit of Fig. 10.75 is selected such that $\mathbf{V} = 5\underline{/120°}$ V.
    (a) Construct a phasor diagram showing $\mathbf{I}_R$, $\mathbf{I}_L$, and $\mathbf{I}_C$. (b) Use the diagram to determine the angle by which $\mathbf{I}_s$ leads $\mathbf{I}_R$, $\mathbf{I}_C$, and $\mathbf{I}_s$.

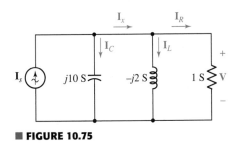

■ **FIGURE 10.75**

72. Let $\mathbf{V}_1 = 100\underline{/0°}$ V, $|\mathbf{V}_2| = 140$ V, and $|\mathbf{V}_1 + \mathbf{V}_2| = 120$ V. Use graphical methods to find two possible values for the angle of $\mathbf{V}_2$.

73. (a) Calculate values for $\mathbf{I}_L$, $\mathbf{I}_R$, $\mathbf{I}_C$, $\mathbf{V}_L$, $\mathbf{V}_R$, and $\mathbf{V}_C$ for the circuit shown in Fig. 10.76. (b) Using scales of 50 V to 1 in and 25 A to 1 in, show all seven quantities on a phasor diagram, and indicate that $\mathbf{I}_L = \mathbf{I}_R + \mathbf{I}_C$ and $\mathbf{V}_s = \mathbf{V}_L + \mathbf{V}_R$.

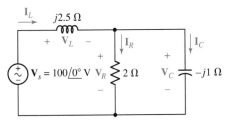

■ **FIGURE 10.76**

74. In the circuit of Fig. 10.77, (a) find values for $\mathbf{I}_1$, $\mathbf{I}_2$, and $\mathbf{I}_3$. (b) Show $\mathbf{V}_s$, $\mathbf{I}_1$, $\mathbf{I}_2$, and $\mathbf{I}_3$ on a phasor diagram (scales of 50 V/in and 2 A/in work fine). (c) Find $\mathbf{I}_s$ graphically and give its amplitude and phase angle.

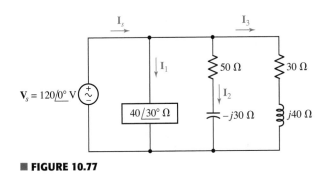

■ **FIGURE 10.77**

75. The voltage source $\mathbf{V}_s$ in Fig. 10.78 is chosen such that $\mathbf{I}_C = 1\underline{/0°}$ A. (a) Draw a phasor diagram showing $\mathbf{V}_1$, $\mathbf{V}_2$, $\mathbf{V}_s$, and $\mathbf{V}_R$. (b) Use the diagram to determine the ratio of $\mathbf{V}_2$ to $\mathbf{V}_1$.

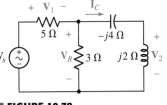

■ **FIGURE 10.78**

Chapter-Integrating Exercises

76. For the circuit shown in Fig. 10.79, (*a*) draw the phasor representation of the circuit; (*b*) determine the Thévenin equivalent seen by the capacitor, and use it to calculate $v_C(t)$. (*c*) Determine the current flowing out of the positive reference terminal of the voltage source. (*d*) Verify your solution with an appropriate PSpice simulation.

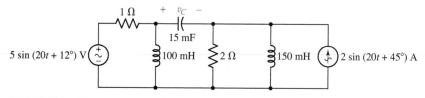

■ **FIGURE 10.79**

77. The circuit of Fig. 10.79 is unfortunately operating differently than specified; the frequency of the current source is only 19 rad/s. Calculate the actual capacitor voltage, and compare it to the expected voltage had the circuit been operating correctly.

78. For the circuit shown in Fig. 10.80, (*a*) draw the corresponding phasor representation; (*b*) obtain an expression for $\mathbf{V}_o/\mathbf{V}_s$; (*c*) Plot $|\mathbf{V}_o/\mathbf{V}_s|$, the magnitude of the phasor voltage ratio, as a function of frequency ω over the range $0.01 \le \omega \le 100$ rad/s (use a logarithmic x axis). (*d*) Does the circuit transfer low frequencies or high frequencies more effectively to the output?

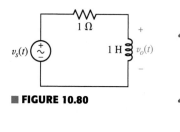

■ **FIGURE 10.80**

79. (*a*) Replace the inductor in the circuit of Fig. 10.80 with a 1 F capacitor and repeat Exercise. 10.78. (*b*) If we design the "corner frequency" of the circuit as the frequency at which the output is reduced to $1/\sqrt{2}$ times its maximum value, redesign the circuit to achieve a corner frequency of 2 kHz.

80. Design a purely passive network (containing only resistors, capacitors, and inductors) which has an impedance of $(22 - j7)/5\underline{/8°}$ Ω at a frequency of $f = 100$ MHz.

AC Circuit Power Analysis

INTRODUCTION

Often an integral part of circuit analysis is the determination of either power delivered or power absorbed (or both). In the context of ac power, we find that the rather simple approach we have taken previously does not provide a convenient picture of how a particular system is operating, so we introduce several different power-related quantities in this chapter.

We begin by considering *instantaneous* power, the product of the time-domain voltage and time-domain current associated with the element or network of interest. The instantaneous power is sometimes quite useful in its own right, because its maximum value might have to be limited to avoid exceeding the safe operating range of a physical device. For example, transistor and vacuum-tube power amplifiers both produce a distorted output, and speakers give a distorted sound, when the peak power exceeds a certain limiting value. However, we are mainly interested in instantaneous power for the simple reason that it provides us with the means to calculate a more important quantity, the *average* power. In a similar way, the progress of a cross-country road trip is best described by the average velocity; our interest in the instantaneous velocity is limited to the avoidance of maximum velocities that will endanger our safety or arouse the highway patrol.

In practical problems we will deal with values of average power which range from the small fraction of a picowatt available in a telemetry signal from outer space, to the few watts of audio power supplied to the speakers in a good stereo system, to the several hundred watts required to run the morning coffeepot, or to the 10 billion watts generated at the Grand Coulee Dam. Still, we will

421

see that even the concept of average power has its limitations, especially when dealing with the energy exchange between reactive loads and power sources. This is easily handled by introducing the concepts of reactive power, complex power, and the power factor—all very common terms in the power industry.

11.1 INSTANTANEOUS POWER

The ***instantaneous power*** delivered to any device is given by the product of the instantaneous voltage across the device and the instantaneous current through it (the passive sign convention is assumed). Thus,[1]

$$p(t) = v(t)i(t) \tag{1}$$

If the device in question is a resistor of resistance R, then the power may be expressed solely in terms of either the current or the voltage:

$$p(t) = v(t)i(t) = i^2(t)R = \frac{v^2(t)}{R} \tag{2}$$

If the voltage and current are associated with a device that is entirely inductive, then

$$p(t) = v(t)i(t) = Li(t)\frac{di(t)}{dt} = \frac{1}{L}v(t)\int_{-\infty}^{t} v(t')\,dt' \tag{3}$$

where we will arbitrarily assume that the voltage is zero at $t = -\infty$. In the case of a capacitor,

$$p(t) = v(t)i(t) = Cv(t)\frac{dv(t)}{dt} = \frac{1}{C}i(t)\int_{-\infty}^{t} i(t')\,dt' \tag{4}$$

where a similar assumption about the current is made.

For example, consider the series RL circuit as shown in Fig. 11.1, excited by a step-voltage source. The familiar current response is

$$i(t) = \frac{V_0}{R}(1 - e^{-Rt/L})u(t)$$

and thus the total power delivered by the source or absorbed by the passive network is

$$p(t) = v(t)i(t) = \frac{V_0^2}{R}(1 - e^{-Rt/L})u(t)$$

since the square of the unit-step function is simply the unit-step function itself.

The power delivered to the resistor is

$$p_R(t) = i^2(t)R = \frac{V_0^2}{R}(1 - e^{-Rt/L})^2 u(t)$$

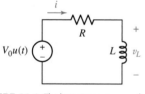

■ **FIGURE 11.1** The instantaneous power that is delivered to R is $p_R(t) = i^2(t)R = (V_0^2/R)(1 - e^{-Rt/L})^2 u(t)$.

---

(1) Earlier, we agreed that lowercase variables in italics were understood to be functions of time, and we have carried on in this spirit up to now. However, in order to emphasize the fact that these quantities must be evaluated at a specific instant in time, we will explicitly denote the time dependence throughout this chapter.

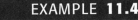

EXAMPLE **11.4**

Find the average power absorbed by each of the three passive elements in Fig. 11.5, as well as the average power supplied by each source.

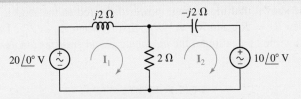

■ **FIGURE 11.5** The average power delivered to each reactive element is zero in the sinusoidal steady state.

Without even analyzing the circuit, we already know that the average power absorbed by the two reactive elements is zero.

The values of $\mathbf{I}_1$ and $\mathbf{I}_2$ are found by any of several methods, such as mesh analysis, nodal analysis, or superposition. They are

$$\mathbf{I}_1 = 5 - j10 = 11.18\underline{/-63.43°} \text{ A}$$
$$\mathbf{I}_2 = 5 - j5 = 7.071\underline{/-45°} \text{ A}$$

The downward current through the 2 Ω resistor is

$$\mathbf{I}_1 - \mathbf{I}_2 = -j5 = 5\underline{/-90°} \text{ A}$$

so that $I_m = 5$ A, and the average power absorbed by the resistor is found most easily by Eq. [12]:

$$P_R = \tfrac{1}{2}I_m^2 R = \tfrac{1}{2}(5^2)2 = 25 \text{ W}$$

This result may be checked by using Eq. [11] or Eq. [13]. We next turn to the left source. The voltage $20\underline{/0°}$ V and associated current $\mathbf{I}_1 = 11.18\underline{/-63.43°}$ A satisfy the *active* sign convention, and thus the power *delivered* by this source is

$$P_{\text{left}} = \tfrac{1}{2}(20)(11.18)\cos[0° - (-63.43°)] = 50 \text{ W}$$

In a similar manner, we find the power *absorbed* by the right source using the *passive* sign convention,

$$P_{\text{right}} = \tfrac{1}{2}(10)(7.071)\cos(0° + 45°) = 25 \text{ W}$$

Since $50 = 25 + 25$, the power relations check.

PRACTICE

11.4 For the circuit of Fig. 11.6, compute the average power delivered to each of the passive elements. Verify your answer by computing the power delivered by the two sources.

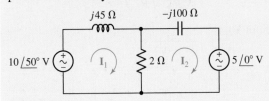

■ **FIGURE 11.6**

Ans: 0, 37.6 mW, 0, 42.0 mW, −4.4 mW.

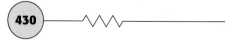

The notation $\mathbf{Z}^*$ denotes the **complex conjugate** of the complex number $\mathbf{Z}$. It is formed by replacing all "j"s with "$-j$"s. See Appendix 5 for more details.

Maximum Power Transfer

We previously considered the maximum power transfer theorem as it applied to resistive loads and resistive source impedances. For a Thévenin source $\mathbf{V}_{th}$ and impedance $\mathbf{Z}_{th} = R_{th} + jX_{th}$ connected to a load $\mathbf{Z}_L = R_L + jX_L$, it may be shown that the average power delivered to the load is a maximum when $R_L = R_{th}$ and $X_L = -X_{th}$, that is, when $\mathbf{Z}_L = \mathbf{Z}_{th}^*$. This result is often dignified by calling it the *maximum power transfer theorem for the sinusoidal steady state:*

> An independent voltage source in *series* with an impedance $\mathbf{Z}_{th}$ or an independent current source in *parallel* with an impedance $\mathbf{Z}_{th}$ delivers a ***maximum average power*** to that load impedance $\mathbf{Z}_L$ which is the conjugate of $\mathbf{Z}_{th}$, or $\mathbf{Z}_L = \mathbf{Z}_{th}^*$.

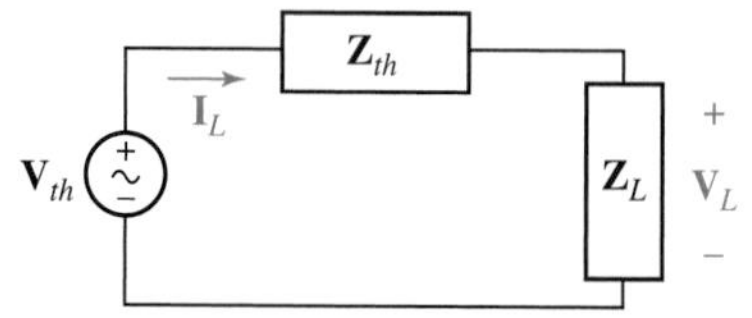

The details of the proof are left to the reader, but the basic approach can be understood by considering the simple loop circuit of Fig. 11.7. The Thévenin equivalent impedance $\mathbf{Z}_{th}$ may be written as the sum of two components, $R_{th} + jX_{th}$, and in a similar fashion the load impedance $\mathbf{Z}_L$ may be written as $R_L + jX_L$. The current flowing through the loop is

$$\mathbf{I}_L = \frac{\mathbf{V}_{th}}{\mathbf{Z}_{th} + \mathbf{Z}_L}$$

$$= \frac{\mathbf{V}_{th}}{R_{th} + jX_{th} + R_L + jX_L} = \frac{\mathbf{V}_{th}}{R_{th} + R_L + j(X_{th} + X_L)}$$

and

$$\mathbf{V}_L = \mathbf{V}_{th} \frac{\mathbf{Z}_L}{\mathbf{Z}_{th} + \mathbf{Z}_L}$$

$$= \mathbf{V}_{th} \frac{R_L + jX_L}{R_{th} + jX_{th} + R_L + jX_L} = \mathbf{V}_{th} \frac{R_L + jX_L}{R_{th} + R_L + j(X_{th} + X_L)}$$

The magnitude of $\mathbf{I}_L$ is

$$\frac{|\mathbf{V}_{th}|}{\sqrt{(R_{th} + R_L)^2 + (X_{th} + X_L)^2}}$$

and the phase angle is

$$\underline{/\mathbf{V}_{th}} - \tan^{-1} \left(\frac{X_{th} + X_L}{R_{th} + R_L} \right)$$

Similarly, the magnitude of $\mathbf{V}_L$ is

$$\frac{|\mathbf{V}_{th}| \sqrt{R_L^2 + X_L^2}}{\sqrt{(R_{th} + R_L)^2 + (X_{th} + X_L)^2}}$$

and its phase angle is

$$\underline{/\mathbf{V}_{th}} + \tan^{-1} \left(\frac{X_L}{R_L} \right) - \tan^{-1} \left(\frac{X_{th} + X_L}{R_{th} + R_L} \right)$$

■ **FIGURE 11.7** A simple loop circuit used to illustrate the derivation of the maximum power transfer theorem as it applies to circuits operating in the sinusoidal steady state.

Referring to Eq. [11], then, we find an expression for the average power P delivered to the load impedance $\mathbf{Z}_L$:

$$P = \frac{\frac{1}{2}|\mathbf{V}_{th}|^2 \sqrt{R_L^2 + X_L^2}}{(R_{th} + R_L)^2 + (X_{th} + X_L)^2} \cos\left(\tan^{-1}\left(\frac{X_L}{R_L}\right)\right) \qquad [14]$$

In order to prove that maximum average power is indeed delivered to the load when $\mathbf{Z}_L = \mathbf{Z}_{th}^*$, we must perform two separate steps. First, the derivative of Eq. [14] with respect to R_L must be set to zero. Second, the derivative of Eq. [14] with respect to X_L must be set to zero. The remaining details are left as an exercise for the avid reader.

EXAMPLE 11.5

A particular circuit is composed of the series combination of a sinusoidal voltage source $3\cos(100t - 3°)$ V, a 500 Ω resistor, a 30 mH inductor, and an unknown impedance. If we are assured that the voltage source is delivering maximum average power to the unknown impedance, what is its value?

The phasor representation of the circuit is sketched in Fig. 11.8. The circuit is easily seen as an unknown impedance $\mathbf{Z}_?$ in series with a Thévenin equivalent consisting of the $3\underline{/-3°}$ V source and a Thévenin impedance $500 + j3$ Ω.

Since the circuit of Fig. 11.8 is already in the form required to employ the maximum average power transfer theorem, we know that maximum average power will be transferred to an impedance equal to the complex conjugate of $\mathbf{Z}_{th}$, or

$$\mathbf{Z}_? = \mathbf{Z}_{th}^* = 500 - j3 \ \Omega$$

This impedance can be constructed in several ways, the simplest being a 500 Ω resistor in series with a capacitor having impedance $-j3$ Ω. Since the operating frequency of the circuit is 100 rad/s, this corresponds to a capacitance of 3.333 mF.

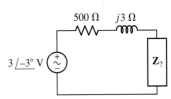

■ **FIGURE 11.8** The phasor representation of a simple series circuit composed of a sinusoidal voltage source, a resistor, an inductor, and an unknown impedance.

PRACTICE
●

11.5 If the 30 mH inductor of Example 11.5 is replaced with a 10 μF capacitor, what is the value of the inductive component of the unknown impedance $\mathbf{Z}_?$ if it is known that $\mathbf{Z}_?$ is absorbing maximum power?

Ans: 10 H.

Average Power for Nonperiodic Functions

We should pay some attention to *nonperiodic* functions. One practical example of a nonperiodic power function for which an average power value is desired is the power output of a radio telescope directed toward a "radio star." Another is the sum of a number of periodic functions, each function having a different period, such that no greater common period can be found for the combination. For example, the current

$$i(t) = \sin t + \sin \pi t \qquad [15]$$

is nonperiodic because the ratio of the periods of the two sine waves is an irrational number. At $t = 0$, both terms are zero and increasing. But the first term is zero and increasing only when $t = 2\pi n$, where n is an integer, and thus periodicity demands that πt or $\pi(2\pi n)$ must equal $2\pi m$, where m is also an integer. No solution (integral values for both m and n) for this equation is possible. It may be illuminating to compare the nonperiodic expression in Eq. [15] with the *periodic* function

$$i(t) = \sin t + \sin 3.14t \qquad [16]$$

where 3.14 is an exact decimal expression and is *not* to be interpreted as 3.141592. . . . With a little effort,[2] it can be shown that the period of this current wave is 100π seconds.

The average value of the power delivered to a $1\ \Omega$ resistor by either a periodic current such as Eq. [16] or a nonperiodic current such as Eq. [15] may be found by integrating over an infinite interval. Much of the actual integration can be avoided because of our thorough knowledge of the average values of simple functions. We therefore obtain the average power delivered by the current in Eq. [15] by applying Eq. [9]:

$$P = \lim_{\tau \to \infty} \frac{1}{\tau} \int_{-\tau/2}^{\tau/2} (\sin^2 t + \sin^2 \pi t + 2\sin t \sin \pi t)\, dt$$

We now consider P as the sum of three average values. The average value of $\sin^2 t$ over an infinite interval is found by replacing $\sin^2 t$ with $(\frac{1}{2} - \frac{1}{2}\cos 2t)$; the average is simply $\frac{1}{2}$. Similarly, the average value of $\sin^2 \pi t$ is also $\frac{1}{2}$. And the last term can be expressed as the sum of two cosine functions, each of which must certainly have an average value of zero. Thus,

$$P = \tfrac{1}{2} + \tfrac{1}{2} = 1\ \text{W}$$

An identical result is obtained for the periodic current of Eq. [16]. Applying this same method to a current function which is the sum of several sinusoids of *different periods* and arbitrary amplitudes,

$$i(t) = I_{m1}\cos \omega_1 t + I_{m2}\cos \omega_2 t + \cdots + I_{mN}\cos \omega_N t \qquad [17]$$

we find the average power delivered to a resistance R,

$$P = \tfrac{1}{2}\left(I_{m1}^2 + I_{m2}^2 + \cdots + I_{mN}^2\right)R \qquad [18]$$

The result is unchanged if an arbitrary phase angle is assigned to each component of the current. This important result is surprisingly simple when we think of the steps required for its derivation: squaring the current function, integrating, and taking the limit. The result is also just plain

---

EXAMPLE 11.6

Find the average power delivered to a $4\ \Omega$ resistor by the current $i_1 = 2\cos 10t - 3\cos 20t$ A.

Since the two cosine terms are at *different* frequencies, the two average-power values may be calculated separately and added. Thus, this current delivers $\frac{1}{2}(2^2)4 + \frac{1}{2}(3^2)4 = 8 + 18 = 26$ W to a $4\ \Omega$ resistor.

---

(2) $T_1 = 2\pi$ and $T_2 = 2\pi/3.14$. Therefore, we seek integral values of m and n such that $2\pi n = 2\pi m/3.14$, or $3.14n = m$, or $\frac{314}{100}n = m$ or $157n = 50m$. Thus, the smallest integral values for n and m are $n = 50$ and $m = 157$. The period is therefore $T = 2\pi n = 100\pi$, or $T = 2\pi(157/3.14) = 100\pi$ s.

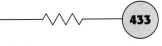

EXAMPLE **11.7**

Find the average power delivered to a 4 Ω resistor by the current $i_2 = 2\cos 10t - 3\cos 10t$ A.

Here, the two components of the current are at the *same* frequency, and they must therefore be combined into a single sinusoid at that frequency. Thus, $i_2 = 2\cos 10t - 3\cos 10t = -\cos 10t$ delivers only $\frac{1}{2}(1^2)4 = 2$ W of average power to a 4 Ω resistor.

PRACTICE

11.6 A voltage source v_s is connected across a 4 Ω resistor. Find the average power absorbed by the resistor if v_s equals (*a*) $8\sin 200t$ V; (*b*) $8\sin 200t - 6\cos(200t - 45°)$ V; (*c*) $8\sin 200t - 4\sin 100t$ V; (*d*) $8\sin 200t - 6\cos(200t - 45°) - 5\sin 100t + 4$ V.

Ans: 8.00 W; 4.01 W; 10.00 W; 11.14 W.

---

surprising, because it shows that, *in this special case of a current such as Eq. [17], where each term has a unique frequency, superposition is applicable to power*. Superposition is *not* applicable for a current which is the sum of two direct currents, nor is it applicable for a current which is the sum of two sinusoids of the same frequency.

11.3 EFFECTIVE VALUES OF CURRENT AND VOLTAGE

In North America, most power outlets deliver a sinusoidal voltage having a frequency of 60 Hz and a "voltage" of 115 V (elsewhere, 50 Hz and 240 V are typically encountered). But what is meant by "115 volts"? This is certainly not the instantaneous value of the voltage, for the voltage is not a constant. The value of 115 V is also not the amplitude which we have been symbolizing as V_m; if we displayed the voltage waveform on a calibrated oscilloscope, we would find that the amplitude of this voltage at one of our ac outlets is $115\sqrt{2}$, or 162.6, volts. We also cannot fit the concept of an average value to the 115 V, because the average value of the sine wave is zero. We might come a little closer by trying the magnitude of the average over a positive or negative half cycle; by using a rectifier-type voltmeter at the outlet, we should measure 103.5 V. As it turns out, however, 115 V is the *effective value* of this sinusoidal voltage; it is a measure of the effectiveness of a voltage source in delivering power to a resistive load.

Effective Value of a Periodic Waveform

Let us arbitrarily define effective value in terms of a current waveform, although a voltage could equally well be selected. The *effective value* of any periodic current is equal to the value of the direct current which, flowing through an R ohm resistor, delivers the same average power to the resistor as does the periodic current.

In other words, we allow the given periodic current to flow through the resistor, determine the instantaneous power $i^2 R$, and then find the average value of $i^2 R$ over a period; this is the average power. We then cause a direct

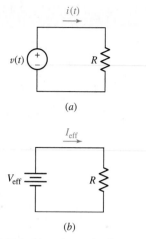

$i(t)$

$v(t)$ R

(a)

I_{eff}

V_{eff} R

(b)

■ **FIGURE 11.9** If the resistor receives the same average power in parts *a* and *b*, then the effective value of *i(t)* is equal to I_{eff}, and the effective value of *v(t)* is equal to V_{eff}.

current to flow through this same resistor and adjust the value of the direct current until the same value of average power is obtained. The resulting magnitude of the direct current is equal to the effective value of the given periodic current. These ideas are illustrated in Fig. 11.9.

The general mathematical expression for the effective value of $i(t)$ is now easily obtained. The average power delivered to the resistor by the periodic current $i(t)$ is

$$P = \frac{1}{T}\int_0^T i^2R\, dt = \frac{R}{T}\int_0^T i^2\, dt$$

where the period of $i(t)$ is T. The power delivered by the direct current is

$$P = I_{\text{eff}}^2 R$$

Equating the power expressions and solving for I_{eff}, we get

$$I_{\text{eff}} = \sqrt{\frac{1}{T}\int_0^T i^2\, dt} \qquad [19]$$

The result is independent of the resistance R, as it must be to provide us with a worthwhile concept. A similar expression is obtained for the effective value of a periodic voltage by replacing i and I_{eff} by v and V_{eff}, respectively.

Notice that the effective value is obtained by first squaring the time function, then taking the average value of the squared function over a period, and finally taking the square root of the average of the squared function. In short, the operation involved in finding an effective value is the (square) *root* of the *mean* of the *square;* for this reason, the effective value is often called the **root-mean-square** value, or simply the **rms** value.

Effective (RMS) Value of a Sinusoidal Waveform

The most important special case is that of the sinusoidal waveform. Let us select the sinusoidal current

$$i(t) = I_m \cos(\omega t + \phi)$$

which has a period

$$T = \frac{2\pi}{\omega}$$

and substitute in Eq. [19] to obtain the effective value

$$I_{\text{eff}} = \sqrt{\frac{1}{T}\int_0^T I_m^2 \cos^2(\omega t + \phi)\, dt}$$

$$= I_m\sqrt{\frac{\omega}{2\pi}\int_0^{2\pi/\omega}\left[\frac{1}{2} + \frac{1}{2}\cos(2\omega t + 2\phi)\right] dt}$$

$$= I_m\sqrt{\frac{\omega}{4\pi}[t]_0^{2\pi/\omega}}$$

$$= \frac{I_m}{\sqrt{2}}$$

Thus the effective value of a sinusoidal current is a real quantity which is independent of the phase angle and numerically equal to $1/\sqrt{2} = 0.707$ times the amplitude of the current. A current $\sqrt{2}\cos(\omega t + \phi)$ A, therefore, has an effective value of 1 A and will deliver the **same** average power to any resistor as will a **direct** current of 1 A.

It should be noted carefully that the $\sqrt{2}$ factor that we obtained as the ratio of the amplitude of the periodic current to the effective value is applicable only when the periodic function is *sinusoidal*. For a sawtooth waveform, for example, the effective value is equal to the maximum value divided by $\sqrt{3}$. The factor by which the maximum value must be divided to obtain the effective value depends on the mathematical form of the given periodic function; it may be either rational or irrational, depending on the nature of the function.

Use of RMS Values to Compute Average Power

The use of the effective value also simplifies slightly the expression for the average power delivered by a sinusoidal current or voltage by avoiding use of the factor $\frac{1}{2}$. For example, the average power delivered to an R ohm resistor by a sinusoidal current is

$$P = \tfrac{1}{2} I_m^2 R$$

Since $I_{\text{eff}} = I_m/\sqrt{2}$, the average power may be written as

$$P = I_{\text{eff}}^2 R \qquad [20]$$

The fact that the effective value is defined in terms of an equivalent dc quantity provides us with average-power formulas for resistive circuits which are identical with those used in dc analysis.

The other power expressions may also be written in terms of effective values:

$$P = V_{\text{eff}} I_{\text{eff}} \cos(\theta - \phi) \qquad [21]$$

$$P = \frac{V_{\text{eff}}^2}{R} \qquad [22]$$

Although we have succeeded in eliminating the factor $\frac{1}{2}$ from our average-power relationships, we must now take care to determine whether a sinusoidal quantity is expressed in terms of its amplitude or its effective value. In practice, the effective value is usually used in the fields of power transmission or distribution and of rotating machinery; in the areas of electronics and communications, the amplitude is more often used. We will assume that the amplitude is specified unless the term "rms" is explicitly used, or we are otherwise instructed.

In the sinusoidal steady state, phasor voltages and currents may be given either as effective values or as amplitudes; the two expressions differ only by a factor of $\sqrt{2}$. The voltage $50\underline{/30^\circ}$ V is expressed in terms of an amplitude; as an rms voltage, we should describe the same voltage as $35.4\underline{/30^\circ}$ V rms.

Effective Value with Multiple-Frequency Circuits

In order to determine the effective value of a periodic or nonperiodic waveform which is composed of the sum of a number of sinusoids of different frequencies, we may use the appropriate average-power relationship of Eq. [18], developed in Sec. 11.2, rewritten in terms of the effective values of the several components:

$$P = \left(I_{1\text{eff}}^2 + I_{2\text{eff}}^2 + \cdots + I_{N\text{eff}}^2 \right) R \qquad [23]$$

From this we see that the effective value of a current which is composed of any number of sinusoidal currents of *different* frequencies can be expressed as

$$I_{\text{eff}} = \sqrt{I_{1\text{eff}}^2 + I_{2\text{eff}}^2 + \cdots + I_{N\text{eff}}^2} \qquad [24]$$

These results indicate that if a sinusoidal current of 5 A rms at 60 Hz flows through a 2 Ω resistor, an average power of $5^2(2) = 50$ W is absorbed by the resistor; if a second current—perhaps 3 A rms at 120 Hz, for example—is also present, the absorbed power is $3^2(2) + 50 = 68$ W. Using Eq. [24] instead, we find that the effective value of the sum of the 60 and 120 Hz currents is 5.831 A. Thus, $P = 5.831^2(2) = 68$ W as before. However, if the second current is also at 60 Hz, the effective value of the sum of the two 60 Hz currents may have any value between 2 and 8 A. Thus, the absorbed power may have *any* value between 8 W and 128 W, depending on the relative phase of the two current components.

PRACTICE

11.7 Calculate the effective value of each of the periodic voltages: (*a*) $6 \cos 25t$; (*b*) $6 \cos 25t + 4 \sin(25t + 30°)$; (*c*) $6 \cos 25t + 5 \cos^2(25t)$; (*d*) $6 \cos 25t + 5 \sin 30t + 4$ V.

Ans: 4.24 V; 6.16 V; 5.23 V; 6.82 V.

> Note that the effective value of a dc quantity K is simply K, not $\dfrac{K}{\sqrt{2}}$.

COMPUTER-AIDED ANALYSIS

Several useful techniques are available through PSpice for calculation of power quantities. In particular, the built-in functions of Probe allow us to both plot the instantaneous power and compute the average power. For example, consider the simple voltage divider circuit of Fig. 11.10, which is being driven by a 60 Hz sine wave with an

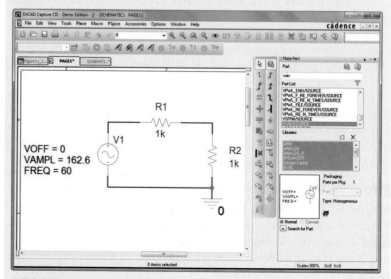

■ **FIGURE 11.10** A simple voltage divider circuit driven by a 115 V rms source operating at 60 Hz.

amplitude of $115\sqrt{2}$ V. We begin by performing a transient simulation over one period of the voltage waveform, $\frac{1}{60}$ s.

The current along with the instantaneous power dissipated in resistor R1 is plotted in Fig. 11.11 by employing the **Add Plot to Window** option under **Plot**. The instantaneous power is periodic, with a nonzero average value and a peak of 6.61 W.

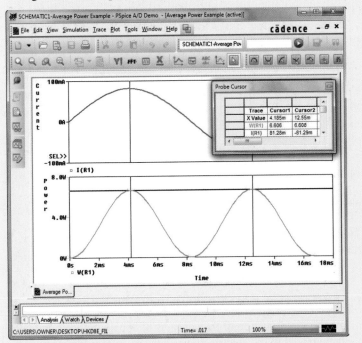

■ FIGURE 11.11 Current and instantaneous power associated with resistor R1 of Fig. 11.10.

The easiest means of using Probe to obtain the average power, which we expect to be $\frac{1}{2}\left(162.6\frac{1000}{1000+1000}\right)(81.3 \times 10^{-3}) = 3.305$ W, is to make use of the built-in "running average" function. Once the **Add Traces** dialog box appears (**Trace**, ⊨ **Add Trace . . .**), type

$$AVG(I(R1) \, {}^*I(R1) \, {}^* 1000)$$

in the **Trace Expression** window.

As can be seen in Fig. 11.12, the average value of the power over either one or two periods is 3.305 W, in agreement with the hand calculation. Note that since PSpice only calculates at specific times, the circuit was not simulated at precisely 8.333 ms and hence **Cursor 1** indicates a slightly higher average power.

Probe also allows us to compute the average over a specific interval using the built-in function **avgx.** For example, to use this function to compute the average power over a single period, which in this case is $1/120 = 8.33$ ms, we would enter

$$AVGX(I(R1) \, {}^*I(R1) \, {}^* 1000, 8.33 \, m)$$

Either approach will result in a value of 3.305 W at the endpoint of the plot.

(Continued on next page)

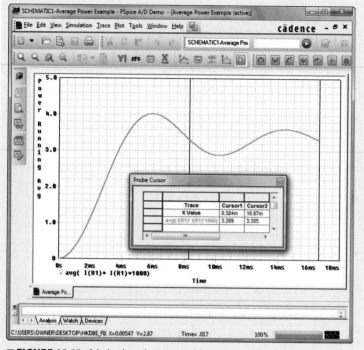

■ FIGURE 11.12 Calculated running average of the power dissipated by resistor R1.

11.4 • APPARENT POWER AND POWER FACTOR

Historically, the introduction of the concepts of apparent power and power factor can be traced to the electric power industry, where large amounts of electric energy must be transferred from one point to another; the efficiency with which this transfer is effected is related directly to the cost of the electric energy, which is eventually paid by the consumer. Customers who provide loads which result in a relatively poor transmission efficiency must pay a greater price for each *kilowatthour* (kWh) of electric energy they actually receive and use. In a similar way, customers who require a costlier investment in transmission and distribution equipment by the power company will also pay more for each kilowatthour unless the company is benevolent and enjoys losing money.

Let us first define *apparent power* and *power factor* and then show briefly how these terms are related to practical economic situations. We assume that the sinusoidal voltage

$$v = V_m \cos(\omega t + \theta)$$

is applied to a network and the resultant sinusoidal current is

$$i = I_m \cos(\omega t + \phi)$$

The phase angle by which the voltage leads the current is therefore $(\theta - \phi)$. The average power delivered to the network, assuming a passive sign convention at its input terminals, may be expressed either in terms of the maximum values:

$$P = \tfrac{1}{2} V_m I_m \cos(\theta - \phi)$$

or in terms of the effective values:

$$P = V_{\text{eff}} I_{\text{eff}} \cos(\theta - \phi)$$

If our applied voltage and current responses had been dc quantities, the average power delivered to the network would have been given simply by the product of the voltage and the current. Applying this dc technique to the sinusoidal problem, we should obtain a value for the absorbed power which is "apparently" given by the familiar product $V_{\text{eff}} I_{\text{eff}}$. However, this product of the *effective* values of the voltage and current is not the average power; we define it as the ***apparent power.*** Dimensionally, apparent power must be measured in the same units as real power, since $\cos(\theta - \phi)$ is dimensionless; but in order to avoid confusion, the term ***volt-amperes,*** or VA, is applied to the apparent power. Since $\cos(\theta - \phi)$ cannot have a magnitude greater than unity, the magnitude of the real power can never be greater than the magnitude of the apparent power.

The ratio of the real or average power to the apparent power is called the ***power factor,*** symbolized by PF. Hence,

$$\text{PF} = \frac{\text{average power}}{\text{apparent power}} = \frac{P}{V_{\text{eff}} I_{\text{eff}}}$$

In the sinusoidal case, the power factor is simply $\cos(\theta - \phi)$, where $(\theta - \phi)$ is the angle by which the voltage leads the current. This relationship is the reason why the angle $(\theta - \phi)$ is often referred to as the ***PF angle.***

For a purely resistive load, the voltage and current are in phase, $(\theta - \phi)$ is zero, and the PF is unity. In other words, the apparent power and the average power are equal. Unity PF, however, may also be achieved for loads that contain both inductance and capacitance if the element values and the operating frequency are carefully selected to provide an input impedance having a zero phase angle. A purely reactive load, that is, one containing no resistance, will cause a phase difference between the voltage and current of either plus or minus 90°, and the PF is therefore zero.

Between these two extreme cases there are the general networks for which the PF can range from zero to unity. A PF of 0.5, for example, indicates a load having an input impedance with a phase angle of either 60° or −60°; the former describes an inductive load, since the voltage leads the current by 60°, while the latter refers to a capacitive load. The ambiguity in the exact nature of the load is resolved by referring to a leading PF or a lagging PF, the terms *leading* or *lagging* referring to the *phase of the current with respect to the voltage.* Thus, an inductive load will have a lagging PF and a capacitive load a leading PF.

Apparent power is not a concept which is limited to sinusoidal forcing functions and responses. It may be determined for any current and voltage waveshapes by simply taking the product of the effective values of the current and voltage.

EXAMPLE 11.8

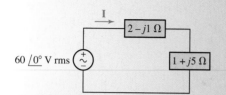

60 $\underline{/0°}$ V rms

$2 - j1\ \Omega$

$1 + j5\ \Omega$

■ FIGURE 11.13 A circuit in which we seek the average power delivered to each element, the apparent power supplied by the source, and the power factor of the combined load.

Calculate values for the average power delivered to each of the two loads shown in Fig. 11.13, the apparent power supplied by the source, and the power factor of the combined loads.

▶ **Identify the goal of the problem.**

The average power refers to the power drawn by the resistive components of the load elements; the apparent power is the product of the effective voltage and the effective current of the load combination.

▶ **Collect the known information.**

The effective voltage is 60 V rms, which appears across a combined load of $2 - j + 1 + j5 = 3 + j4\ \Omega$.

▶ **Devise a plan.**

Simple phasor analysis will provide the current. Knowing voltage and current will enable us to calculate average power and apparent power; these two quantities can be used to obtain the power factor.

▶ **Construct an appropriate set of equations.**

The average power P supplied to a load is given by

$$P = I_{\text{eff}}^2 R$$

where R is the real part of the load impedance. The apparent power supplied by the source is $V_{\text{eff}}I_{\text{eff}}$, where $V_{\text{eff}} = 60$ V rms.

The power factor is calculated as the ratio of these two quantities:

$$\text{PF} = \frac{\text{average power}}{\text{apparent power}} = \frac{P}{V_{\text{eff}}I_{\text{eff}}}$$

▶ **Determine if additional information is required.**

We require I_{eff}:

$$\mathbf{I} = \frac{60\underline{/0°}}{3 + j4} = 12\underline{/-53.13°}\text{ A rms}$$

so $I_{\text{eff}} = 12$ A rms, and *ang* $\mathbf{I} = -53.13°$.

▶ **Attempt a solution.**

The average power delivered to the top load is given by

$$P_{\text{upper}} = I_{\text{eff}}^2 R_{\text{top}} = (12)^2(2) = 288\text{ W}$$

and the average power delivered to the right load is given by

$$P_{\text{lower}} = I_{\text{eff}}^2 R_{\text{right}} = (12)^2(1) = 144\text{ W}$$

The source itself supplies an apparent power of $V_{\text{eff}}I_{\text{eff}} = (60)(12) = 720$ VA.

Finally, the power factor of the combined loads is found by considering the voltage and current associated with the combined loads. This power factor is, of course, identical to the power factor for the source. Thus

$$PF = \frac{P}{V_{eff}I_{eff}} = \frac{432}{60(12)} = 0.6 \; lagging$$

since the combined load is *inductive*.

▶ **Verify the solution. Is it reasonable or expected?**
The total average power delivered to the source is $288 + 144 = 432$ W. The average power supplied by the source is

$$P = V_{eff}I_{eff} \cos(ang\;\mathbf{V} - ang\;\mathbf{I}) = (60)(12)\cos(0 + 53.13°) = 432 \text{ W}$$

so we see the power balance is correct.
    We might also write the combined load impedance as $5\underline{/53.1°}$ Ω, identify 53.1° as the PF angle, and thus have a PF of $\cos 53.1° = 0.6$ *lagging*.

PRACTICE

11.8 For the circuit of Fig. 11.14, determine the power factor of the combined loads if $Z_L = 10$ Ω.

Ans: 0.9966 leading.

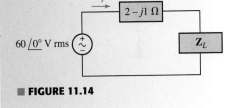

■ **FIGURE 11.14**

11.5 COMPLEX POWER

As we saw in Chap. 10, "complex" numbers do not actually "complicate" analysis. By allowing us to carry two pieces of information together through a series of calculations via the "real" and "imaginary" components, they often greatly simplify what might otherwise be tedious calculations. This is particularly true with power, since we have resistive as well as inductive and capacitive elements in a general load. In this section, we define *complex power* to allow us to calculate the various contributions to the total power in a clean, efficient fashion. The magnitude of the complex power is simply the apparent power. The real part is the average power and—as we are about to see—the imaginary part is a new quantity, termed the *reactive power*, which describes the rate of energy transfer into and out of reactive load components (e.g., inductors and capacitors).
    We define complex power with reference to a general sinusoidal voltage $\mathbf{V}_{eff} = V_{eff}\underline{/\theta}$ across a pair of terminals and a general sinusoidal current $\mathbf{I}_{eff} = I_{eff}\underline{/\phi}$ flowing into one of the terminals in such a way as to satisfy the passive sign convention. The average power P absorbed by the two-terminal network is thus

$$P = V_{eff}I_{eff}\cos(\theta - \phi)$$

Complex nomenclature is next introduced by making use of Euler's formula in the same way as we did in introducing phasors. We express P as

$$P = V_{eff}I_{eff}\,\text{Re}\{e^{j(\theta-\phi)}\}$$

or

$$P = \text{Re}\{V_{\text{eff}}e^{j\theta}I_{\text{eff}}e^{-j\phi}\}$$

The phasor voltage may now be recognized as the first two factors within the brackets in the preceding equation, but the second two factors do not quite correspond to the phasor current, because the angle includes a minus sign, which is not present in the expression for the phasor current. That is, the phasor current is

$$\mathbf{I}_{\text{eff}} = I_{\text{eff}}e^{j\phi}$$

and we therefore must make use of conjugate notation:

$$\mathbf{I}^*_{\text{eff}} = I_{\text{eff}}e^{-j\phi}$$

Hence

$$P = \text{Re}\{\mathbf{V}_{\text{eff}}\mathbf{I}^*_{\text{eff}}\}$$

and we may now let power become complex by defining the **complex power S** as

$$\mathbf{S} = \mathbf{V}_{\text{eff}}\mathbf{I}^*_{\text{eff}} \qquad [25]$$

If we first inspect the polar or exponential form of the complex power,

$$\mathbf{S} = V_{\text{eff}}I_{\text{eff}}\,e^{j(\theta-\phi)}$$

we see that the magnitude of **S**, $V_{\text{eff}}I_{\text{eff}}$, is the apparent power. The angle of **S**, $(\theta - \phi)$, is the PF angle (i.e., the angle by which the voltage leads the current).

In rectangular form, we have

$$\mathbf{S} = P + jQ \qquad [26]$$

where P is the average power, as before. The imaginary part of the complex power is symbolized as Q and is termed the *reactive power*. The dimensions of Q are the same as those of the real power P, the complex power **S**, and the apparent power $|\mathbf{S}|$. In order to avoid confusion with these other quantities, the unit of Q is defined as the **volt-ampere-reactive** (abbreviated VAR). From Eqs. [25] and [26], it is seen that

$$Q = V_{\text{eff}}I_{\text{eff}}\sin(\theta - \phi) \qquad [27]$$

The physical interpretation of reactive power is the time rate of energy flow back and forth between the source (i.e., the utility company) and the reactive components of the load (i.e., inductances and capacitances). These components alternately charge and discharge, which leads to current flow from and to the source, respectively.

The relevant quantities are summarized in Table 11.1 for convenience.

The Power Triangle

A commonly employed graphical representation of complex power is known as the **power triangle,** and is illustrated in Fig. 11.15. The diagram shows that only two of the three power quantities are required, as the third may be obtained by trigonometric relationships. If the power triangle lies in the first quadrant $(\theta - \phi > 0)$; the power factor is *lagging* (corresponding to an inductive load); and if the power triangle lies in the fourth quadrant $(\theta - \phi < 0)$, the power factor is *leading* (corresponding to a capacitive

The sign of the reactive power characterizes the nature of a passive load at which $\mathbf{V}_{\text{eff}}$ and $\mathbf{I}_{\text{eff}}$ are specified. If the load is inductive, then $(\theta - \phi)$ is an angle between 0 and 90°, the sine of this angle is positive, and the reactive power is *positive*. A capacitive load results in a *negative* reactive power.

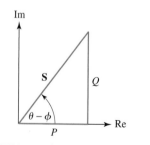

■ **FIGURE 11.15** The power triangle representation of complex power.

READING FURTHER

A good overview of ac power concepts can be found in Chap. 2 of:

B. M. Weedy, *Electric Power Systems,* 3rd ed. Chichester, England: Wiley, 1984.

Contemporary issues pertaining to ac power systems can be found in:

International Journal of Electrical Power & Energy Systems. Guildford, England: IPC Science and Technology Press, 1979–. ISSN: 0142-0615.

EXERCISES

11.1 Instantaneous Power

1. Determine the instantaneous power delivered to the 1 Ω resistor of Fig. 11.23 at $t = 0$, $t = 1$ s, and $t = 2$ s if v_s is equl to (*a*) 9 V; (*b*) 9 sin 2*t* V; (*c*) 9 sin $(2t + 13°)$ V; (*d*) $9e^{-t}$ V.

2. Determine the power absorbed at $t = 1.5$ ms by each of the three elements of the circuit shown in Fig. 11.24 if v_s is equal to (*a*) $30u(-t)$ V; (*b*) $10 + 20u(t)$ V.

3. Calculate the power absorbed at $t = 0^-$, $t = 0^+$, and $t = 200$ ms by each of the elements in the circuit of Fig. 11.25 if v_s is equal to (*a*) $-10u(-t)$ V; (*b*) $20 + 5u(t)$ V.

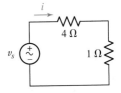

■ **FIGURE 11.23**

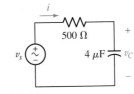

■ **FIGURE 11.24**

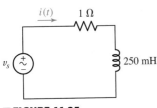

■ **FIGURE 11.25**

4. Three elements are connected in parallel: a 1 k Ω resistor, a 15 mH inductor, and a 100 cos $(2 \times 10^5 t)$ mA sinusoidal source. All transients have long since died out, so the circuit is operating in steady state. Determine the power being absorbed by each element at $t = 10$ μs.

5. Let $i_s = 4u(-t)$ A in the circuit of Fig. 11.26. (*a*) Show that, for all $t > 0$, the instantaneous power absorbed by the resistor is equal in magnitude but opposite in sign to the instantaneous power absorbed by the capacitor. (*b*) Determine the power absorbed by the resistor at $t = 60$ ms.

6. The current source in the circuit of Fig. 11.26 is given by $i_s = 8 - 7u(t)$ A. Compute the power absorbed by all three elements at $t = 0^-$, $t = 0^+$, and $t = 75$ ms.

7. Assuming no transients are present, calcualte the power absorbed by each element shown in the circuit of Fig. 11.27 at $t = 0$, 10, and 20 ms.

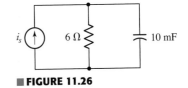

■ **FIGURE 11.26**

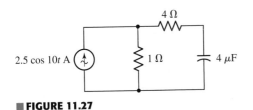

■ **FIGURE 11.27**

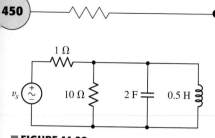

FIGURE 11.28

8. Calculate in Fig. 11.28 the power absorbed by the inductor at $t = 0$ and $t = 1$ s if $v_s = 10u(t)$ V.

9. A 100 mF capacitor is storing 100 mJ of energy up until the point when a conductor of resistance 1.2 Ω falls across its terminals. What is the instantaneous power dissipated in the conductor at $t = 120$ ms? If the specific heat capacity[3] of the conductor is 0.9 kJ/kg · K and its mass is 1 g, estimate the increase in temperature of the conductor in the first second of the capacitor discharge, assuming both elements are initially at 23°C.

10. If we take a typical cloud-to-ground lightning *stroke* to represent a current of 30 kA over an interval of 150 μs, calculate (*a*) the instantaneous power delivered to a copper rod having resistance 1.2 mΩ during the stroke; (*b*) the total energy delivered to the rod.

11.2 Average Power

11. The phasor current $\mathbf{I} = 9\underline{/15°}$ mA (corresponding to a sinusoid operating at 45 rad/s) is applied to the series combination of a 18 kΩ resistor and a 1 μF capacitor. Obtain an expression for (*a*) the instantaneous power and (*b*) the average power absorbed by the combined load.

12. A phasor voltage $\mathbf{V} = 100\underline{/45°}$ V (the sinusoid operates at 155 rad/s) is applied to the parallel combination of a 1 Ω resistor and a 1 mH inductor. (*a*) Obtain an expression for the average power absorbed by each passive element. (*b*) Graph the instantaneous power supplied to the parallel combination, along with the instantaneous power absorbed by each element separately. (Use a single graph.)

13. Calculate the average power delivered by the current $4 - j2$ A to (*a*) $\mathbf{Z} = 9\ \Omega$; (*b*) $\mathbf{Z} = -j1000\ \Omega$; (*c*) $\mathbf{Z} = 1 - j2 + j3\ \Omega$; (*d*) $\mathbf{Z} = 6\underline{/32°}\ \Omega$; (*e*) $\mathbf{Z} = \dfrac{1.5\underline{/-19°}}{2 + j}$ kΩ.

14. With regard to the two-mesh circuit depicted in Fig. 11.29, determine the average power absorbed by each passive element, and the average power supplied by each source, and verify that the total supplied average power = the total absorbed average power.

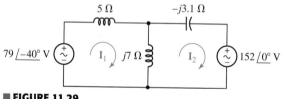

FIGURE 11.29

15. (*a*) Calculate the average power absorbed by each passive element in the circuit of Fig. 11.30, and verify that it equals the average power supplied by the source. (*b*) Check your solution with an appropriate PSpice simulation.

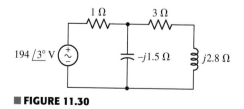

FIGURE 11.30

(3) Assume the specific heat capacity c is given by $c = Q/m \cdot \Delta T$, where $Q =$ the energy delivered to the conductor, m is its mass, and ΔT is the increase in temperature.

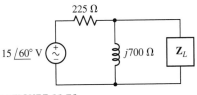

16. (*a*) What load impedance $\mathbf{Z}_L$ will draw the maximum average power from the source shown in Fig. 11.31? (*b*) Calculate the maximum average power supplied to the load.

17. The inductance of Fig. 11.31 is replaced by the impedance $9 - j8$ kΩ. Repeat Exercise 16.

18. Determine the average power supplied by the dependent source in the circuit of Fig. 11.32.

■ **FIGURE 11.31**

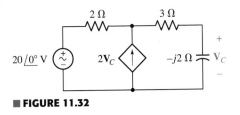

■ **FIGURE 11.32**

19. (*a*) Calculate the average power supplied to each passive element in the circuit of Fig. 11.33. (*b*) Determine the power supplied by each source. (*c*) Replace the 8 Ω resistive load with an impedance capable of drawing maximum average power from the remainder of the circuit. (*d*) Verify your solution with a PSpice simulation.

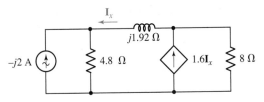

■ **FIGURE 11.33**

20. (*a*) Compute the average value of each waveform shown in Fig. 11.34. (*b*) Square each waveform, and determine the average value of each new periodic waveform.

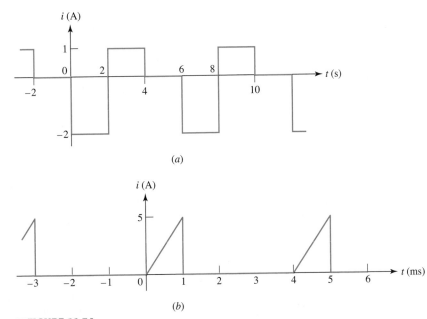

(*a*)

(*b*)

■ **FIGURE 11.34**

21. Calculate the average power delivered to a 2.2 Ω load by the voltage v_s equal to (a) 5 V; (b) 4 cos 80t − 8 sin 80t V; (c) 10 cos 100t + 12.5 cos (100t + 19°) V.

11.3 Effective Values of Current and Voltage

22. Calculate the effective value of the following waveforms: (a) 7 sin 30t V; (b) 100 cos 80t mA; (c) 120$\sqrt{2}$ cos (5000t − 45°) V; (d) $\frac{100}{\sqrt{2}}$ sin (2t + 72°) A.

23. Determine the effective value of the following waveforms: (a) 62.5 cos 100t mV; (b) 1.95 cos 2t A; (c) 208$\sqrt{2}$ cos (100πt + 29°) V; (d) $\frac{400}{\sqrt{2}}$ sin (2000t − 14°) A.

24. Compute the effective value of (a) $i(t) = 3$ sin 4t A; (b) $v(t) = 4$ sin 20t cos 10t; (c) $i(t) = 2 −$ sin 10t mA; (d) the waveform plotted in Fig. 11.35.

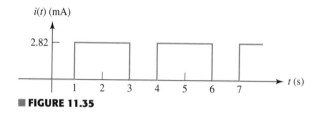

■ **FIGURE 11.35**

25. For each waveform plotted in Fig. 11.34, determine its frequency, period, and rms value.

26. Determine both the average and rms value of each waveform depicted in Fig. 11.36.

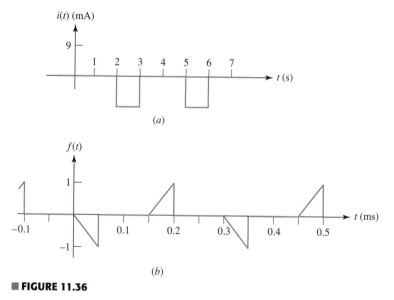

■ **FIGURE 11.36**

27. The series combination of a 1 kΩ resistor and a 2 H inductor must not dissipate more than 250 mW of power at any instant. Assuming a sinusoidal current with $\omega = 500$ rad/s, what is the largest rms current that can be tolerated?

28. For each of the following waveforms, determine its period, frequency and effective value: (a) 5 V; (b) 2 sin 80t − 7 cos 20t + 5 V; (c) 5 cos 50t + 3 sin 50t V; (d) 8 cos² 90t mA. (e) Verify your answers with an appropriate simulation.

29. With regard to the circuit of Fig. 11.37, determine whether a purely real value of R can result in equal rms voltages across the 14 mH inductor and the resistor R. If so, calculate R and the rms voltage across it; if not, explain why not.

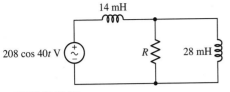

FIGURE 11.37

30. (*a*) Calculate both the average and rms values of the waveform plotted in Fig. 11.38. (*b*) Verify your solutions with appropriate PSpice simulations (*Hint:* you may want to employ two pulse waveforms added together).

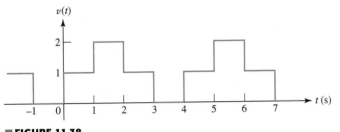

FIGURE 11.38

11.4 Apparent Power and Power Factor

31. For the circuit of Fig. 11.39, compute the average power delivered to each load, the apparent power supplied by the source, and the power factor of the combined loads if (*a*) $\mathbf{Z}_1 = 14\underline{/32°}$ Ω and $\mathbf{Z}_2 = 22$ Ω; (*b*) $\mathbf{Z}_1 = 2\underline{/0°}$ Ω and $\mathbf{Z}_2 = 6 - j$ Ω; (*c*) $\mathbf{Z}_1 = 100\underline{/70°}$ Ω and $\mathbf{Z}_2 = 75\underline{/90°}$ Ω.

32. Calculate the power factor of the combined loads of the circuit depicted in Fig. 11.39 if (*a*) both loads are purely resistive; (*b*) both loads are purely inductive and $\omega = 100$ rad/s; (*c*) both loads are purely capacitive and $\omega = 200$ rad/s; (*d*) $\mathbf{Z}_1 = 2\mathbf{Z}_2 = 5 - j8$ Ω.

33. A given load is connected to an ac power system. If it is known that the load is characterized by resistive losses and either capacitors, inductors, or neither (but not both), which type of reactive element is part of the load if the power factor is measured to be (*a*) unity; (*b*) 0.85 lagging; (*c*) 0.221 leading; (*d*) cos (−90°)?

34. An unknown load is connected to a standard European household outlet (240 V rms, 50 Hz). Determine the phase angle difference between the voltage and current, and whether the voltage leads or lags the current, if (*a*) $\mathbf{V} = 240\underline{/243°}$ V rms and $\mathbf{I} = 3\underline{/9°}$ A rms; (*b*) the power factor of the load is 0.55 lagging; (*c*) the power factor of the load is 0.685 leading; (*d*) the capacitive load draws 100 W average power and 500 volt-amperes apparent power.

35. (*a*) Design a load which draws an average power of 25 W at a leading PF of 0.88 from a standard North American household outlet (120 V rms, 60 Hz). (*b*) Design a capacitor-free load which draws an average power of 150 W and an apparent power of 25 W from a household outlet in eastern Japan (110 V rms, 50 Hz).

36. Assuming an operating frequency of 40 rad/s for the circuit shown in Fig. 11.40, and a load impedance of $50\underline{/-100°}$ Ω, calculate (*a*) the instantaneous power separately delivered to the load and to the 1 kΩ shunt resistance at $t = 20$ ms; (*b*) the average power delivered to both passive elements; (*c*) the apparent power delivered to the load; (*d*) the power factor at which the source is operating.

37. Calculate the power factor at which the source in Fig. 11.40 is operating if the load is (*a*) purely resistive; (*b*) $1000 + j900$ Ω; (*c*) $500\underline{/-5°}$ Ω.

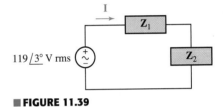

FIGURE 11.39

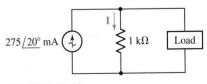

FIGURE 11.40

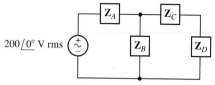

■ **FIGURE 11.41**

38. Determine the load impedance for the circuit depicted in Fig. 11.40 if the source is operating at a PF of (*a*) 0.95 leading; (*b*) unity; (*c*) 0.45 lagging.

39. For the circuit of Fig. 11.41, find the apparent power delivered to each load, and the power factor at which the source operates, if (*a*) $\mathbf{Z}_A = 5 - j2\,\Omega$, $\mathbf{Z}_B = 3\,\Omega$, $\mathbf{Z}_C = 8 + j4\,\Omega$, and $\mathbf{Z}_D = 15\underline{/-30°}\,\Omega$; (*b*) $\mathbf{Z}_A = 2\underline{/-15°}\,\Omega$, $\mathbf{Z}_B = 1\,\Omega$, $\mathbf{Z}_C = 2 + j\,\Omega$, and $\mathbf{Z}_D = 4\underline{/45°}\,\Omega$.

11.5 Complex Power

40. Compute the complex power **S** (in polar form) drawn by a certain load if it is known that (*a*) it draws 100 W average power at a lagging PF of 0.75; (*b*) it draws a current $\mathbf{I} = 9 + j5$ A rms when connected to the voltage $120\underline{/32°}$ V rms; (*c*) it draws 1000 W average power and 10 VAR reactive power at a leading PF; (*d*) it draws an apparent power of 450 W at a lagging PF of 0.65.

41. Calculate the apparent power, power factor, and reactive power associated with a load if it draws complex power **S** equal to (*a*) $1 + j0.5$ kVA; (*b*) 400 VA; (*c*) $150\underline{/-21°}$ VA; (*d*) $75\underline{/25°}$ VA.

42. For each power triangle depicted in Fig. 11.42, determine **S** (in polar form) and the PF.

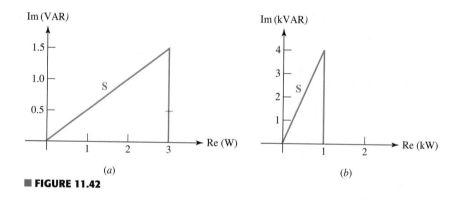

(*a*) (*b*)

■ **FIGURE 11.42**

43. Referring to the network represented in Fig. 11.21, if the motor draws complex power $150\underline{/24°}$ VA, (*a*) determine the PF at which the source is operating; (*b*) determine the impedance of the corrective device required to change the PF of the source to 0.98 lagging. (*c*) Is it physically possible to obtain a leading PF for the source? *Explain*.

44. Determine the complex power absorbed by each passive component in the circuit of Fig. 11.43, and the power factor at which the source is operating.

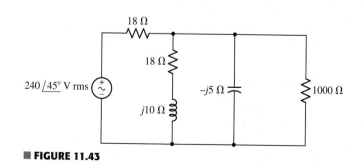

■ **FIGURE 11.43**

45. What value of capacitance must be added in parallel to the 10 Ω resistor of Fig. 11.44 to increase the PF of the source to 0.95 at 50 Hz?

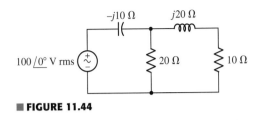

100 /0° V rms

■ **FIGURE 11.44**

46. The kiln operation of a local lumberyard has a monthly average power demand of 175 kW, but associated with that is an average monthly reactive power draw of 205 kVAR. If the lumberyard's utility company charges $0.15 per kVAR for each kVAR above the benchmark value (0.7 times the peak average power demand), (*a*) estimate the annual cost to the lumberyard from PF penalties; (*b*) calculate the money saved in the first and second years, respectively, if 100 kVAR compensation capacitors are available for purchase at $75 each (installed).

47. Calculate the complex power delivered to each passive component of the circuit shown in Fig. 11.45, and determine the power factor of the source.

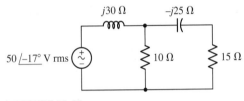

50 /−17° V rms

■ **FIGURE 11.45**

48. Replace the 10 Ω resistor in the circuit of Fig. 11.45 with a 200 mH inductor, assume an operating frequency of 10 rad/s, and calculate (*a*) the PF of the source; (*b*) the apparent power supplied by the source; (*c*) the reactive power delivered by the source.

49. Instead of including a capacitor as indicated in Fig. 11.45, the circuit is erroneously constructed using two identical inductors, each having an impedance of *j*30 W at the operating frequency of 50 Hz. (*a*) Compute the complex power delivered to each passive component. (*b*) Verify your solution by calculating the complex power supplied by the source. (*c*) At what power factor is the source operating?

50. Making use of the general strategy employed in Example 11.9, derive Eq. [28], which enables the corrective value of capacitance to be calculated for a general operating frequency.

Chapter-Integrating Exercises

51. A load is drawing 10 A rms when connected to a 1200 V rms supply running at 50 Hz. If the source is operating at a lagging PF of 0.9, calculate (*a*) the peak voltage magnitude; (*b*) the instantaneous power absorbed by the load at *t* = 1 ms; (*c*) the apparent power supplied by the source; (*d*) the reactive power supplied to the load; (*e*) the load impedance; and (*f*) the complex power supplied by the source (in polar form).

52. For the circuit of Fig. 11.46, assume the source operates at a frequency of 100 rad/s. (*a*) Determine the PF at which the source is operating. (*b*) Calculate the apparent power absorbed by each of the three passive elements. (*c*) Compute the average power supplied by the source. (*d*) Determine the

Thévenin equivalent seen looking into the terminals marked a and b, and calculate the average power delivered to a 100 Ω resistor connected between the same terminals.

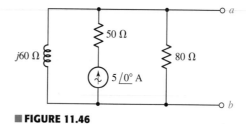

■ **FIGURE 11.46**

53. Remove the 50 Ω resistor in Fig. 11.46, assume an operating frequency of 50 Hz, and (a) determine the power factor at which the load is operating; (b) compute the average power delivered by the source; (c) compute the instantaneous power absorbed by the inductance at $t = 2$ ms; (d) determine the capacitance that must be connected between the terminals marked a and b to increase the PF of the source to 0.95.

54. A source 45 sin 32t V is connected in series with a 5 Ω resistor and a 20 mH inductor. Calculate (a) the reactive power delivered by the source; (b) the apparent power absorbed by each of the three elements; (c) the complex power **S** absorbed by each element; (d) the power factor at which the source is operating.

55. For the circuit of Fig. 11.37, (a) derive an expression for the complex power delivered by the source in terms of the unknown resistance R; (b) compute the necessary capacitance that must be added in parallel to the 28 mH inductor to achieve a unity PF.

12 Polyphase Circuits

INTRODUCTION

The vast majority of power is supplied to consumers in the form of sinusoidal voltages and currents, typically referred to as alternating current or simply *ac*. Although there are exceptions, for example, some types of train motors, most equipment is designed to run on either 50 or 60 Hz. Most 60 Hz systems are now standardized to run on 120 V, whereas 50 Hz systems typically correspond to 240 V (both voltages being quoted in rms units). The actual voltage delivered to an appliance can vary somewhat from these values, and distribution systems employ significantly higher voltages to minimize the current and hence cable size. Originally Thomas Edison advocated a purely dc power distribution network, purportedly due to his preference for the simple algebra required to analyze such circuits. Nikola Tesla and George Westinghouse, two other pioneers in the field of electricity, proposed ac distribution systems as the achievable losses were significantly lower. Ultimately they were more persuasive, despite some rather theatrical demonstrations on the part of Edison.

The transient response of ac power systems is of interest when determining the peak power demand, since most equipment requires more current to start up than it does to run continuously. Often, however, it is the steady-state operation that is of primary interest, so our experience with phasor-based analysis will prove to be handy. In this chapter we introduce a new type of voltage source, the three-phase source, which can be connected in either a three- or four-wire Y configuration or a three-wire Δ configuration. Loads can also be either Y- or Δ-connected, depending on the application.

12.1 • POLYPHASE SYSTEMS

So far, whenever we used the term "sinusoidal source" we pictured a single sinusoidal voltage or current having a particular amplitude, frequency, and phase. In this chapter, we introduce the concept of **polyphase** sources, focusing on three-phase systems in particular. There are distinct advantages in using rotating machinery to generate three-phase power rather than single-phase power, and there are economical advantages in favor of the transmission of power in a three-phase system. Although most of the electrical equipment we have encountered so far is single-phase, three-phase equipment is not uncommon, especially in manufacturing environments. In particular, motors used in large refrigeration systems and in machining facilities are often wired for three-phase power. For the remaining applications, once we have become familiar with the basics of polyphase systems, we will find that it is simple to obtain single-phase power by just connecting to a single "leg" of a polyphase system.

Let us look briefly at the most common polyphase system, a balanced three-phase system. The source has three terminals (not counting a **neutral** or **ground** connection), and voltmeter measurements will show that sinusoidal voltages of equal amplitude are present between any two terminals. However, these voltages are not in phase; each of the three voltages is 120° out of phase with each of the other two, the sign of the phase angle depending on the sense of the voltages. One possible set of voltages is shown in Fig. 12.1. A **balanced load** draws power equally from all three phases. *At no instant does the instantaneous power drawn by the total load reach zero; in fact, the total instantaneous power is constant.* This is an advantage in rotating machinery, for it keeps the torque on the rotor much more constant than it would be if a single-phase source were used. As a result, there is less vibration.

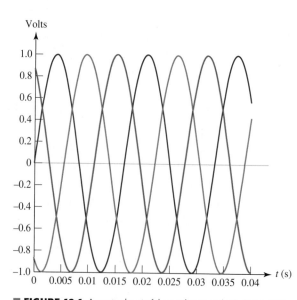

■ **FIGURE 12.1** An example set of three voltages, each of which is 120° out of phase with the other two. As can be seen, only one of the voltages is zero at any particular instant.

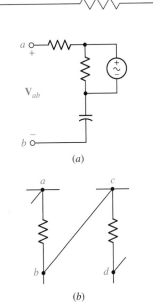

The use of a higher number of phases, such as 6- and 12-phase systems, is limited almost entirely to the supply of power to large **rectifiers.** Rectifiers convert alternating current to direct current by only allowing current to flow to the load in one direction, so that the sign of the voltage across the load remains the same. The rectifier output is a direct current plus a smaller pulsating component, or ripple, which decreases as the number of phases increases.

Almost without exception, polyphase systems in practice contain sources which may be closely approximated by ideal voltage sources or by ideal voltage sources in series with small internal impedances. Three-phase current sources are extremely rare.

Double-Subscript Notation

It is convenient to describe polyphase voltages and currents using **double-subscript notation.** With this notation, a voltage or current, such as $\mathbf{V}_{ab}$ or $\mathbf{I}_{aA}$, has more meaning than if it were indicated simply as $\mathbf{V}_3$ or $\mathbf{I}_x$. By definition, the voltage of point a with respect to point b is $\mathbf{V}_{ab}$. Thus, the plus sign is located at a, as indicated in Fig. 12.2a. We therefore consider the double subscripts to be *equivalent* to a plus-minus sign pair; the use of both would be redundant. With reference to Fig. 12.2b, for example, we see that $\mathbf{V}_{ad} = \mathbf{V}_{ab} + \mathbf{V}_{cd}$. The advantage of the double-subscript notation lies in the fact that Kirchhoff's voltage law requires the voltage between two points to be the same, regardless of the path chosen between the points; thus $\mathbf{V}_{ad} = \mathbf{V}_{ab} + \mathbf{V}_{bd} = \mathbf{V}_{ac} + \mathbf{V}_{cd} = \mathbf{V}_{ab} + \mathbf{V}_{bc} + \mathbf{V}_{cd}$, and so forth. The benefit of this is that KVL may be satisfied without reference to the circuit diagram; correct equations may be written even though a point, or subscript letter, is included which is not marked on the diagram. For example, we might have written $\mathbf{V}_{ad} = \mathbf{V}_{ax} + \mathbf{V}_{xd}$, where x identifies the location of any interesting point of our choice.

One possible representation of a three-phase system of voltages[1] is shown in Fig. 12.3. Let us assume that the voltages $\mathbf{V}_{an}$, $\mathbf{V}_{bn}$, and $\mathbf{V}_{cn}$ are known:

$$\mathbf{V}_{an} = 100\underline{/0°} \text{ V}$$
$$\mathbf{V}_{bn} = 100\underline{/-120°} \text{ V}$$
$$\mathbf{V}_{cn} = 100\underline{/-240°} \text{ V}$$

The voltage $\mathbf{V}_{ab}$ may be found, with an eye on the subscripts, as

$$\mathbf{V}_{ab} = \mathbf{V}_{an} + \mathbf{V}_{nb} = \mathbf{V}_{an} - \mathbf{V}_{bn}$$
$$= 100\underline{/0°} - 100\underline{/-120°} \text{ V}$$
$$= 100 - (-50 - j86.6) \text{ V}$$
$$= 173.2\underline{/30°} \text{ V}$$

The three given voltages and the construction of the phasor $\mathbf{V}_{ab}$ are shown on the phasor diagram of Fig. 12.4.

A double-subscript notation may also be applied to currents. We define the current $\mathbf{I}_{ab}$ as the current flowing from a to b *by the most direct path.* In

---

(1) In keeping with power industry convention, rms values for currents and voltages will be used *implicitly* throughout this chapter.

■ **FIGURE 12.2** (*a*) The definition of the voltage V_{ab}. (*b*) $\mathsf{V}_{ad} = \mathsf{V}_{ab} + \mathsf{V}_{bc} + \mathsf{V}_{cd} = \mathsf{V}_{ab} + \mathsf{V}_{cd}$.

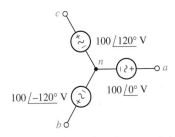

■ **FIGURE 12.3** A network used as a numerical example of double-subscript voltage notation.

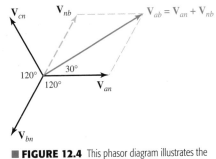

■ **FIGURE 12.4** This phasor diagram illustrates the graphical use of the double-subscript voltage convention to obtain V_{ab} for the network of Fig. 12.3.

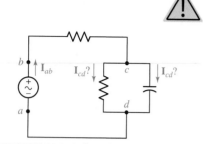

■ **FIGURE 12.5** An illustration of the use and *misuse* of the double-subscript convention for current notation.

every complete circuit we consider, there must of course be at least two possible paths between the points a and b, and we agree that we will not use double-subscript notation unless it is obvious that one path is much shorter, or much more direct. Usually this path is through a single element. Thus, the current I_{ab} is correctly indicated in Fig. 12.5. In fact, we do not even need the direction arrow when talking about this current; the subscripts *tell* us the direction. However, the identification of a current as I_{cd} for the circuit of Fig. 12.5 would cause confusion.

PRACTICE

12.1 Let $V_{ab} = 100\underline{/0°}$ V, $V_{bd} = 40\underline{/80°}$ V, and $V_{ca} = 70\underline{/200°}$ V. Find (a) V_{ad}; (b) V_{bc}; (c) V_{cd}.

12.2 Refer to the circuit of Fig. 12.6 and let $I_{fj} = 3$ A, $I_{de} = 2$ A, and $I_{hd} = -6$ A. Find (a) I_{cd}; (b) I_{ef}; (c) I_{ij}.

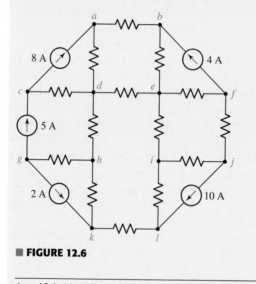

■ **FIGURE 12.6**

Ans: 12.1: $114.0\underline{/20.2°}$ V; $41.8\underline{/145.0°}$ V; $44.0\underline{/20.6°}$ V. 12.2: -3 A; 7 A; 7 A.

12.2 • SINGLE-PHASE THREE-WIRE SYSTEMS

Before studying polyphase systems in detail, it can be helpful first to look at a simple single-phase three-wire system. A *single-phase three-wire source* is defined as a source having three output terminals, such as a, n, and b in Fig. 12.7a, at which the phasor voltages V_{an} and V_{nb} are equal. The source may therefore be represented by the combination of two identical voltage sources; in Fig. 12.7b, $V_{an} = V_{nb} = V_1$. It is apparent that $V_{ab} = 2V_{an} = 2V_{nb}$, and we therefore have a source to which loads operating at either of two voltages may be connected. The normal North American household system is single-phase three-wire, permitting the operation of both 110 V and 220 V appliances. The higher-voltage appliances are normally those drawing larger amounts of power; operation at higher voltage results in a smaller current draw for the same power. Smaller-diameter wire may consequently be used safely in the appliance, the household distribution

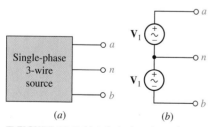

■ **FIGURE 12.7** (a) A single-phase three-wire source. (b) The representation of a single-phase three-wire source by two identical voltage sources.

system, and the distribution system of the utility company, as larger-diameter wire must be used with higher currents to reduce the heat produced due to the resistance of the wire.

The name *single-phase* arises because the voltages $\mathbf{V}_{an}$ and $\mathbf{V}_{nb}$, being equal, must have the same phase angle. From another viewpoint, however, the voltages between the outer wires and the central wire, which is usually referred to as the *neutral*, are exactly 180° out of phase. That is, $\mathbf{V}_{an} = -\mathbf{V}_{bn}$, and $\mathbf{V}_{an} + \mathbf{V}_{bn} = 0$. Later, we will see that balanced polyphase systems are characterized by a set of voltages of equal *amplitude* whose (phasor) sum is zero. From this viewpoint, then, the single-phase three-wire system is really a balanced two-phase system. *Two-phase,* however, is a term that is traditionally reserved for a relatively unimportant unbalanced system utilizing two voltage sources 90° out of phase.

Let us now consider a single-phase three-wire system that contains identical loads $\mathbf{Z}_p$ between each outer wire and the neutral (Fig. 12.8). We first assume that the wires connecting the source to the load are perfect conductors. Since

$$\mathbf{V}_{an} = \mathbf{V}_{nb}$$

then,

$$\mathbf{I}_{aA} = \frac{\mathbf{V}_{an}}{\mathbf{Z}_p} = \mathbf{I}_{Bb} = \frac{\mathbf{V}_{nb}}{\mathbf{Z}_p}$$

and therefore

$$\mathbf{I}_{nN} = \mathbf{I}_{Bb} + \mathbf{I}_{Aa} = \mathbf{I}_{Bb} - \mathbf{I}_{aA} = 0$$

Thus there is no current in the neutral wire, and it could be removed without changing any current or voltage in the system. This result is achieved through the equality of the two loads and of the two sources.

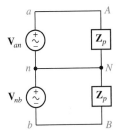

■ **FIGURE 12.8** A simple single-phase three-wire system. The two loads are identical, and the neutral current is zero.

Effect of Finite Wire Impedance

We next consider the effect of a finite impedance in each of the wires. If lines aA and bB each have the same impedance, this impedance may be added to $\mathbf{Z}_p$, resulting in two equal loads once more, and zero neutral current. Now let us allow the neutral wire to possess some impedance $\mathbf{Z}_n$. Without carrying out any detailed analysis, superposition should show us that the symmetry of the circuit will still cause zero neutral current. Moreover, the addition of any impedance connected directly from one of the outer lines to the other outer line also yields a symmetrical circuit and zero neutral current. Thus, zero neutral current is a consequence of a balanced, or symmetrical, load; nonzero impedance in the neutral wire does not destroy the symmetry.

The most general single-phase three-wire system will contain unequal loads between each outside line and the neutral and another load directly between the two outer lines; the impedances of the two outer lines may be expected to be approximately equal, but the neutral impedance is often slightly larger. Let us consider an example of such a system, with particular interest in the current that may flow now through the neutral wire, as well as the overall efficiency with which our system is transmitting power to the unbalanced load.

EXAMPLE 12.1

Analyze the system shown in Fig. 12.9 and determine the power delivered to each of the three loads as well as the power lost in the neutral wire and each of the two lines.

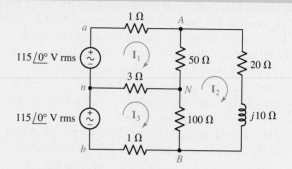

■ **FIGURE 12.9** A typical single-phase three-wire system.

> ### Identify the goal of the problem.

The three loads in the circuit are the 50 Ω resistor, the 100 Ω resistor, and a 20 + j10 Ω impedance. Each of the two lines has a resistance of 1 Ω, and the neutral wire has a resistance of 3 Ω. We need the current through each of these in order to determine power.

> ### Collect the known information.

We have a single-phase three-wire system; the circuit diagram of Fig. 12.9 is completely labeled. The computed currents will be in rms units.

> ### Devise a plan.

The circuit is conducive to mesh analysis, having three clearly defined meshes. The result of the analysis will be a set of mesh currents, which can then be used to compute absorbed power.

> ### Construct an appropriate set of equations.

The three mesh equations are:

$$-115\underline{/0^\circ} + \mathbf{I}_1 + 50(\mathbf{I}_1 - \mathbf{I}_2) + 3(\mathbf{I}_1 - \mathbf{I}_3) = 0$$
$$(20 + j10)\mathbf{I}_2 + 100(\mathbf{I}_2 - \mathbf{I}_3) + 50(\mathbf{I}_2 - \mathbf{I}_1) = 0$$
$$-115\underline{/0^\circ} + 3(\mathbf{I}_3 - \mathbf{I}_1) + 100(\mathbf{I}_3 - \mathbf{I}_2) + \mathbf{I}_3 = 0$$

which can be rearranged to obtain the following three equations

$$\begin{aligned} 54\mathbf{I}_1 \quad\quad -50\mathbf{I}_2 \quad -3\mathbf{I}_3 &= 115\underline{/0^\circ} \\ -50\mathbf{I}_1 + (170 + j10)\mathbf{I}_2 \quad -100\mathbf{I}_3 &= 0 \\ -3\mathbf{I}_1 \quad\quad -100\mathbf{I}_2 \quad +104\mathbf{I}_3 &= 115\underline{/0^\circ} \end{aligned}$$

> ### Determine if additional information is required.

We have a set of three equations in three unknowns, so it is possible to attempt a solution at this point.

▶ **Attempt a solution.**

Solving for the phasor currents $\mathbf{I}_1$, $\mathbf{I}_2$, and $\mathbf{I}_3$ using a scientific calculator, we find

$$\mathbf{I}_1 = 11.24\underline{/-19.83^\circ}\text{ A}$$
$$\mathbf{I}_2 = 9.389\underline{/-24.47^\circ}\text{ A}$$
$$\mathbf{I}_3 = 10.37\underline{/-21.80^\circ}\text{ A}$$

The currents in the outer lines are thus

$$\mathbf{I}_{aA} = \mathbf{I}_1 = 11.24\underline{/-19.83^\circ}\text{ A}$$

and

$$\mathbf{I}_{bB} = -\mathbf{I}_3 = 10.37\underline{/158.20^\circ}\text{ A}$$

and the smaller neutral current is

$$\mathbf{I}_{nN} = \mathbf{I}_3 - \mathbf{I}_1 = 0.9459\underline{/-177.7^\circ}\text{ A}$$

The average power drawn by each load may thus be determined:

$$P_{50} = |\mathbf{I}_1 - \mathbf{I}_2|^2\,(50) = 206\text{ W}$$
$$P_{100} = |\mathbf{I}_3 - \mathbf{I}_2|^2\,(100) = 117\text{ W}$$
$$P_{20+j10} = |\mathbf{I}_2|^2\,(20) = 1763\text{ W}$$

> Note that we do not need to include a factor of $\frac{1}{2}$ since we are working with rms current values.

The total load power is 2086 W. The loss in each of the wires is next found:

$$P_{aA} = |\mathbf{I}_1|^2\,(1) = 126\text{ W}$$
$$P_{bB} = |\mathbf{I}_3|^2\,(1) = 108\text{ W}$$
$$P_{nN} = |\mathbf{I}_{nN}|^2\,(3) = 3\text{ W}$$

giving a total line loss of 237 W. The wires are evidently quite long; otherwise, the relatively high power loss in the two outer lines would cause a dangerous temperature rise.

> Imagine the heat produced by two 100 W light bulbs! These outer wires must dissipate the same amount of power. In order to keep their temperature down, a large surface area is required.

▶ **Verify the solution. Is it reasonable or expected?**

The total absorbed power is $206 + 117 + 1763 + 237$, or 2323 W, which may be checked by finding the power delivered by each voltage source:

$$P_{an} = 115(11.24)\cos 19.83^\circ = 1216\text{ W}$$
$$P_{bn} = 115(10.37)\cos 21.80^\circ = 1107\text{ W}$$

or a total of 2323 W. The *transmission efficiency* for the system is

$$\eta = \frac{\text{total power delivered to load}}{\text{total power generated}} = \frac{2086}{2086 + 237} = 89.8\%$$

This value would be unbelievable for a steam engine or an internal combustion engine, but it is too low for a well-designed distribution system. Larger-diameter wires should be used if the source and the load cannot be placed closer to each other.

(Continued on next page)

A phasor diagram showing the two source voltages, the currents in the outer lines, and the current in the neutral is constructed in Fig. 12.10. The fact that $\mathbf{I}_{aA} + \mathbf{I}_{bB} + \mathbf{I}_{nN} = 0$ is indicated on the diagram.

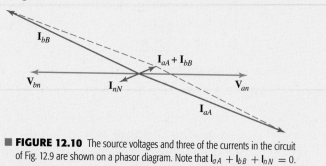

■ **FIGURE 12.10** The source voltages and three of the currents in the circuit of Fig. 12.9 are shown on a phasor diagram. Note that $\mathbf{I}_{aA} + \mathbf{I}_{bB} + \mathbf{I}_{nN} = 0$.

PRACTICE

12.3 Modify Fig. 12.9 by adding a 1.5 Ω resistance to each of the two outer lines, and a 2.5 Ω resistance to the neutral wire. Find the average power delivered to each of the three loads.

Ans: 153.1 W; 95.8 W; 1374 W.

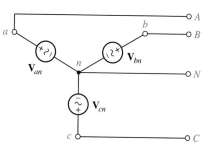

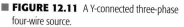

■ **FIGURE 12.11** A Y-connected three-phase four-wire source.

12.3 THREE-PHASE Y-Y CONNECTION

Three-phase sources have three terminals, called the *line* terminals, and they may or may not have a fourth terminal, the *neutral* connection. We will begin by discussing a three-phase source that does have a neutral connection. It may be represented by three ideal voltage sources connected in a Y, as shown in Fig. 12.11; terminals *a*, *b*, *c*, and *n* are available. We will consider only balanced three-phase sources, which may be defined as having

$$|\mathbf{V}_{an}| = |\mathbf{V}_{bn}| = |\mathbf{V}_{cn}|$$

and

$$\mathbf{V}_{an} + \mathbf{V}_{bn} + \mathbf{V}_{cn} = 0$$

These three voltages, each existing between one line and the neutral, are called *phase voltages.* If we arbitrarily choose $\mathbf{V}_{an}$ as the reference, or define

$$\mathbf{V}_{an} = V_p \underline{/0^\circ}$$

where we will consistently use V_p to represent the rms *amplitude* of any of the phase voltages, then the definition of the three-phase source indicates that either

$$\mathbf{V}_{bn} = V_p \underline{/-120^\circ} \quad \text{and} \quad \mathbf{V}_{cn} = V_p \underline{/-240^\circ}$$

or

$$\mathbf{V}_{bn} = V_p \underline{/120^\circ} \quad \text{and} \quad \mathbf{V}_{cn} = V_p \underline{/240^\circ}$$

The former is called *positive phase sequence,* or *abc phase sequence,* and is shown in Fig. 12.12*a*; the latter is termed *negative phase sequence,* or *cba phase sequence,* and is indicated by the phasor diagram of Fig. 12.12*b*.

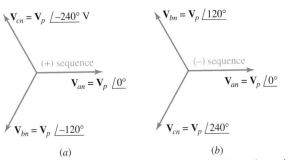

$\mathbf{V}_{cn} = \mathbf{V}_p \underline{/-240°}$ V

$\mathbf{V}_{bn} = \mathbf{V}_p \underline{/120°}$

(+) sequence

$\mathbf{V}_{an} = \mathbf{V}_p \underline{/0°}$

(−) sequence

$\mathbf{V}_{an} = \mathbf{V}_p \underline{/0°}$

$\mathbf{V}_{bn} = \mathbf{V}_p \underline{/-120°}$

$\mathbf{V}_{cn} = \mathbf{V}_p \underline{/240°}$

(a)   (b)

■ **FIGURE 12.12** (a) Positive, or abc, phase sequence. (b) Negative, or cba, phase sequence.

The actual phase sequence of a physical three-phase source depends on the arbitrary choice of the three terminals to be lettered a, b, and c. They may always be chosen to provide positive phase sequence, and we will assume that this has been done in most of the systems we consider.

Line-to-Line Voltages

Let us next find the line-to-line voltages (often simply called the **line voltages**) which are present when the phase voltages are those of Fig. 12.12a. It is easiest to do this with the help of a phasor diagram, since the angles are all multiples of 30°. The necessary construction is shown in Fig. 12.13; the results are

$$\mathbf{V}_{ab} = \sqrt{3}V_p \underline{/30°} \qquad [1]$$

$$\mathbf{V}_{bc} = \sqrt{3}V_p \underline{/-90°} \qquad [2]$$

$$\mathbf{V}_{ca} = \sqrt{3}V_p \underline{/-210°} \qquad [3]$$

Kirchhoff's voltage law requires the sum of these three voltages to be zero; the reader is encouraged to verify this as an exercise.

If the rms amplitude of any of the line voltages is denoted by V_L, then one of the important characteristics of the Y-connected three-phase source may be expressed as

$$\boxed{V_L = \sqrt{3}V_p}$$

Note that with positive phase sequence, $\mathbf{V}_{an}$ leads $\mathbf{V}_{bn}$ and $\mathbf{V}_{bn}$ leads $\mathbf{V}_{cn}$, in each case by 120°, and also that $\mathbf{V}_{ab}$ leads $\mathbf{V}_{bc}$ and $\mathbf{V}_{bc}$ leads $\mathbf{V}_{ca}$, again by 120°. The statement is true for negative phase sequence if "lags" is substituted for "leads."

Now let us connect a balanced Y-connected three-phase load to our source, using three lines and a neutral, as drawn in Fig. 12.14. The load is

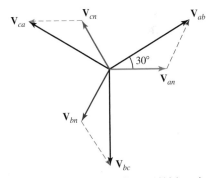

■ **FIGURE 12.13** A phasor diagram which is used to determine the line voltages from the given phase voltages. Or, algebraically, $\mathbf{V}_{ab} = \mathbf{V}_{an} - \mathbf{V}_{bn} = V_p\underline{/0°} - V_p\underline{/-120°} = V_p - V_p\cos(-120°) - jV_p\sin(-120°) = V_p(1 + \frac{1}{2} + j\sqrt{3}/2) = \sqrt{3}V_p\underline{/30°}$.

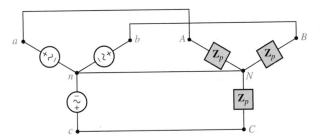

■ **FIGURE 12.14** A balanced three-phase system, connected Y-Y and including a neutral.

represented by an impedance $\mathbf{Z}_p$ between each line and the neutral. The three line currents are found very easily, since we really have three single-phase circuits that possess one common lead:[2]

$$\mathbf{I}_{aA} = \frac{\mathbf{V}_{an}}{\mathbf{Z}_p}$$

$$\mathbf{I}_{bB} = \frac{\mathbf{V}_{bn}}{\mathbf{Z}_p} = \frac{\mathbf{V}_{an}\underline{/-120°}}{\mathbf{Z}_p} = \mathbf{I}_{aA}\underline{/-120°}$$

$$\mathbf{I}_{cC} = \mathbf{I}_{aA}\underline{/-240°}$$

and therefore

$$\mathbf{I}_{Nn} = \mathbf{I}_{aA} + \mathbf{I}_{bB} + \mathbf{I}_{cC} = 0$$

Thus, the neutral carries no current if the source and load are both balanced and if the four wires have zero impedance. How will this change if an impedance $\mathbf{Z}_L$ is inserted in series with each of the three lines and an impedance $\mathbf{Z}_n$ is inserted in the neutral? The line impedances may be combined with the three load impedances; this effective load is still balanced, and a perfectly conducting neutral wire could be removed. Thus, if no change is produced in the system with a short circuit or an open circuit between n and N, any impedance may be inserted in the neutral and the neutral current will remain zero.

It follows that, if we have balanced sources, balanced loads, and balanced line impedances, a neutral wire of any impedance may be replaced by any other impedance, including a short circuit or an open circuit; the replacement will not affect the system's voltages or currents. It is often helpful to *visualize* a short circuit between the two neutral points, whether a neutral wire is actually present or not; the problem is then reduced to three single-phase problems, all identical except for the consistent difference in phase angle. We say that we thus work the problem on a "per-phase" basis.

EXAMPLE 12.2

For the circuit of Fig. 12.15, find both the phase and line currents, and the phase and line voltages throughout the circuit; then calculate the total power dissipated in the load.

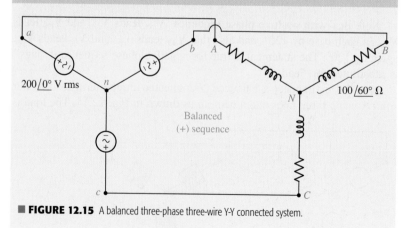

■ **FIGURE 12.15** A balanced three-phase three-wire Y-Y connected system.

---

(2) This can be seen to be true by applying superposition and looking at each phase one at a time.

Since one of the source phase voltages is given and we are told to use the positive phase sequence, the three phase voltages are:

$$\mathbf{V}_{an} = 200\underline{/0°} \text{ V} \qquad \mathbf{V}_{bn} = 200\underline{/-120°} \text{ V} \qquad \mathbf{V}_{cn} = 200\underline{/-240°} \text{ V}$$

The line voltage is $200\sqrt{3} = 346$ V; the phase angle of each line voltage can be determined by constructing a phasor diagram, as we did in Fig. 12.13 (as a matter of fact, the phasor diagram of Fig. 12.13 is applicable), subtracting the phase voltages using a scientific calculator, or by invoking Eqs. [1] to [3]. We find that $\mathbf{V}_{ab}$ is $346\underline{/30°}$ V, $\mathbf{V}_{bc} = 346\underline{/-90°}$ V, and $\mathbf{V}_{ca} = 346\underline{/-210°}$ V.

The line current for phase A is

$$\mathbf{I}_{aA} = \frac{\mathbf{V}_{an}}{\mathbf{Z}_p} = \frac{200\underline{/0°}}{100\underline{/60°}} = 2\underline{/-60°} \text{ A}$$

Since we know this is a balanced three-phase system, we may write the remaining line currents based on $\mathbf{I}_{aA}$:

$$\mathbf{I}_{bB} = 2\underline{/(-60° - 120°)} = 2\underline{/-180°} \text{ A}$$
$$\mathbf{I}_{cC} = 2\underline{/(-60° - 240°)} = 2\underline{/-300°} \text{ A}$$

Finally, the average power absorbed by phase A is $\text{Re}\{\mathbf{V}_{an}\mathbf{I}_{aA}^*\}$, or

$$P_{AN} = 200(2)\cos(0° + 60°) = 200 \text{ W}$$

Thus, the total average power drawn by the three-phase load is 600 W.

The phasor diagram for this circuit is shown in Fig. 12.16. Once we knew any of the line voltage magnitudes and any of the line current magnitudes, the angles for all three voltages and all three currents could have been obtained by simply reading the diagram.

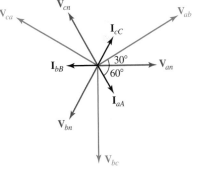

■ **FIGURE 12.16** The phasor diagram that applies to the circuit of Fig. 12.15.

PRACTICE

12.4 A balanced three-phase three-wire system has a Y-connected load. Each phase contains three loads in parallel: $-j100 \ \Omega$, $100 \ \Omega$, and $50 + j50 \ \Omega$. Assume positive phase sequence with $\mathbf{V}_{ab} = 400\underline{/0°}$ V. Find (a) $\mathbf{V}_{an}$; (b) $\mathbf{I}_{aA}$; (c) the total power drawn by the load.

Ans: $231\underline{/-30°}$ V; $4.62\underline{/-30°}$ A; 3200 W.

Before working another example, this would be a good opportunity to quickly explore a statement made in Sec. 12.1, i.e., that even though phase voltages and currents have zero value at specific instants in time (every 1/120 s in North America), the instantaneous power delivered to the *total* load is never zero. Consider phase A of Example 12.2 once more, with the phase voltage and current written in the time domain:

$$v_{AN} = 200\sqrt{2}\cos(120\pi t + 0°) \text{ V}$$

and

$$i_{AN} = 2\sqrt{2}\cos(120\pi t - 60°) \text{ A}$$

The factor of $\sqrt{2}$ is required to convert from rms units.

Thus, the instantaneous power absorbed by phase A is

$$
\begin{aligned}
p_A(t) = v_{AN}i_{AN} &= 800\cos(120\pi t)\cos(120\pi t - 60°) \\
&= 400[\cos(-60°) + \cos(240\pi t - 60°)] \\
&= 200 + 400\cos(240\pi t - 60°)\ \text{W}
\end{aligned}
$$

in a similar fashion,

$$
p_B(t) = 200 + 400\cos(240\pi t - 300°)\ \text{W}
$$

and

$$
p_C(t) = 200 + 400\cos(240\pi t - 180°)\ \text{W}
$$

The instantaneous power absorbed by the *total* load is therefore

$$
p(t) = p_A(t) + p_B(t) + p_C(t) = 600\ \text{W}
$$

independent of time, and the same value as the average power computed in Example 12.2.

EXAMPLE 12.3

A balanced three-phase system with a line voltage of 300 V is supplying a balanced Y-connected load with 1200 W at a leading PF of 0.8. Find the line current and the per-phase load impedance.

The phase voltage is $300/\sqrt{3}$ V and the per-phase power is $1200/3 = 400$ W. Thus the line current may be found from the power relationship

$$
400 = \frac{300}{\sqrt{3}}(I_L)(0.8)
$$

and the line current is therefore 2.89 A. The phase impedance magnitude is given by

$$
|\mathbf{Z}_p| = \frac{V_p}{I_L} = \frac{300/\sqrt{3}}{2.89} = 60\ \Omega
$$

Since the PF is 0.8 leading, the impedance phase angle is $-36.9°$; thus $\mathbf{Z}_p = 60\underline{/-36.9°}\ \Omega$.

PRACTICE

12.5 A balanced three-phase three-wire system has a line voltage of 500 V. Two balanced Y-connected loads are present. One is a capacitive load with $7 - j2\ \Omega$ per phase, and the other is an inductive load with $4 + j2\ \Omega$ per phase. Find (a) the phase voltage; (b) the line current; (c) the total power drawn by the load; (d) the power factor at which the source is operating.

Ans: 289 V; 97.5 A; 83.0 kW; 0.983 lagging.

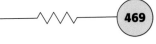

EXAMPLE **12.4**

A balanced 600 W lighting load is added (in parallel) to the system of Example 12.3. Determine the new line current.

We first sketch a suitable per-phase circuit, as shown in Fig. 12.17. The 600 W load is assumed to be a balanced load evenly distributed among the three phases, resulting in an additional 200 W consumed by each phase.

The amplitude of the lighting current (labeled $\mathbf{I}_1$) is determined by

$$200 = \frac{300}{\sqrt{3}} |\mathbf{I}_1| \cos 0°$$

so that

$$|\mathbf{I}_1| = 1.155 \text{ A}$$

In a similar way, the amplitude of the capacitive load current (labeled $\mathbf{I}_2$) is found to be unchanged from its previous value, since the voltage across it has remained the same:

$$|\mathbf{I}_2| = 2.89 \text{ A}$$

If we assume that the phase with which we are working has a phase voltage with an angle of 0°, then since $\cos^{-1}(0.8) = 36.9°$,

$$\mathbf{I}_1 = 1.155 \underline{/0°} \text{ A} \qquad \mathbf{I}_2 = 2.89 \underline{/+36.9°} \text{ A}$$

and the line current is

$$\mathbf{I}_L = \mathbf{I}_1 + \mathbf{I}_2 = 3.87 \underline{/+26.6°} \text{ A}$$

We can check our results by computing the power generated by this phase of the source

$$P_p = \frac{300}{\sqrt{3}} 3.87 \cos(+26.6°) = 600 \text{ W}$$

which agrees with the fact that the individual phase is known to be supplying 200 W to the new lighting load, as well as 400 W to the original load.

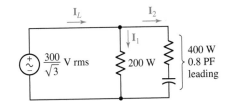

■ **FIGURE 12.17** The per-phase circuit that is used to analyze a *balanced* three-phase example.

PRACTICE

12.6 Three balanced Y-connected loads are installed on a balanced three-phase four-wire system. Load 1 draws a total power of 6 kW at unity PF, load 2 pulls 10 kVA at PF = 0.96 lagging, and load 3 demands 7 kW at 0.85 lagging. If the phase voltage at the loads is 135 V, if each line has a resistance of 0.1 Ω, and if the neutral has a resistance of 1 Ω, find (*a*) the total power drawn by the loads; (*b*) the combined PF of the loads; (*c*) the total power lost in the four lines; (*d*) the phase voltage at the source; (*e*) the power factor at which the source is operating.

Ans: 22.6 kW; 0.954 lag; 1027 W; 140.6 V; 0.957 lagging.

If an *unbalanced* Y-connected load is present in an otherwise balanced three-phase system, the circuit may still be analyzed on a per-phase basis *if* the neutral wire is present and *if* it has zero impedance. If either of these conditions is not met, other methods must be used, such as mesh or nodal analysis. However, engineers who spend most of their time with unbalanced three-phase systems will find the use of *symmetrical components* a great time saver.

We leave this topic for more advanced texts.

12.4 • THE DELTA (Δ) CONNECTION

An alternative to the Y-connected load is the Δ-connected configuration, as shown in Fig. 12.18. This type of configuration is very common, and does not possess a neutral connection.

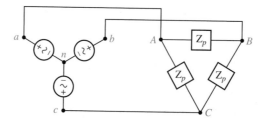

■ **FIGURE 12.18** A balanced Δ-connected load is present on a three-wire three-phase system. The source happens to be Y-connected.

Let us consider a balanced Δ-connected load which consists of an impedance $\mathbf{Z}_p$ inserted between each pair of lines. With reference to Fig. 12.18, let us assume known line voltages

$$V_L = |\mathbf{V}_{ab}| = |\mathbf{V}_{bc}| = |\mathbf{V}_{ca}|$$

or known phase voltages

$$V_p = |\mathbf{V}_{an}| = |\mathbf{V}_{bn}| = |\mathbf{V}_{cn}|$$

where

$$V_L = \sqrt{3}V_p \qquad \text{and} \qquad \mathbf{V}_{ab} = \sqrt{3}V_p \underline{/30°}$$

as we found previously. Because the voltage across each branch of the Δ is known, the *phase currents* are easily found:

$$\mathbf{I}_{AB} = \frac{\mathbf{V}_{ab}}{\mathbf{Z}_p} \qquad \mathbf{I}_{BC} = \frac{\mathbf{V}_{bc}}{\mathbf{Z}_p} \qquad \mathbf{I}_{CA} = \frac{\mathbf{V}_{ca}}{\mathbf{Z}_p}$$

and their differences provide us with the line currents, such as

$$\mathbf{I}_{aA} = \mathbf{I}_{AB} - \mathbf{I}_{CA}$$

Since we are working with a balanced system, the three phase currents are of equal amplitude:

$$I_p = |\mathbf{I}_{AB}| = |\mathbf{I}_{BC}| = |\mathbf{I}_{CA}|$$

The line currents are also equal in amplitude; the symmetry is apparent from the phasor diagram of Fig. 12.19. We thus have

$$I_L = |\mathbf{I}_{aA}| = |\mathbf{I}_{bB}| = |\mathbf{I}_{cC}|$$

and

$$I_L = \sqrt{3}I_p$$

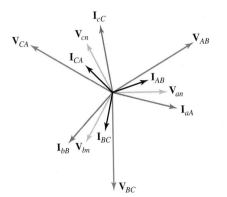

■ FIGURE 12.19 A phasor diagram that could apply to the circuit of Fig. 12.18 if Z_p were an inductive impedance.

Let us disregard the source for the moment and consider only the balanced load. If the load is Δ-connected, then the phase voltage and the line voltage are indistinguishable, but the line current is larger than the phase current by a factor of $\sqrt{3}$; with a Y-connected load, however, the phase current and the line current refer to the same current, and the line voltage is greater than the phase voltage by a factor of $\sqrt{3}$.

EXAMPLE **12.5**

Determine the amplitude of the line current in a three-phase system with a line voltage of 300 V that supplies 1200 W to a Δ-connected load at a lagging PF of 0.8; then find the phase impedance.

Let us again consider a single phase. It draws 400 W, 0.8 lagging PF, at a 300 V line voltage. Thus,

$$400 = 300(I_p)(0.8)$$

and

$$I_p = 1.667 \text{ A}$$

and the relationship between phase currents and line currents yields

$$I_L = \sqrt{3}(1.667) = 2.89 \text{ A}$$

Next, the phase angle of the load is $\cos^{-1}(0.8) = 36.9°$, and therefore the impedance in each phase must be

$$\mathbf{Z}_p = \frac{300}{1.667}\underline{/36.9°} = 180\underline{/36.9°} \ \Omega$$

> Again, keep in mind that we are assuming all voltages and currents are quoted as rms values.

PRACTICE

12.7 Each phase of a balanced three-phase Δ-connected load consists of a 200 mH inductor in series with the parallel combination of a 5 μF capacitor and a 200 Ω resistance. Assume zero line resistance and a phase voltage of 200 V at $\omega = 400$ rad/s. Find (*a*) the phase current; (*b*) the line current; (*c*) the total power absorbed by the load.

Ans: 1.158 A; 2.01 A; 693 W.

EXAMPLE 12.6

Determine the amplitude of the line current in a three-phase system with a 300 V line voltage that supplies 1200 W to a Y-connected load at a lagging PF of 0.8. (*This is the same circuit as in Example 12.5, but with a Y-connected load instead.*)

On a per-phase basis, we now have a phase voltage of $300/\sqrt{3}$ V, a power of 400 W, and a lagging PF of 0.8. Thus,

$$400 = \frac{300}{\sqrt{3}}(I_p)(0.8)$$

and

$$I_p = 2.89 \qquad \text{(and so } I_L = 2.89 \text{ A)}$$

The phase angle of the load is again 36.9°, and thus the impedance in each phase of the Y is

$$\mathbf{Z}_p = \frac{300/\sqrt{3}}{2.89}\underline{/36.9°} = 60\underline{/36.9°}\ \Omega$$

The $\sqrt{3}$ factor not only relates phase and line quantities but also appears in a useful expression for the total power drawn by any balanced three-phase load. If we assume a Y-connected load with a power-factor angle θ, the power taken by any phase is

$$P_p = V_p I_p \cos\theta = V_p I_L \cos\theta = \frac{V_L}{\sqrt{3}}I_L \cos\theta$$

and the total power is

$$P = 3P_p = \sqrt{3}V_L I_L \cos\theta$$

In a similar way, the power delivered to each phase of a Δ-connected load is

$$P_p = V_p I_p \cos\theta = V_L I_p \cos\theta = V_L \frac{I_L}{\sqrt{3}}\cos\theta$$

giving a total power

$$P = 3P_p = \sqrt{3}V_L I_L \cos\theta \qquad [4]$$

Thus Eq. [4] enables us to calculate the total power delivered to a balanced load from a knowledge of the magnitude of the line voltage, of the line current, and of the phase angle of the load impedance (or admittance), regardless of whether the load is Y-connected or Δ-connected. The line current in

PRACTICE

12.8 A balanced three-phase three-wire system is terminated with two Δ-connected loads in parallel. Load 1 draws 40 kVA at a lagging PF of 0.8, while load 2 absorbs 24 kW at a leading PF of 0.9. Assume no line resistance, and let $\mathbf{V}_{ab} = 440\underline{/30°}$ V. Find (*a*) the total power drawn by the loads; (*b*) the phase current $\mathbf{I}_{AB1}$ for the lagging load; (*c*) $\mathbf{I}_{AB2}$; (*d*) $\mathbf{I}_{aA}$.

Ans: 56.0 kW; $30.3\underline{/-6.87°}$ A; $20.2\underline{/55.8°}$ A; $75.3\underline{/-12.46°}$ A.

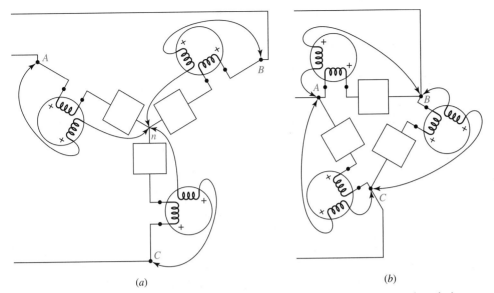

(a)   (b)

■ **FIGURE 12.25** Three wattmeters are connected in such a way that each reads the power taken by one phase of a three-phase load, and the sum of the readings is the total power. (*a*) A Y-connected load. (*b*) A Δ-connected load. Neither the loads nor the source need be balanced.

connect three wattmeters in such a way that each has its current coil in one line and its voltage coil between that line and some common point *x*, as shown in Fig. 12.26. Although a system with a Y-connected load is illustrated, the arguments we present are equally valid for a Δ-connected load. The point *x* may be some unspecified point in the three-phase system, or it may be merely a point in space at which the three potential coils have a common node. The average power indicated by wattmeter *A* must be

$$P_A = \frac{1}{T} \int_0^T v_{Ax} i_{aA}\, dt$$

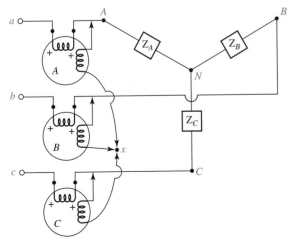

■ **FIGURE 12.26** A method of connecting three wattmeters to measure the total power taken by a three-phase load. Only the three terminals of the load are accessible.

where T is the period of all the source voltages. The readings of the other two wattmeters are given by similar expressions, and the total average power drawn by the load is therefore

$$P = P_A + P_B + P_C = \frac{1}{T} \int_0^T (v_{Ax} i_{aA} + v_{Bx} i_{bB} + v_{Cx} i_{cC}) \, dt$$

Each of the three voltages in the preceding expression may be written in terms of a phase voltage and the voltage between point x and the neutral,

$$v_{Ax} = v_{AN} + v_{Nx}$$
$$v_{Bx} = v_{BN} + v_{Nx}$$
$$v_{Cx} = v_{CN} + v_{Nx}$$

and, therefore,

$$P = \frac{1}{T} \int_0^T (v_{AN} i_{aA} + v_{BN} i_{bB} + v_{CN} i_{cC}) \, dt$$
$$+ \frac{1}{T} \int_0^T v_{Nx} (i_{aA} + i_{bB} + i_{cC}) \, dt$$

However, the entire three-phase load may be considered to be a supernode, and Kirchhoff's current law requires

$$i_{aA} + i_{bB} + i_{cC} = 0$$

Thus

$$P = \frac{1}{T} \int_0^T (v_{AN} i_{aA} + v_{BN} i_{bB} + v_{CN} i_{cC}) \, dt$$

Reference to the circuit diagram shows that this sum is indeed the sum of the average powers taken by each phase of the load, and the sum of the readings of the three wattmeters therefore represents the total average power drawn by the entire load!

Let us illustrate this procedure with a numerical example before we discover that one of these three wattmeters is really superfluous. We will assume a balanced source,

$$\mathbf{V}_{ab} = 100\underline{/0°} \text{ V}$$
$$\mathbf{V}_{bc} = 100\underline{/-120°} \text{ V}$$
$$\mathbf{V}_{ca} = 100\underline{/-240°} \text{ V}$$

or

$$\mathbf{V}_{an} = \frac{100}{\sqrt{3}}\underline{/-30°} \text{ V}$$
$$\mathbf{V}_{bn} = \frac{100}{\sqrt{3}}\underline{/-150°} \text{ V}$$
$$\mathbf{V}_{cn} = \frac{100}{\sqrt{3}}\underline{/-270°} \text{ V}$$

and an unbalanced load,

$$\mathbf{Z}_A = -j10 \ \Omega$$
$$\mathbf{Z}_B = j10 \ \Omega$$
$$\mathbf{Z}_C = 10 \ \Omega$$

Let us assume ideal wattmeters, connected as illustrated in Fig. 12.26, with point x located on the neutral of the source n. The three line currents may be obtained by mesh analysis,

$$\mathbf{I}_{aA} = 19.32\underline{/15°}\ \text{A}$$
$$\mathbf{I}_{bB} = 19.32\underline{/165°}\ \text{A}$$
$$\mathbf{I}_{cC} = 10\underline{/-90°}\ \text{A}$$

The voltage between the neutrals is

$$\mathbf{V}_{nN} = \mathbf{V}_{nb} + \mathbf{V}_{BN} = \mathbf{V}_{nb} + \mathbf{I}_{bB}(j10) = 157.7\underline{/-90°}$$

The average power indicated by each wattmeter may be calculated,

$$P_A = V_p I_{aA}\cos(ang\,\mathbf{V}_{an} - ang\,\mathbf{I}_{aA})$$
$$= \frac{100}{\sqrt{3}} 19.32\cos(-30° - 15°) = 788.7\ \text{W}$$
$$P_B = \frac{100}{\sqrt{3}} 19.32\cos(-150° - 165°) = 788.7\ \text{W}$$
$$P_C = \frac{100}{\sqrt{3}} 10\cos(-270° + 90°) = -577.4\ \text{W}$$

or a total power of 1 kW. Since an rms current of 10 A flows through the *resistive* load, the total power drawn by the load is

$$P = 10^2(10) = 1\ \text{kW}$$

and the two methods agree.

> Note that the reading of one of the wattmeters is negative. Our previous discussion on the basic use of a wattmeter indicates that an upscale reading on that meter can only be obtained after either the potential coil or the current coil is reversed.

The Two-Wattmeter Method

We have proved that point x, the common connection of the three potential coils, may be located any place we wish without affecting the algebraic sum of the three wattmeter readings. Let us now consider the effect of placing point x, this common connection of the three wattmeters, directly on one of the lines. If, for example, one end of each potential coil is returned to B, then there is no voltage across the potential coil of wattmeter B and *this meter must read zero*. It may therefore be removed, and the algebraic sum of the remaining two wattmeter readings is still the total power drawn by the load. When the location of x is selected in this way, we describe the method of power measurement as the **two-wattmeter** method. The sum of the readings indicates the total power, regardless of (1) load unbalance, (2) source unbalance, (3) differences in the two wattmeters, and (4) the waveform of the periodic source. The only assumption we have made is that wattmeter corrections are sufficiently small that we can ignore them. In Fig. 12.26, for example, the current coil of each meter has passing through it the line current drawn by the load plus the current taken by the potential coil. Since the latter current is usually quite small, its effect may be estimated from a knowledge of the resistance of the potential coil and voltage across it. These two quantities enable a close estimate to be made of the power dissipated in the potential coil.

In the numerical example described previously, let us now assume that two wattmeters are used, one with current coil in line A and potential coil between lines A and B, the other with current coil in line C and potential coil

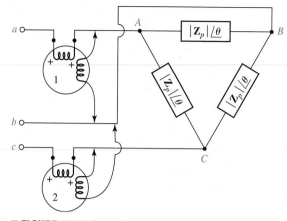

■ **FIGURE 12.27** Two wattmeters connected to read the total power drawn by a balanced three-phase load.

between C and B. The first meter reads

$$P_1 = V_{AB}I_{aA}\cos(ang\ V_{AB} - ang\ I_{aA})$$
$$= 100(19.32)\cos(0° - 15°)$$
$$= 1866\ \text{W}$$

and the second

$$P_2 = V_{CB}I_{cC}\cos(ang\ V_{CB} - ang\ I_{cC})$$
$$= 100(10)\cos(60° + 90°)$$
$$= -866\ \text{W}$$

and, therefore,

$$P = P_1 + P_2 = 1866 - 866 = 1000\ \text{W}$$

as we expect from recent experience with the circuit.

In the case of a balanced load, the two-wattmeter method enables the PF angle to be determined, as well as the total power drawn by the load. Let us assume a load impedance with a phase angle θ; either a Y or Δ connection may be used and we will assume the Δ connection shown in Fig. 12.27. The construction of a standard phasor diagram, such as that of Fig. 12.19, enables us to determine the proper phase angle between the several line voltages and line currents. We therefore determine the readings

$$P_1 = |\mathbf{V}_{AB}|\,|\mathbf{I}_{aA}|\cos(ang\ \mathbf{V}_{AB} - ang\ \mathbf{I}_{aA})$$
$$= V_L I_L \cos(30° + \theta)$$

and

$$P_2 = |\mathbf{V}_{CB}|\,|\mathbf{I}_{cC}|\cos(ang\ \mathbf{V}_{CB} - ang\ \mathbf{I}_{cC})$$
$$= V_L I_L \cos(30° - \theta)$$

The ratio of the two readings is

$$\frac{P_1}{P_2} = \frac{\cos(30° + \theta)}{\cos(30° - \theta)} \qquad [5]$$

If we expand the cosine terms, this equation is easily solved for $\tan\theta$,

$$\tan\theta = \sqrt{3}\frac{P_2 - P_1}{P_2 + P_1} \qquad [6]$$

Thus, equal wattmeter readings indicate a unity PF load; equal and opposite readings indicate a purely reactive load; a reading of P_2 which is (algebraically) greater than P_1 indicates an inductive impedance; and a reading of P_2 which is less than P_1 signifies a capacitive load. How can we tell which wattmeter reads P_1 and which reads P_2? It is true that P_1 is in line A, and P_2 is in line C, and our positive phase-sequence system forces V_{an} to lag V_{cn}. This is enough information to differentiate between two wattmeters, but it is confusing to apply in practice. Even if we were unable to distinguish between the two, we know the magnitude of the phase angle, but not its sign. This is often sufficient information; if the load is an induction motor, the angle must be positive and we do not need to make any tests to determine which reading is which. If no previous knowledge of the load is assumed, then there are several methods of resolving the ambiguity. Perhaps the simplest method is that which involves adding a high-impedance reactive load, say, a three-phase capacitor, across the unknown load. The load must become more capacitive. Thus, if the magnitude of $\tan \theta$ (or the magnitude of θ) decreases, then the load was inductive, whereas an increase in the magnitude of $\tan \theta$ signifies an original capacitive impedance.

EXAMPLE **12.7**

The balanced load in Fig. 12.28 is fed by a balanced three-phase system having $V_{ab} = 230\underline{/0°}$ V rms and positive phase sequence. Find the reading of each wattmeter and the total power drawn by the load.

The potential coil of wattmeter #1 is connected to measure the voltage V_{ac}, and its current coil is measuring the phase current $\mathbf{I}_{aA}$. Since we know to use the positive phase sequence, the line voltages are

$$\mathbf{V}_{ab} = 230\underline{/0°} \text{ V}$$
$$\mathbf{V}_{bc} = 230\underline{/-120°} \text{ V}$$
$$\mathbf{V}_{ca} = 230\underline{/120°} \text{ V}$$

Note that $\mathbf{V}_{ac} = -\mathbf{V}_{ca} = 230\underline{/-60°}$ V.

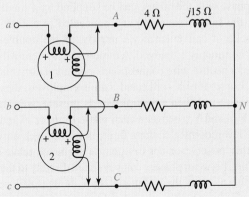

■ **FIGURE 12.28** A balanced three-phase system connected to a balanced three-phase load, the power of which is being measured using the two-wattmeter technique.

(Continued on next page)

The phase current $\mathbf{I}_{aA}$ is given by the phase voltage $\mathbf{V}_{an}$ divided by the phase impedance $4 + j15\ \Omega$,

$$\mathbf{I}_{aA} = \frac{\mathbf{V}_{an}}{4 + j15} = \frac{(230/\sqrt{3})\underline{/-30^\circ}}{4 + j15}\ \text{A}$$

$$= 8.554\underline{/-105.1^\circ}\ \text{A}$$

We may now compute the power measured by wattmeter #1 as

$$P_1 = |\mathbf{V}_{ac}|\,|\mathbf{I}_{aA}|\cos(ang\ \mathbf{V}_{ac} - ang\ \mathbf{I}_{aA})$$

$$= (230)(8.554)\cos(-60^\circ + 105.1^\circ)\ \text{W}$$

$$= 1389\ \text{W}$$

In a similar fashion, we determine that

$$P_2 = |\mathbf{V}_{bc}|\,|\mathbf{I}_{bB}|\cos(ang\ \mathbf{V}_{bc} - ang\ \mathbf{I}_{bB})$$

$$= (230)(8.554)\cos(-120^\circ - 134.9^\circ)\ \text{W}$$

$$= -512.5\ \text{W}$$

Thus, the total average power absorbed by the load is

$$P = P_1 + P_2 = 876.5\ \text{W}$$

> Since this measurement would result in the meter pegged at downscale, one of the coils would need to be reversed in order to take the reading.

PRACTICE

12.10 For the circuit of Fig. 12.26, let the loads be $\mathbf{Z}_A = 25\underline{/60^\circ}\ \Omega$, $\mathbf{Z}_B = 50\underline{/-60^\circ}\ \Omega$, $\mathbf{Z}_C = 50\underline{/60^\circ}\ \Omega$, $\mathbf{V}_{AB} = 600\underline{/0^\circ}$ V rms with (+) phase sequence, and locate point x at C. Find (a) P_A; (b) P_B; (c) P_C.

Ans: 0; 7200 W; 0.

SUMMARY AND REVIEW

Polyphase circuits are not encountered directly by everyone, but are part of almost every large building installation. In this chapter we studied how three voltages, each 120° out of phase with the others, can be supplied by a single generator (and hence have the same frequency) and connected to a three-component load. For the sake of convenience we introduced the double-subscript notation, which is commonly employed. A three-phase system will have at least three terminals; a neutral wire connection is not mandatory but is common at least for the source. If a Δ-connected load is employed, then there is no neutral connection to it. When a neutral wire is present, we can define *phase voltages* $\mathbf{V}_{an}$, $\mathbf{V}_{bn}$, and $\mathbf{V}_{cn}$ between each phase (a, b, or c) and neutral. Kirchhoff's voltage law requires that these three phase voltages sum to zero, regardless of whether positive or negative phase sequence relates their angles. *Line voltages* (i.e., between phases) can be related directly to the phase voltages; for a Δ-connected load they are equal. In a similar fashion, *line currents* and *phase currents* can be directly related to one another; in a Y-connected load, they are equal. At first glance such systems can appear to be somewhat complicated, but symmetry often allows us to perform the analysis on a per-phase basis, simplifying our calculations considerably.

A concise list of key concepts of the chapter is presented below for the convenience of the reader, along with the corresponding example numbers.

- ❏ The majority of electricity production is in the form of three-phase power.
- ❏ Most residential electricity in North America is in the form of single-phase alternating current at a frequency of 60 Hz and an rms voltage of 115 V. Elsewhere, 50 Hz at 240 V rms is most common.
- ❏ Double-subscript notation is commonly employed in power systems for both voltages and currents. (Example 12.1)
- ❏ Three-phase sources can be either Y- or Δ-connected. Both types of sources have three terminals, one for each phase; Y-connected sources have a neutral connection as well. (Example 12.2)
- ❏ In a balanced three-phase system, each phase voltage has the same magnitude, but is 120° out of phase with the other two. (Example 12.2)
- ❏ Loads in a three-phase system may be either Y- or Δ-connected.
- ❏ In a balanced Y-connected source with positive ("*abc*") phase sequence, the line voltages are

$$\mathbf{V}_{ab} = \sqrt{3}V_p\underline{/30^\circ} \qquad \mathbf{V}_{bc} = \sqrt{3}V_p\underline{/-90^\circ}$$
$$\mathbf{V}_{ca} = \sqrt{3}V_p\underline{/-210^\circ}$$

where the phase voltages are

$$\mathbf{V}_{an} = V_p\underline{/0^\circ} \qquad \mathbf{V}_{bn} = V_p\underline{/-120^\circ} \qquad \mathbf{V}_{cn} = V_p\underline{/-240^\circ}$$

(Example 12.2)

- ❏ In a system with a Y-connected load, the line currents are equal to the phase currents. (Examples 12.3, 12.4, 12.6)
- ❏ In a Δ-connected load, the line voltages are equal to the phase voltages. (Example 12.5)
- ❏ In a balanced system with positive phase sequence and a balanced Δ-connected load, the line currents are

$$\mathbf{I}_a = \mathbf{I}_{AB}\sqrt{3}\underline{/-30^\circ} \qquad \mathbf{I}_b = \mathbf{I}_{BC}\sqrt{3}\underline{/-150^\circ} \qquad \mathbf{I}_c = \mathbf{I}_{CA}\sqrt{3}\underline{/+90^\circ}$$

where the phase currents are

$$\mathbf{I}_{AB} = \frac{\mathbf{V}_{AB}}{\mathbf{Z}_\Delta} = \frac{\mathbf{V}_{ab}}{\mathbf{Z}_\Delta} \qquad \mathbf{I}_{BC} = \frac{\mathbf{V}_{BC}}{\mathbf{Z}_\Delta} = \frac{\mathbf{V}_{bc}}{\mathbf{Z}_\Delta} \qquad \mathbf{I}_{CA} = \frac{\mathbf{V}_{CA}}{\mathbf{Z}_\Delta} = \frac{\mathbf{V}_{ca}}{\mathbf{Z}_\Delta}$$

(Example 12.5)

- ❏ Most power calculations are performed on a per-phase basis, assuming a balanced system; otherwise, nodal/mesh analysis is always a valid approach. (Examples 12.3, 12.4, 12.5)
- ❏ The power in a three-phase system (balanced or unbalanced) can be measured with only two wattmeters. (Example 12.7)
- ❏ The instantaneous power in any balanced three-phase system is constant.

READING FURTHER

A good overview of ac power concepts can be found in Chap. 2 of:

> B. M. Weedy, *Electric Power Systems,* 3rd ed. Chichester, England: Wiley, 1984.

A comprehensive book on generation of electrical power from wind is:

> T. Burton, D. Sharpe, N. Jenkins, and E. Bossanyi, *Wind Energy Handbook.* Chichester, England: Wiley, 2001.

EXERCISES

12.1 Polyphase Systems

1. An unknown three terminal device has leads named b, c, and e. When installed in one particular circuit, measurements indicated that $V_{ec} = -9$ V and $V_{eb} = -0.65$ V. (*a*) Calculate V_{cb}. (*b*) Determine the power dissipated in the b-e junction if the current I_b flowing into the terminal marked b is equal to 1 μA.

2. A common type of transistor is known as the MESFET, which is an acronym for **me**tal-semiconductor **f**ield **e**ffect **t**ransistor. It has three terminals, named the gate (g), the source (s), and the drain (d). As an example, consider one particular MESFET operating in a circuit such that $V_{sg} = 0.2$ V and $V_{ds} = 3$ V. (*a*) Calculate V_{gs} and V_{dg}. (*b*) If a gate current $I_g = 100$ pA is found to be flowing into the gate terminal, compute the power lost at the gate-source junction.

3. For a certain Y-connected three-phase source, $\mathbf{V}_{an} = 400\underline{/33°}$ V, $\mathbf{V}_{bn} = 400\underline{/153°}$ V, and $\mathbf{V}_{cx} = 160\underline{/208°}$ V. Determine (*a*) $\mathbf{V}_{cn}$; (*b*) $\mathbf{V}_{an} - \mathbf{V}_{bn}$; (*c*) $\mathbf{V}_{ax}$; (*d*) $\mathbf{V}_{bx}$.

4. Describe what is meant by a "polyphase" source, state one possible advantage of such sources that might outweigh their additional complexity over single-phase sources of power, and explain the difference between "balanced" and "unbalanced" sources.

5. Several of the voltages associated with a certain circuit are given by $\mathbf{V}_{12} = 9\underline{/30°}$ V, $\mathbf{V}_{32} = 3\underline{/130°}$ V, and $\mathbf{V}_{14} = 2\underline{/10°}$ V. Determine $\mathbf{V}_{21}$, $\mathbf{V}_{13}$, $\mathbf{V}_{34}$, and $\mathbf{V}_{24}$.

6. The nodal voltages which describe a particular circuit can be expressed as $\mathbf{V}_{14} = 9 - j$ V, $\mathbf{V}_{24} = 3 + j3$ V, and $\mathbf{V}_{34} = 8$ V. Calculate $\mathbf{V}_{12}$, $\mathbf{V}_{32}$, and $\mathbf{V}_{13}$. Express your answers in phasor form.

7. In the circuit of Fig. 12.29, the resistor markings unfortunately have been omitted, but several of the currents are known. Specifically, $I_{ad} = 1$ A. (*a*) Compute I_{ab}, I_{cd}, I_{de}, I_{fe}, and I_{be}. (*b*) If $V_{ba} = 125$ V, determine the value of the resistor linking nodes a and b.

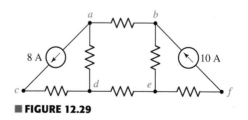

■ **FIGURE 12.29**

8. For the circuit shown in Fig. 12.30, (*a*) determine I_{gh}, I_{cd}, and I_{dh}. (*b*) Calculate I_{ed}, I_{ei}, and I_{jf}. (*c*) If all resistors in the circuit each have a value of 1 Ω, determine the three clockwise-flowing mesh currents.

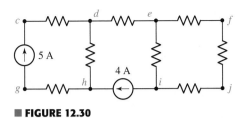

FIGURE 12.30

9. Additional resistors are added in parallel to the resistors between terminals *d* and *e*, and terminals *f* and *j*, respectively, of the circuit in Fig. 12.30. (*a*) Which voltages may still be described using double-subscript notation? (*b*) Which line currents may still be described by double-subscript notation?

12.2 Single-Phase Three-Wire Systems

10. Most consumer electronics are powered by 110 V outlets, but several types of appliances (such as clothes dryers) are powered from 220 V outlets. Lower voltages are generally safer. What, then, motivates manufactures of some pieces of equipment to design them to run on 220 V?

11. The single-phase three-wire system of Fig. 12.31 has three separate load impedances. If the source is balanced and $\mathbf{V}_{an} = 110 + j0$ V rms, (*a*) express $\mathbf{V}_{an}$ and $\mathbf{V}_{bn}$ in phasor notation. (*b*) Determine the phasor voltage which appears across the impedance $\mathbf{Z}_3$. (*c*) Determine the average power delivered by the two sources if $\mathbf{Z}_1 = 50 + j0$ Ω, $\mathbf{Z}_2 = 100 + j45$ Ω, and $\mathbf{Z}_3 = 100 - j90$ Ω. (*d*) Represent load $\mathbf{Z}_3$ by a series connection of two elements, and state their respective values if the sources operate at 60 Hz.

12. For the system represented in Fig. 12.32, the ohmic losses in the neutral wire are so small they can be neglected and it can be adequately modeled as a short circuit. (*a*) Calculate the power lost in the two lines as a result of their nonzero resistance. (*b*) Compute the average power delivered to the load. (*c*) Determine the power factor of the total load.

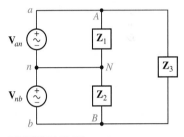

FIGURE 12.31

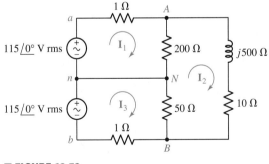

FIGURE 12.32

13. Referring to the balanced load represented in Fig. 12.33, if it is connected to a three-wire balanced source operating at 50 Hz such that $V_{AN} = 115$ V, (*a*) determine the power factor of the load if the capacitor is omitted; (*b*) determine the value of capacitance C that will achieve a unity power factor for the total load.

14. In the three-wire system of Fig. 12.32, (*a*) replace the 50 Ω resistor with a 200 Ω resistor, and calculate the current flowing through the neutral wire. (*b*) Determine a new value for the 50 Ω resistor such that the neutral wire current magnitude is 25% that of line current $\mathbf{I}_{aA}$.

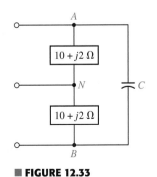

FIGURE 12.33

12.3 Three-Phase Y-Y Connection

15. (a) Show that if $\mathbf{V}_{an} = 400\underline{/33°}$ V, $\mathbf{V}_{bn} = 400\underline{/-87°}$ V, and $\mathbf{V}_{bn} = 400\underline{/-207°}$ V, that $\mathbf{V}_{an} + \mathbf{V}_{bn} + \mathbf{V}_{cn} = 0$. (b) Do the voltages in part (a) represent positive or negative phase sequence? *Explain.*

16. Consider a simple positive phase sequence, three-phase, three-wire system operated at 50 Hz and with a balance load. Each phase voltage of 240 V is connected across a load composed of a series-connected 50 Ω and 500 mH combination. Calculate (a) each line current; (b) the power factor of the load; (c) the total power supplied by the three-phase source.

17. Assume the system shown in Fig. 12.34 is balanced, $R_w = 0$, $\mathbf{V}_{an} = 208\underline{/0°}$ V, and a positive phase sequence applies. Calculate all phase and line currents, and all phase and line voltages, if $\mathbf{Z}_p$ is equal to (a) 1 kΩ; (b) $100 + j48$ Ω; (c) $100 - j48$Ω.

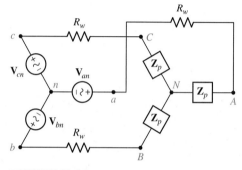

■ **FIGURE 12.34**

18. Repeat Exercise 17 with $R_w = 10$ Ω, and verify your answers with appropriate PSpice simulations if the operating frequency is 60 Hz.

19. Each impedance $\mathbf{Z}_p$ in the balanced three-phase system of Fig. 12.34 is constructed using the parallel combination of a 1 mF capacitance, a 100 mH inductance, and a 10 Ω resistance. The sources have positive phase sequence and operate at 50 Hz. If $\mathbf{V}_{ab} = 208\underline{/0°}$ V, and $R_w = 0$, calculate (a) all phase voltages; (b) all line voltages; (c) all three line currents; (d) the total power drawn by the load.

20. With the assumption that the three-phase system pictured in Fig. 12.34 is balanced with a line voltage of 100 V, calculate the line current and per-phase load impedance if $R_w = 0$ and the load draws (a) 1 kW at a PF of 0.85 lagging; (b) 300 W per phase at a PF of 0.92 leading.

21. The balanced three-phase system of Fig. 12.34 is characterized by a positive phase sequence and a line voltage of 300 V. And $\mathbf{Z}_p$ is given by the parallel combination of a $5 - j3$ Ω capacitive load and a $9 + j2$ Ω inductive load. If $R_w = 0$, calculate (a) the power factor of the source; (b) the total power supplied by the source. (c) Repeat parts (a) and (b) if $R_w = 1$ Ω.

22. A balanced Y-connected load of $100 + j50$ Ω is connected to a balanced three-phase source. If the line current is 42 A and the source supplies 12 kW, determine (a) the line voltage; (b) the phase voltage.

23. A three-phase system is constructed from a balanced Y-connected source operating at 50 Hz and having a line voltage of 210 V, and each phase of the balanced load draws 130 W at a leading power factor of 0.75. (a) Calculate the line current and the total power supplied to the load. (b) If a purely resistive load of 1 Ω is connected in parallel with each existing load, calculate the new line current and total power supplied to the load. (c) Verify your answers using appropriate PSpice simulations.

24. Returning to the balanced three-phase system described in Exercise 21, determine the complex power delivered to the load for both $R_w = 0$ and $R_w = 1\ \Omega$.

25. Each load in the circuit of Fig. 12.34 is composed of a 1.5 H inductor in parallel with a 100 μF capacitor and a 1 kΩ resistor. The resistance is labeled $R_w = 0\ \Omega$. Using positive phase sequence with $\mathbf{V}_{ab} = 115\underline{/0^\circ}$ V at $f = 60$ Hz, determine the rms line current and the total power delivered to the load. Verify your answers with an appropriate PSpice simulation.

12.4 The Delta (Δ) Connection

26. A particular balanced three-phase system is supplying a Δ-connected load with 10 kW at a leading power factor of 0.7. If the phase voltage is 208 V and the source operates at 50 V, (a) compute the line current; (b) determine the phase impedance; (c) calculate the new power factor and new total power delivered to the load if a 2.5 H inductor is connected in parallel with each phase of the load.

27. If each of the three phases in a balanced Δ-connected load is composed of a 10 mF capacitor in parallel with a series-connected 470 Ω resistor and 4 mH inductor combination, assume a phase voltage of 400 V at 50 Hz. (a) Calculate the phase current; (b) the line current; (c) the line voltage; (d) the power factor at which the source operates; (e) the total power delivered to the load.

28. A three-phase load is to be powered by a three-wire three-phase Y-connected source having phase voltage of 400 V and operating at 50 Hz. Each phase of the load consists of a parallel combination of a 500 Ω resistor, 10 mH inductor, and 1 mF capacitor. (a) Compute the line current, line voltage, phase current, and power factor of the load if the load is also Y-connected. (b) Rewire the load so that it is Δ-connected and find the same quantities requested in part (a).

29. For the two situations described in Exercise 28, compute the total power delivered to each of the two loads.

30. Two Δ-connected loads are connected in parallel and powered by a balanced Y-connected system. The smaller of the two loads draws 10 kVA at a lagging PF of 0.75, and the larger draws 25 kVA at a leading PF of 0.80. The line voltage is 400 V. Calculate (a) the power factor at which the source is operating; (b) the total power drawn by the two loads; (c) the phase current of each load.

31. For the balanced three-phase system shown in Fig. 12.35, it is determined that 100 W is lost in each wire. If the phase voltage of the source is 400 V, and the load draws 12 kW at a lagging PF of 0.83, determine the wire resistance R_w.

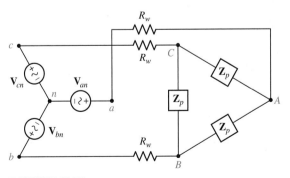

■ FIGURE 12.35

32. The balanced Δ-connected load in Fig. 12.35 is demanding 10 kVA at a lagging PF of 0.91. If line losses are negligible, calculate $\mathbf{I}_{bB}$ and $\mathbf{V}_{an}$ if $\mathbf{V}_{ca} = 160\underline{/30^\circ}$ V and the source voltages are described using positive phase sequence.

33. Repeat Exercise 32 if $R_w = 1\ \Omega$. Verify your solution using an appropriate PSpice simulation.

34. Compute $\mathbf{I}_{aA}$, $\mathbf{I}_{AB}$, and $\mathbf{V}_{an}$ if the Δ-connected load of Fig. 12.35 draws a total complex power of $1800 + j700$ W, $R_w = 1.2\ \Omega$, and the source generates a complex power of $1850 + j700$ W.

35. A balanced three-phase system having line voltage of 240 V rms contains a Δ-connected load of $12 + j$ kΩ per phase and also a Y-connected load of $5 + j3$ kΩ per phase. Find the line current, the power taken by the combined load, and the power factor of the load.

12.5 Power Measurement in Three-Phase Systems

36. Determine the wattmeter reading (stating whether or not the leads had to be reversed to obtain it) in the circuit of Fig. 12.36 if terminals A and B, respectively, are connected to (a) x and y; (b) x and z; (c) y and z.

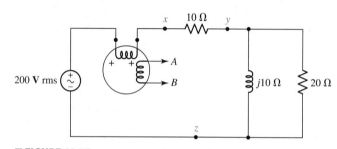

FIGURE 12.36

37. A wattmeter is connected into the circuit of Fig. 12.37 so that $\mathbf{I}_1$ enters the $(+)$ terminal of the current coil, while $\mathbf{V}_2$ is the voltage across the potential coil. Find the wattmeter reading, and verify your solution with an appropriate PSpice simulation.

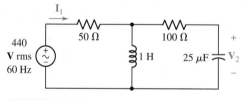

FIGURE 12.37

38. Find the reading of the wattmeter connected in the circuit of Fig. 12.38.

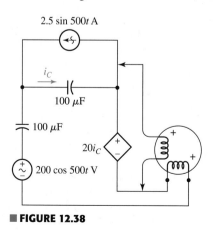

FIGURE 12.38

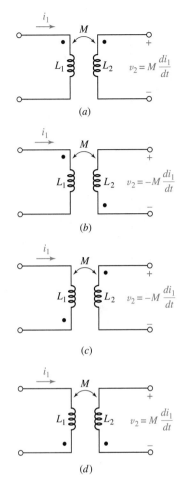

■ **FIGURE 13.2** Current entering the dotted terminal of one coil produces a voltage that is sensed positively at the dotted terminal of the second coil. Current entering the undotted terminal of one coil produces a voltage that is sensed positively at the undotted terminal of the second coil.

Dot Convention

The inductor is a two-terminal element, and we are able to use the passive sign convention in order to select the correct sign for the voltage $L\,di/dt$ or $j\omega L\mathbf{I}$. If the current enters the terminal at which the positive voltage reference is located, then the positive sign is used. Mutual inductance, however, cannot be treated in exactly the same way because four terminals are involved. The choice of a correct sign is established by use of one of several possibilities that include the *"dot convention,"* or by an examination of the particular way in which each coil is wound. We will use the dot convention and merely look briefly at the physical construction of the coils; the use of other special symbols is not necessary when only two coils are coupled.

The dot convention makes use of a large dot placed at one end of each of the two coils which are mutually coupled. We determine the sign of the mutual voltage as follows:

> A current entering the *dotted* terminal of one coil produces an open-circuit voltage with a *positive* voltage reference at the *dotted* terminal of the second coil.

Thus, in Fig. 13.2*a*, i_1 enters the dotted terminal of L_1, v_2 is sensed positively at the dotted terminal of L_2, and $v_2 = M\,di_1/dt$. We have found previously that it is often not possible to select voltages or currents throughout a circuit so that the passive sign convention is everywhere satisfied; the same situation arises with mutual coupling. For example, it may be more convenient to represent v_2 by a positive voltage reference at the undotted terminal, as shown in Fig. 13.2*b*; then $v_2 = -M\,di_1/dt$. Currents that enter the dotted terminal are also not always available, as indicated by Fig. 13.2*c* and *d*. We note then that:

> A current entering the *undotted* terminal of one coil provides a voltage that is *positively* sensed at the *undotted* terminal of the second coil.

Note that the preceding discussion does not include any contribution to the voltage from self-induction, which would occur if i_2 were nonzero. We will consider this important situation in detail, but a quick example first is appropriate.

EXAMPLE 13.1

For the circuit shown in Fig. 13.3, (*a*) determine v_1 if $i_2 = 5\sin 45t$ A and $i_1 = 0$; (*b*) determine v_2 if $i_1 = -8e^{-t}$ A and $i_2 = 0$.

(*a*) Since the current i_2 is entering the *undotted* terminal of the right coil, the positive reference for the voltage induced across the left coil is the undotted terminal. Thus, we have an open-circuit voltage

$$v_1 = -(2)(45)(5\cos 45t) = -450\cos 45t \qquad \text{V}$$

appearing across the terminals of the left coil as a result of the time-varying magnetic flux generated by i_2 flowing into the right coil.

(*Continued on next page*)

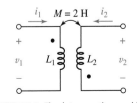

■ **FIGURE 13.3** The dot convention provides a relationship between the terminal at which a current enters one coil, and the positive voltage reference for the other coil.

Since no current flows through the coil on the left, there is no contribution to v_1 from self-induction.

(b) We now have a current entering a *dotted* terminal, but v_2 has its positive reference at the *undotted* terminal. Thus,

$$v_2 = -(2)(-1)(-8e^{-t}) = -16e^{-t} \quad \text{V}$$

PRACTICE

13.1 Assuming $M = 10$ H, coil L_2 is open-circuited, and $i_1 = -2e^{-5t}$ A, find the voltage v_2 for (a) Fig. 13.2a; (b) Fig. 13.2b.

Ans: $100e^{-5t}$ V; $-100e^{-5t}$ V.

Combined Mutual and Self-Induction Voltage

So far, we have considered only a mutual voltage present across an *open-circuited* coil. In general, a nonzero current will be flowing in each of the two coils, and a mutual voltage will be produced in each coil because of the current flowing in the other coil. *This mutual voltage is present independently of and in addition to any voltage of self-induction.* In other words, the voltage across the terminals of L_1 will be composed of two terms, $L_1 \, di_1/dt$ and $M \, di_2/dt$, each carrying a sign depending on the current directions, the assumed voltage sense, and the placement of the two dots. In the portion of a circuit drawn in Fig. 13.4, currents i_1 and i_2 are shown, each entering a dotted terminal. The voltage across L_1 is thus composed of two parts,

$$v_1 = L_1 \frac{di_1}{dt} + M \frac{di_2}{dt}$$

as is the voltage across L_2,

$$v_2 = L_2 \frac{di_2}{dt} + M \frac{di_1}{dt}$$

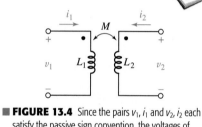

■ **FIGURE 13.4** Since the pairs v_1, i_1 and v_2, i_2 each satisfy the passive sign convention, the voltages of self-induction are both positive; since i_1 and i_2 each enter dotted terminals, and since v_1 and v_2 are both positively sensed at the dotted terminals, the voltages of mutual induction are also both positive.

In Fig. 13.5 the currents and voltages are not selected with the object of obtaining all positive terms for v_1 and v_2. By inspecting only the reference symbols for i_1 and v_1, it is apparent that the passive sign convention is not satisfied and the sign of $L_1 \, di_1/dt$ must therefore be negative. An identical conclusion is reached for the term $L_2 \, di_2/dt$. The mutual term of v_2 is signed by inspecting the direction of i_1 and v_2; since i_1 enters the dotted terminal and v_2 is sensed positive at the dotted terminal, the sign of $M \, di_1/dt$ must be positive. Finally, i_2 enters the undotted terminal of L_2, and v_1 is sensed positive at the undotted terminal of L_1; hence, the mutual portion of v_1, $M \, di_2/dt$, must also be positive. Thus, we have

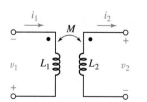

■ **FIGURE 13.5** Since the pairs v_1, i_1 and v_2, i_2 are not sensed according to the passive sign convention, the voltages of self-induction are both negative; since i_1 enters the dotted terminal and v_2 is positively sensed at the dotted terminal, the mutual term of v_2 is positive; and since i_2 enters the undotted terminal and v_1 is positively sensed at the undotted terminal, the mutual term of v_1 is also positive.

$$v_1 = -L_1 \frac{di_1}{dt} + M \frac{di_2}{dt} \qquad v_2 = -L_2 \frac{di_2}{dt} + M \frac{di_1}{dt}$$

The same considerations lead to identical choices of signs for excitation by a sinusoidal source operating at frequency ω

$$\mathbf{V}_1 = -j\omega L_1 \mathbf{I}_1 + j\omega M \mathbf{I}_2 \qquad \mathbf{V}_2 = -j\omega L_2 \mathbf{I}_2 + j\omega M \mathbf{I}_1$$

Physical Basis of the Dot Convention

We can gain a more complete understanding of the dot symbolism by looking at the physical basis for the convention; the meaning of the dots is now interpreted in terms of *magnetic flux*. Two coils are shown wound on a cylindrical form in Fig. 13.6, and the direction of each winding is evident. Let us assume that the current i_1 is positive and increasing with time. The magnetic flux that i_1 produces within the form has a direction which may be found by the right-hand rule: when the right hand is wrapped around the coil with the fingers pointing in the direction of current flow, the thumb indicates the direction of the flux within the coil. Thus i_1 produces a flux which is directed downward; since i_1 is increasing with time, the flux, which is proportional to i_1, is also increasing with time. Turning now to the second coil, let us also think of i_2 as positive and increasing; the application of the right-hand rule shows that i_2 also produces a magnetic flux which is directed downward and is increasing. In other words, the assumed currents i_1 and i_2 produce *additive* fluxes.

The voltage across the terminals of any coil results from the time rate of change of the flux within that coil. The voltage across the terminals of the first coil is therefore greater with i_2 flowing than it would be if i_2 were zero; i_2 induces a voltage in the first coil which has the same sense as the self-induced voltage in that coil. The sign of the self-induced voltage is known from the passive sign convention, and the sign of the mutual voltage is thus obtained.

The dot convention enables us to suppress the physical construction of the coils by placing a dot at one terminal of each coil such that currents entering dot-marked terminals produce additive fluxes. It is apparent that there are always two possible locations for the dots, because both dots may always be moved to the other ends of the coils and additive fluxes will still result.

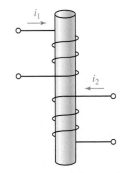

■ **FIGURE 13.6** The physical construction of two mutually coupled coils. From a consideration of the direction of magnetic flux produced by each coil, it is shown that dots may be placed either on the upper terminal of each coil or on the lower terminal of each coil.

EXAMPLE 13.2

For the circuit shown in Fig. 13.7*a*, find the ratio of the output voltage across the 400 Ω resistor to the source voltage, expressed using phasor notation.

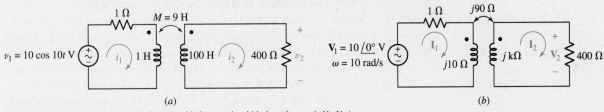

■ **FIGURE 13.7** (*a*) A circuit containing mutual inductance in which the voltage ratio V_2/V_1 is desired. (*b*) Self- and mutual inductances are replaced by the corresponding impedances.

▶ **Identify the goal of the problem.**
We need a numerical value for $\mathbf{V}_2$. We will then divide by $10\underline{/0°}$ V.

▶ **Collect the known information.**
We begin by replacing the 1 H and 100 H inductances by their corresponding impedances, $j10$ Ω and j kΩ, respectively (Fig. 13.7*b*). We also replace the 9 H mutual inductance by $j\omega M = j90$ Ω.

(Continued on next page)

Devise a plan.

Mesh analysis is likely to be a good approach, as we have a circuit with two clearly defined meshes. Once we find I_2, V_2 is simply $400I_2$.

Construct an appropriate set of equations.

In the left mesh, the sign of the mutual term is determined by applying the dot convention. Since I_2 enters the undotted terminal of L_2, the mutual voltage across L_1 must have the positive reference at the undotted terminal. Thus,

$$(1 + j10)I_1 - j90I_2 = 10\underline{/0°}$$

Since I_1 enters the dot-marked terminal, the mutual term in the right mesh has its (+) reference at the dotted terminal of the 100 H inductor. Therefore, we may write

$$(400 + j1000)I_2 - j90I_1 = 0$$

Determine if additional information is required.

We have two equations in two unknowns, I_1 and I_2. Once we solve for the two currents, the output voltage V_2 may be obtained by multiplying I_2 by 400 Ω.

Attempt a solution.

Upon solving these two equations with a scientific calculator, we find that

$$I_2 = 0.172\underline{/-16.70°}\text{ A}$$

Thus,

$$\frac{V_2}{V_1} = \frac{400(0.172\underline{/-16.70°})}{10\underline{/0°}}$$

$$= 6.880\underline{/-16.70°}$$

Verify the solution. Is it reasonable or expected?

We note that the output voltage V_2 is actually larger in magnitude than the input voltage V_1. Should we always expect this result? The answer is no. As we will see in later sections, the transformer can be constructed to achieve *either* a reduction *or* an increase in the voltage. We can perform a quick estimate, however, and at least find an upper and lower bound for our answer. If the 400 Ω resistor is replaced with a short circuit, $V_2 = 0$. If instead we replace the 400 Ω resistor with an open circuit, $I_2 = 0$ and hence

$$V_1 = (1 + j\omega L_1)I_1$$

and

$$V_2 = j\omega M I_1$$

Solving, we find that the maximum value we could expect for V_2/V_1 is $8.955\underline{/5.711°}$. Thus, our answer at least appears reasonable.

The output voltage of the circuit in Fig. 13.7a is greater in magnitude than the input voltage, so that a voltage gain is possible with this type of circuit. It is also interesting to consider this voltage ratio as a function of ω.

To find $\mathbf{I}_2(j\omega)$ for this particular circuit, we write the mesh equations in terms of an unspecified angular frequency ω:

$$(1 + j\omega)\mathbf{I}_1 \qquad\qquad -j\omega 9\mathbf{I}_2 = 10\underline{/0°}$$

and

$$-j\omega 9\mathbf{I}_1 + (400 + j\omega 100)\mathbf{I}_2 = 0$$

Solving by substitution, we find that

$$\mathbf{I}_2 = \frac{j90\omega}{400 + j500\omega - 19\omega^2}$$

Thus, we obtain the ratio of output voltage to input voltage as a function of frequency ω

$$\frac{\mathbf{V}_2}{\mathbf{V}_1} = \frac{400\mathbf{I}_2}{10}$$

$$= \frac{j\omega 3600}{400 + j500\omega - 19\omega^2}$$

The magnitude of this ratio, sometimes referred to as the ***circuit transfer function,*** is plotted in Fig. 13.8 and has a peak magnitude of approximately

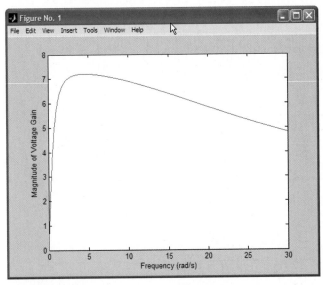

■ **FIGURE 13.8** The voltage gain $|\mathbf{V}_2/\mathbf{V}_1|$ of the circuit shown in Fig. 13.7a is plotted as a function of ω using the following MATLAB script:

```
>> w = linspace(0,30,1000);
>> num = j*w*3600;
>> for indx = 1:1000
den = 400 + j*500*w(indx) − 19*w(indx)*w(indx);
gain(indx) = num(indx)/den;
end
>> plot(w, abs(gain));
>> xlabel('Frequency (rad/s)');
>> ylabel('Magnitude of Voltage Gain');
```

7 near a frequency of 4.6 rad/s. However, for very small or very large frequencies, the magnitude of the transfer function is less than unity.

The circuit is still passive, except for the voltage source, and the *voltage gain* must not be mistakenly interpreted as a *power gain*. At $\omega = 10$ rad/s, the voltage gain is 6.88, but the ideal voltage source, having a terminal voltage of 10 V, delivers a total power of 8.07 W, of which only 5.94 W reaches the 400 Ω resistor. The ratio of the output power to the source power, which we may define as the **power gain,** is thus 0.736.

PRACTICE

13.2 For the circuit of Fig. 13.9, write appropriate mesh equations for the left mesh and the right mesh if $v_s = 20e^{-1000t}$ V.

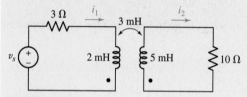

■ **FIGURE 13.9**

Ans: $20e^{-1000t} = 3i_1 + 0.002\,di_1/dt - 0.003\,di_2/dt$; $10i_2 + 0.005\,di_2/dt - 0.003\,di_1/dt = 0$.

EXAMPLE 13.3

Write a complete set of phasor mesh equations for the circuit of Fig. 13.10a.

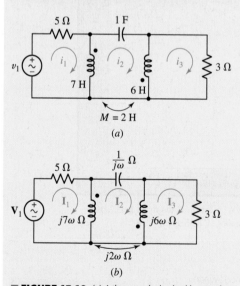

(a)

(b)

■ **FIGURE 13.10** (a) A three-mesh circuit with mutual coupling. (b) The 1 F capacitance as well as the self- and mutual inductances are replaced by their corresponding impedances.

Once again, our first step is to replace both the mutual inductance and the two self-inductances with their corresponding impedances as shown in Fig. 13.10b. Applying Kirchhoff's voltage law to the first mesh, a positive sign for the mutual term is ensured by selecting $(\mathbf{I}_3 - \mathbf{I}_2)$ as the current through the second coil. Thus,

$$5\mathbf{I}_1 + 7j\omega(\mathbf{I}_1 - \mathbf{I}_2) + 2j\omega(\mathbf{I}_3 - \mathbf{I}_2) = \mathbf{V}_1$$

or

$$(5 + 7j\omega)\mathbf{I}_1 - 9j\omega\mathbf{I}_2 + 2j\omega\mathbf{I}_3 = \mathbf{V}_1 \qquad [3]$$

The second mesh requires two self-inductance terms and two mutual inductance terms. Paying close attention to dots, we obtain

$$7j\omega(\mathbf{I}_2 - \mathbf{I}_1) + 2j\omega(\mathbf{I}_2 - \mathbf{I}_3) + \frac{1}{j\omega}\mathbf{I}_2 + 6j\omega(\mathbf{I}_2 - \mathbf{I}_3)$$
$$+ 2j\omega(\mathbf{I}_2 - \mathbf{I}_1) = 0$$

or

$$-9j\omega\mathbf{I}_1 + \left(17j\omega + \frac{1}{j\omega}\right)\mathbf{I}_2 - 8j\omega\mathbf{I}_3 = 0 \qquad [4]$$

Finally, for the third mesh,

$$6j\omega(\mathbf{I}_3 - \mathbf{I}_2) + 2j\omega(\mathbf{I}_1 - \mathbf{I}_2) + 3\mathbf{I}_3 = 0$$

or

$$2j\omega\mathbf{I}_1 - 8j\omega\mathbf{I}_2 + (3 + 6j\omega)\mathbf{I}_3 = 0 \qquad [5]$$

Equations [3] to [5] may be solved by any of the conventional methods.

PRACTICE

13.3 For the circuit of Fig. 13.11, write an appropriate mesh equation in terms of the phasor currents $\mathbf{I}_1$ and $\mathbf{I}_2$ for the (a) left mesh; (b) right mesh.

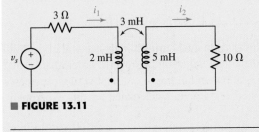

■ FIGURE 13.11

Ans: $\mathbf{V}_s = (3 + j10)\mathbf{I}_1 - j15\mathbf{I}_2$; $0 = -j15\mathbf{I}_1 + (10 + j25)\mathbf{I}_2$.

13.2 • ENERGY CONSIDERATIONS

Let us now consider the energy stored in a pair of mutually coupled inductors. The results will be useful in several different ways. We will first justify our assumption that $M_{12} = M_{21}$, and we may then determine the maximum possible value of the mutual inductance between two given inductors.

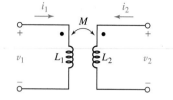

■ **FIGURE 13.12** A pair of coupled coils with a mutual inductance of $M_{12} = M_{21} = M$.

Equality of M_{12} and M_{21}

The pair of coupled coils shown in Fig. 13.12 has currents, voltages, and polarity dots indicated. In order to show that $M_{12} = M_{21}$ we begin by letting all currents and voltages be zero, thus establishing zero initial energy storage in the network. We then open-circuit the right-hand terminal pair and increase i_1 from zero to some constant (dc) value I_1 at time $t = t_1$. The power entering the network from the left at any instant is

$$v_1 i_1 = L_1 \frac{di_1}{dt} i_1$$

and the power entering from the right is

$$v_2 i_2 = 0$$

since $i_2 = 0$.

The energy stored within the network when $i_1 = I_1$ is thus

$$\int_0^{t_1} v_1 i_1 \, dt = \int_0^{I_1} L_1 i_1 \, di_1 = \frac{1}{2} L_1 I_1^2$$

We now hold i_1 constant ($i_1 = I_1$), and we let i_2 change from zero at $t = t_1$ to some constant value I_2 at $t = t_2$. The energy delivered from the right-hand source is thus

$$\int_{t_1}^{t_2} v_2 i_2 \, dt = \int_0^{I_2} L_2 i_2 \, di_2 = \frac{1}{2} L_2 I_2^2$$

However, even though the value of i_1 remains constant, the left-hand source also delivers energy to the network during this time interval:

$$\int_{t_1}^{t_2} v_1 i_1 \, dt = \int_{t_1}^{t_2} M_{12} \frac{di_2}{dt} i_1 \, dt = M_{12} I_1 \int_0^{I_2} di_2 = M_{12} I_1 I_2$$

The total energy stored in the network when both i_1 and i_2 have reached constant values is

$$W_{\text{total}} = \frac{1}{2} L_1 I_1^2 + \frac{1}{2} L_2 I_2^2 + M_{12} I_1 I_2$$

Now, we may establish the same final currents in this network by allowing the currents to reach their final values in the reverse order, that is, first increasing i_2 from zero to I_2 and then holding i_2 constant while i_1 increases from zero to I_1. If the total energy stored is calculated for this experiment, the result is found to be

$$W_{\text{total}} = \frac{1}{2} L_1 I_1^2 + \frac{1}{2} L_2 I_2^2 + M_{21} I_1 I_2$$

The only difference is the interchange of the mutual inductances M_{21} and M_{12}. The initial and final conditions in the network are the same, however, and so the two values of the stored energy must be identical. Thus,

$$M_{12} = M_{21} = M$$

and

$$W = \frac{1}{2} L_1 I_1^2 + \frac{1}{2} L_2 I_2^2 + M I_1 I_2 \qquad [6]$$

conventions: the use of the vertical lines between the two coils to indicate the iron laminations present in many iron-core transformers, the unity value of the coupling coefficient, and the presence of the symbol 1:a, suggesting a turns ratio of N_1 to N_2.

Let us analyze this transformer in the sinusoidal steady state. The two mesh equations are

$$\mathbf{V}_1 = j\omega L_1 \mathbf{I}_1 - j\omega M \mathbf{I}_2 \qquad [22]$$

and

$$0 = -j\omega M \mathbf{I}_1 + (\mathbf{Z}_L + j\omega L_2)\mathbf{I}_2 \qquad [23]$$

First, consider the input impedance of an ideal transformer. By solving Eq. [23] for $\mathbf{I}_2$ and substituting in Eq. [22], we obtain

$$\mathbf{V}_1 = \mathbf{I}_1 j\omega L_1 + \mathbf{I}_1 \frac{\omega^2 M^2}{\mathbf{Z}_L + j\omega L_2}$$

and

$$\mathbf{Z}_{\text{in}} = \frac{\mathbf{V}_1}{\mathbf{I}_1} = j\omega L_1 + \frac{\omega^2 M^2}{\mathbf{Z}_L + j\omega L_2}$$

Since $k = 1$, $M^2 = L_1 L_2$ so

$$\mathbf{Z}_{\text{in}} = j\omega L_1 + \frac{\omega^2 L_1 L_2}{\mathbf{Z}_L + j\omega L_2}$$

Besides a unity coupling coefficient, another characteristic of an ideal transformer is an extremely large impedance for both the primary and secondary coils, regardless of the operating frequency. This suggests that the ideal case would be for both L_1 and L_2 to tend to infinity. Their ratio, however, must remain finite, as specified by the turns ratio. Thus,

$$L_2 = a^2 L_1$$

leads to

$$\mathbf{Z}_{\text{in}} = j\omega L_1 + \frac{\omega^2 a^2 L_1^2}{\mathbf{Z}_L + j\omega a^2 L_1}$$

Now if we let L_1 become infinite, both of the terms on the right-hand side of the preceding equation become infinite, and the result is indeterminate. Thus, it is necessary to first combine these two terms:

$$\mathbf{Z}_{\text{in}} = \frac{j\omega L_1 \mathbf{Z}_L - \omega^2 a^2 L_1^2 + \omega^2 a^2 L_1^2}{\mathbf{Z}_L + j\omega a^2 L_1} \qquad [24]$$

or

$$\mathbf{Z}_{\text{in}} = \frac{j\omega L_1 \mathbf{Z}_L}{\mathbf{Z}_L + j\omega a^2 L_1} = \frac{\mathbf{Z}_L}{\mathbf{Z}_L/j\omega L_1 + a^2} \qquad [25]$$

Now as $L_1 \to \infty$, we see that $\mathbf{Z}_{\text{in}}$ becomes

$$\mathbf{Z}_{\text{in}} = \frac{\mathbf{Z}_L}{a^2} \qquad [26]$$

for finite $\mathbf{Z}_L$.

This result has some interesting implications, and at least one of them appears to contradict one of the characteristics of the linear transformer. The input impedance of an ideal transformer is proportional to the load impedance, the proportionality constant being the reciprocal of the square of the turns ratio. In other words, if the *load* impedance is a capacitive impedance, then the *input* impedance is a capacitive impedance. In the linear transformer, however, the reflected impedance suffered a sign change in its reactive part; a capacitive load led to an inductive contribution to the input impedance. The explanation of this occurrence is achieved by first realizing that $\mathbf{Z}_L/a^2$ is *not* the reflected impedance, although it is often loosely called by that name. The true reflected impedance is infinite in the ideal transformer; otherwise it could not "cancel" the infinite impedance of the primary inductance. This cancellation occurs in the numerator of Eq. [24]. The impedance $\mathbf{Z}_L/a^2$ represents a small term which is the amount by which an exact cancellation does not occur. The true reflected impedance in the ideal transformer does change sign in its reactive part; as the primary and secondary inductances become infinite, however, the effect of the infinite primary-coil reactance and the infinite, but negative, reflected reactance of the secondary coil is one of cancellation.

The first important characteristic of the ideal transformer is therefore its ability to change the magnitude of an impedance, or to change impedance level. An ideal transformer having 100 primary turns and 10,000 secondary turns has a turns ratio of 10,000/100, or 100. Any impedance placed across the secondary then appears at the primary terminals reduced in magnitude by a factor of 100^2, or 10,000. A 20,000 Ω resistor looks like 2 Ω, a 200 mH inductor looks like 20 μH, and a 100 pF capacitor looks like 1 μF. If the primary and secondary windings are interchanged, then $a = 0.01$ and the load impedance is apparently increased in magnitude. In practice, this exact change in magnitude does not always occur, for we must remember that as we took the last step in our derivation and allowed L_1 to become infinite in Eq. [25], it was necessary to neglect $\mathbf{Z}_L$ in comparison with $j\omega L_2$. Since L_2 can never be infinite, it is evident that the ideal transformer model will become invalid if the load impedances are very large.

Use of Transformers for Impedance Matching

A practical example of the use of an iron-core transformer as a device for changing impedance level is in the coupling of an amplifier to a speaker system. In order to achieve maximum power transfer, we know that the resistance of the load should be equal to the internal resistance of the source; the speaker usually has an impedance magnitude (often assumed to be a resistance) of only a few ohms, while an amplifier may possess an internal resistance of several thousand ohms. Thus, an ideal transformer is required in which $N_2 < N_1$. For example, if the amplifier internal impedance is 4000 Ω and the speaker impedance is 8 Ω, then we desire that

$$\mathbf{Z}_g = 4000 = \frac{\mathbf{Z}_L}{a^2} = \frac{8}{a^2}$$

or

$$a = \frac{1}{22.4}$$

and thus

$$\frac{N_1}{N_2} = 22.4$$

Use of Transformers for Current Adjustment

There is a simple relationship between the primary and secondary currents $\mathbf{I}_1$ and $\mathbf{I}_2$ in an ideal transformer. From Eq. [23],

$$\frac{\mathbf{I}_2}{\mathbf{I}_1} = \frac{j\omega M}{\mathbf{Z}_L + j\omega L_2}$$

Once again we allow L_2 to become infinite, and it follows that

$$\frac{\mathbf{I}_2}{\mathbf{I}_1} = \frac{j\omega M}{j\omega L_2} = \sqrt{\frac{L_1}{L_2}}$$

or

$$\boxed{\frac{\mathbf{I}_2}{\mathbf{I}_1} = \frac{1}{a}} \qquad [27]$$

Thus, the ratio of the primary and secondary currents is the turns ratio. If we have $N_2 > N_1$, then $a > 1$, and it is apparent that the larger current flows in the winding with the smaller number of turns. In other words,

$$N_1 \mathbf{I}_1 = N_2 \mathbf{I}_2$$

It should also be noted that the current ratio is the *negative* of the turns ratio if either current is reversed or if either dot location is changed.

In our example in which an ideal transformer was used to change the impedance level to efficiently match a speaker to an amplifier, an rms current of 50 mA at 1000 Hz in the primary causes an rms current of 1.12 A at 1000 Hz in the secondary. The power delivered to the speaker is $(1.12)^2(8)$, or 10 W, and the power delivered to the transformer by the power amplifier is $(0.05)^2 4000$, or 10 W. The result is comforting, since the ideal transformer contains neither an active device which can generate power nor any resistor which can absorb power.

Use of Transformers for Voltage Level Adjustment

Since the power delivered to the ideal transformer is identical with that delivered to the load, whereas the primary and secondary currents are related by the turns ratio, it should seem reasonable that the primary and secondary voltages must also be related to the turns ratio. If we define the secondary voltage, or load voltage, as

$$\mathbf{V}_2 = \mathbf{I}_2 \mathbf{Z}_L$$

and the primary voltage as the voltage across L_1, then

$$\mathbf{V}_1 = \mathbf{I}_1 \mathbf{Z}_{\text{in}} = \mathbf{I}_1 \frac{\mathbf{Z}_L}{a^2}$$

The ratio of the two voltages then becomes

$$\frac{\mathbf{V}_2}{\mathbf{V}_1} = a^2 \frac{\mathbf{I}_2}{\mathbf{I}_1}$$

or

$$\frac{\mathbf{V}_2}{\mathbf{V}_1} = a = \frac{N_2}{N_1} \qquad [28]$$

The ratio of the secondary to the primary voltage is equal to the turns ratio. We should take care to note that this equation is opposite that of Eq. [27], and this is a common source of error for students. This ratio may also be negative if either voltage is reversed or either dot location is changed.

Simply by choosing the turns ratio, therefore, we now have the ability to change any ac voltage to any other ac voltage. If $a > 1$, the secondary voltage will be greater than the primary voltage, and we have what is commonly referred to as a *step-up transformer*. If $a < 1$, the secondary voltage will be less than the primary voltage, and we have a *step-down transformer*. Utility companies typically generate power at a voltage in the range of 12 to 25 kV. Although this is a rather large voltage, transmission losses over long distances can be reduced by increasing the level to several hundred thousand volts using a step-up transformer (Fig. 13.26a). This voltage is then reduced to several tens of kilovolts at substations for local power distribution using step-down transformers (Fig. 13.26b). Additional step-down transformers are located outside buildings to reduce the voltage from the transmission voltage to the 110 or 220 V level required to operate machinery (Fig. 13.26c).

Combining the voltage and current ratios, Eqs. [27] and [28],

$$\mathbf{V}_2\mathbf{I}_2 = \mathbf{V}_1\mathbf{I}_1$$

and we see that the primary and secondary complex voltamperes are equal. The magnitude of this product is usually specified as a maximum allowable value on power transformers. If the load has a phase angle θ, or

$$\mathbf{Z}_L = |\mathbf{Z}_L|\underline{/\theta}$$

then $\mathbf{V}_2$ leads $\mathbf{I}_2$ by an angle θ. Moreover, the input impedance is $\mathbf{Z}_L/a^2$, and thus $\mathbf{V}_1$ also leads $\mathbf{I}_1$ by the same angle θ. If we let the voltage and current represent rms values, then $|\mathbf{V}_2|\,|\mathbf{I}_2|\cos\theta$ must equal $|\mathbf{V}_1|\,|\mathbf{I}_1|\cos\theta$, and all the power delivered to the primary terminals reaches the load; none is absorbed by or delivered to the ideal transformer.

The characteristics of the ideal transformer that we have obtained have all been determined by phasor analysis. They are certainly true in the sinusoidal steady state, but we have no reason to believe that they are correct for the *complete* response. Actually, they are applicable in general, and the demonstration that this statement is true is much simpler than the phasor-based analysis we have just completed. Our analysis, however, has served to point out the specific approximations that must be made on a more exact model of an actual transformer in order to obtain an ideal transformer. For example, we have seen that the reactance of the secondary winding must be much greater in magnitude than the impedance of any load that is connected to the secondary. Some feeling for those operating conditions under which a transformer ceases to behave as an ideal transformer is thus achieved.

(a)

(b)

(c)

■ **FIGURE 13.26** (a) A step-up transformer used to increase the generator output voltage for transmission. (b) Substation transformer used to reduce the voltage from the 220 kV transmission level to several tens of kilovolts for local distribution. (c) Step-down transformer used to reduce the distribution voltage level to 240 V for power consumption.
(*Photos courtesy of Dr. Wade Enright, Te Kura Pukaha Vira O Te Whare Wananga O Waitaha, Aotearoa.*)

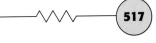

EXAMPLE **13.7**

For the circuit given in Fig. 13.27, determine the average power
dissipated in the 10 kΩ resistor.

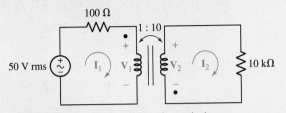

■ **FIGURE 13.27** A simple ideal transformer circuit.

The average power dissipated by the 10 kΩ resistor is simply

$$P = 10,000|\mathbf{I}_2|^2$$

The 50 V rms source "sees" a transformer input impedance of $\mathbf{Z}_L/a^2$
or 100 Ω. Thus, we obtain

$$\mathbf{I}_1 = \frac{50}{100 + 100} = 250 \text{ mA rms}$$

From Eq. [27], $\mathbf{I}_2 = (1/a)\mathbf{I}_1 = 25$ mA rms, so we find that the 10 kΩ
resistor dissipates 6.25 W.

> The phase angles can be ignored in this example as
> they do not impact the calculation of average power
> dissipated by a purely resistive load.

PRACTICE

13.8 Repeat Example 13.7 using voltages to compute the dissipated
power.

Ans: 6.25 W.

Voltage Relationship in the Time Domain

Let us now determine how the time-domain quantities v_1 and v_2 are related
in the ideal transformer. Returning to the circuit shown in Fig. 13.17 and the
two equations, [16] and [17], describing it, we may solve the second equation for di_2/dt and substitute in the first equation:

$$v_1 = L_1 \frac{di_1}{dt} + \frac{M}{L_2}v_2 - \frac{M^2}{L_2}\frac{di_1}{dt}$$

However, for unity coupling, $M^2 = L_1 L_2$, and so

$$v_1 = \frac{M}{L_2}v_2 = \sqrt{\frac{L_1}{L_2}}v_2 = \frac{1}{a}v_2$$

The relationship between primary and secondary voltage therefore does
apply to the complete time-domain response.

PRACTICAL APPLICATION

Superconducting Transformers

For the most part, we have neglected the various types of losses that may be present in a particular transformer. When dealing with large power transformers, however, close attention must be paid to such nonidealities, despite overall efficiencies of typically 97 percent or more. Although such a high efficiency may seem nearly ideal, it can represent a great deal of wasted energy when the transformer is handling several thousand amperes. So-called i^2R (pronounced "eye-squared-R") losses represent power dissipated as heat, which can increase the temperature of the transformer coils. Wire resistance increases with temperature, so heating only leads to greater losses. High temperatures can also lead to degradation of the wire insulation, resulting in shorter transformer life. As a result, many modern power transformers employ a liquid oil bath to remove excess heat from the transformer coils. Such an approach has its drawbacks, however, including environmental impact and fire danger from leaking oil as a result of corrosion over time (Fig. 13.28).

One possible means of improving the performance of such transformers is to make use of superconducting wire to replace the resistive coils of a standard transformer design. Superconductors are materials that are resistive at high temperature, but suddenly show no resistance to the flow of current below a critical temperature. Most elements are superconducting only near

■ **FIGURE 13.28** Fire that broke out in 2004 at the 340,000 V American Electric Power Substation near Mishawaka, Indiana. (© *AP/Wide World Photos*)

An expression relating primary and secondary current in the time domain is most quickly obtained by dividing Eq. [16] throughout by L_1,

$$\frac{v_1}{L_1} = \frac{di_1}{dt} + \frac{M}{L_1}\frac{di_2}{dt} = \frac{di_1}{dt} + a\frac{di_2}{dt}$$

and then invoking one of the hypotheses underlying the ideal transformer: L_1 must be infinite. If we assume that v_1 is not infinite, then

$$\frac{di_1}{dt} = -a\frac{di_2}{dt}$$

Integrating,

$$i_1 = -ai_2 + A$$

where A is a constant of integration that does not vary with time. Thus, if we neglect any direct currents in the two windings and fix our attention only on absolute zero, requiring expensive liquid helium–based cryogenic cooling. With the discovery in the 1980s of ceramic superconductors having critical temperatures of 90 K ($-183°C$) and higher, it became possible to replace helium–based equipment with significantly cheaper liquid nitrogen systems.

Figure 13.29 shows a prototype partial-core superconducting transformer being developed at the University of Canterbury. This design employs environmentally benign liquid nitrogen in place of an oil bath, and is also significantly smaller than a comparably rated conventional transformer. The result is a measurable improvement in overall transformer efficiency, which translates into operational cost savings for the owner.

Still, all designs have disadvantages that must be weighed against their potential advantages, and superconducting transformers are no exception. The most significant obstacle at present is the relatively high cost of fabricating superconducting wire several kilometers in length compared to copper wire. Part of this is due to the challenge of fabricating long wires from ceramic materials, but part of it is also due to the silver tubing used to surround the superconductor to provide a low-resistance current path in the event of a cooling system failure (although less expensive than silver, copper reacts with the ceramic and is therefore not a viable alternative). The net result is that although a superconducting transformer is likely to save a utility money over a long period of time—many transformers see over 30 years of service—

■ **FIGURE 13.29** Prototype 15 kVA partial core superconducting power transformer.
(*Photo courtesy of Department of Electrical and Computer Engineering, University of Canterbury.*)

the initial cost is much higher than for a traditional resistive transformer. At present, many companies (including utilities) are driven by short-term cost considerations and are not always eager to make large capital investments with only long-term cost benefits.

the time-varying portion of the response, then

$$i_1 = -ai_2$$

The minus sign arises from the placement of the dots and selection of the current directions in Fig. 13.17.

The same current and voltage relationships are therefore obtained in the time domain as were obtained previously in the frequency domain, provided that dc components are ignored. The time-domain results are more general, but they have been obtained by a less informative process.

Equivalent Circuits

The characteristics of the ideal transformer which we have established may be utilized to simplify circuits in which ideal transformers appear. Let us assume, for purposes of illustration, that everything to the left of the primary terminals has been replaced by its Thévenin equivalent, as has the network to the right of the secondary terminals. We thus consider the circuit shown in Fig. 13.30. Excitation at any frequency ω is assumed.

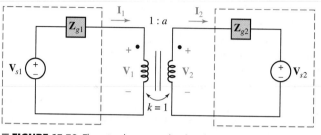

■ **FIGURE 13.30** The networks connected to the primary and secondary terminals of an ideal transformer are represented by their Thévenin equivalents.

Thévenin's or Norton's theorem may be used to achieve an equivalent circuit that does not contain a transformer. For example, let us determine the Thévenin equivalent of the network to the left of the secondary terminals. Open-circuiting the secondary, $I_2 = 0$ and therefore $I_1 = 0$ (remember that L_1 is infinite). No voltage appears across Z_{g1}, and thus $V_1 = V_{s1}$ and $V_{2oc} = aV_{s1}$. The Thévenin impedance is obtained by setting V_{s1} to zero and utilizing the square of the turns ratio, being careful to use the reciprocal turns ratio, since we are looking in at the secondary terminals. Thus, $Z_{th2} = Z_{g1}a^2$.

As a check on our equivalent, let us also determine the short-circuit secondary current I_{2sc}. With the secondary short-circuited, the primary generator faces an impedance of Z_{g1}, and, thus, $I_1 = V_{s1}/Z_{g1}$. Therefore, $I_{2sc} = V_{s1}/aZ_{g1}$. The ratio of the open-circuit voltage to the short-circuit current is a^2Z_{g1}, as it should be. The Thévenin equivalent of the transformer and primary circuit is shown in the circuit of Fig. 13.31.

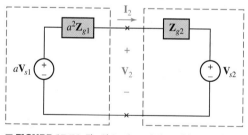

■ **FIGURE 13.31** The Thévenin equivalent of the network to the left of the secondary terminals in Fig. 13.30 is used to simplify that circuit.

Each primary voltage may therefore be multiplied by the turns ratio, each primary current divided by the turns ratio, and each primary impedance multiplied by the square of the turns ratio; and then these modified voltages, currents, and impedances replace the given voltages, currents, and impedances plus the transformer. If either dot is interchanged, the equivalent may be obtained by using the negative of the turns ratio.

Note that this equivalence, as illustrated by Fig. 13.31, is possible only if the network connected to the two primary terminals, and that connected to the two secondary terminals, can be replaced by their Thévenin equivalents. That is, each must be a two-terminal network. For example, if we cut the two primary leads at the transformer, the circuit must be divided into two separate networks; there can be no element or network bridging across the transformer between primary and secondary.

A similar analysis of the transformer and the secondary network shows that everything to the right of the primary terminals may be replaced by an identical network without the transformer, each voltage being divided by a, each current being multiplied by a, and each impedance being divided by a^2. A reversal of either winding requires the use of a turns ratio of $-a$.

EXAMPLE 13.8

For the circuit given in Fig. 13.32, determine the equivalent circuit in which the transformer and the secondary circuit are replaced, and also that in which the transformer and the primary circuit are replaced.

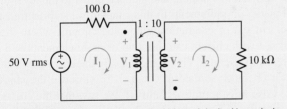
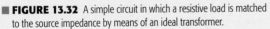

■ **FIGURE 13.32** A simple circuit in which a resistive load is matched to the source impedance by means of an ideal transformer.

This is the same circuit we analyzed in Example 13.7. As before, the input impedance is $10,000/(10)^2$, or 100 Ω and so $|\mathbf{I}_1| = 250$ mA rms. We can also compute the voltage across the primary coil

$$|\mathbf{V}_1| = |50 - 100\mathbf{I}_1| = 25 \text{ V rms}$$

and thus find that the source delivers $(25 \times 10^{-3})(50) = 12.5$ W, of which $(25 \times 10^{-3})^2(100) = 6.25$ W is dissipated in the internal resistance of the source and $12.5 - 6.25 = 6.25$ W is delivered to the load. This is the condition for maximum power transfer to the load.

If the secondary circuit and the ideal transformer are removed by the use of the Thévenin equivalent, the 50 V source and 100 Ω resistor simply see a 100 Ω impedance, and the simplified circuit of Fig. 13.33a is obtained. The primary current and voltage are now immediately evident.

If, instead, the network to the left of the secondary terminals is replaced by its Thévenin equivalent, we find (keeping in mind the location of the dots) $\mathbf{V}_{th} = -10(50) = -500$ V rms, and $\mathbf{Z}_{th} = (-10)^2(100) = 10$ kΩ; the resulting circuit is shown in Fig. 13.33b.

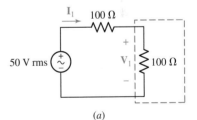

(a)

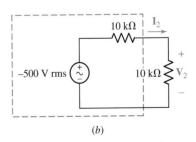

(b)

■ **FIGURE 13.33** The circuit of Fig. 13.32 is simplified by replacing (a) the transformer and secondary circuit by the Thévenin equivalent or (b) the transformer and primary circuit by the Thévenin equivalent.

PRACTICE

13.9 Let $N_1 = 1000$ turns and $N_2 = 5000$ turns in the ideal transformer shown in Fig. 13.34. If $\mathbf{Z}_L = 500 - j400 \ \Omega$, find the average power delivered to $\mathbf{Z}_L$ for (a) $\mathbf{I}_2 = 1.4\underline{/20°}$ A rms; (b) $\mathbf{V}_2 = 900\underline{/40°}$ V rms; (c) $\mathbf{V}_1 = 80\underline{/100°}$ V rms; (d) $\mathbf{I}_1 = 6\underline{/45°}$ A rms; (e) $\mathbf{V}_s = 200\underline{/0°}$ V rms.

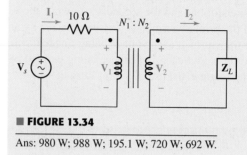

■ **FIGURE 13.34**

Ans: 980 W; 988 W; 195.1 W; 720 W; 692 W.

SUMMARY AND REVIEW

Transformers play a critical role in the power industry, allowing voltages to be stepped up for transmission, and stepped down to the level required for individual pieces of equipment. In this chapter, we studied transformers in the broader context of magnetically coupled circuits, where the magnetic flux associated with current can link two or more elements in a circuit (or even neighboring circuits). This is most easily understood by extending the concept of inductance studied in Chap. 7 to introduce the idea of mutual inductance (also having units of henrys). We saw that the coefficient M of mutual inductance is limited to less than the geometric mean of the two inductances being coupled (i.e., $M \le \sqrt{L_1 L_2}$), and made use of the dot convention to determine the polarity of the voltage induced across one inductance as a result of current flowing through the other. When the two inductances are not particularly close, M might be rather small. However, in the case of a well-designed transformer, it might approach its maximum value. To describe such situations, we introduced the concept of the coupling coefficient k. When dealing with a linear transformer, analysis may be assisted by representing the element with an equivalent T (or, less commonly, Π) network, but a great deal of circuit analysis is performed assuming an ideal transformer. In such instances we no longer concern ourselves with M or k, but rather the turns ratio a. We saw that the voltages across the primary and secondary coils, as well as their individual currents, are related by this parameter. This approximation is very useful for both analysis and design. We concluded the chapter with a brief discussion of how Thévenin's theorem can be applied to circuits with ideal transformers.

We could continue, as the study of inductively coupled circuits is an interesting and important topic, but at this point it might be appropriate to list some of the key concepts we have already discussed, along with corresponding example numbers.

❏ Mutual inductance describes the voltage induced at the ends of a coil due to the magnetic field generated by a second coil. (Example 13.1)

46. Calculate $\mathbf{I}_x$ and $\mathbf{V}_2$ as labeled in Fig. 13.68.

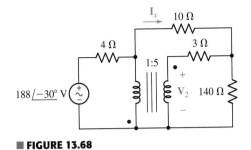

■ **FIGURE 13.68**

47. The ideal transformer of the circuit in Fig. 13.68 is removed, flipped across its vertical axis, and reconnected such that the same terminals remain connected to the negative terminal of the source. (*a*) Calculate $\mathbf{I}_x$ and $\mathbf{V}_2$. (*b*) Repeat part (*a*) if both dots are placed at the bottom terminals of the transformer.

48. For the circuit of Fig. 13.69, $v_s = 117 \sin 500t$ V. Calculate v_2 if the terminals marked *a* and *b* are (*a*) left open-circuited; (*b*) short-circuited; (*c*) bridged by a 2 Ω resistor.

49. The turns ratio of the ideal transformer in Fig. 13.69 is changed from 30:1 to 1:3. Take $v_s = 720 \cos 120\pi t$ V, and calculate v_2 if terminals *a* and *b* are (*a*) short-circuited; (*b*) bridged by a 10 Ω resistor; (*c*) bridged by a 1 MΩ resistor.

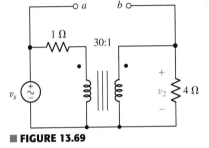

■ **FIGURE 13.69**

DP 50. For the circuit of Fig. 13.70, $R_1 = 1$ Ω, $R_2 = 4$ Ω, and $R_L = 1$ Ω. Select *a* and *b* to achieve a peak voltage of 200 V magnitude across R_L.

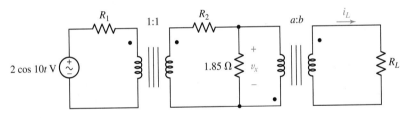

■ **FIGURE 13.70**

51. Calculate v_x for the circuit of Fig. 13.70 if $a = 0.01b = 1$, $R_1 = 300$ Ω, $R_2 = 14$ Ω, and $R_L = 1$ kΩ.

52. (*a*) Referring to the ideal transformer circuit in Fig. 13.70, determine the load current i_L if $b = 0.25a = 1$, $R_1 = 2.2$ Ω, $R_2 = 3.1$ Ω, and $R_L = 200$ Ω. (*b*) Verify your solution with an appropriate PSpice simulation.

53. Determine the Thévenin equivalent of the network in Fig. 13.71 as seen looking into terminals *a* and *b*.

54. Calculate $\mathbf{V}_2$ and the average power delivered to the 8 Ω resistor of Fig. 13.72 if $\mathbf{V}_s = 10/\underline{15°}$ V, and the control parameter *c* is equal to (*a*) 0; (*b*) 1 mS.

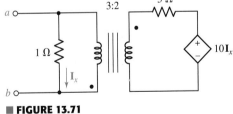

■ **FIGURE 13.71**

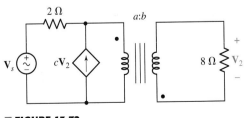

■ **FIGURE 13.72**

55. (*a*) For the circuit of Fig. 13.72, take $c = -2.5$ mS and select values of a and b such that 100 W average power is delivered to the 8 Ω load when $\mathbf{V}_s = 5\underline{/-35°}$ V, (*b*) Verify your solution with an appropriate PSpice simulation.

Chapter-Integrating Exercises

56. A transformer whose nameplate reads $\boxed{2300/230 \text{ V}, 25 \text{ kVA}}$ operates with primary and secondary voltages of 2300 V and 230 V rms, respectively, and can supply 25 kVA from its secondary winding. If this transformer is supplied with 2300 V rms and is connected to secondary loads requiring 8 kW at unity PF and 15 kVA at 0.8 PF lagging, (*a*) what is the primary current? (*b*) How many kilowatts can the transformer still supply to a load operating at 0.95 PF lagging? (*c*) Verify your answers with PSpice.

57. A friend brings a vintage stereo system back from a recent trip to Warnemünde, unaware that it was designed to operate on twice the supply voltage (240 VAC) available at American household outlets. Design a circuit to allow your friend to listen to the stereo in the United States, assuming the operating frequency (50 Hz in Germany, 60 Hz in the United States) difference can be neglected.

58. The friend referred to in Exercise 57 attempts to justify the erroneous assumption made regarding the stereo by pointing out that the wall outlet in the W.C. (bathroom) had a socket for his U.S. electric razor, clearly marked 120 VAC. He failed to notice that the small sign below the outlet clearly stated "Razors only." With the knowledge that all power lines running into the room operated at 240 VAC, draw the likely circuit built into the bathroom wall outlet, and explain why it is limited to "razors only."

59. Obtain an expression for $\mathbf{V}_2/\mathbf{V}_s$ in the circuit of Fig. 13.73 if (*a*) $L_1 = 100$ mH, $L_2 = 500$ mH, and M is its maximum possible value; (*b*) $L_1 = 5L_2 = 1.4$ H and $k = 87\%$ of its maximum possible value; (*c*) the two coils can be treated as an ideal transformer, the left-hand coil having 500 turns and the right-hand coil having 10,000 turns.

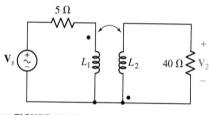

■ **FIGURE 13.73**

60. You notice your neighbor has installed a large coil of wire in close proximity to the power line coming into your house (underground cables are not available in your neighborhood). (*a*) What is the likely intention of your neighbor? (*b*) Is the plan likely to succeed? *Explain.* (*c*) When confronted, your neighbor simply shrugs and claims there's no way it can cost you anything, anyway, since nothing of his is touching anything on your property. True or not? *Explain.*

Complex Frequency and the Laplace Transform

14

INTRODUCTION

When faced with time-varying sources, or a circuit with switches installed, we have several choices with respect to the analysis approach. Chapters 7 through 9 detail direct differential equation–based analysis, which is particularly useful when examining turn-on or turn-off transients. In contrast, Chaps. 10 to 13 describe analysis situations where sinusoidal excitation is assumed, and transients are of little or no interest. Unfortunately, not all sources are sinusoidal, and there are times when both transient and steady-state responses are required. In such instances, the Laplace transform proves to be a highly valuable tool.

Many textbooks simply launch straight into the Laplace transform integral, but this approach conveys no intuitive understanding. For this reason, we have chosen to first introduce what may strike the reader at first as a somewhat odd concept—the notion of a "complex" frequency. Simply a mathematical convenience, complex frequency allows us to manipulate both periodic and nonperiodic time-varying quantities in parallel, greatly simplifying the analysis. After getting a feel for the basic technique, we develop it as a specific circuit analysis tool in Chap. 15.

14.1 COMPLEX FREQUENCY

We introduce the notion of **complex frequency** by considering a (purely real) exponentially damped sinusoidal function, such as the voltage

$$v(t) = V_m e^{\sigma t} \cos(\omega t + \theta) \qquad [1]$$

533

where σ (sigma) is a real quantity and is usually negative. Although we often refer to this function as being "damped," it is conceivable that we might occasionally encounter a situation where $\sigma > 0$ and hence the sinusoidal amplitude is increasing. (In Chap. 9, our study of the natural response of the *RLC* circuit also indicates that σ is the negative of the exponential damping coefficient.)

Note that we may construct a constant voltage from Eq. [1] by letting $\sigma = \omega = 0$:

$$v(t) = V_m \cos \theta = V_0 \qquad [2]$$

If we set only σ equal to zero, then we obtain a general sinusoidal voltage

$$v(t) = V_m \cos(\omega t + \theta) \qquad [3]$$

and if $\omega = 0$, we have the exponential voltage

$$v(t) = V_m \cos \theta \; e^{\sigma t} = V_0 e^{\sigma t} \qquad [4]$$

Thus, the damped sinusoid of Eq. [1] includes as special cases the dc Eq. [2], sinusoidal Eq. [3], and exponential Eq. [4] functions.

Some additional insight into the significance of σ can be obtained by comparing the exponential function of Eq. [4] with the complex representation of a sinusoidal function with a zero-degree phase angle,

$$v(t) = V_0 e^{j\omega t} \qquad [5]$$

It is apparent that the two functions, Eqs. [4] and [5], have much in common. The only difference is that the exponent in Eq. [4] is real and the one in Eq. [5] is imaginary. The similarity between the two functions is emphasized by describing σ as a "frequency." This choice of terminology will be discussed in detail in the following sections, but for now we need merely note that σ is specifically termed the *real part* of the complex frequency. It should not be called the "real frequency," however, for this is a term that is more suitable for f (or, loosely, for ω). We will also refer to σ as the **neper frequency,** the name arising from the dimensionless unit of the exponent of e. Thus, given e^{7t}, the dimensions of $7t$ are **nepers** (Np), and 7 is the neper frequency in nepers per second.

The neper itself was named after the Scottish philosopher and mathematician John Napier (1550–1617) and his napierian logarithm system; the spelling of his name is historically uncertain (see, for example, H. A. Wheeler, *IRE Transactions on Circuit Theory* **2**, 1955, p. 219).

The General Form

The forced response of a network to a general forcing function of the form of Eq. [1] can be found very simply by using a method almost identical with that used in phasor-based analysis. Once we are able to find the forced response to this damped sinusoid, we will also have found the forced response to a dc voltage, an exponential voltage, and a sinusoidal voltage. First we consider σ and ω as the real and imaginary parts of a complex frequency.

We suggest that any function that may be written in the form

$$f(t) = \mathbf{K}e^{st} \qquad [6]$$

where **K** and **s** are complex constants (independent of time) is characterized by the complex frequency **s.** The complex frequency **s** is therefore simply

Before we actually carry out the details of an analysis problem and see how the procedure resembles what we used in sinusoidal analysis, it is worthwhile to outline the steps of the basic method.

1. We first characterize the circuit with a set of loop or nodal integrodifferential equations.

2. The given forcing functions, in complex form, and the assumed forced responses, also in complex form, are substituted in the equations and the indicated integrations and differentiations are performed.

3. Each term in every equation will then contain the same factor e^{st}. We divide throughout by this factor, or "suppress e^{st}," understanding that it must be reinserted if a time-domain description of any response function is desired.

With the Re{ } notation and the e^{st} factor gone, we have converted all the voltages and currents from the *time domain* to the *frequency domain*. The integrodifferential equations become algebraic equations, and their solution is obtained just as easily as in the sinusoidal steady state. Let us illustrate the basic method by a numerical example.

EXAMPLE 14.1

Apply the forcing function $v(t) = 60e^{-2t} \cos(4t + 10°)$ V to the series *RLC* circuit shown in Fig. 14.1, and specify the forced response by finding values for I_m and ϕ in the time-domain expression $i(t) = I_m e^{-2t} \cos(4t + \phi)$.

We first express the forcing function in Re{ } notation:

$$v(t) = 60e^{-2t} \cos(4t + 10°) = \text{Re}\{60e^{-2t} e^{j(4t+10°)}\}$$
$$= \text{Re}\{60e^{j10°} e^{(-2+j4)t}\}$$

or

$$v(t) = \text{Re}\{\mathbf{V}e^{st}\}$$

where

$$\mathbf{V} = 60\underline{/10°} \qquad \text{and} \qquad s = -2 + j4$$

After dropping Re{ }, we are left with the complex forcing function

$$60\underline{/10°}e^{st}$$

Similarly, we represent the unknown response by the complex quantity $\mathbf{I}e^{st}$, where $\mathbf{I} = I_m\underline{/\phi}$.

Our next step must be the integrodifferential equation for the circuit. From Kirchhoff's voltage law, we obtain

$$v(t) = Ri + L\frac{di}{dt} + \frac{1}{C}\int i\, dt = 2i + 3\frac{di}{dt} + 10\int i\, dt$$

(Continued on next page)

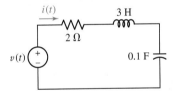

■ **FIGURE 14.1** A series *RLC* circuit to which a damped sinusoidal forcing function is applied. A frequency-domain solution for $i(t)$ is desired.

If the notation here seems unfamiliar, the reader may wish to pause and read Appendix 5, particularly section 4, which deals with the polar form of complex number representation.

and we substitute the given complex forcing function and the assumed complex forced response in this equation:

$$60\underline{/10°}e^{st} = 2\mathbf{I}e^{st} + 3s\mathbf{I}e^{st} + \frac{10}{s}\mathbf{I}e^{st}$$

The common factor e^{st} is next suppressed:

$$60\underline{/10°} = 2\mathbf{I} + 3s\mathbf{I} + \frac{10}{s}\mathbf{I}$$

and thus

$$\mathbf{I} = \frac{60\underline{/10°}}{2 + 3s + 10/s}$$

We now let $s = -2 + j4$ and solve for the complex current $\mathbf{I}$:

$$\mathbf{I} = \frac{60\underline{/10°}}{2 + 3(-2 + j4) + 10/(-2 + j4)}$$

After manipulating the complex numbers, we find

$$\mathbf{I} = 5.37\underline{/-106.6°}$$

Thus, I_m is 5.37 A, ϕ is $-106.6°$, and the forced response can be written directly (recalling that $s = -2 + j4$) as

$$i(t) = 5.37e^{-2t}\cos(4t - 106.6°) \text{ A}$$

We have thus solved the problem by reducing a *calculus*-based expression to an *algebraic* expression. This is only a small indication of the power of the technique we are about to study.

PRACTICE

14.3 Give the phasor current that is equivalent to the time-domain current: (a) $24\sin(90t + 60°)$ A; (b) $24e^{-10t}\cos(90t + 60°)$ A; (c) $24e^{-10t}\cos 60° \times \cos 90t$ A. If $\mathbf{V} = 12\underline{/35°}$ V, find $v(t)$ for s equal to (d) 0; (e) -20 s^{-1}; (f) $-20 + j5$ s^{-1}.

Ans: $24\underline{/-30°}$ A; $24\underline{/60°}$ A; $12\underline{/0°}$ A; 9.83 V; $9.83e^{-20t}$ V; $12e^{-20t}\cos(5t + 35°)$ V.

14.3 • DEFINITION OF THE LAPLACE TRANSFORM

Our constant goal has been one of analysis: given some forcing function at one point in a linear circuit, determine the response at some other point. For the first several chapters, we played only with dc forcing functions and responses of the form V_0e^0. However, after the introduction of inductance and capacitance, the sudden dc excitation of simple *RL* and *RC* circuits produced responses varying exponentially with time: $V_0e^{\sigma t}$. When we considered the *RLC* circuit, the responses took on the form of the exponentially varying sinusoid, $V_0e^{\sigma t}\cos(\omega t + \theta)$. All this work was accomplished in the time domain, and the dc forcing function was the only one we considered.

As we advanced to the use of the sinusoidal forcing function, the tedium and complexity of solving the integrodifferential equations caused us to begin casting about for an easier way to work problems. The phasor transform was the result, and we might remember that we were led to it through consideration of a complex forcing function of the form $V_0 e^{j\theta} e^{j\omega t}$. As soon as we concluded that we did not need the factor containing t, we were left with the phasor $V_0 e^{j\theta}$; we had arrived at the *frequency domain.*

Now a little flexing of our cerebral cortex has caused us to apply a forcing function of the form $V_0 e^{j\theta} e^{(\sigma + j\omega)t}$, leading to the invention of the complex frequency $\mathbf{s}$, and thereby relegating all our previous functional forms to special cases: dc ($\mathbf{s} = 0$), exponential ($\mathbf{s} = \sigma$), sinusoidal ($\mathbf{s} = j\omega$), and exponential sinusoid ($\mathbf{s} = \sigma + j\omega$). By analogy to our previous experience with phasors, we saw that in these cases we may omit the factor containing t, and once again obtain a solution by working in the frequency domain.

The Two-Sided Laplace Transform

We know that sinusoidal forcing functions lead to sinusoidal responses, and also that exponential forcing functions lead to exponential responses. However, as practicing engineers we will encounter many waveforms that are neither sinusoidal nor exponential, such as square waves, sawtooth waveforms, and pulses beginning at arbitrary instants of time. When such forcing functions are applied to a linear circuit, we will see that the response is neither similar to the form of the excitation waveform nor exponential. As a result, we are not able to eliminate the terms containing t to form a frequency-domain response. This is rather unfortunate, as working in the frequency domain has proved to be rather useful.

There is a solution, however, which makes use of a technique that allows us to expand any function into a *sum* of exponential waveforms, each with its own complex frequency. Since we are considering linear circuits, we know that the total response of our circuit can be obtained by simply adding the individual response to each exponential waveform. And, in dealing with each exponential waveform, we may once again neglect any terms containing t, and work instead in the *frequency* domain. It unfortunately takes an infinite number of exponential terms to accurately represent a general time function, so that taking a brute-force approach and applying superposition to the exponential series might be somewhat insane. Instead, we will sum these terms by performing an integration, leading to a frequency-domain function.

We formalize this approach using what is known as a ***Laplace transform,*** defined for a general function $f(t)$ as

$$\mathbf{F}(\mathbf{s}) = \int_{-\infty}^{\infty} e^{-\mathbf{s}t} f(t)\, dt \qquad [12]$$

The mathematical derivation of this integral operation requires an understanding of Fourier series and the Fourier transform, which are discussed in Chap. 18. The fundamental concept behind the Laplace transform, however, can be understood based on our discussion of complex frequency and our prior experience with phasors and converting back and forth between the time domain and the frequency domain. In fact, that is precisely what the Laplace transform does: it converts the general time-domain function $f(t)$ into a corresponding frequency-domain representation, $\mathbf{F}(\mathbf{s})$.

The Two-Sided Inverse Laplace Transform

Equation [12] defines the *two-sided,* or *bilateral,* Laplace transform of $f(t)$. The term *two-sided* or *bilateral* is used to emphasize the fact that both positive and negative values of t are included in the range of integration. The inverse operation, often referred to as the **inverse Laplace transform,** is also defined as an integral expression[1]

$$f(t) = \frac{1}{2\pi j} \int_{\sigma_0 - j\infty}^{\sigma_0 + j\infty} e^{st} \mathbf{F(s)} \, ds \qquad [13]$$

where the real constant σ_0 is included in the limits to ensure convergence of this improper integral; the two equations [12] and [13] constitute the two-sided Laplace transform pair. The good news is that Eq. [13] need never be invoked in the study of circuit analysis: there is a quick and easy alternative to look forward to learning.

The One-Sided Laplace Transform

In many of our circuit analysis problems, the forcing and response functions do not exist forever in time, but rather they are initiated at some specific instant that we usually select as $t = 0$. Thus, for time functions that do not exist for $t < 0$, or for those time functions whose behavior for $t < 0$ is of no interest, the time-domain description can be thought of as $v(t)u(t)$. The defining integral for the Laplace transform is taken with the lower limit at $t = 0^-$ in order to include the effect of any discontinuity at $t = 0$, such as an impulse or a higher-order singularity. The corresponding Laplace transform is then

$$\mathbf{F(s)} = \int_{-\infty}^{\infty} e^{-st} f(t) u(t) \, dt = \int_{0^-}^{\infty} e^{-st} f(t) \, dt$$

This defines the *one-sided* Laplace transform of $f(t)$, or simply the *Laplace transform* of $f(t)$, one-sided being understood. The inverse transform expression remains unchanged, but when evaluated, it is understood to be valid only for $t > 0$. Here then is the definition of the Laplace transform pair that we will use from now on:

$$\boxed{\mathbf{F(s)} = \int_{0^-}^{\infty} e^{-st} f(t) \, dt} \qquad [14]$$

$$\boxed{\begin{aligned} f(t) &= \frac{1}{2\pi j} \int_{\sigma_0 - j\infty}^{\sigma_0 + j\infty} e^{st} \mathbf{F(s)} \, ds \\ f(t) &\Leftrightarrow \mathbf{F(s)} \end{aligned}} \qquad [15]$$

The script $\mathcal{L}$ may also be used to indicate the direct or inverse Laplace transform operation:

$$\mathbf{F(s)} = \mathcal{L}\{f(t)\} \qquad \text{and} \qquad f(t) = \mathcal{L}^{-1}\{\mathbf{F(s)}\}$$

---

(1) If we ignore the distracting factor of $1/2\pi j$ and view the integral as a summation over all frequencies such that $f(t) \propto \Sigma[\mathbf{F(s)}\,ds]e^{st}$, this reinforces the notion that $f(t)$ is indeed a sum of complex frequency terms having a magnitude proportional to $\mathbf{F(s)}$.

EXAMPLE **14.2**

Compute the Laplace transform of the function $f(t) = 2u(t - 3)$.

In order to find the one-sided Laplace transform of $f(t) = 2u(t - 3)$, we must evaluate the integral

$$\mathbf{F(s)} = \int_{0-}^{\infty} e^{-st} f(t)\, dt$$

$$= \int_{0-}^{\infty} e^{-st} 2u(t - 3)\, dt$$

$$= 2 \int_{3}^{\infty} e^{-st}\, dt$$

Simplifying, we find

$$\mathbf{F(s)} = \frac{-2}{\mathbf{s}} e^{-st} \Big|_{3}^{\infty} = \frac{-2}{\mathbf{s}} (0 - e^{-3s}) = \frac{2}{\mathbf{s}} e^{-3s}$$

PRACTICE

14.4 Let $f(t) = -6e^{-2t}[u(t + 3) - u(t - 2)]$. Find the (*a*) two-sided $\mathbf{F(s)}$; (*b*) one-sided $\mathbf{F(s)}$.

Ans: $\frac{6}{2+s}[e^{-4-2s} - e^{6+3s}]$; $\frac{6}{2+s}[e^{-4-2s} - 1]$.

14.4 LAPLACE TRANSFORMS OF SIMPLE TIME FUNCTIONS

In this section we will begin to build up a catalog of Laplace transforms for those time functions most frequently encountered in circuit analysis; we will assume for now that the function of interest is a voltage, although such a choice is strictly arbitrary. We will create this catalog, at least initially, by utilizing the definition,

$$\mathbf{V(s)} = \int_{0-}^{\infty} e^{-st} v(t)\, dt = \mathcal{L}\{v(t)\}$$

which, along with the expression for the inverse transform,

$$v(t) = \frac{1}{2\pi j} \int_{\sigma_0 - j\infty}^{\sigma_0 + j\infty} e^{st} \mathbf{V(s)}\, d\mathbf{s} = \mathcal{L}^{-1}\{\mathbf{V(s)}\}$$

establishes a one-to-one correspondence between $v(t)$ and $\mathbf{V(s)}$. That is, for every $v(t)$ for which $\mathbf{V(s)}$ exists, there is a unique $\mathbf{V(s)}$. At this point, we may be looking with some trepidation at the rather ominous form given for the inverse transform. Fear not! As we will see shortly, *an introductory study of Laplace transform theory does not require actual evaluation of this integral*. By going from the time domain to the frequency domain and taking advantage of the uniqueness just mentioned, we will be able to generate a catalog of transform pairs that will already contain the corresponding time function for nearly every transform that we wish to invert.

Before we continue, however, we should pause to consider whether there is any chance that the transform may not even exist for some $v(t)$ that concerns us. A set of conditions sufficient to ensure the absolute convergence of the Laplace integral for $\text{Re}\{s\} > \sigma_0$ is

1. The function $v(t)$ is integrable in every finite interval $t_1 < t < t_2$, where $0 \le t_1 < t_2 < \infty$.

2. $\lim\limits_{t \to \infty} e^{-\sigma_0 t} |v(t)|$ exists for some value of σ_0.

Time functions that do not satisfy these conditions are seldom encountered by the circuit analyst.[2]

The Unit-Step Function $u(t)$

Now let us look at some specific transforms. We first examine the Laplace transform of the unit-step function $u(t)$. From the defining equation, we may write

$$\mathscr{L}\{u(t)\} = \int_{0^-}^{\infty} e^{-st} u(t)\, dt = \int_{0}^{\infty} e^{-st}\, dt$$

$$= -\frac{1}{s} e^{-st} \Big|_0^{\infty} = \frac{1}{s}$$

for $\text{Re}\{s\} > 0$, to satisfy condition 2. Thus,

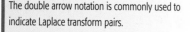

The double arrow notation is commonly used to indicate Laplace transform pairs.

$$u(t) \Leftrightarrow \frac{1}{s} \tag{16}$$

and our first Laplace transform pair has been established with great ease.

The Unit-Impulse Function $\delta(t - t_0)$

A singularity function whose transform is of considerable interest is the unit-impulse function $\delta(t - t_0)$. This function, plotted in Fig. 14.2, seems rather strange at first but is enormously useful in practice. The unit-impulse function is defined to have an area of unity, so that

$$\delta(t - t_0) = 0 \qquad t \ne t_0$$

$$\int_{t_0 - \varepsilon}^{t_0 + \varepsilon} \delta(t - t_0)\, dt = 1$$

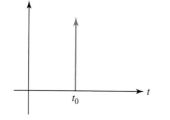

■ **FIGURE 14.2** The unit-impulse function $\delta(t - t_0)$. This function is often used to approximate a signal pulse whose duration is very short compared to circuit time constants.

where ε is a small constant. Thus, this "function" (a naming that makes many purist mathematicians cringe) has a nonzero value only at the point t_0. For $t_0 > 0^-$, we therefore find the Laplace transform to be

$$\mathscr{L}\{\delta(t - t_0)\} = \int_{0^-}^{\infty} e^{-st} \delta(t - t_0)\, dt = e^{-st_0}$$

$$\delta(t - t_0) \Leftrightarrow e^{-st_0} \tag{17}$$

---

(2) Examples of such functions are e^{t^2} and e^{e^t}, but not t^n or n^t. For a somewhat more detailed discussion of the Laplace transform and its applications, refer to Clare D. McGillem and George R. Cooper, *Continuous and Discrete Signal and System Analysis,* 3d ed. Oxford University Press, North Carolina: 1991, Chap. 5.

In particular, note that we obtain

$$\delta(t) \Leftrightarrow 1 \qquad [18]$$

for $t_0 = 0$.

An interesting feature of the unit-impulse function is known as the **sifting property**. Consider the integral of the impulse function multiplied by an arbitrary function $f(t)$:

$$\int_{-\infty}^{\infty} f(t)\delta(t - t_0)\,dt$$

Since the function $\delta(t - t_0)$ is zero everywhere except at $t = t_0$, the value of this integral is simply $f(t_0)$. The property turns out to be *very* useful in simplifying integral expressions containing the unit-impulse function.

The Exponential Function $e^{-\alpha t}$

Recalling our past interest in the exponential function, we examine its transform,

$$\mathcal{L}\{e^{-\alpha t}u(t)\} = \int_{0^-}^{\infty} e^{-\alpha t}e^{-st}\,dt$$

$$= -\frac{1}{s + \alpha}e^{-(s+\alpha)t}\Big|_0^{\infty} = \frac{1}{s + \alpha}$$

and therefore,

$$e^{-\alpha t}u(t) \Longleftrightarrow \frac{1}{s + \alpha} \qquad [19]$$

It is understood that $\text{Re}\{s\} > -\alpha$.

The Ramp Function $tu(t)$

As a final example, for the moment, let us consider the ramp function $tu(t)$. We obtain

$$\mathcal{L}\{tu(t)\} = \int_{0^-}^{\infty} te^{-st}\,dt = \frac{1}{s^2}$$

$$tu(t) \Leftrightarrow \frac{1}{s^2} \qquad [20]$$

either by a straightforward integration by parts or from a table of integrals.

So what of the function $te^{-\alpha t}u(t)$? We leave it to the reader to show that

$$te^{-\alpha t}u(t) \Leftrightarrow \frac{1}{(s + \alpha)^2} \qquad [21]$$

There are, of course, quite a few additional time functions worth considering, but it may be best if we pause for the moment to consider the reverse of the process—the inverse Laplace transform—before returning to add to our list.

14.5 INVERSE TRANSFORM TECHNIQUES

The Linearity Theorem

Although we mentioned that Eq. [13] can be applied to convert an s-domain expression into a time-domain expression, we also alluded to the fact that this is more work than required—if we're willing to exploit the uniqueness of any Laplace transform pair. In order to fully capitalize on this fact, we must first introduce one of several helpful and well-known Laplace transform theorems—the *linearity theorem.* This theorem states that the Laplace transform of the sum of two or more time functions is equal to the sum of the transforms of the individual time functions. For two time functions we have

$$\mathcal{L}\{f_1(t) + f_2(t)\} = \int_{0^-}^{\infty} e^{-st}[f_1(t) + f_2(t)]\,dt$$
$$= \int_{0^-}^{\infty} e^{-st}f_1(t)\,dt + \int_{0^-}^{\infty} e^{-st}f_2(t)\,dt$$
$$= \mathbf{F}_1(s) + \mathbf{F}_2(s)$$

This is known as the "additive property" of the Laplace transform.

As an example of the use of this theorem, suppose that we have a Laplace transform $\mathbf{V}(s)$ and want to know the corresponding time function $v(t)$. It will often be possible to decompose $\mathbf{V}(s)$ into the sum of two or more functions, e.g., $\mathbf{V}_1(s)$ and $\mathbf{V}_2(s)$, whose inverse transforms, $v_1(t)$ and $v_2(t)$, are already tabulated. It then becomes a simple matter to apply the linearity theorem and write

$$v(t) = \mathcal{L}^{-1}\{\mathbf{V}(s)\} = \mathcal{L}^{-1}\{\mathbf{V}_1(s) + \mathbf{V}_2(s)\}$$
$$= \mathcal{L}^{-1}\{\mathbf{V}_1(s)\} + \mathcal{L}^{-1}\{\mathbf{V}_2(s)\} = v_1(t) + v_2(t)$$

Another important consequence of the linearity theorem is evident by studying the definition of the Laplace transform. Since we are working with an integral, *the Laplace transform of a constant times a function is equal to the constant times the Laplace transform of the function.* In other words,

$$\mathcal{L}\{kv(t)\} = k\mathcal{L}\{v(t)\}$$

or

$$kv(t) \Leftrightarrow k\mathbf{V}(s) \qquad [22]$$

This is known as the "homogeneity property" of the Laplace transform.

where k is a constant of proportionality. This result is extremely handy in many situations that arise from circuit analysis, as we will see.

EXAMPLE **14.3**

Given the function G(s) = (7/s) − 31/(s + 17), obtain g(t).

This **s**-domain function is composed of the sum of two terms, $7/\mathbf{s}$ and $−31/(\mathbf{s} + 17)$. Through the linearity theorem we know that $g(t)$ will be composed of two terms as well, each the inverse Laplace transform of one of the two **s**-domain terms:

$$g(t) = \mathcal{L}^{-1}\left\{\frac{7}{\mathbf{s}}\right\} - \mathcal{L}^{-1}\left\{\frac{31}{\mathbf{s} + 17}\right\}$$

Let's begin with the first term. The homogeneity property of the Laplace transform allows us to write that

$$\mathcal{L}^{-1}\left\{\frac{7}{\mathbf{s}}\right\} = 7\mathcal{L}^{-1}\left\{\frac{1}{\mathbf{s}}\right\} = 7u(t)$$

Thus, we have made use of the known transform pair $u(t) \Leftrightarrow 1/\mathbf{s}$ and the homogeneity property to find this first component of $g(t)$. In a similar fashion, we find that $\mathcal{L}^{-1}\left\{\dfrac{31}{\mathbf{s} + 17}\right\} = 31e^{-17t}u(t)$. Putting these two terms together,

$$g(t) = [7 - 31e^{-17t}]u(t)$$

PRACTICE

14.7 Given $\mathbf{H}(\mathbf{s}) = \dfrac{2}{\mathbf{s}} - \dfrac{4}{\mathbf{s}^2} + \dfrac{3.5}{(\mathbf{s} + 10)(\mathbf{s} + 10)}$, obtain $h(t)$.

Ans: $h(t) = [2 - 4t + 3.5te^{-10t}]u(t)$.

Inverse Transform Techniques for Rational Functions

In analyzing circuits with multiple energy storage elements, we will often encounter **s**-domain expressions that are ratios of **s**-polynomials. We thus expect to routinely encounter expressions of the form

$$\mathbf{V}(\mathbf{s}) = \frac{\mathbf{N}(\mathbf{s})}{\mathbf{D}(\mathbf{s})}$$

where $\mathbf{N}(\mathbf{s})$ and $\mathbf{D}(\mathbf{s})$ are polynomials in **s**. The values of **s** which lead to $\mathbf{N}(\mathbf{s}) = 0$ are referred to as *zeros* of $\mathbf{V}(\mathbf{s})$, and those values of **s** which lead to $\mathbf{D}(\mathbf{s}) = 0$ are referred to as *poles* of $\mathbf{V}(\mathbf{s})$.

Rather than rolling up our sleeves and invoking Eq. [13] each time we need to find an inverse transform, it is often possible to decompose these expressions using the method of residues into simpler terms whose inverse transforms are already known. The criterion for this is that $\mathbf{V}(\mathbf{s})$ must be a *rational function* for which the degree of the numerator $\mathbf{N}(\mathbf{s})$ must be less than that of the denominator $\mathbf{D}(\mathbf{s})$. If it is not, we must first perform a simple division step, as shown in the following example. The result will include an impulse function (assuming the degree of the numerator is the same as that of the denominator) and a rational function. The inverse transform of the first is simple; the straightforward method of residues applies to the rational function if its inverse transform is not already known.

In practice, it is seldom necessary to ever invoke Eq. [13] for functions encountered in circuit analysis, provided that we are clever in using the various techniques presented in this chapter.

EXAMPLE 14.4

Calculate the inverse transform of $F(s) = 2\dfrac{s+2}{s}$.

Since the degree of the numerator is equal to the degree of the denominator, $F(s)$ is *not* a rational function. Thus, we begin by performing long division:

$$F(s) = s\overline{)\begin{array}{l} 2 \\ 2s+4 \\ \underline{2s} \\ 4 \end{array}}$$

so that $F(s) = 2 + (4/s)$. By the linearity theorem,

$$\mathcal{L}^{-1}\{F(s)\} = \mathcal{L}^{-1}\{2\} + \mathcal{L}^{-1}\left\{\frac{4}{s}\right\} = 2\delta(t) + 4u(t)$$

(It should be noted that this particular function can be simplified without the process of long division; such a route was chosen to provide an example of the basic process.)

PRACTICE

14.8 Given the function $Q(s) = \dfrac{3s^2 - 4}{s^2}$, find $q(t)$.

Ans: $q(t) = 3\delta(t) - 4tu(t)$.

In employing the method of residues, essentially performing a partial fraction expansion of $V(s)$, we focus our attention on the roots of the denominator. Thus, it is first necessary to factor the s-polynomial that comprises $D(s)$ into a product of binomial terms. The roots of $D(s)$ may be any combination of distinct or repeated roots, and may be real or complex. It is worth noting, however, that complex roots always occur as conjugate pairs, provided that the coefficients of $D(s)$ are real.

Distinct Poles and the Method of Residues

As a specific example, let us determine the inverse Laplace transform of

$$V(s) = \frac{1}{(s+\alpha)(s+\beta)}$$

The denominator has been factored into two distinct roots, $-\alpha$ and $-\beta$. Although it is possible to substitute this expression in the defining equation for the inverse transform, it is much easier to utilize the linearity theorem. Using partial-fraction expansion, we can split the given transform into the sum of two simpler transforms,

$$V(s) = \frac{A}{s+\alpha} + \frac{B}{s+\beta}$$

where A and B may be found by any of several methods. Perhaps the quickest solution is obtained by recognizing that

$$A = \lim_{s \to -\alpha}\left[(s+\alpha)V(s) - \frac{(s+\alpha)}{(s+\beta)}B\right]$$

$$= \lim_{s \to -\alpha}\left[\frac{1}{s+\beta} - 0\right] = \frac{1}{\beta - \alpha}$$

In this equation, we use the single-fraction (i.e., nonexpanded) version of $V(s)$.

Recognizing that the second term is always zero, in practice we would simply write

$$A = (s + \alpha)\mathbf{V}(s)|_{s=-\alpha}$$

Similarly,

$$B = (s + \beta)\mathbf{V}(s)|_{s=-\beta} = \frac{1}{\alpha - \beta}$$

and therefore,

$$\mathbf{V}(s) = \frac{1/(\beta - \alpha)}{s + \alpha} + \frac{1/(\alpha - \beta)}{s + \beta}$$

We have already evaluated inverse transforms of this form, and so

$$v(t) = \frac{1}{\beta - \alpha}e^{-\alpha t}u(t) + \frac{1}{\alpha - \beta}e^{-\beta t}u(t)$$

$$= \frac{1}{\beta - \alpha}(e^{-\alpha t} - e^{-\beta t})u(t)$$

If we wished, we could now include this as a new entry in our catalog of Laplace pairs,

$$\frac{1}{\beta - \alpha}(e^{-\alpha t} - e^{-\beta t})u(t) \Leftrightarrow \frac{1}{(s + \alpha)(s + \beta)}$$

This approach is easily extended to functions whose denominators are higher-order **s**-polynomials, although the operations can become somewhat tedious. It should also be noted that we did not specify that the constants A and B must be real. However, in situations where α and β are complex, we will find that α and β are also complex conjugates (this is not required mathematically, but is required for physical circuits). In such instances, we will also find that $A = B^*$; in other words, the coefficients will be complex conjugates as well.

EXAMPLE 14.5

Determine the inverse transform of

$$\mathbf{P}(s) = \frac{7s + 5}{s^2 + s}$$

We see that $\mathbf{P}(s)$ is a rational function (the degree of the numerator is *one,* whereas the degree of the denominator is *two*), so we begin by factoring the denominator and write:

$$\mathbf{P}(s) = \frac{7s + 5}{s(s + 1)} = \frac{a}{s} + \frac{b}{s + 1}$$

where our next step is to determine values for a and b. Applying the method of residues,

$$a = \frac{7s + 5}{s + 1}\bigg|_{s=0} = 5 \quad \text{and} \quad b = \frac{7s + 5}{s}\bigg|_{s=-1} = 2$$

(Continued on next page)

We may now write $\mathbf{P(s)}$ as

$$\mathbf{P(s)} = \frac{5}{s} + \frac{2}{s+1}$$

the inverse transform of which is simply $p(t) = [5 + 2e^{-t}]u(t)$.

PRACTICE

14.9 Given the function $\mathbf{Q(s)} = \dfrac{11s + 30}{s^2 + 3s}$, find $q(t)$.

Ans: $q(t) = [10 + e^{-3t}]u(t)$.

Repeated Poles

A closely related situation is that of repeated poles. Consider the function

$$\mathbf{V(s)} = \frac{\mathbf{N(s)}}{(\mathbf{s} - p)^n}$$

which we want to expand into

$$\mathbf{V(s)} = \frac{a_n}{(\mathbf{s} - p)^n} + \frac{a_{n-1}}{(\mathbf{s} - p)^{n-1}} + \cdots + \frac{a_1}{(\mathbf{s} - p)}$$

To determine each constant, we first multiply the nonexpanded version of $\mathbf{V(s)}$ by $(\mathbf{s} - p)^n$. The constant a_n is found by evaluating the resulting expression at $\mathbf{s} = p$. The remaining constants are found by differentiating the expression $(\mathbf{s} - p)^n \mathbf{V(s)}$ the appropriate number of times prior to evaluating at $\mathbf{s} = p$, and dividing by a factorial term. The differentiation procedure removes the constants previously found, and evaluating at $\mathbf{s} = p$ removes the remaining constants.

For example, a_{n-2} is found by evaluating

$$\frac{1}{2!}\frac{d^2}{d\mathbf{s}^2}[(\mathbf{s} - p)^n \mathbf{V(s)}]_{\mathbf{s}=p}$$

and the term a_{n-k} is found by evaluating

$$\frac{1}{k!}\frac{d^k}{d\mathbf{s}^k}[(\mathbf{s} - p)^n \mathbf{V(s)}]_{\mathbf{s}=p}$$

To illustrate the basic procedure, let's find the inverse Laplace transform of a function having a combination of both situations: one pole at $\mathbf{s} = 0$ and two poles at $\mathbf{s} = -6$.

EXAMPLE 14.6

Compute the inverse transform of the function
$$\mathbf{V(s)} = \frac{2}{s^3 + 12s^2 + 36s}$$

We note that the denominator can be easily factored, leading to

$$\mathbf{V(s)} = \frac{2}{s(s+6)(s+6)} = \frac{2}{s(s+6)^2}$$

We assume that both $f(t)$ and its first derivative are transformable. Now, the last term of this equation is readily expressed as a limit,

$$\int_{0^-}^{\infty} \frac{df}{dt}\, dt = \lim_{t\to\infty} \int_{0^-}^{t} \frac{df}{dt}\, dt$$

$$= \lim_{t\to\infty} [f(t) - f(0^-)]$$

By recognizing that $f(0^-)$ is a constant, a comparison of the last two equations shows us that

$$\lim_{t\to\infty} f(t) = \lim_{\mathbf{s}\to 0}[\mathbf{sF}(\mathbf{s})] \qquad\qquad [31]$$

which is the *final-value theorem.* In applying this theorem, it is necessary to know that $f(\infty)$, the limit of $f(t)$ as t becomes infinite, exists or—what amounts to the same thing—that the poles of $\mathbf{F}(\mathbf{s})$ all lie *within* the left half of the $\mathbf{s}$ plane except for (possibly) a simple pole at the origin. The product $\mathbf{sF}(\mathbf{s})$ thus has all of its poles lying within the left half plane.

EXAMPLE **14.11**

Use the final-value theorem to determine $f(\infty)$ for the function $(1 - e^{-at})u(t)$, where $a > 0$.

Without even using the final-value theorem, we see immediately that $f(\infty) = 1$. The transform of $f(t)$ is

$$\mathbf{F}(\mathbf{s}) = \frac{1}{\mathbf{s}} - \frac{1}{\mathbf{s} + a}$$

$$= \frac{a}{\mathbf{s}(\mathbf{s} + a)}$$

The poles of $\mathbf{F}(\mathbf{s})$ are $\mathbf{s} = 0$ and $\mathbf{s} = -a$. Thus, the nonzero pole of $\mathbf{F}(\mathbf{s})$ lies in the left-hand $\mathbf{s}$-plane, as we were assured that $a > 0$; we find that we may indeed apply the final-value theorem to this function. Multiplying by $\mathbf{s}$ and letting $\mathbf{s}$ approach zero, we obtain

$$\lim_{\mathbf{s}\to 0}[\mathbf{sF}(\mathbf{s})] = \lim_{\mathbf{s}\to 0} \frac{a}{\mathbf{s} + a} = 1$$

which agrees with $f(\infty)$.

If $f(t)$ is a sinusoid, however, so that $\mathbf{F}(\mathbf{s})$ has poles on the $j\omega$ axis, then a blind use of the final-value theorem might lead us to conclude that the final value is zero. We know, however, that the final value of either $\sin \omega_0 t$ or $\cos \omega_0 t$ is indeterminate. So, beware of $j\omega$-axis poles!

PRACTICE

14.14 Without finding $f(t)$ first, determine $f(0^+)$ and $f(\infty)$ for each of the following transforms: (*a*) $4e^{-2\mathbf{s}}(\mathbf{s} + 50)/\mathbf{s}$; (*b*) $(\mathbf{s}^2 + 6)/(\mathbf{s}^2 + 7)$; (*c*) $(5\mathbf{s}^2 + 10)/[2\mathbf{s}(\mathbf{s}^2 + 3\mathbf{s} + 5)]$.

Ans: 0, 200; ∞, indeterminate (poles lie on the $j\omega$ axis); 2.5, 1.

SUMMARY AND REVIEW

The primary topic of this chapter was the Laplace transform, a mathematical tool for converting well-behaved time-domain functions into frequency-domain expressions. Before introducing the transform, we first considered the notion of a complex frequency, which we referred to as **s**. This convenient term has both a real (σ) and imaginary (ω) component, so can be written as $\mathbf{s} = \sigma + j\omega$. In reality, this is shorthand for an exponentially damped sinusoid, and we noted that several common functions are actually special cases of this function. Limited circuit analysis can be performed with this generalized function, but its real purpose was to simply acquaint the reader with the idea of a so-called complex frequency.

One of the most surprising things is that day-to-day circuit analysis does not require direct implementation of either the Laplace transform integral or its corresponding inverse integral! Instead, look-up tables are routinely employed, and the **s**-polynomials which result from analyzing circuits in the **s** domain are factored into smaller, easily recognizable terms. This works because each Laplace transform pair is unique. There are several theorems associated with Laplace transforms which do see daily usage, however. These include the linearity theorem, the time-differentiation theorem, and the time-integration theorem. The time-shift as well as initial-value and final-value theorems are also commonly employed.

The Laplace technique is not restricted to circuit analysis, or even electrical engineering for that matter. Any system which is described by *integrodifferential* equations can make use of the concepts studied in this chapter. At this stage, however, it is probably best to review the key concepts already discussed, highlighting appropriate examples.

❑ The concept of complex frequency allows us to consider the exponentially damped and oscillatory components of a function simultaneously. (Example 14.1)

❑ The complex frequency $\mathbf{s} = \sigma + j\omega$ is the general case; dc ($\mathbf{s} = 0$), exponential ($\omega = 0$), and sinusoidal ($\sigma = 0$) functions are special cases.

❑ Analyzing circuits in the **s** domain results in the conversion of time-domain *integrodifferential* equations into frequency-domain algebraic equations. (Example 14.1)

❑ In circuit analysis problems, we convert time-domain functions into the frequency domain using the one-sided Laplace transform: $\mathbf{F(s)} = \int_{0^-}^{\infty} e^{-\mathbf{s}t} f(t)\, dt$. (Example 14.2)

❑ The inverse Laplace transform converts frequency-domain expressions into the time domain. However, it is seldom needed due to the existence of tables listing Laplace transform pairs. (Example 14.3)

❑ The unit-impulse function is a common approximation to pulses with very narrow widths compared to circuit time constants. It is nonzero only at a single point, and has unity area.

❑ $\mathcal{L}\{f_1(t) + f_2(t)\} = \mathcal{L}\{f_1(t)\} + \mathcal{L}\{f_2(t)\}$ (*additive property*)

❑ $\mathcal{L}\{kf(t)\} = k\mathcal{L}\{f(t)\}, k = $ constant   (*homogeneity property*)

❑ Inverse transforms are typically found using a combination of partial-fraction expansion techniques and various operations (Table 14.2) to simplify **s**-domain quantities into expressions that can be found in transform tables (such as Table 14.1). (Examples 14.4, 14.5, 14.6, 14.10)

❑ The differentiation and integration theorems allow us to convert integrodifferential equations in the time domain into simple algebraic equations in the frequency domain. (Examples 14.7, 14.8, 14.9)

❑ The initial-value and final-value theorems are useful when only the specific values $f(t = 0^+)$ or $f(t \to \infty)$ are desired. (Example 14.11)

READING FURTHER

An easily readable development of the Laplace transform and some of its key properties can be found in Chap. 4 of:

A. Pinkus and S. Zafrany, *Fourier Series and Integral Transforms.* Cambridge, United Kingdom: Cambridge University Press, 1997.

A much more detailed treatment of integral transforms and their application to science and engineering problems can be found in:

B. Davies, *Integral Transforms and Their Applications,* 3rd ed. New York: Springer-Verlag, 2002.

Stability and the Routh test are discussed in Chap. 5 of:

K. Ogata, *Modern Control Engineering,* 4th ed. Englewood Cliffs, N.J.: Prentice-Hall, 2002.

EXERCISES

14.1 Complex Frequency

1. Determine the conjugate of each of the following: (a) $8 - j$; (b) $8e^{-9t}$; (c) 22.5; (d) $4e^{j9}$; (e) $j2e^{-j11}$.

2. Compute the complex conjugate of each of the following expressions: (a) -1; (b) $\dfrac{-j}{5\underline{/20°}}$; (c) $5e^{-j5} + 2e^{j3}$; (d) $(2 + j)(8\underline{/30°})e^{j2t}$.

3. Several real voltages are written down on a piece of paper, but coffee spills across half of each one. Complete the voltage expression if the legible part is (a) $5e^{-j50t}$; (b) $(2 + j)e^{j9t}$; (c) $(1 - j)e^{j78t}$; (d) $-je^{-5t}$. Assume the units of each voltage are volts (V).

4. State the complex frequency or frequencies associated with each function: (a) $f(t) = \sin 100t$; (b) $f(t) = 10$; (c) $g(t) = 5e^{-7t} \cos 80t$; (d) $f(t) = 5e^{8t}$; (e) $g(t) = (4e^{-2t} - e^{-t}) \cos(4t - 95°)$.

5. For each of the following functions, determine both the complex frequency **s** as well as **s***: (a) $7e^{-9t} \sin(100t + 9°)$; (b) $\cos 9t$; (c) $2 \sin 45t$; (d) $e^{7t} \cos 7t$.

6. Use real constants A, B, θ, ϕ, etc. to construct the general form of a real time function characterized by the following frequency components: (a) $10 - j3$ s^{-1}; (b) 0.25 s^{-1}; (c) $0, 1, -j, 1 + j$ (all s^{-1}).

7. The following voltage sources $Ae^{Bt} \cos(Ct + \theta)$ are connected (one at a time) to a 280 Ω resistor. Calculate the resulting current at $t = 0, 0.1,$ and 0.5 s, assuming the passive sign convention: (a) $A = 1$ V, $B = 0.2$ Hz, $C = 0$, $\theta = 45°$; (b) $A = 285$ mV, $B = -1$ Hz, $C = 2$ rad/s, $\theta = -45°$.

8. Your neighbor's cell phone interferes with your laptop speaker system whenever the phone is connecting to the local network. Connecting an oscilloscope to the output jack of your computer, you observe a voltage waveform that can

be described by a complex frequency $s = -1 + j200\pi$ s^{-1}. (a) What can you deduce about your neighbor's movements? (b) The imaginary part of the complex frequency starts to decrease suddenly. Alter your deduction as appropriate.

9. Compute the real part of each of the following complex functions: (a) $\mathbf{v}(t) = 9e^{-j4t}$ V; (b) $\mathbf{v}(t) = 12 - j9$ V; (c) $5\cos 100t - j43\sin 100t$ V; (d) $(2 + j)e^{j3t}$ V.

10. Your new assistant has measured the signal coming from a piece of test equipment, writing $v(t) = \mathbf{V}_x e^{(-2+j60)t}$, where $\mathbf{V}_x = 8 - j100$ V. (a) There is a missing term. What is it, and how can you tell it's missing? (b) What is the complex frequency of the signal? (c) What is the significance of the fact that Im$\{\mathbf{V}_x\}$ > Re$\{\mathbf{V}_x\}$? (d) What is the significance of the fact that $|$Re$\{\mathbf{s}\}|$ < $|$Im$\{\mathbf{s}\}|$?

14.2 The Damped Sinusoidal Forcing Function

11. State the time-domain voltage $v(t)$ which corresponds to the voltage $\mathbf{V} = 19\underline{/84°}$ V if $\mathbf{s}$ is equal to (a) 5 s^{-1}; (b) 0; (c) $-4 + j$ s^{-1}.

12. For the circuit of Fig. 14.10, the voltage source is chosen such that it can be represented by the complex frequency domain function $\mathbf{V}e^{\mathbf{s}t}$, with $\mathbf{V} = 2.5\underline{/-20°}$ V and $\mathbf{s} = -1 + j100$ s^{-1}. Calculate (a) $\mathbf{s}^*$; (b) $v(t)$, the time-domain representation of the voltage source; (c) the current $i(t)$.

13. With regard to the circuit depicted in Fig. 14.10, determine the time-domain voltage $v(t)$ which corresponds to a frequency-domain current $\mathbf{i}(t) = 5\underline{/30°}$ A for a complex frequency of (a) $\mathbf{s} = -2 + j2$ s^{-1}; (b) $\mathbf{s} = -3 + j$ s^{-1}.

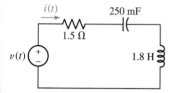

■ **FIGURE 14.10**

14. For the circuit depicted in Fig. 14.11, take $\mathbf{s} = -200 + j150$ s^{-1}. Determine the ratio of the frequency-domain voltages $\mathbf{V}_2$ and $\mathbf{V}_1$, which correspond to $v_2(t)$ and $v_1(t)$, respectively.

15. If the complex frequency describing the circuit of Fig. 14.11 is $\mathbf{s} = -150 + j100$ s^{-1}, determine the time-domain voltage which corresponds to a frequency-domain voltage $\mathbf{V}_2 = 5\underline{/-25°}$ V.

16. Calculate the time-domain voltage v in the circuit of Fig. 14.12 if the frequency-domain representation of the current source is $2.3\underline{/5°}$ A at a complex frequency of $\mathbf{s} = -1 + j2$ s^{-1}.

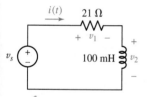

■ **FIGURE 14.11**

17. The circuit of Fig. 14.12 is operated for an extended period of time without interruption. The frequency-domain voltage which develops across the three elements can be represented as $1.8\underline{/75°}$ V at a complex frequency of $\mathbf{s} = -2 + j1.5$ s^{-1}. Determine the time-domain current i_s.

18. The circuit of Fig. 14.13 is driven by $v_S(t) = 10\cos 5t$ V. (a) Determine the complex frequency of the source. (b) Determine the frequency-domain representation of the source: (c) Compute the frequency-domain representation of i_x. (d) Obtain the time-domain expression for i_x.

19. The frequency-domain current $\mathbf{I}_x$ which flows through the 2.2 Ω resistor of Fig. 14.13 can be represented as $2\underline{/10°}$ A at a complex frequency of $\mathbf{s} = -1 + j0.5$ s^{-1}. Determine the time-domain voltage v_s.

20. Let $i_{s1} = 20e^{-3t}\cos 4t$ A and $i_{s2} = 30e^{-3t}\sin 4t$ A in the circuit of Fig. 14.14. (a) Work in the frequency domain to find $\mathbf{V}_x$. (b) Find $v_x(t)$.

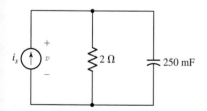

■ **FIGURE 14.12**

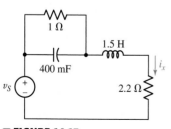

■ **FIGURE 14.13**

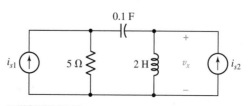

■ **FIGURE 14.14**

14.3 Definition of the Laplace Transform

21. Calculate, with the assistance of Eq. [14] (and showing intermediate steps), the Laplace transform of the following: (a) $2.1u(t)$; (b) $2u(t - 1)$; (c) $5u(t - 2) - 2u(t)$; (d) $3u(t - b)$, where $b > 0$.

22. Employ the one-sided Laplace transform integral (with intermediate steps explicitly included) to compute the **s**-domain expressions which correspond to the following: (a) $5u(t - 6)$; (b) $2e^{-t}u(t)$; (c) $2e^{-t}u(t - 1)$; (d) $e^{-2t}\sin 5t u(t)$.

23. With the assistance of Eq. [14] and showing appropriate intermediate steps, compute the one-sided Laplace transform of the following: (a) $(t - 1)u(t - 1)$; (b) $t2u(t)$; (c) $\sin 2tu(t)$; (d) $\cos 100t\, u(t)$.

24. The Laplace transform of $tf(t)$, assuming $\mathcal{L}\{f(t)\} = \mathbf{F(s)}$, is given by $-\dfrac{d}{d\mathbf{s}}\,\mathbf{F(s)}$. Test this by comparing the predicted result to what is found by directly employing Eq. [14] for (a) $tu(t)$; (b) $t^2u(t)$; (c) $t^3u(t)$; (d) $te^{-t}u(t)$.

14.4 Laplace Transforms of Simple Time Functions

25. For the following functions, specify the range of σ_0 for which the one-sided Laplace transform exists: (a) $t + 4$; (b) $(t + 1)(t - 2)$; (c) $e^{-t/2}u(t)$; (d) $\sin 10t\, u(t + 1)$.

26. Show, with the assistance of Eq. [14], that $\mathcal{L}\{f(t) + g(t) + h(t)\} = \mathcal{L}\{f(t)\} + \mathcal{L}\{g(t)\} + \mathcal{L}\{h(t)\}$.

27. Determine $\mathbf{F(s)}$ if $f(t)$ is equal to (a) $3u(t - 2)$; (b) $3e^{-2t}u(t) + 5u(t)$; (c) $\delta(t) + u(t) - tu(t)$; (d) $5\delta(t)$.

28. Obtain an expression for $\mathbf{G(s)}$ if $g(t)$ is given by (a) $[5u(t)]^2 - u(t)$; (b) $2u(t) - 2u(t - 2)$; (c) $tu(2t)$; (d) $2e^{-t}u(t) + 3u(t)$.

29. Without recourse to Eq. [15], obtain an expression for $f(t)$ if $\mathbf{F(s)}$ is given by (a) $\dfrac{1}{\mathbf{s}}$; (b) $1.55 - \dfrac{2}{\mathbf{s}}$; (c) $\dfrac{1}{\mathbf{s} + 1.5}$; (d) $\dfrac{5}{\mathbf{s}^2} + \dfrac{5}{\mathbf{s}} + 5$. (Provide some brief explanation of how you arrived at your solution.)

30. Obtain an expression for $g(t)$ without employing the inverse Laplace transform integral, if $\mathbf{G(s)}$ is known to be (a) $\dfrac{1.5}{(\mathbf{s} + 9)^2}$; (b) $\dfrac{2}{\mathbf{s}} - 0$; (c) π; (d) $\dfrac{a}{(\mathbf{s} + 1)^2} - a, a > 0$. (Provide some brief explanation of your solution process for each.)

31. Evaluate the following: (a) $\delta(t)$ at $t = 1$; (b) $5\delta(t + 1) + u(t + 1)$ at $t = 0$; (c) $\int_{-1}^{2} \delta(t)\, dt$; (d) $3 - \int_{-1}^{2} 2\delta(t)\, dt$.

32. Evaluate the following: (a) $[\delta(2t)]^2$ at $t = 1$; (b) $2\delta(t - 1) + u(-t + 1)$ at $t = 0$; (c) $\frac{1}{3}\int_{-0.001}^{0.003} \delta(t)\, dt$; (d) $\overline{\left[\frac{1}{2}\int_{0}^{2} \delta(t - 1)\, dt\right]}^{2}$.

33. Evaluate the following expressions at $t = 0$:
(a) $\int_{-\infty}^{+\infty} 2\delta(t - 1)\, dt$; (b) $\dfrac{\int_{-\infty}^{+\infty} \delta(t + 1)\, dt}{u(t + 1)}$; (c) $\dfrac{\sqrt{3\int_{-\infty}^{+\infty} \delta(t - 2)\, dt}}{[u(1 - t)]^3} - \sqrt{u(t + 2)}$;
(d) $\left[\dfrac{\int_{-\infty}^{+\infty} \delta(t - 1)\, dt}{\int_{-\infty}^{+\infty} \delta(t + 1)\, dt}\right]^{2}$.

34. Evaluate the following:
(a) $\int_{-\infty}^{+\infty} e^{-100}\delta\left(t - \dfrac{1}{5}\right) dt$; (b) $\int_{-\infty}^{+\infty} 4t\delta(t - 2)\, dt$; (c) $\int_{-\infty}^{+\infty} 4t^2\delta(t - 1.5)\, dt$;
(d) $\dfrac{\int_{-\infty}^{+\infty}(4 - t)\delta(t - 1)\, dt}{\int_{-\infty}^{+\infty}(4 - t)\delta(t + 1)\, dt}$.

14.5 Inverse Transform Techniques

35. Determine the inverse transform of $\mathbf{F(s)}$ equal to $(a)\, 5 + \dfrac{5}{s^2} - \dfrac{5}{(s+1)}$;

    $(b)\, \dfrac{1}{s} + \dfrac{5}{(0.1s+4)} - 3$; $(c)\, -\dfrac{1}{2s} + \dfrac{1}{(0.5s)^2} + \dfrac{4}{(s+5)(s+5)} + 2$;

    $(d)\, \dfrac{4}{(s+5)(s+5)} + \dfrac{2}{s+1} + \dfrac{1}{s+3}$.

36. Obtain an expression for $g(t)$ if $\mathbf{G(s)}$ is given by $(a)\, \dfrac{3(s+1)}{(s+1)^2} + \dfrac{2s}{s^2} - \dfrac{1}{(s+2)^2}$;

    $(b)\, -\dfrac{10}{(s+3)^3}$; $(c)\, 19 - \dfrac{8}{(s+3)^2} + \dfrac{18}{s^2+6s+9}$.

37. Reconstruct the time-domain function if its transform is $(a)\, \dfrac{s}{s(s+2)}$; $(b)\, 1$;

    $(c)\, 3\dfrac{s+2}{(s^2+2s+4)}$; $(d)\, 4\dfrac{s}{2s+3}$.

38. Determine the inverse transform of $\mathbf{V(s)}$ equal to $(a)\, \dfrac{s^2+2}{s} + 1$;

    $(b)\, \dfrac{s+8}{s} + \dfrac{2}{s^2}$; $(c)\, \dfrac{s+1}{s(s+2)} + \dfrac{2s^2-1}{s^2}$; $(d)\, \dfrac{s^2+4s+4}{s}$.

39. Obtain the time-domain expression which corresponds to each of the following
    s-domain functions: $(a)\, 2\dfrac{3s+\frac{1}{2}}{s^2+3s}$; $(b)\, 7 - \dfrac{s+\frac{1}{s}}{s^2+3s+1}$;

    $(c)\, \dfrac{2}{s^2} + \dfrac{1}{s} + \dfrac{s+2}{\left(\frac{s}{2}\right)^2+4s+6}$; $(d)\, \dfrac{2}{(s+1)(s+1)}$; $(e)\, \dfrac{14}{(s+1)^2(s+4)(s+5)}$.

40. Find the inverse Laplace transform of the following: $(a)\, \dfrac{1}{s^2+9s+20}$;

    $(b)\, \dfrac{4}{s^3+18s^2+17s} + \dfrac{1}{s}$; $(c)\, (0.25)\dfrac{1}{\left(\frac{s}{2}\right)^2+1.75s+2.5}$;

    $(d)\, \dfrac{3}{s(s+1)(s+4)(s+5)(s+2)}$. (e) Verify your answers with MATLAB.

41. Determine the inverse Laplace transform of each of the following s-domain
    expressions:

    $(a)\, \dfrac{1}{(s+2)^2(s+1)}$; $(b)\, \dfrac{s}{(s^2+4s+4)(s+2)}$; $(c)\, \dfrac{8}{s^3+8s^2+21s+18}$.

    (d) Verify your answers with MATLAB.

42. Given the following expressions in the s-domain, determine the corresponding
    time-domain functions: $(a)\, \dfrac{1}{3s} - \dfrac{1}{2s+1} + \dfrac{3}{s^3+8s^2+16s} - 1$;

    $(b)\, \dfrac{1}{3s+5} + \dfrac{3}{s^3/8+0.25s^2}$; $(c)\, \dfrac{2s}{(s+a)^2}$.

43. Compute $\mathscr{L}^{-1}\{\mathbf{G(s)}\}$ if $\mathbf{G(s)}$ is given by $(a)\, \dfrac{3s}{(s/2+2)^2(s+2)}$;

    $(b)\, 3 - 3\dfrac{s}{(2s^2+24s+70)(s+5)}$; $(c)\, 2 - \dfrac{1}{s+100} + \dfrac{s}{s^2+100}$; $(d)\, \mathscr{L}\{tu(2t)\}$.

44. Obtain the time-domain expression which corresponds to the following
    s-domain functions: $(a)\, \dfrac{s}{(s+2)^3}$; $(b)\, \dfrac{4}{(s+1)^4(s+1)^2}$;

    $(c)\, \dfrac{1}{s^2(s+4)^2(s+6)^3} - \dfrac{2s^2}{s} + 9$. (d) Verify your solutions with MATLAB.

14.6 Basic Theorems for the Laplace Transform

45. Take the Laplace transform of the following equations:

    $(a)\, 5\,di/dt - 7\,d^2i/dt^2 + 9i = 4$; $(b)\, m\dfrac{d^2p}{dt^2} + \mu_f\dfrac{dp}{dt} + kp(t) = 0$,

    the equation that describes the "force-free" response of a simple shock

The Convolution Integral

If the input to our system N is the forcing function $x(t)$, we know the output must be the function $y(t)$ as depicted in Fig. 15.20a. Thus, from Fig. 15.20f we conclude that

$$y(t) = \int_{-\infty}^{\infty} x(\lambda) h(t - \lambda) \, d\lambda \qquad [9]$$

where $h(t)$ is the impulse response of N. This important relationship is known far and wide as the ***convolution integral.*** In words, this last equation states that *the output is equal to the input convolved with the impulse response.* It is often abbreviated by means of

$$y(t) = x(t) * h(t)$$

where the asterisk is read "convolved with."

Equation [9] sometimes appears in a slightly different but equivalent form. If we let $z = t - \lambda$, then $d\lambda = -dz$, and the expression for $y(t)$ becomes

$$y(t) = \int_{\infty}^{-\infty} -x(t - z) h(z) \, dz = \int_{-\infty}^{\infty} x(t - z) h(z) \, dz$$

and since the symbol that we use for the variable of integration is unimportant, we can modify Eq. [9] to write

$$y(t) = x(t) * h(t) = \int_{-\infty}^{\infty} x(z) h(t - z) \, dz$$

$$= \int_{-\infty}^{\infty} x(t - z) h(z) \, dz \qquad [10]$$

> Be careful not to confuse this new notation with multiplication!

Convolution and Realizable Systems

The result that we have in Eq. [10] is very general; it applies to any linear system. However, we are usually interested in ***physically realizable systems,*** those that *do* exist or *could* exist, and such systems have a property that modifies the convolution integral slightly. That is, *the response of the system cannot begin before the forcing function is applied.* In particular, $h(t)$ is the response of the system resulting from the application of a unit impulse at $t = 0$. Therefore, $h(t)$ cannot exist for $t < 0$. It follows that, in the second integral of Eq. [10], the integrand is zero when $z < 0$; in the first integral, the integrand is zero when $(t - z)$ is negative, or when $z > t$. Therefore, for *realizable* systems the limits of integration change in the convolution integrals:

$$y(t) = x(t) * h(t) = \int_{-\infty}^{t} x(z) h(t - z) \, dz$$

$$= \int_{0}^{\infty} x(t - z) h(z) \, dz \qquad [11]$$

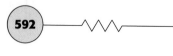

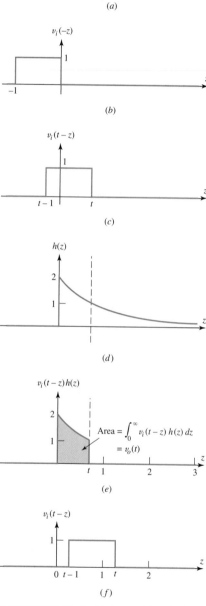

$v_i(z)$

(a)

$v_i(-z)$

(b)

$v_i(t-z)$

(c)

$h(z)$

(d)

$v_i(t-z)h(z)$

Area $= \int_0^{\infty} v_i(t-z)\, h(z)\, dz$
$= v_o(t)$

(e)

$v_i(t-z)$

(f)

■ **FIGURE 15.21** Graphical concepts in evaluating a convolution integral.

Equations [10] and [11] are both valid, but the latter is more specific when we are speaking of *realizable* linear systems, and well worth memorizing.

Graphical Method of Convolution

Before discussing the significance of the impulse response of a circuit any further, let us consider a numerical example that will give us some insight into just how the convolution integral can be evaluated. Although the expression itself is simple enough, the evaluation is sometimes troublesome, especially with regard to the values used as the limits of integration.

Suppose that the input is a rectangular voltage pulse that starts at $t = 0$, has a duration of 1 second, and is 1 V in amplitude:

$$x(t) = v_i(t) = u(t) - u(t-1)$$

Suppose also that this voltage pulse is applied to a circuit whose impulse response is known to be an exponential function of the form:

$$h(t) = 2e^{-t}u(t)$$

We wish to evaluate the output voltage $v_o(t)$, and we can write the answer immediately in integral form,

$$y(t) = v_o(t) = v_i(t) * h(t) = \int_0^{\infty} v_i(t-z)h(z)\, dz$$
$$= \int_0^{\infty} [u(t-z) - u(t-z-1)][2e^{-z}u(z)]\, dz$$

Obtaining this expression for $v_o(t)$ is simple enough, but the presence of the many unit-step functions tends to make its evaluation confusing and possibly even a little obnoxious. Careful attention must be paid to the determination of those portions of the range of integration in which the integrand is zero.

Let us use some graphical assistance to help us understand what the convolution integral says. We begin by drawing several z axes lined up one above the other, as shown in Fig. 15.21. We know what $v_i(t)$ looks like, and so we know what $v_i(z)$ looks like also; this is plotted as Fig. 15.21a. The function $v_i(-z)$ is simply $v_i(z)$ run backward with respect to z, or rotated about the ordinate axis; it is shown in Fig. 15.21b. Next we wish to represent $v_i(t-z)$, which is $v_i(-z)$ after it is shifted to the right by an amount $z = t$ as shown in Fig. 15.21c. On the next z axis, in Fig. 15.21d, our impulse response $h(z) = 2e^{-z}u(z)$ is plotted.

The next step is to multiply the two functions $v_i(t-z)$ and $h(z)$; the result for an arbitrary value of $t < 1$ is shown in Fig. 15.21e. We are after a value for the output $v_o(t)$, which is given by the *area* under the product curve (shown shaded in the figure).

First consider $t < 0$. There is no overlap between $v_i(t-z)$ and $h(z)$, so $v_o = 0$. As we increase t, we slide the pulse shown in Fig. 15.21c to the right, leading to an overlap with $h(z)$ once $t > 0$. The area under the corresponding curve of Fig. 15.21e continues to increase as we increase the value of t until we reach $t = 1$. As t increases above this value, a gap opens up between $z = 0$ and the leading edge of the pulse, as shown in Fig. 15.21f. As a result, the overlap with $h(z)$ decreases.

In other words, for values of t that lie between zero and unity, we must integrate from $z = 0$ to $z = t$; for values of t that exceed unity, the range of integration is $t - 1 < z < t$. Thus, we may write

$$
v_o(t) = \begin{cases} 0 & t < 0 \\ \displaystyle\int_0^t 2e^{-z}\,dz = 2(1 - e^{-t}) & 0 \le t \le 1 \\ \displaystyle\int_{t-1}^t 2e^{-z}\,dz = 2(e - 1)e^{-t} & t > 1 \end{cases}
$$

This function is shown plotted versus the time variable t in Fig. 15.22, and our solution is completed.

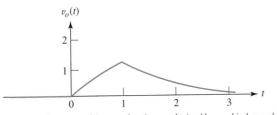

■ **FIGURE 15.22** The output function v_o obtained by graphical convolution.

EXAMPLE **15.8**

Apply a unit-step function, $x(t) = u(t)$, as the input to a system whose impulse response is $h(t) = u(t) - 2u(t - 1) + u(t - 2)$, and determine the corresponding output $y(t) = x(t) * h(t)$.

Our first step is to plot both $x(t)$ and $h(t)$, as shown in Fig. 15.23.

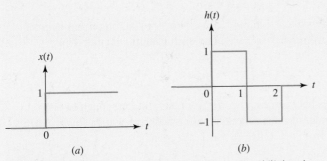

(a)   (b)

■ **FIGURE 15.23** Sketches of (a) the input signal $x(t) = u(t)$ and (b) the unit-impulse response $h(t) = u(t) - 2u(t - 1) + u(t - 2)$, for a linear system.

We arbitrarily choose to evaluate the first integral of Eq. [11],

$$
y(t) = \int_{-\infty}^t x(z)h(t - z)\,dz
$$

and prepare a sequence of sketches to help select the correct limits of integration. Figure 15.24 shows these functions in order: the input $x(z)$ as a function of z; the impulse response $h(z)$; the curve of $h(-z)$,

(Continued on next page)

which is just $h(z)$ rotated about the vertical axis; and $h(t-z)$, obtained by sliding $h(-z)$ to the right t units. For this sketch, we have selected t in the range $0 < t < 1$.

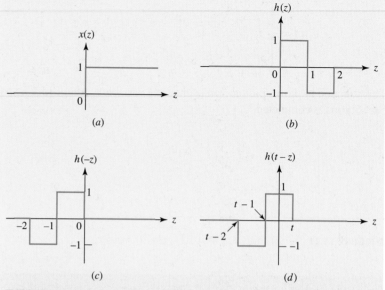

■ **FIGURE 15.24** (a) The input signal and (b) the impulse response are plotted as functions of z. (c) $h(-z)$ is obtained by flipping $h(z)$ about the vertical axis, and (d) $h(t-z)$ results when $h(-z)$ is slid t units to the right.

It is now possible to visualize the product of the first graph, $x(z)$, and the last, $h(t-z)$, for the various ranges of t. When t is less than zero, there is no overlap, and

$$y(t) = 0 \qquad t < 0$$

For the case sketched in Fig. 15.24d, $h(t-z)$ has a nonzero overlap with $x(z)$ from $z = 0$ to $z = t$, and each is unity in value. Thus,

$$y(t) = \int_0^t (1 \times 1)\, dz = t \qquad 0 < t < 1$$

When t lies between 1 and 2, $h(t-z)$ has slid far enough to the right to bring under the step function that part of the negative square wave extending from $z = 0$ to $z = t - 1$. We then have

$$y(t) = \int_0^{t-1} [1 \times (-1)]\, dz + \int_{t-1}^t (1 \times 1)\, dz = -z \Big|_{z=0}^{z=t-1} + z \Big|_{z=t-1}^{z=t}$$

Therefore,

$$y(t) = -(t-1) + t - (t-1) = 2 - t \qquad 1 < t < 2$$

Finally, when t is greater than 2, $h(t-z)$ has slid far enough to the right that it lies entirely to the right of $z = 0$. The intersection with the unit step is complete, and

$$y(t) = \int_{t-2}^{t-1} [1 \times (-1)]\, dz + \int_{t-1}^t (1 \times 1)\, dz = -z \Big|_{z=t-2}^{z=t-1} + z \Big|_{z=t-1}^{z=t}$$

or

$$y(t) = -(t-1) + (t-2) + t - (t-1) = 0 \qquad t > 2$$

These four segments of $y(t)$ are collected as a continuous curve in Fig. 15.25.

PRACTICE

15.8 Repeat Example 15.8 using the *second* integral of Eq. [11].

15.9 The impulse response of a network is given by $h(t) = 5u(t-1)$. If an input signal $x(t) = 2[u(t) - u(t-3)]$ is applied, determine the output $y(t)$ at t equal to (a) -0.5; (b) 0.5; (c) 2.5; (d) 3.5.

Ans: 15.9: 0, 0, 15, 25.

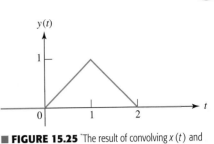

■ **FIGURE 15.25** The result of convolving $x(t)$ and $h(t)$ as shown in Fig. 15.23.

Convolution and the Laplace Transform

Convolution has applications in a wide variety of disciplines beyond linear circuit analysis, including image processing, communications, and semiconductor transport theory. It is often helpful therefore to have a graphical intuition of the basic process, even if the integral expressions of Eqs. [10] and [11] are not always the best solution route. One powerful alternative approach makes use of properties of the Laplace transform—hence our introduction to convolution in this chapter.

Let $\mathbf{F}_1(\mathbf{s})$ and $\mathbf{F}_2(\mathbf{s})$ be the Laplace transforms of $f_1(t)$ and $f_2(t)$, respectively, and consider the Laplace transform of $f_1(t) * f_2(t)$,

$$\mathcal{L}\{f_1(t) * f_2(t)\} = \mathcal{L}\left\{\int_{-\infty}^{\infty} f_1(\lambda) f_2(t-\lambda)\, d\lambda\right\}$$

One of these time functions will typically be the forcing function that is applied at the input terminals of a linear circuit, and the other will be the unit-impulse response of the circuit.

Since we are now dealing with time functions that do not exist prior to $t = 0^-$ (the definition of the Laplace transform forces us to assume this), the lower limit of integration can be changed to 0^-. Then, using the definition of the Laplace transform, we get

$$\mathcal{L}\{f_1(t) * f_2(t)\} = \int_{0^-}^{\infty} e^{-st} \left[\int_{0^-}^{\infty} f_1(\lambda) f_2(t-\lambda)\, d\lambda\right] dt$$

Since e^{-st} does not depend upon λ, we can move this factor inside the inner integral. If we do this and also reverse the order of integration, the result is

$$\mathcal{L}\{f_1(t) * f_2(t)\} = \int_{0^-}^{\infty} \left[\int_{0^-}^{\infty} e^{-st} f_1(\lambda) f_2(t-\lambda)\, dt\right] d\lambda$$

Continuing with the same type of trickery, we note that $f_1(\lambda)$ does not depend upon t, and so it can be moved outside the inner integral:

$$\mathcal{L}\{f_1(t) * f_2(t)\} = \int_{0^-}^{\infty} f_1(\lambda) \left[\int_{0^-}^{\infty} e^{-st} f_2(t-\lambda)\, dt\right] d\lambda$$

We then make the substitution $x = t - \lambda$ in the bracketed integral (where we may treat λ as a constant):

$$
\begin{aligned}
\mathcal{L}\{f_1(t) * f_2(t)\} &= \int_{0^-}^{\infty} f_1(\lambda) \left[\int_{-\lambda}^{\infty} e^{-s(x+\lambda)} f_2(x)\, dx \right] d\lambda \\
&= \int_{0^-}^{\infty} f_1(\lambda) e^{-s\lambda} \left[\int_{-\lambda}^{\infty} e^{-sx} f_2(x)\, dx \right] d\lambda \\
&= \int_{0^-}^{\infty} f_1(\lambda) e^{-s\lambda} [\mathbf{F}_2(\mathbf{s})]\, d\lambda \\
&= \mathbf{F}_2(\mathbf{s}) \int_{0^-}^{\infty} f_1(\lambda) e^{-s\lambda}\, d\lambda
\end{aligned}
$$

Since the remaining integral is simply $\mathbf{F}_1(\mathbf{s})$, we find that

$$
\boxed{\mathcal{L}\{f_1(t) * f_2(t)\} = \mathbf{F}_1(\mathbf{s}) \cdot \mathbf{F}_2(\mathbf{s})}
$$

[12]

Stated slightly differently, we may conclude that the inverse transform of the product of two transforms is the convolution of the individual inverse transforms, a result that is sometimes useful in obtaining inverse transforms.

EXAMPLE **15.9**

Find $v(t)$ by applying convolution techniques, given that $V(s) = 1/[(s + \alpha)\,(s + \beta)]$.

We obtained the inverse transform of this particular function in Sec. 14.5 using a partial-fraction expansion. We now identify $\mathbf{V}(\mathbf{s})$ as the product of two transforms,

$$
\mathbf{V}_1(\mathbf{s}) = \frac{1}{\mathbf{s} + \alpha}
$$

and

$$
\mathbf{V}_2(\mathbf{s}) = \frac{1}{\mathbf{s} + \beta}
$$

where

$$
v_1(t) = e^{-\alpha t} u(t)
$$

and

$$
v_2(t) = e^{-\beta t} u(t)
$$

The desired $v(t)$ can be expressed as

$$
\begin{aligned}
v(t) &= \mathcal{L}^{-1}\{\mathbf{V}_1(\mathbf{s})\mathbf{V}_2(\mathbf{s})\} = v_1(t) * v_2(t) = \int_{0^-}^{\infty} v_1(\lambda) v_2(t - \lambda)\, d\lambda \\
&= \int_{0^-}^{\infty} e^{-\alpha\lambda} u(\lambda) e^{-\beta(t-\lambda)} u(t - \lambda)\, d\lambda = \int_{0^-}^{t} e^{-\alpha\lambda} e^{-\beta t} e^{\beta\lambda}\, d\lambda \\
&= e^{-\beta t} \int_{0^-}^{t} e^{(\beta-\alpha)\lambda}\, d\lambda = e^{-\beta t} \frac{e^{(\beta-\alpha)t} - 1}{\beta - \alpha} u(t)
\end{aligned}
$$

or, more compactly,

$$v(t) = \frac{1}{\beta - \alpha}(e^{-\alpha t} - e^{-\beta t})u(t)$$

which is the same result that we obtained before using partial-fraction expansion. Note that it is necessary to insert the unit step $u(t)$ in the result because all (one-sided) Laplace transforms are valid only for nonnegative time.

Was the result easier to obtain by this method? Not unless one is in love with convolution integrals! The partial-fraction-expansion method is usually simpler, assuming that the expansion itself is not too cumbersome. However, the operation of convolution is easier to perform in the **s**-domain, since it only requires multiplication.

PRACTICE

15.10 Repeat Example 15.8, performing the convolution in the **s**-domain.

Further Comments on Transfer Functions

As we have noted several times before, the output $v_o(t)$ at some point in a linear circuit can be obtained by convolving the input $v_i(t)$ with the unit-impulse response $h(t)$. However, we must remember that the impulse response results from the application of a unit impulse at $t = 0$ *with all initial conditions zero*. Under these conditions, the Laplace transform of $v_o(t)$ is

$$\mathcal{L}\{v_o(t)\} = \mathbf{V}_o(\mathbf{s}) = \mathcal{L}\{v_i(t) * h(t)\} = \mathbf{V}_i(\mathbf{s})[\mathcal{L}\{h(t)\}]$$

Thus, the ratio $\mathbf{V}_o(\mathbf{s})/\mathbf{V}_i(\mathbf{s})$ is equal to the transform of the impulse response, which we shall denote by $\mathbf{H}(\mathbf{s})$,

$$\mathcal{L}\{h(t)\} = \mathbf{H}(\mathbf{s}) = \frac{\mathbf{V}_o(\mathbf{s})}{\mathbf{V}_i(\mathbf{s})} \qquad [13]$$

From Eq. [13] we see that the impulse response and the transfer function make up a Laplace transform pair,

$$h(t) \Leftrightarrow \mathbf{H}(\mathbf{s})$$

This is an important fact that we shall explore further in Sec. 15.7, after becoming familiar with the concept of pole-zero plots and the complex-frequency plane. At this point, however, we are already able to exploit this new concept of convolution for circuit analysis.

EXAMPLE 15.10

Determine the impulse response of the circuit in Fig. 15.26a, and use this to compute the forced response $v_o(t)$ if the input $v_{in}(t) = 6e^{-t}u(t)$ V.

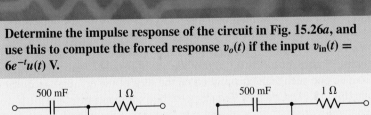

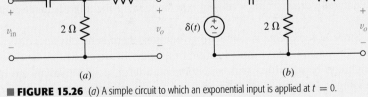

(a) (b)

■ **FIGURE 15.26** (a) A simple circuit to which an exponential input is applied at $t = 0$. (b) Circuit used to determine $h(t)$.

(Continued on next page)

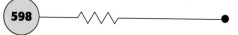

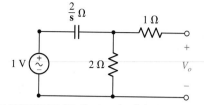

■ **FIGURE 15.27** Circuit used to find **H(s)**.

We first connect an impulse voltage pulse $\delta(t)$ V to the circuit as shown in Fig. 15.26*b*. Although we may work in either the time domain with $h(t)$ or the s-domain with **H(s)**, we choose the latter, so we next consider the s-domain representation of Fig. 15.26*b* as depicted in Fig. 15.27.

The impulse response **H(s)** is given by

$$\mathbf{H(s)} = \frac{\mathbf{V}_o}{1}$$

so our immediate goal is to find $\mathbf{V}_o$—a task easily performed by simple voltage division:

$$\mathbf{V}_o\bigg|_{v_{in}=\partial(t)} = \frac{2}{2/\mathbf{s}+2} = \frac{\mathbf{s}}{\mathbf{s}+1} = \mathbf{H(s)}$$

We may now find $v_o(t)$ when $v_{in} = 6e^{-t}u(t)$ using convolution, as

$$v_{in} = \mathcal{L}^{-1}\{\mathbf{V}_{in}(\mathbf{s}) \cdot \mathbf{H(s)}\}$$

Since $\mathbf{V}_{in}(\mathbf{s}) = 6/(\mathbf{s}+1)$,

$$\mathbf{V}_o = \frac{6\mathbf{s}}{(\mathbf{s}+1)^2} = \frac{6}{\mathbf{s}+1} - \frac{6}{(\mathbf{s}+1)^2}$$

Taking the inverse Laplace transform, we find that

$$v_o(t) = 6e^{-t}(1-t)u(t) \qquad \text{V}$$

PRACTICE

15.11 Referring to the circuit of Fig. 15.26*a*, use convolution to obtain $v_o(t)$ if $v_{in} = tu(t)$ V.

Ans: $v_o(t) = (1 - e^{-t})u(t)$ V.

15.6 • THE COMPLEX-FREQUENCY PLANE

As evident in the last several examples, circuits with even a comparatively small number of elements can lead to rather unwieldy s-domain expressions. In such instances, a graphical representation of a particular circuit response or transfer function can provide useful insights. In this section, we introduce one such technique, based on the idea of the complex-frequency plane (Fig. 15.28). Complex frequency has two components (σ and ω), so we are naturally drawn to representing our functions using a three-dimensional model.

Since ω represents an oscillating function, there is no physical distinction between a positive and negative frequency. In the case of σ, however, which can be identified with an exponential term, positive values are increasing in magnitude, whereas negative values are decaying. The origin of the s plane corresponds to dc (no time variation). A pictorial summary of these ideas is presented in Fig. 15.29.

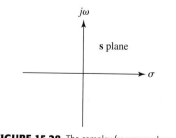

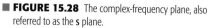

■ **FIGURE 15.28** The complex-frequency plane, also referred to as the **s** plane.

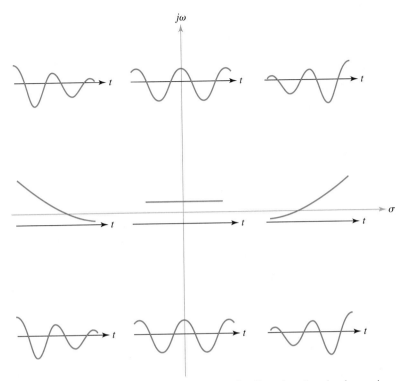

■ **FIGURE 15.29** An illustration of the physical meaning of positive and negative values for σ and ω, as would be represented on the complex-frequency plane. When $\omega = 0$, a function will have no oscillatory component; when $\sigma = 0$, the function is purely sinusoidal except when ω is also zero.

To construct an appropriate three-dimensional representation of some function $\mathbf{F(s)}$, we first note that we have its magnitude in mind, although the phase will have a strong complex-frequency dependence as well and may be graphed in a similar fashion. Thus, we will begin by substituting $\sigma + j\omega$ for $\mathbf{s}$ in our function $\mathbf{F(s)}$, then proceed to determine an expression for $|\mathbf{F(s)}|$. We next draw an axis perpendicular to the $\mathbf{s}$ plane, and use this to plot $|\mathbf{F(s)}|$ for each value of σ and ω. The basic process is illustrated in the following example.

EXAMPLE 15.11

Sketch the admittance of the series combination of a 1 H inductor and a 3 Ω resistor as a function of both $j\omega$ and σ.

The admittance of these two series elements is given by

$$\mathbf{Y(s)} = \frac{1}{\mathbf{s} + 3}$$

Substituting $\mathbf{s} = \sigma + j\omega$, we find the magnitude of the function is

$$|\mathbf{Y(s)}| = \frac{1}{\sqrt{(\sigma + 3)^2 + \omega^2}}$$

(Continued on next page)

When $\mathbf{s} = -3 + j0$, the response magnitude is infinite; when $\mathbf{s}$ is infinite, the magnitude of $\mathbf{Y(s)}$ is zero. Thus our model must have infinite height over the point $(-3 + j0)$, and it must have zero height at all points infinitely far away from the origin. A cutaway view of such a model is shown in Fig. 15.30*a*.

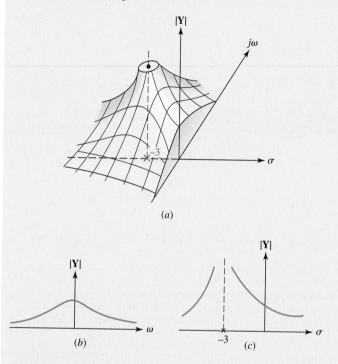

(a)

(b)   (c)

■ **FIGURE 15.30** (*a*) A cutaway view of a clay model whose top surface represents $|\mathbf{Y(s)}|$ for the series combination of a 1 H inductor and a 3 Ω resistor. (*b*) $|\mathbf{Y(s)}|$ as a function of ω. (*c*) $|\mathbf{Y(s)}|$ as a function of σ.

Once the model is constructed, it is simple to visualize the variation of $|\mathbf{Y}|$ as a function of ω (with $\sigma = 0$) by cutting the model with a perpendicular plane containing the $j\omega$ axis. The model shown in Fig. 15.30*a* happens to be cut along this plane, and the desired plot of $|\mathbf{Y}|$ versus ω can be seen; the curve is also drawn in Fig. 15.30*b*. In a similar manner, a vertical plane containing the σ axis enables us to obtain $|\mathbf{Y}|$ versus σ (with $\omega = 0$), as shown in Fig. 15.30*c*.

PRACTICE

15.12 Sketch the magnitude of the impedance $\mathbf{Z(s)} = 2 + 5\mathbf{s}$ as a function of σ and $j\omega$.

Ans: See Fig. 15.31.

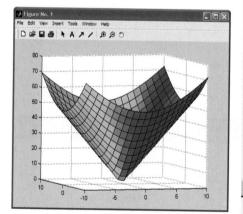

■ **FIGURE 15.31** Solution for Practice Problem 15.12, generated with the following code:

```
EDU» sigma = linspace(−10,10,21);
EDU» omega = linspace(−10,10,21);
EDU» [X, Y] = meshgrid(sigma,omega);
EDU» Z = abs(2 + 5*X + j*5*Y);
EDU» colormap(hsv);
EDU» s = [−5 3 8];
EDU» surfl(X,Y,Z,s);
EDU» view (−20,5)
```

Pole-Zero Constellations

This approach works well for relatively simple functions, but a more practical method is needed in general. Let us visualize the **s** plane once again as the floor and then imagine a large elastic sheet laid on it. We now fix our attention on all the poles and zeros of the response. At each zero, the response is zero, the height of the sheet must be zero, and we therefore tack the sheet to the floor. At the value of **s** corresponding to each pole, we may prop up the sheet with a thin vertical rod. Zeros and poles at infinity must be treated by using a large-radius clamping ring or a high circular fence, respectively. If we have used an infinitely large, weightless, perfectly elastic sheet, tacked down with vanishingly small tacks, and propped up with infinitely long, zero-diameter rods, then the elastic sheet assumes a height that is exactly proportional to the magnitude of the response.

These comments may be illustrated by considering the configuration of the poles and zeros, sometimes called a ***pole-zero constellation,*** that locates all the critical frequencies of a frequency-domain quantity, for example, an impedance **Z(s)**. A pole-zero constellation for an example impedance is shown in Fig. 15.32; in such a diagram, poles are denoted by crosses and zeros by circles. If we visualize an elastic-sheet model, tacked down at **s** $= -2 + j0$ and propped up at **s** $= -1 + j5$ and at **s** $= -1 - j5$, we should see a terrain whose distinguishing features are two mountains and one conical crater or depression. The portion of the model for the upper LHP is shown in Fig. 15.32*b*.

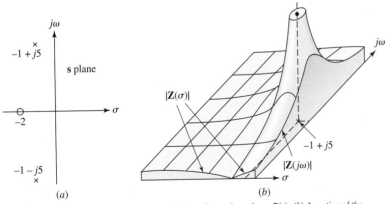

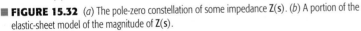

■ **FIGURE 15.32** (*a*) The pole-zero constellation of some impedance Z(s). (*b*) A portion of the elastic-sheet model of the magnitude of Z(s).

Let us now build up the expression for **Z(s)** that leads to this pole-zero configuration. The zero requires a factor of $(s + 2)$ in the numerator, and the two poles require the factors $(s + 1 - j5)$ and $(s + 1 + j5)$ in the denominator. Except for a multiplying constant k, we now know the form of **Z(s)**:

$$\mathbf{Z(s)} = k \frac{s + 2}{(s + 1 - j5)(s + 1 + j5)}$$

or

$$\mathbf{Z(s)} = k \frac{s + 2}{s^2 + 2s + 26} \qquad [14]$$

In order to determine k, we require a value for **Z(s)** at some **s** other than a critical frequency. For this function, let us suppose we are told **Z(0)** $= 1$.

By direct substitution in Eq. [14], we find that k is 13, and therefore

$$\mathbf{Z}(\mathbf{s}) = 13\frac{\mathbf{s}+2}{\mathbf{s}^2 + 2\mathbf{s} + 26} \qquad [15]$$

The plots $|\mathbf{Z}(\sigma)|$ versus σ and $|\mathbf{Z}(j\omega)|$ versus ω may be obtained exactly from Eq. [15], but the general form of the function is apparent from the pole-zero configuration and the elastic-sheet analogy. Portions of these two curves appear at the sides of the model shown in Fig. 15.32b.

PRACTICE

15.13 The parallel combination of 0.25 mH and 5 Ω is in series with the parallel combination of 40 μF and 5 Ω. (a) Find $\mathbf{Z}_{in}(\mathbf{s})$, the input impedance of the series combination. (b) Specify all the zeros of $\mathbf{Z}_{in}(\mathbf{s})$. (c) Specify all the poles of $\mathbf{Z}_{in}(\mathbf{s})$. (d) Draw the pole-zero configuration.

Ans: $5(\mathbf{s}^2 + 10{,}000\mathbf{s} + 10^8)/(\mathbf{s}^2 + 25{,}000\mathbf{s} + 10^8)$ Ω; $-5 \pm j8.66$ krad/s; $-5, -20$ krad/s.

15.7 • NATURAL RESPONSE AND THE s PLANE

At the beginning of this chapter, we explored how working in the frequency domain through the Laplace transform allows us to consider a broad range of time-varying circuits, discarding integrodifferential equations and working algebraically instead. This approach is very powerful, but it does suffer from not being a very visual process. In contrast, there is a *tremendous* amount of information contained in the pole-zero plot of a forced response. In this section, we consider how such plots can be used to obtain the *complete* response of a circuit—natural plus forced—provided the initial conditions are known. The advantage of such an approach is a more *intuitive* linkage between the location of the critical frequencies, easily visualized through the pole-zero plot, and the desired response.

Let us introduce the method by considering the simplest example, a series RL circuit as shown in Fig. 15.33. A general voltage source $v_s(t)$ causes the current $i(t)$ to flow after closure of the switch at $t = 0$. The complete response $i(t)$ for $t > 0$ is composed of a natural response and a forced response:

$$i(t) = i_n(t) + i_f(t)$$

We may find the forced response by working in the frequency domain, assuming, of course, that $v_s(t)$ has a functional form that we can transform to the frequency domain; if $v_s(t) = 1/(1+t^2)$, for example, we must proceed as best we can from the basic differential equation for the circuit. For the circuit of Fig. 15.33, we have

$$\mathbf{I}_f(\mathbf{s}) = \frac{\mathbf{V}_s}{R + \mathbf{s}L}$$

or

$$\mathbf{I}_f(\mathbf{s}) = \frac{1}{L}\frac{\mathbf{V}_s}{\mathbf{s} + R/L} \qquad [16]$$

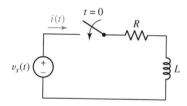

■ **FIGURE 15.33** An example that illustrates the determination of the complete response through a knowledge of the critical frequencies of the impedance faced by the source.

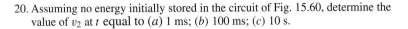

20. Assuming no energy initially stored in the circuit of Fig. 15.60, determine the value of v_2 at t equal to (*a*) 1 ms; (*b*) 100 ms; (*c*) 10 s.

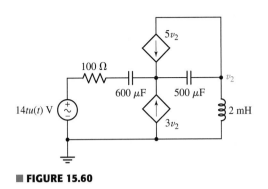

■ **FIGURE 15.60**

15.3 Additional Circuit Analysis Techniques

21. Using repeated source transformations, obtain an **s**-domain expression for the Thévenin equivalent seen by the element labeled **Z** in the circuit of Fig. 15.61.

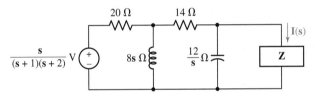

■ **FIGURE 15.61**

22. Calculate **I(s)** as labeled in the circuit of Fig. 15.61 if the element **Z** has impedance of (*a*) 2 Ω; (*b*) $\frac{1}{2s}$ Ω; (*c*) $s + \frac{1}{2s} + 3$ Ω.

23. For the circuit shown in Fig. 15.62, determine the **s**-domain Thévenin equivalent seen by the (*a*) 2 Ω resistor; (*b*) 4 Ω resistor; (*c*) 1.2 F capacitor; (*d*) current source.

24. Calculate both currents labeled in the circuit of Fig. 15.62.

25. For the circuit of Fig. 15.63, take $i_s(t) = 5u(t)$ A and determine (*a*) the Thévenin equivalent impedance seen by the 10 Ω resistor; (*b*) the inductor current $i_L(t)$.

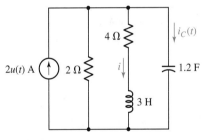

■ **FIGURE 15.62**

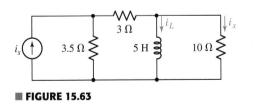

■ **FIGURE 15.63**

26. If the current source of Fig. 15.63 is $1.5e^{-2t}u(t)$ A, and $i_L(0^-) = 1$ A, determine $i_x(t)$.

27. For the **s**-domain circuit of Fig. 15.64, determine the Thévenin equivalent seen looking into the terminals marked *a* and *b*.

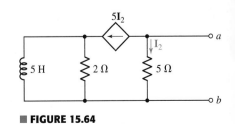

■ **FIGURE 15.64**

28. (*a*) Use superposition in the **s**-domain to find an expression for $\mathbf{V}_1(\mathbf{s})$ as labeled in Fig. 15.65. (*b*) Find $v_1(t)$.

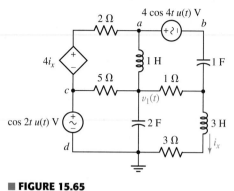

■ **FIGURE 15.65**

29. If the top right voltage source of Fig. 15.65 is open-circuited, determine the Thévenin equivalent seen looking into the terminals marked *a* and *b*.

30. If the bottom left voltage source of Fig. 15.65 is open-circuited, determine the Thévenin equivalent seen looking into the terminals marked *c* and *d*.

15.4 Poles, Zeros, and Transfer Functions

31. Determine the poles and zeros of the following **s**-domain functions:

(a) $\dfrac{s}{s+12.5}$; (b) $\dfrac{s(s+1)}{(s+5)(s+3)}$; (c) $\dfrac{s+4}{s^2+8s+7}$; (d) $\dfrac{s^2-s-2}{3s^3+24s^2+21s}$.

32. Use appropriate means to ascertain the poles and zeros of

(a) $s+4$; (b) $\dfrac{2s}{s^2-8s+16}$; (c) $\dfrac{4}{s^3+8s+7}$; (d) $\dfrac{s-5}{s^3-7s+6}$.

33. Consider the following expressions and determine the critical frequencies of each:

(a) $5+s^{-1}$; (b) $\dfrac{s(s+1)(s+4)}{(s+5)(s+3)^2}$; (c) $\dfrac{1}{s^2+4}$; (d) $\dfrac{0.5s^2-18}{s^2+1}$.

34. For the network represented schematically in Fig. 15.66, (*a*) write the transfer function $\mathbf{H}(\mathbf{s}) \equiv \mathbf{V}_{out}(\mathbf{s})/\mathbf{V}_{in}(\mathbf{s})$; (*b*) determine the poles and zeros of $\mathbf{H}(\mathbf{s})$.

35. For each of the two networks represented schematically in Fig. 15.67, (*a*) write the transfer function $\mathbf{H}(\mathbf{s}) \equiv \mathbf{V}_{out}(\mathbf{s})/\mathbf{V}_{in}(\mathbf{s})$; (*b*) determine the poles and zeros of $\mathbf{H}(\mathbf{s})$.

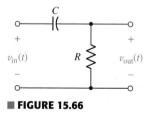

■ **FIGURE 15.66**

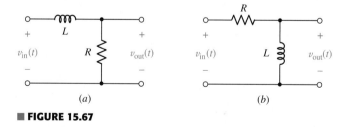

(a)   (b)

■ **FIGURE 15.67**

36. Determine the criticial frequencies of $\mathbf{Z}_{in}$ as defined in Fig. 15.50.

37. Specify the poles and zeros of $\mathbf{Y}(\mathbf{s})$ as defined by Fig. 15.51.

38. If a network is found to have the transfer function $\mathbf{H}(\mathbf{s}) = \dfrac{\mathbf{s}}{s^2+8s+7}$, determine the **s**-domain output voltage for $v_{in}(t)$ equal to (*a*) $3u(t)$ V; (*b*) $25e^{-2t}u(t)$ V; (*c*) $4u(t+1)$ V; (*d*) $2\sin 5t\,u(t)$ V.

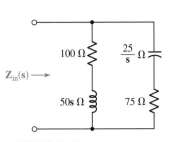

FIGURE 15.68

39. A particular network is known to be characterized by the transfer function $\mathbf{H(s)} = \mathbf{s} + 1/(\mathbf{s}^2 + 23\mathbf{s} + 60)$. Determine the critical frequencies of the output if the input is (a) $2u(t) + 4\delta(t)$; (b) $-5e^{-t}u(t)$; (c) $4te^{-2t}u(t)$; (d) $5\sqrt{2}e^{-10t}\cos 5t\ u(t)$ V.

40. For the network represented in Fig. 15.68, determine the critical frequencies of $\mathbf{Z}_{in}(\mathbf{s})$.

15.5 Convolution

41. Referring to Fig. 15.69, employ Eq. [11] to obtain $x(t) * y(t)$.

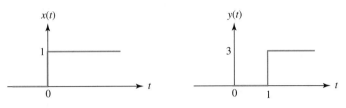

FIGURE 15.69

42. With respect to the functions $x(t)$ and $y(t)$ as plotted in Fig. 15.69, use Eq. [11] to obtain (a) $x(t) * x(t)$; (b) $y(t) * \delta(t)$.

43. Employ graphical convolution techniques to determine $f * g$ if $f(t) = 5u(t)$ and $g(t) = 2u(t) - 2u(t-2) + 2u(t-4) - 2u(t-6)$.

44. Let $h(t) = 2e^{-3t}u(t)$ and $x(t) = u(t) - \delta(t)$. Find $y(t) = h(t) * x(t)$ by (a) using convolution in the time domain; (b) finding $\mathbf{H(s)}$ and $\mathbf{X(s)}$ and then obtaining $\mathcal{L}^{-1}\{\mathbf{H(s)X(s)}\}$.

45. (a) Determine the impulse response $h(t)$ of the network shown in Fig. 15.70. (b) Use convolution to determine $v_o(t)$ if $v_{in}(t) = 8u(t)$ V.

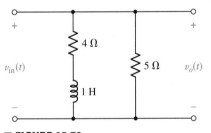

FIGURE 15.70

15.6 The Complex-Frequency Plane

 46. A 2 Ω resistor is placed in series with a 250 mF capacitor. Sketch the magnitude of the equivalent impedance as a function of (a) σ; (b) ω; (c) σ and ω, using an elastic-sheet type approach. (d) Verify your solutions using MATLAB.

47. Sketch the magnitude of $\mathbf{Z(s)} = \mathbf{s}^2 + \mathbf{s}$ as a function of (a) σ; (b) ω; (c) σ, and ω, using an elastic-sheet type approach. (d) Verify your solutions using MATLAB.

48. Sketch the pole-zero constellation of each of the following: (a) $\dfrac{\mathbf{s(s+4)}}{\mathbf{(s+5)(s+2)}}$; (b) $\dfrac{\mathbf{s-1}}{\mathbf{s}^2 + 8\mathbf{s} + 7}$; (c) $\dfrac{\mathbf{s}^2 + 1}{\mathbf{s(s}^2 + 10\mathbf{s} + 16)}$; (d) $\dfrac{5}{\mathbf{s}^2 + 2\mathbf{s} + 5}$.

49. The partially labeled pole-zero constellation of a particular transfer function $\mathbf{H(s)}$ is shown in Fig. 15.71. Obtain an expression for $\mathbf{H(s)}$ if $\mathbf{H(0)}$ is equal to (a) 1; (b) -5. (c) Is the system $\mathbf{H(s)}$ represents expected to be stable or unstable? *Explain.*

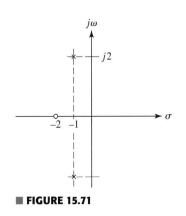

FIGURE 15.71

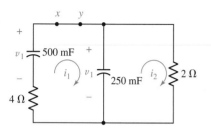

■ **FIGURE 15.72**

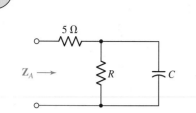

■ **FIGURE 15.73**

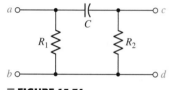

■ **FIGURE 15.74**

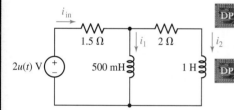

■ **FIGURE 15.75**

50. The three-element network shown in Fig. 15.72 has an input impedance $\mathbf{Z}_A(\mathbf{s})$ that has a zero at $\mathbf{s} = -10 + j0$. If a 20 Ω resistor is placed in series with the network, the zero of the new impedance shifts to $\mathbf{s} = -3.6 + j0$. Calculate R and C.

51. Let $\mathbf{H}(\mathbf{s}) = 100(\mathbf{s}+2)/(\mathbf{s}^2 + 2\mathbf{s} + 5)$ and (*a*) show the pole-zero plot for $\mathbf{H}(\mathbf{s})$; (*b*) find $\mathbf{H}(j\omega)$; (*c*) find $|\mathbf{H}(j\omega)|$; (*d*) sketch $|\mathbf{H}(j\omega)|$ versus ω; (*e*) find $\omega_{\max}$, the frequency at which $|\mathbf{H}(j\omega)|$ is a maximum.

15.7 Natural Response and the s Plane

52. Determine expressions for $i_1(t)$ and $i_2(t)$ for the circuit of Fig. 15.73, assuming $v_1(0^-) = 2$ V and $v_2(0^-) = 0$ V.

53. The 250 mF capacitor in the circuit of Fig. 15.73 is replaced with a 2 H inductor. If $v_1(t) = 0$ V and $i_1(0^-) - i_2(0^-) = 1$ A, obtain an expression for $i_2(t)$.

54. In the network of Fig. 15.74, a current source $i_x(t) = 2u(t)$ A is connected between terminals c and d such that the arrow of the source points upward. Determine the natural frequencies present in the voltage $v_{ab}(t)$ which results.

55. With regard to the circuit shown in Fig. 15.75, let $i_1(0^-) = 1$ A and $i_2(0^-) = 0$. (*a*) Determine the poles of $\mathbf{I}_{in}(\mathbf{s})/\mathbf{V}_{in}(\mathbf{s})$; (*b*) use this information to obtain expressions for $i_1(t)$ and $i_2(t)$.

15.8 A Technique for Synthesizing the Voltage Ratio H(s) = V_out/V_in

 56. Design a circuit which produces the transfer function $\mathbf{H}(\mathbf{s}) = \mathbf{V}_{out}/\mathbf{V}_{in}$ equal to (*a*) $5(\mathbf{s}+1)$; (*b*) $\dfrac{5}{(\mathbf{s}+1)}$; (*c*) $5\dfrac{\mathbf{s}+1}{\mathbf{s}+2}$.

 57. Design a circuit which produces the transfer function $\mathbf{H}(\mathbf{s}) = \mathbf{V}_{out}/\mathbf{V}_{in}$ equal to (*a*) $2(\mathbf{s}+1)^2$; (*b*) $\dfrac{3}{(\mathbf{s}+500)(\mathbf{s}+100)}$.

 58. Design a circuit which produces the transfer function
$$\mathbf{H}(\mathbf{s}) = \frac{\mathbf{V}_{out}}{\mathbf{V}_{in}} = 5\frac{\mathbf{s}+10^4}{\mathbf{s}+2\times10^5}.$$

59. Design a circuit which produces the transfer function
$$\mathbf{H}(\mathbf{s}) = \frac{\mathbf{V}_{out}}{\mathbf{V}_{in}} = 3\frac{\mathbf{s}+50}{(\mathbf{s}+75)^2}.$$

 60. Find $\mathbf{H}(\mathbf{s}) = \mathbf{V}_{out}/\mathbf{V}_{in}$ as a ratio of polynomials in $\mathbf{s}$ for the op-amp circuit of Fig. 15.42, given the impedance values (in Ω): (*a*) $\mathbf{Z}_1(\mathbf{s}) = 10^3 + (10^8/\mathbf{s})$, $\mathbf{Z}_f(\mathbf{s}) = 5000$; (*b*) $\mathbf{Z}_1(\mathbf{s}) = 5000$, $\mathbf{Z}_f(\mathbf{s}) = 10^3 + (10^8/\mathbf{s})$; (*c*) $\mathbf{Z}_1(\mathbf{s}) = 10^3 + (10^8/\mathbf{s})$, $\mathbf{Z}_f(\mathbf{s}) = 10^4 + (10^8/\mathbf{s})$.

Chapter-Integrating Exercises

61. Design a circuit that provides a frequency of 16 Hz, which is near the lower end of the human hearing range. Verify your design with an appropriate simulation.

62. Design a circuit which provides a dual-tone multifrequency (DTMF) signal corresponding to the number 9, which is a voltage output composed of a 1477 Hz signal and an 852 Hz signal.

63. (*a*) Design a circuit which provides a signal at 261.6 Hz, which is approximately middle "C". Use only standard 5% tolerance resistance values. (*b*) Estimate the likely frequency range of your signal generator based on the range of possible resistor values which can be used in construction.

64. (*a*) Many people with partial hearing loss, especially the elderly, have difficulty in detecting standard smoke detectors. An alternative is to lower the frequency to approximately 500 Hz. Design a circuit which provides such a signal, using only standard 10% tolerance resistor and capacitor values. (*b*) Estimate the actual frequency range expected from your design if it is manufactured based on the possible range of component values.

65. Design a circuit which provides either a 200 Hz signal or a 400 Hz signal by closing appropriate switches.

curve, are also identified. Let us first show that the maximum impedance magnitude is R and that this maximum occurs at resonance.

The admittance, as specified by Eq. [1], possesses a constant conductance and a susceptance which has a minimum magnitude (zero) at resonance. The minimum admittance magnitude therefore occurs at resonance, and it is $1/R$. Hence, the maximum impedance magnitude is R, *and it occurs at resonance*.

At the resonant frequency, therefore, the voltage across the parallel resonant circuit of Fig. 16.1 is simply $\mathbf{I}R$, and the *entire* source current $\mathbf{I}$ flows through the resistor. However, current is also present in L and C. For the inductor, $\mathbf{I}_{L,0} = \mathbf{V}_{L,0}/j\omega_0 L = \mathbf{I}R/j\omega_0 L$, and the capacitor current at resonance is $\mathbf{I}_{C,0} = (j\omega_0 C)\mathbf{V}_{C,0} = j\omega_0 CR\mathbf{I}$. Since $1/\omega_0 C = \omega_0 L$ at resonance, we find that

$$\mathbf{I}_{C,0} = -\mathbf{I}_{L,0} = j\omega_0 CR\mathbf{I} \qquad [5]$$

and

$$\mathbf{I}_{C,0} + \mathbf{I}_{L,0} = \mathbf{I}_{LC} = 0$$

Thus, the *net* current flowing into the LC combination is zero. The maximum value of the response magnitude and the frequency at which it occurs are not always found so easily. In less standard resonant circuits, we may find it necessary to express the magnitude of the response in analytical form, usually as the square root of the sum of the real part squared and the imaginary part squared; then we should differentiate this expression with respect to frequency, equate the derivative to zero, solve for the frequency of maximum response, and finally substitute this frequency in the magnitude expression to obtain the maximum-amplitude response. The procedure may be carried out for this simple case merely as a corroborative exercise; but, as we have seen, it is not necessary.

Quality Factor

It should be emphasized that, although the *height* of the response curve of Fig. 16.3 depends only upon the value of R for constant-amplitude excitation, the width of the curve or the steepness of the sides depends upon the other two element values also. We will shortly relate the "width of the response curve" to a more carefully defined quantity, the *bandwidth*, but it is helpful to express this relationship in terms of a very important parameter, the ***quality factor Q.***

We will find that the sharpness of the response curve of any resonant circuit is determined by the maximum amount of energy that can be stored in the circuit, compared with the energy that is lost during one complete period of the response.

We define Q as

> We should be very careful not to confuse the quality factor with charge or reactive power, all of which unfortunately are represented by the letter Q.

$$\boxed{Q = \text{quality factor} \equiv 2\pi \frac{\text{maximum energy stored}}{\text{total energy lost per period}}} \qquad [6]$$

The proportionality constant 2π is included in the definition in order to simplify the more useful expressions for Q which we will now obtain. Since energy can be stored only in the inductor and the capacitor, and can be lost only in the resistor, we may express Q in terms of the instantaneous energy associated with each of the reactive elements and the average power P_R dissipated in the resistor:

$$Q = 2\pi \frac{[w_L(t) + w_C(t)]_{\text{max}}}{P_R T}$$

where T is the period of the sinusoidal frequency at which Q is evaluated.

Now let us apply this definition to the parallel *RLC* circuit of Fig. 16.1 and determine the value of Q at the resonant frequency; this value of Q is denoted by Q_0. We select the current forcing function

$$i(t) = \mathbf{I}_m \cos \omega_0 t$$

and obtain the corresponding voltage response at resonance,

$$v(t) = Ri(t) = R\mathbf{I}_m \cos \omega_0 t$$

The energy stored in the capacitor is then

$$w_C(t) = \frac{1}{2}Cv^2 = \frac{\mathbf{I}_m^2 R^2 C}{2} \cos^2 \omega_0 t$$

and the instantaneous energy stored in the inductor is given by

$$w_L(t) = \frac{1}{2}Li_L^2 = \frac{1}{2}L\left(\frac{1}{L}\int v\, dt\right)^2 = \frac{1}{2L}\left[\frac{R\mathbf{I}_m}{\omega_0}\sin \omega_0 t\right]^2$$

so that

$$w_L(t) = \frac{\mathbf{I}_m^2 R^2 C}{2}\sin^2 \omega_0 t$$

The total *instantaneous* stored energy is therefore constant:

$$w(t) = w_L(t) + w_C(t) = \frac{\mathbf{I}_m^2 R^2 C}{2}$$

and this constant value must also be the maximum value. In order to find the energy lost in the resistor in one period, we take the average power absorbed by the resistor (see Sec. 11.2),

$$P_R = \tfrac{1}{2}\mathbf{I}_m^2 R$$

and multiply by one period, obtaining

$$P_R T = \frac{1}{2f_0}\mathbf{I}_m^2 R$$

We thus find the quality factor at resonance:

$$Q_0 = 2\pi \frac{\mathbf{I}_m^2 R^2 C/2}{\mathbf{I}_m^2 R/2f_0}$$

or

$$Q_0 = 2\pi f_0 RC = \omega_0 RC \qquad\qquad [7]$$

This equation (as well as any expression in Eq. [8]) holds only for the simple parallel *RLC* circuit of Fig. 16.1. Equivalent expressions for Q_0 which are often quite useful may be obtained by simple substitution:

$$Q_0 = R\sqrt{\frac{C}{L}} = \frac{R}{|X_{C,0}|} = \frac{R}{|X_{L,0}|} \qquad [8]$$

So we see that for this specific circuit, decreasing the resistance decreases Q_0; the lower the resistance, the greater the amount of energy lost in the element. Intriguingly, increasing the capacitance *increases* Q_0, but increasing the inductance leads to a *reduction* in Q_0. These statements, of course, apply to operation of the circuit at the resonant frequency.

Other Interpretations of Q

Another useful interpretation of Q is obtained when we inspect the inductor and capacitor currents at resonance, as given by Eq. [5],

$$\mathbf{I}_{C,0} = -\mathbf{I}_{L,0} = j\omega_0 C R \mathbf{I} = jQ_0\mathbf{I} \qquad [9]$$

Note that each is Q_0 times the source current in amplitude and that each is 180° out of phase with the other. Thus, if we apply 2 mA at the resonant frequency to a parallel resonant circuit with a Q_0 of 50, we find 2 mA in the resistor, and 100 mA in both the inductor and the capacitor. A parallel resonant circuit can therefore act as a current amplifier, but not, of course, as a power amplifier, since it is a passive network.

Resonance, by definition, is fundamentally associated with the forced response, since it is defined in terms of a (purely resistive) input impedance, a sinusoidal steady-state concept. The two most important parameters of a resonant circuit are perhaps the resonant frequency ω_0 and the quality factor Q_0. Both the exponential damping coefficient and the natural resonant frequency may be expressed in terms of ω_0 and Q_0:

$$\alpha = \frac{1}{2RC} = \frac{1}{2(Q_0/\omega_0 C)C}$$

or

$$\alpha = \frac{\omega_0}{2Q_0} \qquad [10]$$

and

$$\omega_d = \sqrt{\omega_0^2 - \alpha^2}$$

or

$$\omega_d = \omega_0\sqrt{1 - \left(\frac{1}{2Q_0}\right)^2} \qquad [11]$$

Damping Factor

For future reference it may be helpful to note one additional relationship involving ω_0 and Q_0. The quadratic factor appearing in the numerator of Eq. [4],

$$s^2 + \frac{1}{RC}s + \frac{1}{LC}$$

may be written in terms of α and ω_0:

$$\mathbf{s}^2 + 2\alpha\mathbf{s} + \omega_0^2$$

In the field of system theory or automatic control theory, it is traditional to write this factor in a slightly different form that utilizes the dimensionless parameter ζ (zeta), called the ***damping factor:***

$$\mathbf{s}^2 + 2\zeta\omega_0\mathbf{s} + \omega_0^2$$

Comparison of these expressions allows us to relate ζ to other parameters:

$$\zeta = \frac{\alpha}{\omega_0} = \frac{1}{2Q_0} \qquad\qquad [12]$$

EXAMPLE 16.1

Consider a parallel *RLC* circuit such that $L = 2$ mH, $Q_0 = 5$, and $C = 10$ nF. Determine the value of R and the magnitude of the steady-state admittance at $0.1\omega_0$, ω_0, and $1.1\omega_0$.

We derived several expressions for Q_0, a parameter directly related to energy loss, and hence the resistance in our circuit. Rearranging the expression in Eq. [8], we calculate

$$R = Q_0\sqrt{\frac{L}{C}} = 2.236 \text{ k}\Omega$$

Next, we compute ω_0, a term we may recall from Chap. 9,

$$\omega_0 = \frac{1}{\sqrt{LC}} = 223.6 \text{ krad/s}$$

or, alternatively, we may exploit Eq. [7] and obtain the same answer,

$$\omega_0 = \frac{Q_0}{RC} = 223.6 \text{ krad/s}$$

The admittance of any parallel *RLC* network is simply

$$\mathbf{Y} = \frac{1}{R} + j\omega C + \frac{1}{j\omega L}$$

and hence

$$|\mathbf{Y}| = \frac{1}{R} + j\omega C + \frac{1}{j\omega L}$$

evaluated at the three designated frequencies is equal to

$$|\mathbf{Y}(0.9\omega_0)| = 6.504 \times 10^{-4} \text{ S} \qquad |\mathbf{Y}(\omega_0)| = 4.472 \times 10^{-4} \text{ S}$$

$$|\mathbf{Y}(1.1\omega_0)| = 6.182 \times 10^{-4} \text{ S}$$

We thus obtain a minimum impedance at the resonant frequency, or a *maximum voltage response* to a particular input current. If we quickly compute the reactance at these three frequencies, we find

$$X(0.9\omega_0) = -4.72 \times 10^{-4} \text{ S} \qquad X(1.1\omega_0) = 4.72 \times 10^{-4} \text{ S}$$

$$X(\omega_0) = -1.36 \times 10^{-7}$$

We leave it to the reader to show that our value for $X(\omega_0)$ is nonzero only as a result of rounding error.

line for $\omega > a$, and a broken line for $\omega < a$. Note that the two asymptotes intersect at $\omega = a$, the frequency of the zero. This frequency is also described as the **corner, break, 3 dB,** or **half-power frequency.**

Smoothing Bode Plots

Let us see how much error is embodied in our asymptotic response curve. At the corner frequency ($\omega = a$),

$$H_{\text{dB}} = 20 \log \sqrt{1 + \frac{a^2}{a^2}} = 3 \text{ dB}$$

Note that we continue to abide by the convention of taking $\sqrt{2}$ as corresponding to 3 dB.

as compared with an asymptotic value of 0 dB. At $\omega = 0.5a$, we have

$$H_{\text{dB}} = 20 \log \sqrt{1.25} \approx 1 \text{ dB}$$

Thus, the exact response is represented by a smooth curve that lies 3 dB above the asymptotic response at $\omega = a$, and 1 dB above it at $\omega = 0.5a$ (and also at $\omega = 2a$). This information can always be used to smooth out the corner if a more exact result is desired.

Multiple Terms

Most transfer functions will consist of more than a simple zero (or simple pole). This, however, is easily handled by the Bode method, since we are in fact working with logarithms. For example, consider a function

$$\mathbf{H}(\mathbf{s}) = K \left(1 + \frac{\mathbf{s}}{s_1}\right) \left(1 + \frac{\mathbf{s}}{s_2}\right)$$

where $K = $ constant, and $-s_1$ and $-s_2$ represent the two zeros of our function $\mathbf{H}(\mathbf{s})$. For this function H_{dB} may be written as

$$H_{\text{dB}} = 20 \log \left| K \left(1 + \frac{j\omega}{s_1}\right) \left(1 + \frac{j\omega}{s_2}\right) \right|$$

$$= 20 \log \left[K \sqrt{1 + \left(\frac{\omega}{s_1}\right)^2} \sqrt{1 + \left(\frac{\omega}{s_2}\right)^2} \right]$$

or

$$H_{\text{dB}} = 20 \log K + 20 \log \sqrt{1 + \left(\frac{\omega}{s_1}\right)^2} + 20 \log \sqrt{1 + \left(\frac{\omega}{s_2}\right)^2}$$

which is simply the sum of a constant (frequency-independent) term $20 \log K$ and two simple zero terms of the form previously considered. In other words, *we may construct a sketch of H_{dB} by simply graphically adding the plots of the separate terms.* We explore this in the following example.

EXAMPLE 16.7

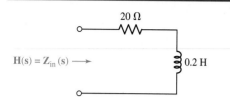

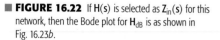

H(s) = Z_{in} (s) →

20 Ω

0.2 H

■ **FIGURE 16.22** If **H(s)** is selected as Z_{in}(s) for this network, then the Bode plot for H_{dB} is as shown in Fig. 16.23*b*.

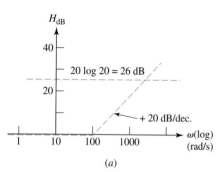

(*a*)

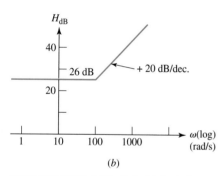

(*b*)

■ **FIGURE 16.23** (*a*) The Bode plots for the factors of **H(s)** = 20(1 + s/100) are sketched individually. (*b*) The composite Bode plot is shown as the sum of the plots of part (*a*).

Obtain the Bode plot of the input impedance of the network shown in Fig. 16.22.

We have the input impedance,

$$\mathbf{Z}_{in}(\mathbf{s}) = \mathbf{H}(\mathbf{s}) = 20 + 0.2\mathbf{s}$$

Putting this in standard form, we obtain

$$\mathbf{H}(\mathbf{s}) = 20\left(1 + \frac{\mathbf{s}}{100}\right)$$

The two factors constituting **H(s)** are a zero at **s** = −100, leading to a break frequency of ω = 100 rad/s, and a constant equivalent to $20\log 20 = 26$ dB. Each of these is sketched lightly in Fig. 16.23*a*. Since we are working with the logarithm of $|\mathbf{H}(j\omega)|$, we next add together the Bode plots corresponding to the individual factors. The resultant magnitude plot appears as Fig. 16.23*b*. No attempt has been made to smooth out the corner with a +3 dB correction at ω = 100 rad/s; this is left to the reader as a quick exercise.

PRACTICE

16.11 Construct a Bode magnitude plot for **H(s)** = 50 + s.

Ans: 34 dB, ω < 50 rad/s; slope = +20 dB/decade ω > 50 rad/s.

Phase Response

Returning to the transfer function of Eq. [26], we would now like to determine the ***phase response*** for the simple zero,

$$\text{ang } \mathbf{H}(j\omega) = \text{ang}\left(1 + \frac{j\omega}{a}\right) = \tan^{-1}\frac{\omega}{a}$$

This expression is also represented by its asymptotes, although three straight-line segments are required. For $\omega \ll a$, ang $\mathbf{H}(j\omega) \approx 0°$, and we use this as our asymptote when $\omega < 0.1a$:

$$\text{ang } \mathbf{H}(j\omega) = 0° \qquad (\omega < 0.1a)$$

At the high end, $\omega \gg a$, we have ang $\mathbf{H}(j\omega) \approx 90°$, and we use this above $\omega = 10a$:

$$\text{ang } \mathbf{H}(j\omega) = 90° \qquad (\omega > 10a)$$

Since the angle is 45° at $\omega = a$, we now construct the straight-line asymptote extending from 0° at $\omega = 0.1a$, through 45° at $\omega = a$, to 90° at $\omega = 10a$. This straight line has a slope of 45°/decade. It is shown as a solid curve in Fig. 16.24, while the exact angle response is shown as a broken line. The maximum differences between the asymptotic and true responses are ±5.71° at $\omega = 0.1a$ and 10a. Errors of ∓5.29° occur at $\omega = 0.394a$ and

$2.54a$; the error is zero at $\omega = 0.159a$, a, and $6.31a$. The phase angle plot is typically left as a straight-line approximation, although smooth curves can also be drawn in a manner similar to that depicted in Fig. 16.24 by the dashed line.

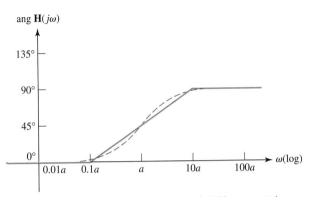

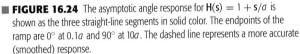

■ **FIGURE 16.24** The asymptotic angle response for $H(s) = 1 + s/a$ is shown as the three straight-line segments in solid color. The endpoints of the ramp are $0°$ at $0.1a$ and $90°$ at $10a$. The dashed line represents a more accurate (smoothed) response.

It is worth pausing briefly here to consider what the phase plot is telling us. In the case of a simple zero at $s = a$, we see that for frequencies much less than the corner frequency, the phase of the response function is $0°$. For high frequencies, however ($\omega \gg a$), the phase is $90°$. In the vicinity of the corner frequency, the phase of the transfer function varies somewhat rapidly. The actual phase angle imparted to the response can therefore be selected through the design of the circuit (which determines a).

PRACTICE

16.12 Draw the Bode phase plot for the transfer function of Example 16.7.

Ans: $0°$, $\omega \le 10$; $90°$, $\omega \ge 1000$; $45°$, $\omega = 100$; $45°$/dec slope, $10 < \omega < 1000$. (ω in rad/s).

Additional Considerations in Creating Bode Plots

We next consider a simple pole,

$$H(s) = \frac{1}{1 + s/a} \qquad [27]$$

Since this is the reciprocal of a zero, the logarithmic operation leads to a Bode plot which is the *negative* of that obtained previously. The amplitude is 0 dB up to $\omega = a$, and then the slope is -20 dB/decade for $\omega > a$. The angle plot is $0°$ for $\omega < 0.1a$, $-90°$ for $\omega > 10a$, and $-45°$ at $\omega = a$, and it has a slope of $-45°$/decade when $0.1a < \omega < 10a$. The reader is encouraged to generate the Bode plot for this function by working directly with Eq. [27].

Another term that can appear in $\mathbf{H}(\mathbf{s})$ is a factor of $\mathbf{s}$ in the numerator or denominator. If $\mathbf{H}(\mathbf{s}) = \mathbf{s}$, then

$$H_{dB} = 20 \log |\omega|$$

Thus, we have an infinite straight line passing through 0 dB at $\omega = 1$ and having a slope everywhere of 20 dB/decade. This is shown in Fig. 16.25a. If the $\mathbf{s}$ factor occurs in the denominator, a straight line is obtained having a slope of -20 dB/decade and passing through 0 dB at $\omega = 1$, as shown in Fig. 16.25b.

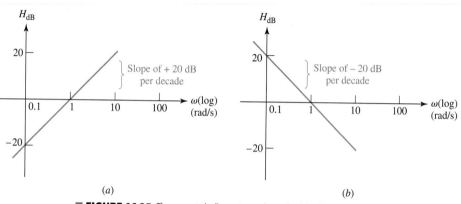

(a)   (b)

■ **FIGURE 16.25** The asymptotic diagrams are shown for (a) $\mathbf{H}(\mathbf{s}) = \mathbf{s}$ and (b) $\mathbf{H}(\mathbf{s}) = 1/\mathbf{s}$. Both are infinitely long straight lines passing through 0 dB at $\omega = 1$ and having slopes of ± 20 dB/decade.

Another simple term found in $\mathbf{H}(\mathbf{s})$ is the multiplying constant K. This yields a Bode plot which is a horizontal straight line lying $20 \log |K|$ dB above the abscissa. It will actually be below the abscissa if $|K| < 1$.

EXAMPLE 16.8

Obtain the Bode plot for the gain of the circuit shown in Fig. 16.26.

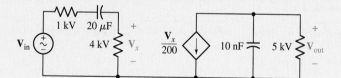

■ **FIGURE 16.26** If $\mathbf{H}(\mathbf{s}) = \mathbf{V}_{out}/\mathbf{V}_{in}$, this amplifier is found to have the Bode amplitude plot shown in Fig. 16.27b, and the phase plot shown in Fig. 16.28.

We work from left to right through the circuit and write the expression for the voltage gain,

$$\mathbf{H}(\mathbf{s}) = \frac{\mathbf{V}_{out}}{\mathbf{V}_{in}} = \frac{4000}{5000 + 10^6/20\mathbf{s}} \left(-\frac{1}{200} \right) \frac{5000(10^8/\mathbf{s})}{5000 + 10^8/\mathbf{s}}$$

which simplifies (mercifully) to

$$\mathbf{H}(\mathbf{s}) = \frac{-2\mathbf{s}}{(1 + \mathbf{s}/10)(1 + \mathbf{s}/20{,}000)} \qquad [28]$$

We see a constant, $20 \log |-2| = 6$ dB, break points at $\omega = 10$ rad/s and $\omega = 20{,}000$ rad/s, and a linear factor $\mathbf{s}$. Each of these is sketched in Fig. 16.27a, and the four sketches are added to give the Bode magnitude plot in Fig. 16.27b.

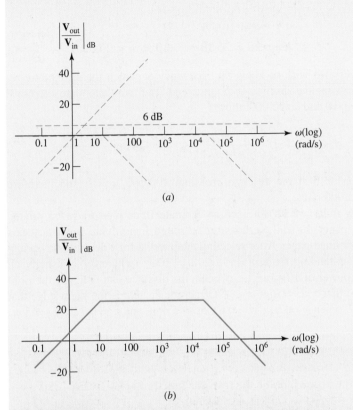

(a)

(b)

■ **FIGURE 16.27** (a) Individual Bode magnitude sketches are made for the factors (-2), $(\mathbf{s})$, $(1 + \mathbf{s}/10)^{-1}$, and $(1 + \mathbf{s}/20{,}000)^{-1}$. (b) The four separate plots of part (a) are added to give the Bode magnitude plots for the amplifier of Fig. 16.26.

PRACTICE

16.13 Construct a Bode magnitude plot for $\mathbf{H(s)}$ equal to
(a) $50/(\mathbf{s} + 100)$; (b) $(\mathbf{s} + 10)/(\mathbf{s} + 100)$; (c) $(\mathbf{s} + 10)/\mathbf{s}$.

Ans: (a) -6 dB, $\omega < 100$; -20 dB/decade, $\omega > 100$; (b) -20 dB, $\omega < 10$; $+20$ dB/decade, $10 < \omega < 100$; 0 dB, $\omega > 100$; (c) 0 dB, $\omega > 10$; -20 dB/decade, $\omega < 10$.

Before we construct the phase plot for the amplifier of Fig. 16.26, let us take a few moments to investigate several of the details of the magnitude plot.

First, it is wise not to rely too heavily on graphical addition of the individual magnitude plots. Instead, the exact value of the combined magnitude plot may be found easily at selected points by considering the asymptotic value of each factor of $\mathbf{H(s)}$ at the point in question. For example, in the flat region of Fig. 16.27a between $\omega = 10$ and $\omega = 20{,}000$, we are below the corner at $\omega = 20{,}000$, and so we represent $(1 + s/20{,}000)$ by 1; but we are above $\omega = 10$, so $(1 + s/10)$ is represented as $\omega/10$. Hence,

$$H_{\text{dB}} = 20 \log \left| \frac{-2\omega}{(\omega/10)(1)} \right|$$

$$= 20 \log 20 = 26 \text{ dB} \qquad (10 < \omega < 20{,}000)$$

We might also wish to know the frequency at which the asymptotic response crosses the abscissa at the high end. The two factors are expressed here as $\omega/10$ and $\omega/20{,}000$; thus

$$H_{\text{dB}} = 20 \log \left| \frac{-2\omega}{(\omega/10)(\omega/20{,}000)} \right| = 20 \log \left| \frac{400{,}000}{\omega} \right|$$

Since $H_{\text{dB}} = 0$ at the abscissa crossing, $400{,}000/\omega = 1$, and therefore $\omega = 400{,}000$ rad/s.

Many times we do not need an accurate Bode plot drawn on printed semilog paper. Instead we construct a rough logarithmic frequency axis on simple lined paper. After selecting the interval for a decade—say, a distance L extending from $\omega = \omega_1$ to $\omega = 10\omega_1$ (where ω_1 is usually an integral power of 10)—we let x locate the distance that ω lies to the right of ω_1, so that $x/L = \log(\omega/\omega_1)$. Of particular help is the knowledge that $x = 0.3L$ when $\omega = 2\omega_1$, $x = 0.6L$ at $\omega = 4\omega_1$, and $x = 0.7L$ at $\omega = 5\omega_1$.

EXAMPLE 16.9

Draw the phase plot for the transfer function given by Eq. [28], $\mathbf{H(s)} = -2s/[(1 + s/10)(1 + s/20{,}000)]$.

We begin by inspecting $\mathbf{H}(j\omega)$:

$$\mathbf{H}(j\omega) = \frac{-j2\omega}{(1 + j\omega/10)(1 + j\omega/20{,}000)} \qquad [29]$$

The angle of the numerator is a constant, $-90°$.

The remaining factors are represented as the sum of the angles contributed by break points at $\omega = 10$ and $\omega = 20{,}000$. These three terms appear as broken-line asymptotic curves in Fig. 16.28, and their sum is shown as the solid curve. An equivalent representation is obtained if the curve is shifted upward by $360°$.

Exact values can also be obtained for the asymptotic phase response. For example, at $\omega = 10^4$ rad/s, the angle in Fig. 16.28 is obtained from the numerator and denominator terms in Eq. [29]. The numerator angle is $-90°$. The angle for the pole at $\omega = 10$ is $-90°$, since ω is greater than 10 times the corner frequency. Between 0.1 and 10 times the corner frequency, we recall that the slope is $-45°$ per decade for a simple pole. For the break point at 20,000 rad/s, we therefore calculate the angle, $-45° \log(\omega/0.1a) = -45° \log[10{,}000/(0.1 \times 20{,}000)] = -31.5°$.

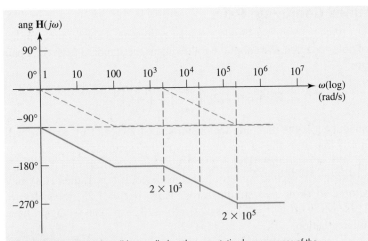

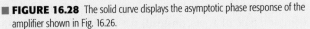

■ FIGURE 16.28 The solid curve displays the asymptotic phase response of the amplifier shown in Fig. 16.26.

The algebraic sum of these three contributions is $-90° - 90°$ $-31.5° = -211.5°$, a value that appears to be moderately near the asymptotic phase curve of Fig. 16.28.

PRACTICE

16.14 Draw the Bode phase plot for $\mathbf{H(s)}$ equal to (*a*) $50/(\mathbf{s} + 100)$; (*b*) $(\mathbf{s} + 10)/(\mathbf{s} + 100)$; (*c*) $(\mathbf{s} + 10)/\mathbf{s}$.

Ans: (*a*) $0°$, $\omega < 10$; $-45°$/decade, $10 < \omega < 1000$; $-90°$, $\omega > 1000$; (*b*) $0°$, $\omega < 1$; $+45°$/decade, $1 < \omega < 10$; $45°$, $10 < \omega < 100$; $-45°$/decade, $100 < \omega < 1000$; $0°$, $\omega > 1000$; (*c*) $-90°$, $\omega < 1$; $+45°$/decade, $1 < \omega < 100$; $0°$, $\omega > 100$.

Higher-Order Terms

The zeros and poles that we have been considering are all first-order terms, such as $\mathbf{s}^{\pm 1}$, $(1 + 0.2\mathbf{s})^{\pm 1}$, and so forth. We may extend our analysis to higher-order poles and zeros very easily, however. A term $\mathbf{s}^{\pm n}$ yields a magnitude response that passes through $\omega = 1$ with a slope of $\pm 20n$ dB/decade; the phase response is a constant angle of $\pm 90n°$. Also, a multiple zero, $(1 + \mathbf{s}/a)^n$, must represent the sum of n of the magnitude-response curves, or n of the phase-response curves of the simple zero. We therefore obtain an asymptotic magnitude plot that is 0 dB for $\omega < a$ and has a slope of $20n$ dB/decade when $\omega > a$; the error is $-3n$ dB at $\omega = a$, and $-n$ dB at $\omega = 0.5a$ and $2a$. The phase plot is $0°$ for $\omega < 0.1a$, $90n°$ for $\omega > 10a$, $45n°$ at $\omega = a$, and a straight line with a slope of $45n°/$ decade for $0.1a < \omega < 10a$, and it has errors as large as $\pm 5.71n°$ at two frequencies.

The asymptotic magnitude and phase curves associated with a factor such as $(1 + \mathbf{s}/20)^{-3}$ may be drawn quickly, but the relatively large errors associated with the higher powers should be kept in mind.

Complex Conjugate Pairs

The last type of factor we should consider represents a conjugate complex pair of poles or zeros. We adopt the following as the standard form for a pair of zeros:

$$\mathbf{H}(\mathbf{s}) = 1 + 2\zeta\left(\frac{\mathbf{s}}{\omega_0}\right) + \left(\frac{\mathbf{s}}{\omega_0}\right)^2$$

The quantity ζ is the damping factor introduced in Sec. 16.1, and we will see shortly that ω_0 is the corner frequency of the asymptotic response.

If $\zeta = 1$, we see that $\mathbf{H}(\mathbf{s}) = 1 + 2(\mathbf{s}/\omega_0) + (\mathbf{s}/\omega_0)^2 = (1 + \mathbf{s}/\omega_0)^2$, a second-order zero such as we have just considered. If $\zeta > 1$, then $\mathbf{H}(\mathbf{s})$ may be factored to show two simple zeros. Thus, if $\zeta = 1.25$, then $\mathbf{H}(\mathbf{s}) = 1 + 2.5(\mathbf{s}/\omega_0) + (\mathbf{s}/\omega_0)^2 = (1 + \mathbf{s}/2\omega_0)(1 + \mathbf{s}/0.5\omega_0)$, and we again have a familiar situation.

A new case arises when $0 \leq \zeta \leq 1$. There is no need to find values for the conjugate complex pair of roots. Instead, we determine the low- and high-frequency asymptotic values for both the magnitude and phase response, and then apply a correction that depends on the value of ζ.

For the magnitude response, we have

$$H_{\text{dB}} = 20\log|\mathbf{H}(j\omega)| = 20\log\left|1 + j2\zeta\left(\frac{\omega}{\omega_0}\right) - \left(\frac{\omega}{\omega_0}\right)^2\right| \qquad [30]$$

When $\omega \ll \omega_0$, $H_{\text{dB}} = 20\log|1| = 0$ dB. This is the low-frequency asymptote. Next, if $\omega \gg \omega_0$, only the squared term is important, and $H_{\text{dB}} = 20\log|-(\omega/\omega_0)^2| = 40\log(\omega/\omega_0)$. We have a slope of $+40$ dB/decade. This is the high-frequency asymptote, and the two asymptotes intersect at 0 dB, $\omega = \omega_0$. The solid curve in Fig. 16.29 shows this asymptotic representation of the magnitude response. However, a correction must be applied

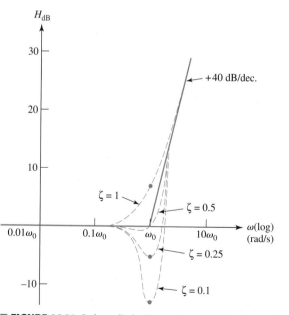

■ **FIGURE 16.29** Bode amplitude plots are shown for $\mathbf{H}(\mathbf{s}) = 1 + 2\zeta(\mathbf{s}/\omega_0) + (\mathbf{s}/\omega_0)^2$ for several values of the damping factor ζ.

Comparing Eqs. [41], [42], and [43], we determine that our design must ensure that

$$1 = \frac{1}{R^2 C^2}$$

and

$$1 = \frac{3 - G}{RC}$$

Consequently, $RC = 1$ and $G = 4$. If we choose $R_A = 3$ kΩ, it follows that $R_B = 1$ kΩ. We may scale these values later if we choose when adjusting for operation at 2000 rad/s, but this is unnecessary as the dc gain is set by the ratio of the two resistors.

Initially we design for $R = 1$ Ω and $C = 1$ F as this automatically satifies the $RC = 1$ requirement. Neither value being easy to locate, however, we select a more reasonable capacitor value of 0.1 μF, which combined with our frequency scaling factor $k_f = 2000$, results in a resistor scaling factor $k_m = 5000$. Thus, $R = 5$ kΩ in our final design.

All that remains is to select values for R_1, R_f, and C_f in our front-end stage. Recall that the transfer function of this stage is

$$-\frac{1/R_1 C_f}{\mathbf{s} + (1/R_f C_f)}$$

Setting $R_f = 1$ Ω and $C_f = 1$ F initially allows the pole to be properly located prior to scaling operations, which dictate that we build the circuit with $R_f = 5$ kΩ and $C_f = 0.1$ μF. Our only remaining choice, then, is to ensure that R_1 allows us to meet our gain requirement of 4. Since we have already achieved this with our Sallen-Key stage, R_1 must be equal to R_f, or 5 kΩ.

Design of Chebyshev filters proceeds along the same lines as that of Butterworth filters, except we have more choices now with the ripple factor. Also, for filters not having a 3 dB ripple factor, the critical frequency is where the ripple channel in the passband terminates, which is slightly different than what we have specified previously. Filters with order $n > 2$ are constructed by cascading stages, either multiple Sallen-Key stages for even orders, or a simple stage such as Fig. 15.49a in conjunction with the appropriate number of Sallen-Key stages for odd orders. For filters with a specific gain requirement, an op amp stage containing only resistors is typically required at the output.

SUMMARY AND REVIEW

We began this chapter with a short discussion of *resonance*. Of course the reader was likely to already have an intuitive understanding of the basic concept—timing when to kick our legs on a swing as a child; watching videos of crystal glasses shattering under the power of a trained soprano's voice; instinctively slowing down when driving over a corrugated surface. In the context of linear circuit analysis, we found (perhaps surprisingly) that a frequency can be chosen even for networks with capacitors and inductors such that the voltage and current are in phase (hence the network appears purely resistive at that particular frequency). How quickly our circuit response changes as we move "off resonance" was related to a new term—the *quality factor* (*Q*) of our circuit. After defining what is meant by *critical frequencies* for our circuit response, we introduced the concept of *bandwidth,* and discovered that our expressions may be simplified rather dramatically for high-*Q* (*Q* > 5) circuits. We briefly extended this discussion to consider the differences between series and parallel circuits near resonance, along with more practical networks which cannot be classed as either.

The remainder of this chapter dealt with the analysis and design of filter circuits. Prior to launching into that discussion, the topic of "scaled" circuit components dealt with both *frequency* and *magnitude scaling* as a convenient design tool. We also introduced the handy method of *Bode plots,* which allows us to quickly sketch a reasonable approximation to the response of a filter circuit as a function of frequency. We next considered both *passive* and *active filters,* starting with simple designs using a single capacitor to achieve either low-pass or high-pass behavior. Shortly thereafter, *bandpass filter* design was studied. Although they are straightforward to work with, the response of such simple circuits is not particularly abrupt. As an alternative, filter designs based on either Butterworth or Chebyshev polynomials were examined, with higher-order filters yielding sharper magnitude response at the expense of increased complexity.

❑ Resonance is the condition in which a fixed-amplitude sinusoidal forcing function produces a response of maximum amplitude. (Example 16.1)

❑ An electrical network is in resonance when the voltage and current at the network input terminals are in phase. (Example 16.1)

❑ The quality factor is proportional to the maximum energy stored in a network divided by the total energy lost per period.

❑ A half-power frequency is defined as the frequency at which the magnitude of a circuit response function is reduced to $1/\sqrt{2}$ times its maximum value.

❑ A high-*Q* circuit is a resonant circuit in which the quality factor is ≥ 5. (Example 16.2)

❑ The bandwidth of a resonant circuit is defined as the difference between the upper and lower half-power frequencies.

❑ In a high-*Q* circuit, each half-power frequency is located approximately one-half bandwidth from the resonant frequency. (Example 16.2)

❑ A series resonant circuit is characterized by a *low* impedance at resonance, whereas a parallel resonant circuit is characterized by a *high* impedance at resonance. (Examples 16.1 and 16.3)

❑ A series resonant circuit and a parallel resonant circuit are equivalent if $R_p = R_s(1 + Q^2)$ and $X_p = X_s(1 + Q^{-2})$. (Examples 16.4, 16.5)

❑ Impractical values for components often make design easier. The transfer function of a network may be scaled in magnitude or frequency using appropriate replacement values for components. (Example 16.6)

- ❏ Bode diagrams allow the rough shape of a transfer function to be plotted quickly from the poles and zeros. (Examples 16.7, 16.8, 16.9, 16.10)

- ❏ The four basic types of filters are low-pass, high-pass, bandpass, and bandstop. (Examples 16.11, 16.12)

- ❏ Passive filters use only resistors, capacitors, and inductors; active filters are based on op amps or other active elements. (Example 16.13)

- ❏ Butterworth and Chebyshev filters can be designed based on the simple Sallen-Key amplifier. Filter gain typically must be adjusted by adding a purely resistor-based amplifier circuit at the output.

READING FURTHER

A good discussion of a large variety of filters can be found in:

J. T. Taylor and Q. Huang, eds., *CRC Handbook of Electrical Filters*. Boca Raton, Fla.: CRC Press, 1997.

A comprehensive compilation of various active filter circuits and design procedures is given in:

D. Lancaster, *Lancaster's Active Filter Cookbook,* 2nd ed. Burlington, Mass.: Newnes, 1996.

Additional filter references which the reader might find useful include:

D. E. Johnson and J. L. Hilburn, *Rapid Practical Design of Active Filters.* New York: John Wiley & Sons, Inc., 1975.

J. V. Wait, L. P. Huelsman, and G. A. Korn, *Introduction to Operational Amplifier Theory and Applications,* 2nd ed. New York: McGraw-Hill, 1992.

EXERCISES

16.1 Parallel Resonance

1. Compute Q_0 and ζ for a simple parallel *RLC* network if (*a*) $R = 1$ kΩ, $C = 10$ mF, and $L = 1$ H; (*b*) $R = 1$ Ω, $C = 10$ mF, and $L = 1$ H; (*c*) $R = 1$ kΩ, $C = 1$ F, and $L = 1$ H; (*d*) $R = 1$ Ω, $C = 1$ F, and $L = 1$ H.

2. For the circuit shown in Fig. 16.1, let $R = 1$ kΩ, $C = 22$ mF, and $L = 12$ mH. (*a*) Calculate α, ω_0, ζ, f_0, and ω_d for the circuit. (*b*) If $\mathbf{I} = 1\underline{/0°}$ A, plot $\mathbf{V}$, $\mathbf{I}_{LC}$, $\mathbf{I}_L$, and $\mathbf{I}_C$ as a function of frequency, and verify that $\mathbf{I}$ and $\mathbf{V}$ are in phase at ω_0. (*c*) What is the relationship of $\mathbf{I}_L$ to $\mathbf{I}_C$ at ω_0?

3. A certain parallel *RLC* circuit is built using component values $L = 50$ mH and $C = 33$ mF. If $Q_0 = 10$, determine the value of R, and the sketch the magnitude of the steady-state impedance over the range of $2 < \omega < 40$ rad/s.

4. A parallel *RLC* network is constructed using $R = 5$ Ω, $L = 100$ mH, and $C = 1$ mF. (*a*) Compute Q_0. (*b*) Determine at which frequencies the impedance magnitude drops to 90% of its maximum value.

5. For the network of Fig. 16.50, derive an expression for the steady-state input impedance and determine the frequency at which it has maximum amplitude.

6. Plot the input admittance of the network depicted in Fig. 16.50 using a logarithmic frequency scale over the range $0.01\omega_0 < \omega_0 < 100\omega_0$, and determine the resonant frequency and the bandwidth of the network.

7. Delete the 2 Ω resistor in the network of Fig. 16.50 and determine (*a*) the magnitude of the input impedance at resonance; (*b*) the resonant frequency; (*c*) Q_0.

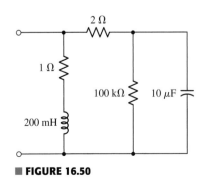

■ **FIGURE 16.50**

8. Delete the 1 Ω resistor in the network of Fig. 16.50 and determine (*a*) the magnitude of the input impedance at resonance; (*b*) the resonant frequency; (*c*) Q_0.

9. A varactor is a semiconductor device whose reactance may be varied by applying a bias voltage. The quality factor can be expressed[3] as

$$Q \approx \frac{\omega C_J R_P}{1 + \omega^2 C_J^2 R_P R_S}$$

where C_J is the junction capacitance (which depends on the voltage applied to the device), R_S is the series resistance of the device, and R_P is an equivalent parallel resistance term. (*a*) If $C_J = 3.77$ pF at 1.5 V, $R_P = 1.5$ MΩ, and $R_S = 2.8\ \Omega$, plot the quality factor as a function of frequency ω. (*b*) Differentiate the expression for Q to obtain both ω_0 and Q_{max}.

16.2 Bandwidth and High-Q Circuits

10. The circuit of Fig. 16.1 is built using component values $L = 1$ mH and $C = 100\ \mu$F. If $Q_0 = 15$, determine the bandwidth and estimate the magnitude and angle of the input impedance for operation at (*a*) 3162 rad/s; (*b*) 3000 rad/s; (*c*) 3200 rad/s; (*d*) 2000 rad/s. (*e*) Verify your estimates using an exact expression for $\mathbf{Y}(j\omega)$.

11. A parallel *RLC* network is constructed with a 5 mH inductor, and the remaining component values are chosen such that $Q_0 = 6.5$ and $\omega_0 = 1000$ rad/s. Determine the approximate value of the input impedance magnitude for operation at (*a*) 500 rad/s; (*b*) 750 rad/s; (*c*) 900 rad/s; (*d*) 1100 rad/s. (*e*) Plot your estimates along with the exact result using a linear frequency (rad/s) axis.

12. A parallel *RLC* network is constructed with a 200 μH inductor, and the remaining component values are chosen such that $Q_0 = 8$ and $\omega_0 = 5000$ rad/s. Use approximate expressions to estimate the input impedance angle for operation at (*a*) 2000 rad/s; (*b*) 3000 rad/s; (*c*) 4000 rad/s; (*d*) 4500 rad/s. (*e*) Plot your estimates along with the exact result using a linear frequency (rad/s) axis.

13. Find the bandwidth of each of the response curves shown in Fig. 16.51.

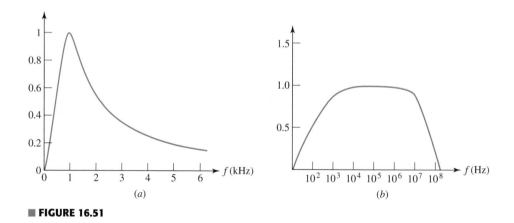

■ **FIGURE 16.51**

14. A parallel *RLC* circuit is constructed such that it has the impedance magnitude characteristic plotted in Fig. 16.52. (*a*) Determine the resistor value. (*b*) Determine the capacitor value if a 1 H inductor was used. (*c*) Obtain values for the bandwidth, Q_0, and both the low half-power frequency and the high half-power frequency.

(3) S. M. Sze, *Physics of Semiconductor Devices*, 2d ed. New York: Wiley, 1981, p. 116.

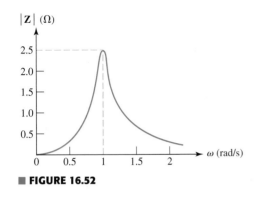

■ FIGURE 16.52

16.3 Series Resonance

15. A series *RLC* circuit is constructed employing component values $R = 100\ \Omega$ and $L = 1.5$ mH along with a sinusoidal voltage source v_s. If $Q_0 = 7$, determine (*a*) the magnitude of the impedance at 500 Mrad/s; (*b*) the current which flows in response to a voltage $v_s = 2.5\cos(425 \times 10^6 t)$ V.

16. With regard to the series *RLC* circuit described in Exercise 15, adjust the resistor value such that Q_0 is reduced to 5, and (*a*) estimate the angle of the impedance at 90 krad/s, 100 krad/s, and 110 krad/s. (*b*) Determine the percent error in the estimated values, compared to the exact expression.

17. An *RLC* circuit is constructed using $R = 5\ \Omega$, $L = 20$ mH, and $C = 1$ mF. Calculate Q_0, the bandwidth, and the magnitude of the impedance at $0.95\omega_0$ if the circuit is (*a*) parallel-connected; (*b*) series-connected. (*c*) Verify your solutions using appropriate PSpice simulations. (*Hint: a large resistor in parallel with the capacitor will avoid error messages associated with no dc path to ground, and a small resistance in series with the VAC source will avoid shorting by the inductor during dc bias point determination.*)

18. Inspect the circuit of Fig. 16.53, noting the amplitude of the source voltage. Now decide whether you would be willing to put your bare hands across the capacitor if the circuit were actually built in the lab. Plot $|\mathbf{V}_C|$ versus ω to justify your answer.

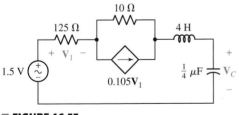

■ FIGURE 16.53

19. After deriving $\mathbf{Z}_{\text{in}}(\mathbf{s})$ in Fig. 16.54, find (*a*) ω_0; (*b*) Q_0.

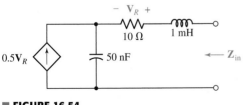

■ FIGURE 16.54

16.4 Other Resonant Forms

20. For the network of Fig. 16.9a, $R_1 = 100\ \Omega$, $R_2 = 150\ \Omega$, $L = 30$ mH, and C is chosen so that $\omega_0 = 750$ rad/s. Calculate the impedance magnitude at (a) the frequency corresponding to resonance when $R_1 = 0$; (b) 700 rad/s; (c) 800 rad/s.

21. Assuming an operating frequency of 200 rad/s, find a series equivalent of the parallel combination of a 500 Ω resistor and (a) a 1.5 μF capacitor; (b) a 200 mH inductor.

22. If the frequency of operation is either 40 rad/s or 80 rad/s, find a parallel equivalent of the series combination of a 2 Ω resistor and (a) a 100 mF capacitor; (b) a 3 mH inductor.

23. For the network represented in Fig. 16.55, determine the resonant frequency and the corresponding value of $|\mathbf{Z}_{in}|$.

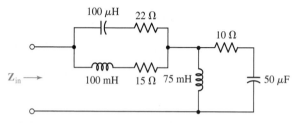

■ **FIGURE 16.55**

24. For the circuit shown in Fig. 16.56, the voltage source has magnitude 1 V and phase angle 0°. Determine the resonant frequency ω_0 and the value of $\mathbf{V}_x$ at $0.95\omega_0$.

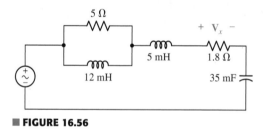

■ **FIGURE 16.56**

16.5 Scaling

25. A parallel RLC circuit is constructed using component values $R = 1\ \Omega$, $C = 3$ F, and $L = \frac{1}{3}$ H. Determine the required component values if the network is to have (a) a resonant frequency of 200 kHz; (b) a peak impedance of 500 kΩ; (c) a resonant frequency of 750 kHz and an impedance magnitude at resonance of 25 Ω.

26. A series RLC circuit is constructed using component values $R = 1\ \Omega$, $C = 5$ F, and $L = \frac{1}{5}$ H. Determine the required component values if the network is to have (a) a resonant frequency of 430 Hz; (b) a peak impedance of 100 Ω; (c) a resonant frequency of 75 kHz and an impedance magnitude at resonance of 15 kΩ.

27. Scale the network shown in Fig. 16.57 by $K_m = 200$ and $K_f = 700$, and obtain an expression for the new impedance $\mathbf{Z}_{in}(s)$.

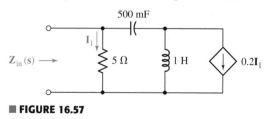

■ **FIGURE 16.57**

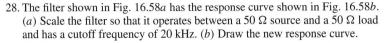

28. The filter shown in Fig. 16.58a has the response curve shown in Fig. 16.58b.
(a) Scale the filter so that it operates between a 50 Ω source and a 50 Ω load and has a cutoff frequency of 20 kHz. (b) Draw the new response curve.

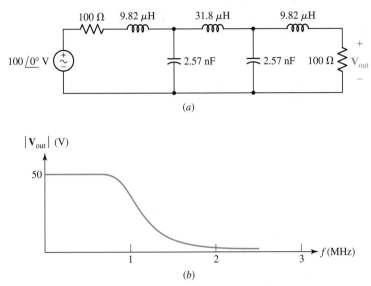

(a)

(b)

■ **FIGURE 16.58**

29. (a) Draw the new configuration for Fig. 16.59 after the network is scaled by $K_m = 250$ and $K_f = 400$. (b) Determine the Thévenin equivalent of the scaled network at $\omega = 1$ krad/s.

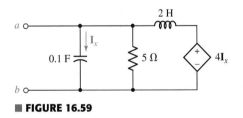

■ **FIGURE 16.59**

16.6 Bode Diagrams

30. Sketch the Bode magnitude and phase plots for the following functions:
(a) $3 + 4s$; (b) $\dfrac{1}{3 + 4s}$.

31. For the following functions, sketch the Bode magnitude and phase plots:
(a) $25\left(1 + \dfrac{s}{3}\right)(5 + s)$; (b) $\dfrac{0.1}{(1 + 5s)(2 + s)}$.

32. Use the Bode approach to sketch the magnitude of each of the following responses, then verify your solutions with appropriate MATLAB simulations:
(a) $3\dfrac{s}{s^2 + 7s + 10}$; (b) $\dfrac{4}{s^3 + 7s^2 + 12s}$.

33. If a particular network is described by transfer function $\mathbf{H}(s)$, plot the magnitude of $\mathbf{H}(s)$ as a function of frequency for $\mathbf{H}(s)$ equal to
(a) $\dfrac{s + 300}{s(5s + 8)}$; (b) $\dfrac{s(s^2 + 7s + 7)}{s(2s + 4)^2}$.

34. Sketch the phase plot of each of the following transfer functions:
(a) $\dfrac{s + 1}{s(s + 2)^2}$; (b) $5\dfrac{s^2 + s}{s + 2}$.

35. Determine the Bode magnitude plot for the following transfer functions, and compare to what is predicted using MATLAB: (a) $\mathbf{s}^2 + 0.2\mathbf{s} + 1$;

(b) $\left(\dfrac{\mathbf{s}}{4}\right)^2 + 0.1\left(\dfrac{\mathbf{s}}{4}\right) + 1.$

 36. Determine the phase plot corresponding to each of the transfer functions in Exercises 33 and 35, and compare your sketches to what is predicted using MATLAB.

37. Determine the Bode magnitude plot for each of the following:

(a) $\dfrac{3 + 0.1\mathbf{s} + \mathbf{s}^2/3}{\mathbf{s}^2 + 1}$; (b) $2\dfrac{\mathbf{s}^2 + 9\mathbf{s} + 20}{\mathbf{s}^2(\mathbf{s}+1)^3}$.

38. For the circuit of Fig. 16.60, (a) derive an expression for the transfer function $\mathbf{H}(\mathbf{s}) = \mathbf{V}_{\text{out}}/\mathbf{V}_{\text{in}}$. (b) Sketch the corresponding Bode magnitude and phase plots.

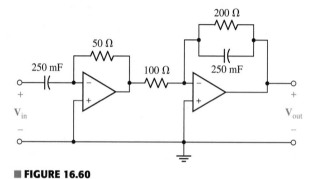

■ **FIGURE 16.60**

39. (a) Modify the circuit shown in Fig. 16.60 to add a double pole at 0.05 rad/s and a zero at 0.01 rad/s. (b) Sketch the corresponding Bode magnitude and phase plots.

16.7 Basic Filter Design

 40. (a) Design a high-pass filter with a corner frequency of 100 rad/s. (b) Verify your design with an appropriate PSpice simulation.

41. (a) Design a low-pass filter with a break frequency of 1450 rad/s. (b) Sketch the Bode magnitude and phase plots for your design. (c) Verify your filter performance with an appropriate simulation.

42. (a) Design a bandpass filter characterized by a bandwidth of 1000 rad/s and a low-frequency corner of 250 Hz. (b) Verify your design with an appropriate PSpice simulation.

43. Design a bandpass filter having a low-frequency cutoff of 500 Hz and a high-frequency cutoff of 1580 Hz.

44. Design a notch filter which removes 60 Hz "noise" from power line influences on a particular signal by taking the output across the inductor-capacitor series connection in the circuit of Fig. 16.39.

45. Design a low-pass filter characterized by a voltage gain of 25 dB and a corner frequency of 5000 rad/s.

46. Design a high-pass filter characterized by a voltage gain of 30 dB and a corner frequency of 50 rad/s.

 47. (a) Design a two-stage op amp filter circuit with a bandwidth of 1000 rad/s, a low-frequency cutoff of 100 rad/s, and a voltage gain of 20 dB. (b) Verify your design with an appropriate PSpice simulation.

 48. Design a circuit which removes the entire audio frequency range (approximately 20 Hz to 20 kHz, for human hearing), but amplifies the signal voltage of all other frequencies by a factor of 15.

49. Depending on which song you're listening to, your MP3 player sometimes provides too little bass, even when the appropriate setting is maximized. Design a filter which allows you to vary the gain real time of all signals less than 500 Hz prior to reaching your earphones. Include a diagram of the overall system.

16.8 Advanced Filter Design

50. Show that the circuit represented by Eq. [36] cannot be implemented as either a Butterworth or a Chebyshev low-pass filter.

51. Design a second-order low-pass filter having a voltage gain of 5 dB and a cutoff frequency of 1700 kHz based on (a) Butterworth polynomials; (b) Chebyshev polynomials for 3 dB ripple factor.

52. If a high-pass filter is required having gain of 6 dB and a cutoff frequency of 350 Hz, design a suitable second-order Butterworth-based solution.

53. (a) Design a second-order low-pass Butterworth filter with a cutoff frequency of 890 rad/s and a voltage gain of 8 dB. (b) Verify your design with an appropriate PSpice simulation.

54. (a) Design a second-order high-pass Butterworth filter with a cutoff frequency of 2000 Hz and a voltage gain of 4.5 dB. (b) Verify your design with an appropriate PSpice simulation.

55. A third-order low-pass Butterworth filter is required having a cutoff frequency of 1200 Hz and a voltage gain of at least 3 dB. Design a suitable circuit.

56. (a) Design a third-order low-pass Butterworth filter having a gain of 13 dB and a corner frequency at 1800 Hz. (b) Compare your filter response to that of a Chebyshev filter with the same specifications.

57. Design a fourth-order high-pass Butterworth filter having a minimum gain of 15 dB and a corner frequency of 1100 rad/s.

58. Choose parameters for the circuit described by Eq. [36] such that it has a cutoff frequency at 450 rad/s, and compare its performance to a comparable second-order Butterworth filter.

Chapter-Integrating Exercises

59. Design a parallel resonant circuit for an AM radio so that a variable inductor can adjust the resonant frequency over the AM broadcast band, 535 to 1605 kHz, with $Q_0 = 45$ at one end of the band and $Q_0 \le 45$ throughout the band. Let $R = 20$ kΩ, and specify values for C, L_{min}, and L_{max}.

60. Derive an expression for the transfer function $\mathbf{V}_{out}/\mathbf{V}_{in}$ which describes the circuit shown in Fig. 16.61, and sketch its magnitude as a function of frequency.

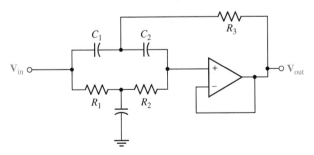

■ **FIGURE 16.61**

61. The network of Fig. 16.36 was implemented as a low-pass filter designed with a corner frequency of 1250 rad/s. Its performance is inadequate in two respects: (1) a voltage gain of at least 2 dB is required, and (2) the magnitude of the output voltage does not decrease quickly enough in the stopband.

Design a better alternative if only one op amp is available and only two 1 μF capacitors can be located.

62. Determine the effect of component tolerance on the circuit designed in Example 16.14 if each component is specified to be only within 10% of its stated value.

63. Derive an expression for the transfer function $\mathbf{V}_{out}/\mathbf{V}_{in}$ which describes the circuit shown in Fig. 16.62, and sketch its magnitude as a function of frequency.

64. For the circuit shown in Fig. 16.62, select component values to design for corner frequencies at 500 rad/s and 1500 rad/s. Verify your design.

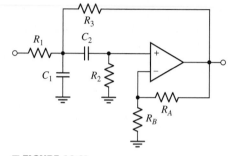

■ **FIGURE 16.62**

65. Design a bandpass filter that spans the portion of the audio spectrum from 200 Hz to 2 kHz which has a minimum gain of 5 dB, and a steeper magnitude characteristic on the high-frequency side than on the low-frequency side. Verify your design using a suitable simulation.

Two-Port Networks

KEY CONCEPTS

INTRODUCTION

A general network having two pairs of terminals, one often labeled the "input terminals" and the other the "output terminals," is a very important building block in electronic systems, communication systems, automatic control systems, transmission and distribution systems, or other systems in which an electrical signal or electric energy enters the input terminals, is acted upon by the network, and leaves via the output terminals. The output terminal pair may very well connect with the input terminal pair of another network. When we studied the concept of Thévenin and Norton equivalent networks in Chap. 5, we were introduced to the idea that it is not always necessary to know the detailed workings of part of a circuit. This chapter extends such concepts to linear networks, resulting in parameters that allow us to predict how any network will interact with other networks.

17.1 ● ONE-PORT NETWORKS

A pair of terminals at which a signal may enter or leave a network is called a *port,* and a network having only one such pair of terminals is called a *one-port network,* or simply a *one-port.* No connections may be made to any other nodes internal to the one-port, and it is therefore evident that i_a must equal i_b in the one-port shown in Fig. 17.1*a*. When more than one pair of terminals is present, the network is known as a *multiport network.* The two-port network to which this chapter is principally devoted is shown in Fig. 17.1*b*. The currents in the two leads making up each port must be equal, and so it follows that $i_a = i_b$ and $i_c = i_d$ in the two-port shown in Fig. 17.1*b*. Sources and loads must be connected directly across the two terminals of a port if the methods of this chapter are to be used. In other

687

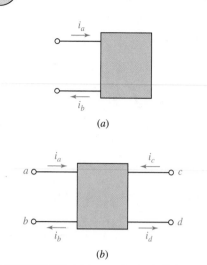

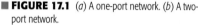

(a)

(b)

■ **FIGURE 17.1** (a) A one-port network. (b) A two-port network.

words, each port can be connected only to a one-port network or to a port of another multiport network. For example, no device may be connected between terminals *a* and *c* of the two-port network in Fig. 17.1*b*. If such a circuit must be analyzed, general loop or nodal equations should be written.

Some of the introductory study of one- and two-port networks is accomplished best by using a generalized network notation and the abbreviated nomenclature for determinants introduced in Appendix 2. Thus, if we write a set of loop equations for a passive network,

$$
\begin{aligned}
\mathbf{Z}_{11}\mathbf{I}_1 + \mathbf{Z}_{12}\mathbf{I}_2 + \mathbf{Z}_{13}\mathbf{I}_3 + \cdots + \mathbf{Z}_{1N}\mathbf{I}_N &= \mathbf{V}_1 \\
\mathbf{Z}_{21}\mathbf{I}_1 + \mathbf{Z}_{22}\mathbf{I}_2 + \mathbf{Z}_{23}\mathbf{I}_3 + \cdots + \mathbf{Z}_{2N}\mathbf{I}_N &= \mathbf{V}_2 \\
\mathbf{Z}_{31}\mathbf{I}_1 + \mathbf{Z}_{32}\mathbf{I}_2 + \mathbf{Z}_{33}\mathbf{I}_3 + \cdots + \mathbf{Z}_{3N}\mathbf{I}_N &= \mathbf{V}_3 \\
&\cdots\cdots\cdots \\
\mathbf{Z}_{N1}\mathbf{I}_1 + \mathbf{Z}_{N2}\mathbf{I}_2 + \mathbf{Z}_{N3}\mathbf{I}_3 + \cdots + \mathbf{Z}_{NN}\mathbf{I}_N &= \mathbf{V}_N
\end{aligned}
\qquad [1]
$$

then the coefficient of each current will be an impedance $\mathbf{Z}_{ij}(\mathbf{s})$, and the circuit determinant, or determinant of the coefficients, is

$$
\Delta_{\mathbf{z}} =
\begin{vmatrix}
\mathbf{Z}_{11} & \mathbf{Z}_{12} & \mathbf{Z}_{13} & \cdots & \mathbf{Z}_{1N} \\
\mathbf{Z}_{21} & \mathbf{Z}_{22} & \mathbf{Z}_{23} & \cdots & \mathbf{Z}_{2N} \\
\mathbf{Z}_{31} & \mathbf{Z}_{32} & \mathbf{Z}_{33} & \cdots & \mathbf{Z}_{3N} \\
\cdots & \cdots & \cdots & \cdots & \cdots \\
\mathbf{Z}_{N1} & \mathbf{Z}_{N2} & \mathbf{Z}_{N3} & \cdots & \mathbf{Z}_{NN}
\end{vmatrix}
\qquad [2]
$$

Here *N* loops have been assumed, the currents appear in subscript order in each equation, and the order of the equations is the same as that of the currents. We also assume that KVL is applied so that the sign of each $\mathbf{Z}_{ii}$ term $(\mathbf{Z}_{11}, \mathbf{Z}_{22}, \ldots, \mathbf{Z}_{NN})$ is positive; the sign of any $\mathbf{Z}_{ij}(i \neq j)$ or mutual term may be either positive or negative, depending on the reference directions assigned to $\mathbf{I}_i$ and $\mathbf{I}_j$.

If there are dependent sources within the network, then it is possible that not all the coefficients in the loop equations must be resistances or impedances. Even so, we will continue to refer to the circuit determinant as $\Delta_{\mathbf{z}}$.

The use of minor notation (Appendix 2) allows for the input or driving-point impedance at the terminals of a *one-port* network to be expressed very concisely. The result is also applicable to a *two-port* network if one of the two ports is terminated in a passive impedance, including an open or a short circuit.

Let us suppose that the one-port network shown in Fig. 17.2 is composed entirely of passive elements and dependent sources; linearity is also assumed. An ideal voltage source $\mathbf{V}_1$ is connected to the port, and the source current is identified as the current in loop 1. Employing Cramer's rule, then,

Cramer's rule is reviewed in Appendix 2.

$$
\mathbf{I}_1 = \frac{
\begin{vmatrix}
\mathbf{V}_1 & \mathbf{Z}_{12} & \mathbf{Z}_{13} & \cdots & \mathbf{Z}_{1N} \\
0 & \mathbf{Z}_{22} & \mathbf{Z}_{23} & \cdots & \mathbf{Z}_{2N} \\
0 & \mathbf{Z}_{32} & \mathbf{Z}_{33} & \cdots & \mathbf{Z}_{3N} \\
\cdots & \cdots & \cdots & \cdots & \cdots \\
0 & \mathbf{Z}_{N2} & \mathbf{Z}_{N3} & \cdots & \mathbf{Z}_{NN}
\end{vmatrix}
}{
\begin{vmatrix}
\mathbf{Z}_{11} & \mathbf{Z}_{12} & \mathbf{Z}_{13} & \cdots & \mathbf{Z}_{1N} \\
\mathbf{Z}_{21} & \mathbf{Z}_{22} & \mathbf{Z}_{23} & \cdots & \mathbf{Z}_{2N} \\
\mathbf{Z}_{31} & \mathbf{Z}_{32} & \mathbf{Z}_{33} & \cdots & \mathbf{Z}_{3N} \\
\cdots & \cdots & \cdots & \cdots & \cdots \\
\mathbf{Z}_{N1} & \mathbf{Z}_{N2} & \mathbf{Z}_{N3} & \cdots & \mathbf{Z}_{NN}
\end{vmatrix}
}
$$

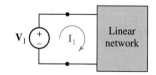

■ **FIGURE 17.2** An ideal voltage source $\mathbf{V}_1$ is connected to the single port of a linear one-port network containing no independent sources; $\mathbf{Z}_{in} = \Delta_{\mathbf{z}}/\Delta_{11}$.

or, more concisely,

$$\mathbf{I}_1 = \frac{\mathbf{V}_1 \Delta_{11}}{\Delta_{\mathbf{z}}}$$

Thus,

$$\mathbf{Z}_{in} = \frac{\mathbf{V}_1}{\mathbf{I}_1} = \frac{\Delta_{\mathbf{z}}}{\Delta_{11}} \qquad\qquad [3]$$

EXAMPLE **17.1**

Calculate the input impedance for the one-port resistive network shown in Fig. 17.3.

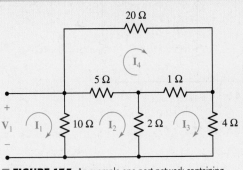

FIGURE 17.3 An example one-port network containing only resistive elements.

We first assign the four mesh currents as shown and write the corresponding mesh equations by inspection:

$$\mathbf{V}_1 = 10\mathbf{I}_1 - 10\mathbf{I}_2$$
$$0 = -10\mathbf{I}_1 + 17\mathbf{I}_2 - 2\mathbf{I}_3 - 5\mathbf{I}_4$$
$$0 = \phantom{-10\mathbf{I}_1 +} - 2\mathbf{I}_2 + 7\mathbf{I}_3 - \mathbf{I}_4$$
$$0 = \phantom{-10\mathbf{I}_1 +} - 5\mathbf{I}_2 - \mathbf{I}_3 + 26\mathbf{I}_4$$

The circuit determinant is then given by

$$\Delta_{\mathbf{z}} = \begin{vmatrix} 10 & -10 & 0 & 0 \\ -10 & 17 & -2 & -5 \\ 0 & -2 & 7 & -1 \\ 0 & -5 & -1 & 26 \end{vmatrix}$$

and has the value $9680\ \Omega^4$. Eliminating the first row and first column, we have

$$\Delta_{11} = \begin{vmatrix} 17 & -2 & -5 \\ -2 & 7 & -1 \\ -5 & -1 & 26 \end{vmatrix} = 2778\ \Omega^3$$

(Continued on next page)

Thus, Eq. [3] provides the value of the input impedance,

$$\mathbf{Z}_{\text{in}} = \frac{9680}{2778} = 3.485 \ \Omega$$

PRACTICE

17.1 Find the input impedance of the network shown in Fig. 17.4 if it is formed into a one-port network by breaking it at terminals (a) a and a'; (b) b and b'; (c) c and c'.

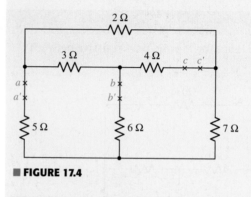

■ **FIGURE 17.4**

Ans: 9.47 Ω; 10.63 Ω; 7.58 Ω.

EXAMPLE **17.2**

Find the input impedance of the network shown in Fig. 17.5.

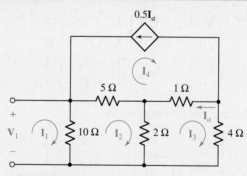

■ **FIGURE 17.5** A one-port network containing a dependent source.

The four mesh equations are written in terms of the four assigned mesh currents:

$$10\mathbf{I}_1 - 10\mathbf{I}_2 \qquad\qquad = \mathbf{V}_1$$
$$-10\mathbf{I}_1 + 17\mathbf{I}_2 - 2\mathbf{I}_3 - 5\mathbf{I}_4 = 0$$
$$-\ 2\mathbf{I}_2 + 7\mathbf{I}_3 -\ \mathbf{I}_4 = 0$$

and

$$\mathbf{I}_4 = -0.5\mathbf{I}_a = -0.5(\mathbf{I}_4 - \mathbf{I}_3)$$

or

$$-0.5\mathbf{I}_3 + 1.5\mathbf{I}_4 = 0$$

Thus we can write

$$\Delta_{\mathbf{z}} = \begin{vmatrix} 10 & -10 & 0 & 0 \\ -10 & 17 & -2 & -5 \\ 0 & -2 & 7 & -1 \\ 0 & 0 & -0.5 & 1.5 \end{vmatrix} = 590 \ \Omega^4$$

while

$$\Delta_{11} = \begin{vmatrix} 17 & -2 & -5 \\ -2 & 7 & -1 \\ 0 & -0.5 & 1.5 \end{vmatrix} = 159 \ \Omega^3$$

giving

$$\mathbf{Z}_{\text{in}} = \tfrac{590}{159} = 3.711 \ \Omega$$

We may also select a similar procedure using nodal equations, yielding the input admittance:

$$\mathbf{Y}_{\text{in}} = \frac{1}{\mathbf{Z}_{\text{in}}} = \frac{\Delta_{\mathbf{Y}}}{\Delta_{11}} \qquad [4]$$

where Δ_{11} now refers to the minor of $\Delta_{\mathbf{Y}}$.

PRACTICE

17.2 Write a set of nodal equations for the circuit of Fig. 17.6, calculate $\Delta_{\mathbf{Y}}$, and then find the input admittance seen between (a) node 1 and the reference node; (b) node 2 and the reference.

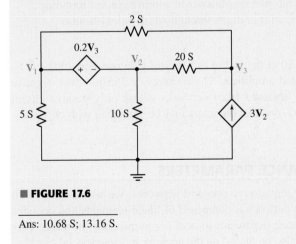

■ FIGURE 17.6

Ans: 10.68 S; 13.16 S.

EXAMPLE 17.3

Use Eq. [4] to again determine the input impedance of the network shown in Fig. 17.3, repeated here as Fig. 17.7.

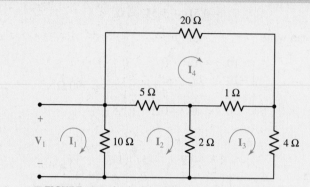

■ **FIGURE 17.7** The circuit from Example 17.1, repeated for convenience.

We first order the node voltages $\mathbf{V}_1$, $\mathbf{V}_2$, and $\mathbf{V}_3$ from left to right, select the reference at the bottom node, and then write the system admittance matrix by inspection:

$$\Delta_{\mathbf{Y}} = \begin{vmatrix} 0.35 & -0.2 & -0.05 \\ -0.2 & 1.7 & -1 \\ -0.05 & -1 & 1.3 \end{vmatrix} = 0.3473 \text{ S}^3$$

$$\Delta_{11} = \begin{vmatrix} 1.7 & -1 \\ -1 & 1.3 \end{vmatrix} = 1.21 \text{ S}^2$$

so that

$$\mathbf{Y}_{\text{in}} = \frac{0.3473}{1.21} = 0.2870 \text{ S}$$

which corresponds to

$$\mathbf{Z}_{\text{in}} = \frac{1}{0.287} = 3.484 \text{ }\Omega$$

which agrees with our previous answer to within expected rounding error (we only retained four digits throughout the calculations).

Exercises 9 and 10 at the end of the chapter give one-ports that can be built using operational amplifiers. These exercises illustrate that *negative* resistances may be obtained from networks whose only passive circuit elements are resistors, and that inductors may be simulated with only resistors and capacitors.

17.2 • ADMITTANCE PARAMETERS

Let us now turn our attention to two-port networks. We will assume in all that follows that the network is composed of linear elements and contains no independent sources; dependent sources *are* permissible, however. Further conditions will also be placed on the network in some special cases.

We will consider the two-port as it is shown in Fig. 17.8; the voltage and current at the input terminals are $\mathbf{V}_1$ and $\mathbf{I}_1$, and $\mathbf{V}_2$ and $\mathbf{I}_2$ are specified at the output port. The directions of $\mathbf{I}_1$ and $\mathbf{I}_2$ are both customarily selected as *into* the network at the upper conductors (and out at the lower conductors). Since the network is linear and contains no independent sources within it, $\mathbf{I}_1$ may be considered to be the superposition of two components, one caused by $\mathbf{V}_1$ and the other by $\mathbf{V}_2$. When the same argument is applied to $\mathbf{I}_2$, we may begin with the set of equations

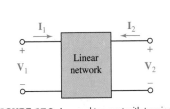

$$\mathbf{I}_1 = \mathbf{y}_{11}\mathbf{V}_1 + \mathbf{y}_{12}\mathbf{V}_2 \qquad [5]$$

$$\mathbf{I}_2 = \mathbf{y}_{21}\mathbf{V}_1 + \mathbf{y}_{22}\mathbf{V}_2 \qquad [6]$$

■ FIGURE 17.8 A general two-port with terminal voltages and currents specified. The two-port is composed of linear elements, possibly including dependent sources, but not containing any independent sources.

where the $\mathbf{y}$'s are no more than proportionality constants, or unknown coefficients, for the present. However, it should be clear that their dimensions must be A/V, or S. They are therefore called the $\mathbf{y}$ (or admittance) parameters, and are defined by Eqs. [5] and [6].

The $\mathbf{y}$ parameters, as well as other sets of parameters we will define later in the chapter, are represented concisely as matrices. Here, we define the (2×1) column matrix $\mathbf{I}$,

$$\mathbf{I} = \begin{bmatrix} \mathbf{I}_1 \\ \mathbf{I}_2 \end{bmatrix} \qquad [7]$$

the (2×2) square matrix of the $\mathbf{y}$ parameters,

$$\mathbf{y} = \begin{bmatrix} \mathbf{y}_{11} & \mathbf{y}_{12} \\ \mathbf{y}_{21} & \mathbf{y}_{22} \end{bmatrix} \qquad [8]$$

and the (2×1) column matrix $\mathbf{V}$,

$$\mathbf{V} = \begin{bmatrix} \mathbf{V}_1 \\ \mathbf{V}_2 \end{bmatrix} \qquad [9]$$

The notation adopted in this text to represent a matrix is standard, but also can be easily confused with our previous notation for phasors or general complex quantities. The nature of any such symbol should be clear from the context in which it is used.

Thus, we may write the matrix equation $\mathbf{I} = \mathbf{y}\mathbf{V}$, or

$$\begin{bmatrix} \mathbf{I}_1 \\ \mathbf{I}_2 \end{bmatrix} = \begin{bmatrix} \mathbf{y}_{11} & \mathbf{y}_{12} \\ \mathbf{y}_{21} & \mathbf{y}_{22} \end{bmatrix} \begin{bmatrix} \mathbf{V}_1 \\ \mathbf{V}_2 \end{bmatrix}$$

and matrix multiplication of the right-hand side gives us the equality

$$\begin{bmatrix} \mathbf{I}_1 \\ \mathbf{I}_2 \end{bmatrix} = \begin{bmatrix} \mathbf{y}_{11}\mathbf{V}_1 + \mathbf{y}_{12}\mathbf{V}_2 \\ \mathbf{y}_{21}\mathbf{V}_1 + \mathbf{y}_{22}\mathbf{V}_2 \end{bmatrix}$$

These (2×1) matrices must be equal, element by element, and thus we are led to the defining equations, [5] and [6].

The most useful and informative way to attach a physical meaning to the $\mathbf{y}$ parameters is through a direct inspection of Eqs. [5] and [6]. Consider Eq. [5], for example; if we let $\mathbf{V}_2$ be zero, then we see that $\mathbf{y}_{11}$ must be given by the ratio of $\mathbf{I}_1$ to $\mathbf{V}_1$. We therefore describe $\mathbf{y}_{11}$ as the admittance measured at the input terminals with the output terminals *short-circuited* ($\mathbf{V}_2 = 0$). Since there can be no question which terminals are short-circuited, $\mathbf{y}_{11}$ is best described as the *short-circuit input admittance*. Alternatively, we might describe $\mathbf{y}_{11}$ as the reciprocal of the input impedance measured with the output terminals short-circuited, but a description as an admittance is obviously more direct. It is not the *name* of the parameter that is important. Rather, it is the conditions which must be applied to Eq. [5] or [6], and hence to the network, that are most meaningful; when the conditions are determined, the parameter can be found directly from an analysis of the circuit (or by experiment on the physical circuit). Each of the **y** parameters may be described as a current-voltage ratio with either $V_1 = 0$ (the input terminals short-circuited) or $V_2 = 0$ (the output terminals short-circuited):

$$\mathbf{y}_{11} = \frac{\mathbf{I}_1}{\mathbf{V}_1}\bigg|_{\mathbf{V}_2=0} \quad\quad [10]$$

$$\mathbf{y}_{12} = \frac{\mathbf{I}_1}{\mathbf{V}_2}\bigg|_{\mathbf{V}_1=0} \quad\quad [11]$$

$$\mathbf{y}_{21} = \frac{\mathbf{I}_2}{\mathbf{V}_1}\bigg|_{\mathbf{V}_2=0} \quad\quad [12]$$

$$\mathbf{y}_{22} = \frac{\mathbf{I}_2}{\mathbf{V}_2}\bigg|_{\mathbf{V}_1=0} \quad\quad [13]$$

Because each parameter is an admittance which is obtained by short-circuiting either the output or the input port, the y parameters are known as the **short-circuit admittance parameters.** The specific name of $\mathbf{y}_{11}$ is the **short-circuit input admittance,** $\mathbf{y}_{22}$ is the **short-circuit output admittance,** and $\mathbf{y}_{12}$ and $\mathbf{y}_{21}$ are the **short-circuit transfer admittances.**

EXAMPLE 17.4

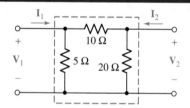

■ FIGURE 17.9 A resistive two-port.

Find the four short-circuit admittance parameters for the resistive two-port shown in Fig. 17.9.

The values of the parameters may be easily established by applying Eqs. [10] to [13], which we obtained directly from the defining equations, [5] and [6]. To determine $\mathbf{y}_{11}$, we short-circuit the output and find the ratio of $\mathbf{I}_1$ to $\mathbf{V}_1$. This may be done by letting $\mathbf{V}_1 = 1$ V, for then $\mathbf{y}_{11} = \mathbf{I}_1$. By inspection of Fig. 17.9, it is apparent that 1 V applied at the input with the output short-circuited will cause an input current of $(\frac{1}{5} + \frac{1}{10})$, or 0.3 A. Hence,

$$\mathbf{y}_{11} = 0.3 \text{ S}$$

In order to find $\mathbf{y}_{12}$, we short-circuit the input terminals and apply 1 V at the output terminals. The input current flows through the short circuit and is $\mathbf{I}_1 = -\frac{1}{10}$ A. Thus

$$\mathbf{y}_{12} = -0.1 \text{ S}$$

By similar methods,

$$\mathbf{y}_{21} = -0.1 \text{ S} \quad\quad \mathbf{y}_{22} = 0.15 \text{ S}$$

The describing equations for this two-port in terms of the admittance parameters are, therefore,

$$\mathbf{I}_1 = 0.3\mathbf{V}_1 - 0.1\mathbf{V}_2 \quad\quad [14]$$

$$\mathbf{I}_2 = -0.1\mathbf{V}_1 + 0.15\mathbf{V}_2 \quad\quad [15]$$

and

$$y = \begin{bmatrix} 0.3 & -0.1 \\ -0.1 & 0.15 \end{bmatrix} \quad \text{(all S)}$$

It is not necessary to find these parameters one at a time by using Eqs. [10] to [13], however. We may find them all at once—as shown in the next example.

EXAMPLE 17.5

Assign node voltages V_1 and V_2 in the two-port of Fig. 17.9 and write the expressions for I_1 and I_2 in terms of them.

We have

$$I_1 = \frac{V_1}{5} + \frac{V_1 - V_2}{10} = 0.3V_1 - 0.1V_2$$

and

$$I_2 = \frac{V_2 - V_1}{10} + \frac{V_2}{20} = -0.1V_1 + 0.15V_2$$

These equations are identical with Eqs. [14] and [15], and the four **y** parameters may be read from them *directly*.

PRACTICE

17.3 By applying the appropriate 1 V sources and short circuits to the circuit shown in Fig. 17.10, find (*a*) y_{11}; (*b*) y_{21}; (*c*) y_{22}; (*d*) y_{12}.

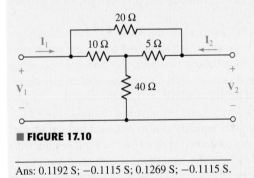

■ **FIGURE 17.10**

Ans: 0.1192 S; −0.1115 S; 0.1269 S; −0.1115 S.

In general, it is easier to use Eq. [10], [11], [12], or [13] when only one parameter is desired. If we need all of them, however, it is usually easier to assign V_1 and V_2 to the input and output nodes, to assign other node-to-reference voltages at any interior nodes, and then to carry through with the general solution.

In order to see what use might be made of such a system of equations, let us now terminate each port with some specific one-port network.

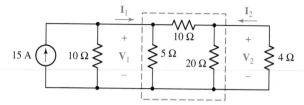

■ FIGURE 17.11 The resistive two-port network of Fig. 17.9, terminated with specific one-port networks.

Consider the simple two-port network of Example 17.4, shown in Fig. 17.11 with a practical current source connected to the input port and a resistive load connected to the output port. A relationship must now exist between V_1 and I_1 that is independent of the two-port network. This relationship may be determined solely from this external circuit. If we apply KCL (or write a single nodal equation) at the input,

$$I_1 = 15 - 0.1V_1$$

For the output, Ohm's law yields

$$I_2 = -0.25V_2$$

Substituting these expressions for I_1 and I_2 in Eqs. [14] and [15], we have

$$15 = \quad 0.4V_1 - 0.1V_2$$
$$0 = -0.1V_1 + 0.4V_2$$

from which are obtained

$$V_1 = 40 \text{ V} \qquad V_2 = 10 \text{ V}$$

The input and output currents are also easily found:

$$I_1 = 11 \text{ A} \qquad I_2 = -2.5 \text{ A}$$

and the complete terminal characteristics of this resistive two-port are then known.

The advantages of two-port analysis do not show up very strongly for such a simple example, but it should be apparent that once the **y** parameters are determined for a more complicated two-port, the performance of the two-port for different terminal conditions is easily determined; it is necessary only to relate V_1 to I_1 at the input and V_2 to I_2 at the output.

In the example just concluded, y_{12} and y_{21} were both found to be -0.1 S. It is not difficult to show that this equality is also obtained if three general impedances Z_A, Z_B, and Z_C are contained in this Π network. It is somewhat more difficult to determine the specific conditions which are necessary in order that $y_{12} = y_{21}$, but the use of determinant notation is of some help. Let us see if the relationships of Eqs. [10] to [13] can be expressed in terms of the impedance determinant and its minors.

Since our concern is with the two-port and not with the specific networks with which it is terminated, we will let V_1 and V_2 be represented by two ideal voltage sources. Equation [10] is applied by letting $V_2 = 0$ (thus short-circuiting the output) and finding the input admittance. The network now, however, is simply a one-port, and the input impedance of a one-port was found in Sec. 17.1. We select loop **1** to include the input terminals, and let I_1 be that loop's current; we identify $(-I_2)$ as the loop current in loop **2** and assign the remaining loop currents in any convenient manner. Thus,

$$\mathbf{Z}_{in}|_{\mathbf{V}_2=0} = \frac{\Delta_{\mathbf{Z}}}{\Delta_{11}}$$

and, therefore,

$$\mathbf{y}_{11} = \frac{\Delta_{11}}{\Delta_{\mathbf{Z}}}$$

Similarly,

$$\mathbf{y}_{22} = \frac{\Delta_{22}}{\Delta_{\mathbf{Z}}}$$

In order to find $\mathbf{y}_{12}$, we let $\mathbf{V}_1 = 0$ and find $\mathbf{I}_1$ as a function of $\mathbf{V}_2$. We find that $\mathbf{I}_1$ is given by the ratio

$$\mathbf{I}_1 = \frac{\begin{vmatrix} 0 & \mathbf{Z}_{12} & \cdots & \mathbf{Z}_{1N} \\ -\mathbf{V}_2 & \mathbf{Z}_{22} & \cdots & \mathbf{Z}_{2N} \\ 0 & \mathbf{Z}_{32} & \cdots & \mathbf{Z}_{3N} \\ \cdots & \cdots & \cdots & \cdots \\ 0 & \mathbf{Z}_{N2} & \cdots & \mathbf{Z}_{NN} \end{vmatrix}}{\begin{vmatrix} \mathbf{Z}_{11} & \mathbf{Z}_{12} & \cdots & \mathbf{Z}_{1N} \\ \mathbf{Z}_{21} & \mathbf{Z}_{22} & \cdots & \mathbf{Z}_{2N} \\ \mathbf{Z}_{31} & \mathbf{Z}_{32} & \cdots & \mathbf{Z}_{3N} \\ \cdots & \cdots & \cdots & \cdots \\ \mathbf{Z}_{N1} & \mathbf{Z}_{N2} & \cdots & \mathbf{Z}_{NN} \end{vmatrix}}$$

Thus,

$$\mathbf{I}_1 = -\frac{(-\mathbf{V}_2)\Delta_{21}}{\Delta_{\mathbf{Z}}}$$

and

$$\mathbf{y}_{12} = \frac{\Delta_{21}}{\Delta_{\mathbf{Z}}}$$

In a similar manner, we may show that

$$\mathbf{y}_{21} = \frac{\Delta_{12}}{\Delta_{\mathbf{Z}}}$$

The equality of $\mathbf{y}_{12}$ and $\mathbf{y}_{21}$ is thus contingent on the equality of the two minors of $\Delta_{\mathbf{Z}}$—Δ_{12} and Δ_{21}. These two minors are

$$\Delta_{21} = \begin{vmatrix} \mathbf{Z}_{12} & \mathbf{Z}_{13} & \mathbf{Z}_{14} & \cdots & \mathbf{Z}_{1N} \\ \mathbf{Z}_{32} & \mathbf{Z}_{33} & \mathbf{Z}_{34} & \cdots & \mathbf{Z}_{3N} \\ \mathbf{Z}_{42} & \mathbf{Z}_{43} & \mathbf{Z}_{44} & \cdots & \mathbf{Z}_{4N} \\ \cdots & \cdots & \cdots & \cdots & \cdots \\ \mathbf{Z}_{N2} & \mathbf{Z}_{N3} & \mathbf{Z}_{N4} & \cdots & \mathbf{Z}_{NN} \end{vmatrix}$$

and

$$\Delta_{12} = \begin{vmatrix} \mathbf{Z}_{21} & \mathbf{Z}_{23} & \mathbf{Z}_{24} & \cdots & \mathbf{Z}_{2N} \\ \mathbf{Z}_{31} & \mathbf{Z}_{33} & \mathbf{Z}_{34} & \cdots & \mathbf{Z}_{3N} \\ \mathbf{Z}_{41} & \mathbf{Z}_{43} & \mathbf{Z}_{44} & \cdots & \mathbf{Z}_{4N} \\ \cdots & \cdots & \cdots & \cdots & \cdots \\ \mathbf{Z}_{N1} & \mathbf{Z}_{N3} & \mathbf{Z}_{N4} & \cdots & \mathbf{Z}_{NN} \end{vmatrix}$$

Their equality is shown by first interchanging the rows and columns of one minor (for example, Δ_{21}), an operation which any college algebra book proves is valid, and then letting every mutual impedance $\mathbf{Z}_{ij}$ be replaced by $\mathbf{Z}_{ji}$. Thus, we set

$$\mathbf{Z}_{12} = \mathbf{Z}_{21} \qquad \mathbf{Z}_{23} = \mathbf{Z}_{32} \qquad \text{etc.}$$

This equality of $\mathbf{Z}_{ij}$ and $\mathbf{Z}_{ji}$ is evident for the three familiar passive elements—the resistor, capacitor, and inductor—and it is also true for mutual inductance. However, it is *not* true for *every* type of device which we may wish to include inside a two-port network. Specifically, it is not true in general for a dependent source, and it is not true for the gyrator, a useful model for Hall-effect devices and for waveguide sections containing ferrites. Over a narrow range of radian frequencies, the gyrator provides an additional phase shift of 180° for a signal passing from the output to the input over that for a signal in the forward direction, and thus $\mathbf{y}_{12} = -\mathbf{y}_{21}$. A common type of passive element leading to the inequality of $\mathbf{Z}_{ij}$ and $\mathbf{Z}_{ji}$, however, is a nonlinear element.

Any device for which $\mathbf{Z}_{ij} = \mathbf{Z}_{ji}$ is called a *bilateral element,* and a circuit which contains only bilateral elements is called a *bilateral circuit.* We have therefore shown that an important property of a bilateral two-port is

$$\mathbf{y}_{12} = \mathbf{y}_{21}$$

and this property is glorified by stating it as the *reciprocity theorem:*

Another way of stating the theorem is to say that the interchange of an ideal voltage source and an ideal ammeter in any passive, linear, bilateral circuit will not change the ammeter reading.

> In any passive linear bilateral network, if the single voltage source $\mathbf{V}_x$ in branch x produces the current response $\mathbf{I}_y$ in branch y, then the removal of the voltage source from branch x and its insertion in branch y will produce the current response $\mathbf{I}_y$ in branch x.

If we had been working with the admittance determinant of the circuit and had proved that the minors Δ_{21} and Δ_{12} of the admittance determinant $\Delta_\mathbf{Y}$ were equal, then we should have obtained the reciprocity theorem in its dual form:

In other words, the interchange of an ideal current source and an ideal voltmeter in any passive linear bilateral circuit will not change the voltmeter reading.

> In any passive linear bilateral network, if the single current source $\mathbf{I}_x$ between nodes x and x' produces the voltage response $\mathbf{V}_y$ between nodes y and y', then the removal of the current source from nodes x and x' and its insertion between nodes y and y' will produce the voltage response $\mathbf{V}_y$ between nodes x and x'.

PRACTICE

17.4 In the circuit of Fig. 17.10, let $\mathbf{I}_1$ and $\mathbf{I}_2$ represent ideal current sources. Assign the node voltage $\mathbf{V}_1$ at the input, $\mathbf{V}_2$ at the output, and $\mathbf{V}_x$ from the central node to the reference node. Write three nodal equations, eliminate $\mathbf{V}_x$ to obtain two equations, and then rearrange these equations into the form of Eqs. [5] and [6] so that all four $\mathbf{y}$ parameters may be read directly from the equations.

17.5 Find **y** for the two-port shown in Fig. 17.12.

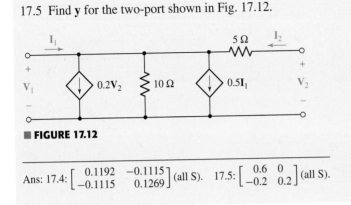

■ **FIGURE 17.12**

Ans: 17.4: $\begin{bmatrix} 0.1192 & -0.1115 \\ -0.1115 & 0.1269 \end{bmatrix}$ (all S).   17.5: $\begin{bmatrix} 0.6 & 0 \\ -0.2 & 0.2 \end{bmatrix}$ (all S).

17.3 SOME EQUIVALENT NETWORKS

When analyzing electronic circuits, it is often necessary to replace the active device (and perhaps some of its associated passive circuitry) with an equivalent two-port containing only three or four impedances. The validity of the equivalent may be restricted to small signal amplitudes and a single frequency, or perhaps a limited range of frequencies. The equivalent is also a linear approximation of a nonlinear circuit. However, if we are faced with a network containing a number of resistors, capacitors, and inductors, plus a transistor labeled 2N3823, then we cannot analyze the circuit by any of the techniques we have studied previously; the transistor must first be replaced by a linear model, just as we replaced the op amp by a linear model in Chap. 6. The **y** parameters provide one such model in the form of a two-port network that is often used at high frequencies. Another common linear model for a transistor appears in Sec. 17.5.

The two equations that determine the short-circuit admittance parameters,

$$\mathbf{I}_1 = \mathbf{y}_{11}\mathbf{V}_1 + \mathbf{y}_{12}\mathbf{V}_2 \qquad [16]$$

$$\mathbf{I}_2 = \mathbf{y}_{21}\mathbf{V}_1 + \mathbf{y}_{22}\mathbf{V}_2 \qquad [17]$$

have the form of a pair of nodal equations written for a circuit containing two nonreference nodes. The determination of an equivalent circuit that leads to Eqs. [16] and [17] is made more difficult by the inequality, in general, of $\mathbf{y}_{12}$ and $\mathbf{y}_{21}$; it helps to resort to a little trickery in order to obtain a pair of equations that possess equal mutual coefficients. Let us both add and subtract $\mathbf{y}_{12}\mathbf{V}_1$ (the term we would like to see present on the right side of Eq. [17]):

$$\mathbf{I}_2 = \mathbf{y}_{12}\mathbf{V}_1 + \mathbf{y}_{22}\mathbf{V}_2 + (\mathbf{y}_{21} - \mathbf{y}_{12})\mathbf{V}_1 \qquad [18]$$

or

$$\mathbf{I}_2 - (\mathbf{y}_{21} - \mathbf{y}_{12})\mathbf{V}_1 = \mathbf{y}_{12}\mathbf{V}_1 + \mathbf{y}_{22}\mathbf{V}_2 \qquad [19]$$

The right-hand sides of Eqs. [16] and [19] now show the proper symmetry for a bilateral circuit; the left-hand side of Eq. [19] may be interpreted as the algebraic sum of two current sources, one an independent source $\mathbf{I}_2$ entering node **2**, and the other a dependent source $(\mathbf{y}_{21} - \mathbf{y}_{12})\mathbf{V}_1$ leaving node **2**.

Let us now "read" the equivalent network from Eqs. [16] and [19]. We first provide a reference node, and then a node labeled $\mathbf{V}_1$ and one labeled $\mathbf{V}_2$.

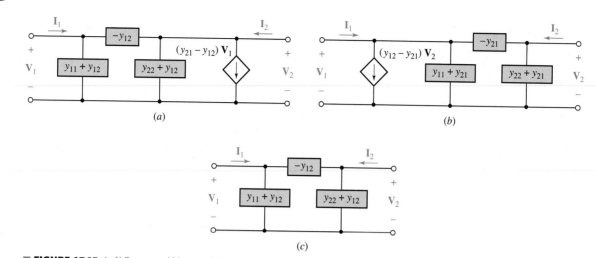

■ FIGURE 17.13 (a, b) Two-ports which are equivalent to any general linear two-port. The dependent source in part (a) depends on V_1, and that in part (b) depends on V_2. (c) An equivalent for a bilateral network.

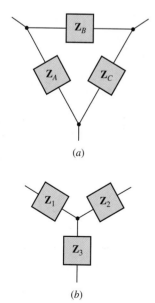

■ FIGURE 17.14 The three-terminal Δ network (a) and the three-terminal Y network (b) are equivalent if the six impedances satisfy the conditions of the Y-Δ (or Π-T) transformation, Eqs. [20] to [25].

From Eq. [16], we establish the current $\mathbf{I}_1$ flowing into node **1**, we supply a mutual admittance $(-\mathbf{y}_{12})$ between nodes **1** and **2**, and we supply an admittance of $(\mathbf{y}_{11} + \mathbf{y}_{12})$ between node **1** and the reference node. With $\mathbf{V}_2 = 0$, the ratio of $\mathbf{I}_1$ to $\mathbf{V}_1$ is then $\mathbf{y}_{11}$, as it should be. Now consider Eq. [19]; we cause the current $\mathbf{I}_2$ to flow into the second node, we cause the current $(\mathbf{y}_{21} - \mathbf{y}_{12})\mathbf{V}_1$ to leave the node, we note that the proper admittance $(-\mathbf{y}_{12})$ exists between the nodes, and we complete the circuit by installing the admittance $(\mathbf{y}_{22} + \mathbf{y}_{12})$ from node **2** to the reference node. The completed circuit is shown in Fig. 17.13a.

Another form of equivalent network is obtained by subtracting and adding $\mathbf{y}_{21}\mathbf{V}_2$ in Eq. [16]; this equivalent circuit is shown in Fig. 17.13b. If the two-port is bilateral, then $\mathbf{y}_{12} = \mathbf{y}_{21}$, and either of the equivalents reduces to a simple passive Π network. The dependent source disappears. This equivalent of the bilateral two-port is shown in Fig. 17.13c.

There are several uses to which these equivalent circuits may be put. In the first place, we have succeeded in showing that an equivalent of any complicated linear two-port *exists*. It does not matter how many nodes or loops are contained within the network; the equivalent is no more complex than the circuits of Fig. 17.13. One of these may be much simpler to use than the given circuit if we are interested only in the terminal characteristics of the given network.

The three-terminal network shown in Fig. 17.14a is often referred to as a Δ of impedances, while that in Fig. 17.14b is called a Y. One network may be replaced by the other if certain specific relationships between the impedances are satisfied, and these interrelationships may be established by use of the **y** parameters. We find that

$$\mathbf{y}_{11} = \frac{1}{\mathbf{Z}_A} + \frac{1}{\mathbf{Z}_B} = \frac{1}{\mathbf{Z}_1 + \mathbf{Z}_2\mathbf{Z}_3/(\mathbf{Z}_2 + \mathbf{Z}_3)}$$

$$\mathbf{y}_{12} = \mathbf{y}_{21} = -\frac{1}{\mathbf{Z}_B} = \frac{-\mathbf{Z}_3}{\mathbf{Z}_1\mathbf{Z}_2 + \mathbf{Z}_2\mathbf{Z}_3 + \mathbf{Z}_3\mathbf{Z}_1}$$

$$\mathbf{y}_{22} = \frac{1}{\mathbf{Z}_C} + \frac{1}{\mathbf{Z}_B} = \frac{1}{\mathbf{Z}_2 + \mathbf{Z}_1\mathbf{Z}_3/(\mathbf{Z}_1 + \mathbf{Z}_3)}$$

These equations may be solved for $\mathbf{Z}_A$, $\mathbf{Z}_B$, and $\mathbf{Z}_C$ in terms of $\mathbf{Z}_1$, $\mathbf{Z}_2$, and $\mathbf{Z}_3$:

$$\mathbf{Z}_A = \frac{\mathbf{Z}_1\mathbf{Z}_2 + \mathbf{Z}_2\mathbf{Z}_3 + \mathbf{Z}_3\mathbf{Z}_1}{\mathbf{Z}_2} \qquad [20]$$

$$\mathbf{Z}_B = \frac{\mathbf{Z}_1\mathbf{Z}_2 + \mathbf{Z}_2\mathbf{Z}_3 + \mathbf{Z}_3\mathbf{Z}_1}{\mathbf{Z}_3} \qquad [21]$$

$$\mathbf{Z}_C = \frac{\mathbf{Z}_1\mathbf{Z}_2 + \mathbf{Z}_2\mathbf{Z}_3 + \mathbf{Z}_3\mathbf{Z}_1}{\mathbf{Z}_1} \qquad [22]$$

or, for the inverse relationships:

$$\mathbf{Z}_1 = \frac{\mathbf{Z}_A\mathbf{Z}_B}{\mathbf{Z}_A + \mathbf{Z}_B + \mathbf{Z}_C} \qquad [23]$$

$$\mathbf{Z}_2 = \frac{\mathbf{Z}_B\mathbf{Z}_C}{\mathbf{Z}_A + \mathbf{Z}_B + \mathbf{Z}_C} \qquad [24]$$

$$\mathbf{Z}_3 = \frac{\mathbf{Z}_C\mathbf{Z}_A}{\mathbf{Z}_A + \mathbf{Z}_B + \mathbf{Z}_C} \qquad [25]$$

The reader may recall these useful relationships from Chap. 5, where their derivation was described.

These equations enable us to transform easily between the equivalent Y and Δ networks, a process known as the Y-Δ transformation (or Π-T transformation if the networks are drawn in the forms of those letters). In going from Y to Δ, Eqs. [20] to [22], first find the value of the common numerator as the sum of the products of the impedances in the Y taken two at a time. Each impedance in the Δ is then found by dividing the numerator by the impedance of that element in the Y which has no common node with the desired Δ element. Conversely, given the Δ, first take the sum of the three impedances around the Δ; then divide the product of the two Δ impedances having a common node with the desired Y element by that sum.

These transformations are often useful in simplifying passive networks, particularly resistive ones, thus avoiding the need for any mesh or nodal analysis.

EXAMPLE **17.6**

Find the input resistance of the circuit shown in Fig. 17.15a.

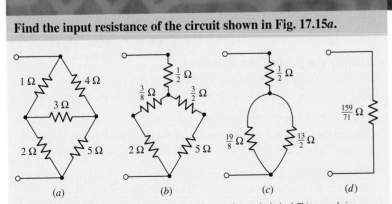

(a) (b) (c) (d)

■ **FIGURE 17.15** (*a*) A resistive network whose input resistance is desired. This example is repeated from Chap. 5. (*b*) The upper Δ is replaced by an equivalent Y. (*c*, *d*) Series and parallel combinations give the equivalent input resistance $\frac{159}{71}$ Ω.

(Continued on next page)

We first make a Δ-Y transformation on the upper Δ appearing in Fig. 17.15a. The sum of the three resistances forming this Δ is $1 + 4 + 3 = 8\ \Omega$. The product of the two resistors connected to the top node is $1 \times 4 = 4\ \Omega^2$. Thus, the upper resistor of the Y is $\frac{4}{8}$, or $\frac{1}{2}\ \Omega$. Repeating this procedure for the other two resistors, we obtain the network shown in Fig. 17.15b.

We next make the series and parallel combinations indicated, obtaining in succession Fig. 17.15c and d. Thus, the input resistance of the circuit in Fig. 17.15a is found to be $\frac{159}{71}$, or 2.24 Ω.

Now let us tackle a slightly more complicated example, shown as Fig. 17.16. We note that the circuit contains a dependent source, and thus the Y-Δ transformation is not applicable.

EXAMPLE **17.7**

The circuit shown in Fig. 17.16 is an approximate linear equivalent of a transistor amplifier in which the emitter terminal is the bottom node, the base terminal is the upper input node, and the collector terminal is the upper output node. A 2000 Ω resistor is connected between collector and base for some special application and makes the analysis of the circuit more difficult. Determine the y parameters for this circuit.

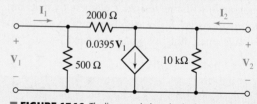

■ **FIGURE 17.16** The linear equivalent circuit of a transistor in common-emitter configuration with resistive feedback between collector and base.

▷ **Identify the goal of the problem.**
Cutting through the problem-specific jargon, we realize that we have been presented with a two-port network and require the **y** parameters.

▷ **Collect the known information.**
Figure 17.16 shows a two-port network with V_1, I_1, V_2, and I_2 already indicated, and a value for each component has been provided.

▷ **Devise a plan.**
There are several ways we might think about this circuit. If we recognize it as being in the form of the equivalent circuit shown in Fig. 17.13a, then we may immediately determine the values of the **y** parameters. If recognition is not immediate, then the **y** parameters

may be determined for the two-port by applying the relationships of Eqs. [10] to [13]. We also might avoid any use of two-port analysis methods and write equations directly for the circuit as it stands. The first option seems best in this case.

▶ **Construct an appropriate set of equations.**

By inspection, we find that $-\mathbf{y}_{21}$ corresponds to the admittance of our 2 kΩ resistor, that $\mathbf{y}_{11} + \mathbf{y}_{12}$ corresponds to the admittance of the 500 Ω resistor, the gain of the dependent current source corresponds to $\mathbf{y}_{21} - \mathbf{y}_{12}$, and finally that $\mathbf{y}_{22} + \mathbf{y}_{12}$ corresponds to the admittance of the 10 kΩ resistor. Hence we may write

$$\mathbf{y}_{12} = -\tfrac{1}{2000}$$

$$\mathbf{y}_{11} = \tfrac{1}{500} - \mathbf{y}_{12}$$

$$\mathbf{y}_{21} = 0.0395 + \mathbf{y}_{12}$$

$$\mathbf{y}_{22} = \tfrac{1}{10,000} - \mathbf{y}_{12}$$

▶ **Determine if additional information is required.**

With the equations written as they are, we see that once $\mathbf{y}_{12}$ is computed, the remaining $\mathbf{y}$ parameters may also be obtained.

▶ **Attempt a solution.**

Plugging the numbers into a calculator, we find that

$$\mathbf{y}_{12} = -\tfrac{1}{2000} = -0.5 \text{ mS}$$

$$\mathbf{y}_{11} = \tfrac{1}{500} - \left(-\tfrac{1}{2000}\right) = 2.5 \text{ mS}$$

$$\mathbf{y}_{22} = \tfrac{1}{10,000} - \left(-\tfrac{1}{2000}\right) = 0.6 \text{ mS}$$

and

$$\mathbf{y}_{21} = 0.0395 + \left(-\tfrac{1}{2000}\right) = 39 \text{ mS}$$

The following equations must then apply:

$$\mathbf{I}_1 = 2.5\mathbf{V}_1 - 0.5\mathbf{V}_2 \qquad\qquad [26]$$

$$\mathbf{I}_2 = 39\mathbf{V}_1 + 0.6\mathbf{V}_2 \qquad\qquad [27]$$

where we are now using units of mA, V, and mS or kΩ.

▶ **Verify the solution. Is it reasonable or expected?**

Writing two nodal equations directly from the circuit, we find

$$\mathbf{I}_1 = \frac{\mathbf{V}_1 - \mathbf{V}_2}{2} + \frac{\mathbf{V}_1}{0.5} \qquad \text{or} \qquad \mathbf{I}_1 = 2.5\mathbf{V}_1 - 0.5\mathbf{V}_2$$

and

$$-39.5\mathbf{V}_1 + \mathbf{I}_2 = \frac{\mathbf{V}_2 - \mathbf{V}_1}{2} + \frac{\mathbf{V}_2}{10} \qquad \text{or} \qquad \mathbf{I}_2 = 39\mathbf{V}_1 + 0.6\mathbf{V}_2$$

which agree with Eqs. [26] and [27] obtained directly from the $\mathbf{y}$ parameters.

Now let us make use of Eqs. [26] and [27] by analyzing the performance of the two-port in Fig. 17.16 under several different operating conditions. We first provide a current source of $1\underline{/0°}$ mA at the input and connect a 0.5 kΩ (2 mS) load to the output. The terminating networks are therefore both one-ports and give us the following specific information relating $\mathbf{I}_1$ to $\mathbf{V}_1$ and $\mathbf{I}_2$ to $\mathbf{V}_2$:

$$\mathbf{I}_1 = 1 \text{ (for any } \mathbf{V}_1) \qquad \mathbf{I}_2 = -2\mathbf{V}_2$$

We now have four equations in the four variables, $\mathbf{V}_1$, $\mathbf{V}_2$, $\mathbf{I}_1$, and $\mathbf{I}_2$. Substituting the two one-port relationships in Eqs. [26] and [27], we obtain two equations relating $\mathbf{V}_1$ and $\mathbf{V}_2$:

$$1 = 2.5\mathbf{V}_1 - 0.5\mathbf{V}_2 \qquad 0 = 39\mathbf{V}_1 + 2.6\mathbf{V}_2$$

Solving, we find that

$$\mathbf{V}_1 = 0.1 \text{ V} \qquad \mathbf{V}_2 = -1.5 \text{ V}$$
$$\mathbf{I}_1 = 1 \text{ mA} \qquad \mathbf{I}_2 = 3 \text{ mA}$$

These four values apply to the two-port operating with a prescribed input ($\mathbf{I}_1 = 1$ mA) and a specified load ($R_L = 0.5$ kΩ).

The performance of an amplifier is often described by giving a few specific values. Let us calculate four of these values for this two-port with its terminations. We will define and evaluate the voltage gain, the current gain, the power gain, and the input impedance.

The *voltage gain* $\mathbf{G}_V$ is

$$\mathbf{G}_V = \frac{\mathbf{V}_2}{\mathbf{V}_1}$$

From the numerical results, it is easy to see that $\mathbf{G}_V = -15$.

The *current gain* $\mathbf{G}_I$ is defined as

$$\mathbf{G}_I = \frac{\mathbf{I}_2}{\mathbf{I}_1}$$

and we have

$$\mathbf{G}_I = 3$$

Let us define and calculate the *power gain* G_P for an assumed sinusoidal excitation. We have

$$G_P = \frac{P_{\text{out}}}{P_{\text{in}}} = \frac{\text{Re}\left\{-\frac{1}{2}\mathbf{V}_2\mathbf{I}_2^*\right\}}{\text{Re}\left\{\frac{1}{2}\mathbf{V}_1\mathbf{I}_1^*\right\}} = 45$$

The device might be termed either a voltage, a current, or a power amplifier, since all the gains are greater than unity. If the 2 kΩ resistor were removed, the power gain would rise to 354.

The input and output impedances of the amplifier are often desired in order that maximum power transfer may be achieved to or from an adjacent two-port. We define the *input impedance* $\mathbf{Z}_{\text{in}}$ as the ratio of input voltage to current:

$$\mathbf{Z}_{\text{in}} = \frac{\mathbf{V}_1}{\mathbf{I}_1} = 0.1 \text{ k}\Omega$$

This is the impedance offered to the current source when the 500 Ω load is connected to the output. (With the output short-circuited, the input impedance is necessarily $1/\mathbf{y}_{11}$, or 400 Ω.)

It should be noted that the input impedance *cannot* be determined by replacing every source with its internal impedance and then combining resistances or conductances. In the given circuit, such a procedure would yield a value of 416 Ω. The error, of course, comes from treating the *dependent* source as an *independent* source. If we think of the input impedance as being numerically equal to the input voltage produced by an input current of 1 A, the application of the 1 A source produces some input voltage $\mathbf{V}_1$, and the strength of the dependent source ($0.0395\mathbf{V}_1$) cannot be zero. We should recall that when we obtain the Thévenin equivalent impedance of a circuit containing a dependent source along with one or more independent sources, we must replace the independent sources with short circuits or open circuits, but a dependent source must not be deactivated. Of course, if the voltage or current on which the dependent source depends is zero, then the dependent source will itself be inactive; occasionally a circuit may be simplified by recognizing such an occurrence.

Besides $\mathbf{G}_V$, $\mathbf{G}_I$, G_P, and $\mathbf{Z}_{in}$, there is one other performance parameter that is quite useful. This is the *output impedance* $\mathbf{Z}_{out}$, and it is determined for a different circuit configuration.

The output impedance is just another term for the Thévenin impedance appearing in the Thévenin equivalent circuit of that portion of the network faced by the load. In our circuit, which we have assumed is driven by a $1\underline{/0°}$ mA current source, we therefore replace this independent source with an open circuit, leave the dependent source alone, and seek the *input* impedance seen looking to the left from the output terminals (with the load removed). Thus, we define

$$\mathbf{Z}_{out} = \mathbf{V}_2|_{\mathbf{I}_2=1 \text{ A with all other independent sources deactivated } and \ R_L \text{ removed}}$$

We therefore remove the load resistor, apply $1\underline{/0°}$ mA (since we are working in V, mA, and kΩ) at the output terminals, and determine $\mathbf{V}_2$. We place these requirements on Eqs. [26] and [27], and obtain

$$0 = 2.5\mathbf{V}_1 - 0.5\mathbf{V}_2 \qquad 1 = 39\mathbf{V}_1 + 0.6\mathbf{V}_2$$

Solving,

$$\mathbf{V}_2 = 0.1190 \text{ V}$$

and thus

$$\mathbf{Z}_{out} = 0.1190 \text{ k}\Omega$$

An alternative procedure might be to find the open-circuit output voltage and the short-circuit output current. That is, the Thévenin impedance is the output impedance:

$$\mathbf{Z}_{out} = \mathbf{Z}_{th} = -\frac{\mathbf{V}_{2oc}}{\mathbf{I}_{2sc}}$$

Carrying out this procedure, we first rekindle the independent source so that $\mathbf{I}_1 = 1$ mA, and then open-circuit the load so that $\mathbf{I}_2 = 0$. We have

$$1 = 2.5\mathbf{V}_1 - 0.5\mathbf{V}_2 \qquad 0 = 39\mathbf{V}_1 + 0.6\mathbf{V}_2$$

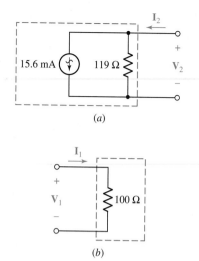

(a)

(b)

■ **FIGURE 17.17** (a) The Norton equivalent of the network in Fig. 17.16 to the left of the output terminal, $I_1 = 1\underline{/0°}$ mA. (b) The Thévenin equivalent of that portion of the network to the right of the input terminals, if $I_2 = -2V_2$ mA.

and thus

$$\mathbf{V}_{2oc} = -1.857 \text{ V}$$

Next, we apply short-circuit conditions by setting $\mathbf{V}_2 = 0$ and again let $\mathbf{I}_1 = 1$ mA. We find that

$$\mathbf{I}_1 = 1 = 2.5\mathbf{V}_1 - 0 \qquad \mathbf{I}_2 = 39\mathbf{V}_1 + 0$$

and thus

$$\mathbf{I}_{2sc} = 15.6 \text{ mA}$$

The assumed directions of $\mathbf{V}_2$ and $\mathbf{I}_2$ therefore result in a Thévenin or output impedance

$$\mathbf{Z}_{\text{out}} = -\frac{\mathbf{V}_{2oc}}{\mathbf{I}_{2sc}} = -\frac{-1.857}{15.6} = 0.1190 \text{ k}\Omega$$

as before.

We now have enough information to enable us to draw the Thévenin or Norton equivalent of the two-port of Fig. 17.16 when it is driven by a $1\underline{/0°}$ mA current source and terminated in a 500 Ω load. Thus, the Norton equivalent presented to the load must contain a current source equal to the short-circuit current $\mathbf{I}_{2sc}$ in parallel with the output impedance; this equivalent is shown in Fig. 17.17a. Also, the Thévenin equivalent offered to the $1\underline{/0°}$ mA input source must consist solely of the input impedance, as drawn in Fig. 17.17b.

Before leaving the **y** parameters, we should recognize their usefulness in describing the parallel connection of two-ports, as indicated in Fig. 17.18. When we first defined a port in Sec. 17.1, we noted that the currents entering and leaving the two terminals of a port had to be equal, and there could be no external connections made that bridged between ports. Apparently the parallel connection shown in Fig. 17.18 violates this condition. However, if each two-port has a reference node that is common to its input and output port, and if the two-ports are connected in parallel so that they have a common reference node, then all ports remain ports after the connection. Thus, for the A network,

$$\mathbf{I}_A = \mathbf{y}_A \mathbf{V}_A$$

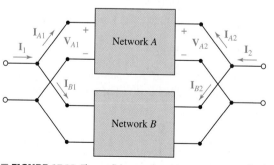

■ **FIGURE 17.18** The parallel connection of two two-port networks. If both inputs and outputs have the same reference node, then the admittance matrix $\mathbf{y} = \mathbf{y}_A + \mathbf{y}_B$.

where

$$\mathbf{I}_A = \begin{bmatrix} \mathbf{I}_{A1} \\ \mathbf{I}_{A2} \end{bmatrix} \qquad \text{and} \qquad \mathbf{V}_A = \begin{bmatrix} \mathbf{V}_{A1} \\ \mathbf{V}_{A2} \end{bmatrix}$$

and for the B network

$$\mathbf{I}_B = \mathbf{y}_B \mathbf{V}_B$$

But

$$\mathbf{V}_A = \mathbf{V}_B = \mathbf{V} \qquad \text{and} \qquad \mathbf{I} = \mathbf{I}_A + \mathbf{I}_B$$

Thus,

$$\mathbf{I} = (\mathbf{y}_A + \mathbf{y}_B)\mathbf{V}$$

and we see that each **y** parameter of the parallel network is given as the sum of the corresponding parameters of the individual networks,

$$\mathbf{y} = \mathbf{y}_A + \mathbf{y}_B \qquad\qquad [28]$$

This may be extended to any number of two-ports connected in parallel.

PRACTICE

17.6 Find **y** and $\mathbf{Z}_{\text{out}}$ for the terminated two-port shown in Fig. 17.19.
17.7 Use Δ-Y and Y-Δ transformations to determine R_{in} for the network shown in (a) Fig. 17.20a; (b) Fig. 17.20b.

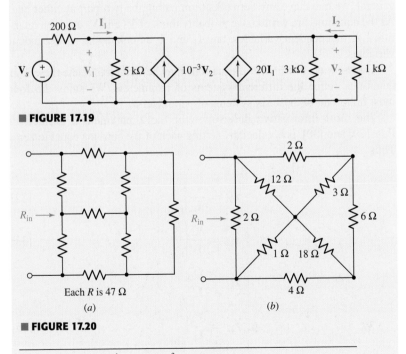

■ FIGURE 17.19

■ FIGURE 17.20

Each R is 47 Ω

(a)   (b)

Ans: 17.6: $\begin{bmatrix} 2 \times 10^{-4} & -10^{-3} \\ -4 \times 10^{-3} & 20.3 \times 10^{-3} \end{bmatrix}$ (S); 51.1 Ω. 17.7: 53.71 Ω, 1.311 Ω.

17.4 IMPEDANCE PARAMETERS

The concept of two-port parameters has been introduced in terms of the short-circuit admittance parameters. There are other sets of parameters, however, and each set is associated with a particular class of networks for which its use provides the simplest analysis. We will consider three other types of parameters, the open-circuit impedance parameters, which are the subject of this section; and the hybrid and the transmission parameters, which are discussed in following sections.

We begin again with a general linear two-port that does not contain any independent sources; the currents and voltages are assigned as before (Fig. 17.8). Now let us consider the voltage $\mathbf{V}_1$ as the response produced by two current sources $\mathbf{I}_1$ and $\mathbf{I}_2$. We thus write for $\mathbf{V}_1$

$$\mathbf{V}_1 = \mathbf{z}_{11}\mathbf{I}_1 + \mathbf{z}_{12}\mathbf{I}_2 \qquad [29]$$

and for $\mathbf{V}_2$

$$\mathbf{V}_2 = \mathbf{z}_{21}\mathbf{I}_1 + \mathbf{z}_{22}\mathbf{I}_2 \qquad [30]$$

or

$$\mathbf{V} = \begin{bmatrix} \mathbf{V}_1 \\ \mathbf{V}_2 \end{bmatrix} = \mathbf{z}\mathbf{I} = \begin{bmatrix} \mathbf{z}_{11} & \mathbf{z}_{12} \\ \mathbf{z}_{21} & \mathbf{z}_{22} \end{bmatrix} \begin{bmatrix} \mathbf{I}_1 \\ \mathbf{I}_2 \end{bmatrix} \qquad [31]$$

Of course, in using these equations it is not necessary that $\mathbf{I}_1$ and $\mathbf{I}_2$ be current sources; nor is it necessary that $\mathbf{V}_1$ and $\mathbf{V}_2$ be voltage sources. In general, we may have any networks terminating the two-port at either end. As the equations are written, we probably think of $\mathbf{V}_1$ and $\mathbf{V}_2$ as given quantities, or independent variables, and $\mathbf{I}_1$ and $\mathbf{I}_2$ as unknowns, or dependent variables.

The six ways in which two equations may be written to relate these four quantities define the different systems of parameters. We study the four most important of these six systems of parameters.

The most informative description of the $\mathbf{z}$ parameters, defined in Eqs. [29] and [30], is obtained by setting each of the currents equal to zero. Thus

$$\mathbf{z}_{11} = \left.\frac{\mathbf{V}_1}{\mathbf{I}_1}\right|_{\mathbf{I}_2=0} \qquad [32]$$

$$\mathbf{z}_{12} = \left.\frac{\mathbf{V}_1}{\mathbf{I}_2}\right|_{\mathbf{I}_1=0} \qquad [33]$$

$$\mathbf{z}_{21} = \left.\frac{\mathbf{V}_2}{\mathbf{I}_1}\right|_{\mathbf{I}_2=0} \qquad [34]$$

$$\mathbf{z}_{22} = \left.\frac{\mathbf{V}_2}{\mathbf{I}_2}\right|_{\mathbf{I}_1=0} \qquad [35]$$

Since zero current results from an open-circuit termination, the $\mathbf{z}$ parameters are known as the *open-circuit impedance parameters*. They are easily related to the short-circuit admittance parameters by solving Eqs. [29] and

[30] for $\mathbf{I}_1$ and $\mathbf{I}_2$:

$$\mathbf{I}_1 = \frac{\begin{vmatrix} \mathbf{V}_1 & \mathbf{z}_{12} \\ \mathbf{V}_2 & \mathbf{z}_{22} \end{vmatrix}}{\begin{vmatrix} \mathbf{z}_{11} & \mathbf{z}_{12} \\ \mathbf{z}_{21} & \mathbf{z}_{22} \end{vmatrix}}$$

or

$$\mathbf{I}_1 = \left(\frac{\mathbf{z}_{22}}{\mathbf{z}_{11}\mathbf{z}_{22} - \mathbf{z}_{12}\mathbf{z}_{21}} \right) \mathbf{V}_1 - \left(\frac{\mathbf{z}_{12}}{\mathbf{z}_{11}\mathbf{z}_{22} - \mathbf{z}_{12}\mathbf{z}_{21}} \right) \mathbf{V}_2$$

Using determinant notation, and being careful that the subscript is a lower-case $\mathbf{z}$, we assume that $\Delta_{\mathbf{z}} \neq 0$ and obtain

$$\mathbf{y}_{11} = \frac{\Delta_{11}}{\Delta_{\mathbf{z}}} = \frac{\mathbf{z}_{22}}{\Delta_{\mathbf{z}}} \qquad \mathbf{y}_{12} = -\frac{\Delta_{21}}{\Delta_{\mathbf{z}}} = -\frac{\mathbf{z}_{12}}{\Delta_{\mathbf{z}}}$$

and from solving for $\mathbf{I}_2$,

$$\mathbf{y}_{21} = -\frac{\Delta_{12}}{\Delta_{\mathbf{z}}} = -\frac{\mathbf{z}_{21}}{\Delta_{\mathbf{z}}} \qquad \mathbf{y}_{22} = \frac{\Delta_{22}}{\Delta_{\mathbf{z}}} = \frac{\mathbf{z}_{11}}{\Delta_{\mathbf{z}}}$$

In a similar manner, the $\mathbf{z}$ parameters may be expressed in terms of the admittance parameters. Transformations of this nature are possible between any of the various parameter systems, and quite a collection of occasionally useful formulas may be obtained. Transformations between the $\mathbf{y}$ and $\mathbf{z}$ parameters (as well as the $\mathbf{h}$ and $\mathbf{t}$ parameters which we will consider in the following sections) are given in Table 17.1 as a helpful reference.

TABLE 17.1 Transformations Between y, z, h, and t Parameters

	y		z		h		t	
y	$\mathbf{y}_{11}$	$\mathbf{y}_{12}$	$\dfrac{\mathbf{z}_{22}}{\Delta_{\mathbf{z}}}$	$\dfrac{-\mathbf{z}_{12}}{\Delta_{\mathbf{z}}}$	$\dfrac{1}{\mathbf{h}_{11}}$	$\dfrac{-\mathbf{h}_{12}}{\mathbf{h}_{11}}$	$\dfrac{\mathbf{t}_{22}}{\mathbf{t}_{12}}$	$\dfrac{-\Delta_{\mathbf{t}}}{\mathbf{t}_{12}}$
	$\mathbf{y}_{21}$	$\mathbf{y}_{22}$	$\dfrac{-\mathbf{z}_{21}}{\Delta_{\mathbf{z}}}$	$\dfrac{\mathbf{z}_{11}}{\Delta_{\mathbf{z}}}$	$\dfrac{\mathbf{h}_{21}}{\mathbf{h}_{11}}$	$\dfrac{\Delta_{\mathbf{h}}}{\mathbf{h}_{11}}$	$\dfrac{-1}{\mathbf{t}_{12}}$	$\dfrac{\mathbf{t}_{11}}{\mathbf{t}_{12}}$
z	$\dfrac{\mathbf{y}_{22}}{\Delta_{\mathbf{y}}}$	$\dfrac{-\mathbf{y}_{12}}{\Delta_{\mathbf{y}}}$	$\mathbf{z}_{11}$	$\mathbf{z}_{12}$	$\dfrac{\Delta_{\mathbf{h}}}{\mathbf{h}_{22}}$	$\dfrac{\mathbf{h}_{12}}{\mathbf{h}_{22}}$	$\dfrac{\mathbf{t}_{11}}{\mathbf{t}_{21}}$	$\dfrac{\Delta_{\mathbf{t}}}{\mathbf{t}_{21}}$
	$\dfrac{-\mathbf{y}_{21}}{\Delta_{\mathbf{y}}}$	$\dfrac{\mathbf{y}_{11}}{\Delta_{\mathbf{y}}}$	$\mathbf{z}_{21}$	$\mathbf{z}_{22}$	$\dfrac{-\mathbf{h}_{21}}{\mathbf{h}_{22}}$	$\dfrac{1}{\mathbf{h}_{22}}$	$\dfrac{1}{\mathbf{t}_{21}}$	$\dfrac{\mathbf{t}_{22}}{\mathbf{t}_{21}}$
h	$\dfrac{1}{\mathbf{y}_{11}}$	$\dfrac{-\mathbf{y}_{12}}{\mathbf{y}_{11}}$	$\dfrac{\Delta_{\mathbf{z}}}{\mathbf{z}_{22}}$	$\dfrac{\mathbf{z}_{12}}{\mathbf{z}_{22}}$	$\mathbf{h}_{11}$	$\mathbf{h}_{12}$	$\dfrac{\mathbf{t}_{12}}{\mathbf{t}_{22}}$	$\dfrac{\Delta_{\mathbf{t}}}{\mathbf{t}_{22}}$
	$\dfrac{\mathbf{y}_{21}}{\mathbf{y}_{11}}$	$\dfrac{\Delta_{\mathbf{y}}}{\mathbf{y}_{11}}$	$\dfrac{-\mathbf{z}_{21}}{\mathbf{z}_{22}}$	$\dfrac{1}{\mathbf{z}_{22}}$	$\mathbf{h}_{21}$	$\mathbf{h}_{22}$	$\dfrac{-1}{\mathbf{t}_{22}}$	$\dfrac{\mathbf{t}_{21}}{\mathbf{t}_{22}}$
t	$\dfrac{-\mathbf{y}_{22}}{\mathbf{y}_{21}}$	$\dfrac{-1}{\mathbf{y}_{21}}$	$\dfrac{\mathbf{z}_{11}}{\mathbf{z}_{21}}$	$\dfrac{\Delta_{\mathbf{z}}}{\mathbf{z}_{21}}$	$\dfrac{-\Delta_{\mathbf{h}}}{\mathbf{h}_{21}}$	$\dfrac{-\mathbf{h}_{11}}{\mathbf{h}_{21}}$	$\mathbf{t}_{11}$	$\mathbf{t}_{12}$
	$\dfrac{-\Delta_{\mathbf{y}}}{\mathbf{y}_{21}}$	$\dfrac{-\mathbf{y}_{11}}{\mathbf{y}_{21}}$	$\dfrac{1}{\mathbf{z}_{21}}$	$\dfrac{\mathbf{z}_{22}}{\mathbf{z}_{21}}$	$\dfrac{-\mathbf{h}_{22}}{\mathbf{h}_{21}}$	$\dfrac{-1}{\mathbf{h}_{21}}$	$\mathbf{t}_{21}$	$\mathbf{t}_{22}$

For all parameter sets: $\Delta_{\mathbf{p}} = \mathbf{p}_{11}\mathbf{p}_{22} - \mathbf{p}_{12}\mathbf{p}_{21}$.

If the two-port is a bilateral network, reciprocity is present; it is easy to show that this results in the equality of $\mathbf{z}_{12}$ and $\mathbf{z}_{21}$.

Equivalent circuits may again be obtained from an inspection of Eqs. [29] and [30]; their construction is facilitated by adding and subtracting either $\mathbf{z}_{12}\mathbf{I}_1$ in Eq. [30] or $\mathbf{z}_{21}\mathbf{I}_2$ in Eq. [29]. Each of these equivalent circuits contains a dependent voltage source.

Let us leave the derivation of such an equivalent to some leisure moment, and consider next an example of a rather general nature. Can we construct a general Thévenin equivalent of the two-port, as viewed from the output terminals? It is necessary first to assume a specific input circuit configuration, and we will select an independent voltage source $\mathbf{V}_s$ (positive sign at top) in series with a generator impedance $\mathbf{Z}_g$. Thus

$$\mathbf{V}_s = \mathbf{V}_1 + \mathbf{I}_1\mathbf{Z}_g$$

Combining this result with Eqs. [29] and [30], we may eliminate $\mathbf{V}_1$ and $\mathbf{I}_1$ and obtain

$$\mathbf{V}_2 = \frac{\mathbf{z}_{21}}{\mathbf{z}_{11} + \mathbf{Z}_g}\mathbf{V}_s + \left(\mathbf{z}_{22} - \frac{\mathbf{z}_{12}\mathbf{z}_{21}}{\mathbf{z}_{11} + \mathbf{Z}_g}\right)\mathbf{I}_2$$

The Thévenin equivalent circuit may be drawn directly from this equation; it is shown in Fig. 17.21. The output impedance, expressed in terms of the $\mathbf{z}$ parameters, is

$$\mathbf{Z}_{\text{out}} = \mathbf{z}_{22} - \frac{\mathbf{z}_{12}\mathbf{z}_{21}}{\mathbf{z}_{11} + \mathbf{Z}_g}$$

If the generator impedance is zero, the simpler expression

$$\mathbf{Z}_{\text{out}} = \frac{\mathbf{z}_{11}\mathbf{z}_{22} - \mathbf{z}_{12}\mathbf{z}_{21}}{\mathbf{z}_{11}} = \frac{\Delta_\mathbf{z}}{\Delta_{22}} = \frac{1}{\mathbf{y}_{22}} \qquad (\mathbf{Z}_g = 0)$$

is obtained. For this special case, the output *admittance* is identical to $\mathbf{y}_{22}$, as indicated by the basic relationship of Eq. [13].

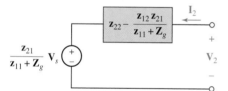

■ **FIGURE 17.21** The Thévenin equivalent of a general two-port, as viewed from the output terminals, expressed in terms of the open-circuit impedance parameters.

EXAMPLE **17.8**

Given the set of impedance parameters

$$\mathbf{z} = \begin{bmatrix} 10^3 & 10 \\ -10^6 & 10^4 \end{bmatrix} \qquad \textbf{(all } \Omega\text{)}$$

which is representative of a bipolar junction transistor operating in the common-emitter configuration, determine the voltage, current, and power gains, as well as the input and output impedances. The two-port is driven by an ideal sinusoidal voltage source $\mathbf{V}_s$ in series with a 500 Ω resistor, and terminated in a 10 kΩ load resistor.

The two describing equations for the two-port are

$$\mathbf{V}_1 = 10^3\mathbf{I}_1 + 10\mathbf{I}_2 \qquad [36]$$

$$\mathbf{V}_2 = -10^6\mathbf{I}_1 + 10^4\mathbf{I}_2 \qquad [37]$$

current, as $(-\mathbf{I}_2)$. Thus, both $\mathbf{I}_1$ and $-\mathbf{I}_2$ are directed to the right, the direction of energy or signal transmission.

Other widely used nomenclature for this set of parameters is

$$\begin{bmatrix} \mathbf{t}_{11} & \mathbf{t}_{12} \\ \mathbf{t}_{21} & \mathbf{t}_{22} \end{bmatrix} = \begin{bmatrix} \mathbf{A} & \mathbf{B} \\ \mathbf{C} & \mathbf{D} \end{bmatrix} \qquad [46]$$

Note that there are no minus signs in the **t** or **ABCD** matrices.

Looking again at Eqs. [43] to [45], we see that the quantities on the left, often thought of as the given or independent variables, are the input voltage and current, $\mathbf{V}_1$ and $\mathbf{I}_1$; the dependent variables, $\mathbf{V}_2$ and $\mathbf{I}_2$, are the output quantities. Thus, the transmission parameters provide a direct relationship between input and output. Their major use arises in transmission-line analysis and in cascaded networks.

Let us find the **t** parameters for the bilateral resistive two-port of Fig. 17.30*a*. To illustrate one possible procedure for finding a single parameter, consider

$$\mathbf{t}_{12} = \left. \frac{\mathbf{V}_1}{-\mathbf{I}_2} \right|_{\mathbf{V}_2=0}$$

We therefore short-circuit the output ($\mathbf{V}_2 = 0$) and set $\mathbf{V}_1 = 1$ V, as shown in Fig. 17.30*b*. Note that we cannot set the denominator equal to unity by placing a 1 A current source at the output; we already have a short circuit there. The equivalent resistance offered to the 1 V source is $R_{\text{eq}} = 2 + (4\|10)\ \Omega$, and we then use current division to get

$$-\mathbf{I}_2 = \frac{1}{2 + (4\|10)} \times \frac{10}{10+4} = \frac{5}{34}\ \text{A}$$

Hence,

$$\mathbf{t}_{12} = \frac{1}{-\mathbf{I}_2} = \frac{34}{5} = 6.8\ \Omega$$

If it is necessary to find all four parameters, we write any convenient pair of equations using all four terminal quantities, $\mathbf{V}_1$, $\mathbf{V}_2$, $\mathbf{I}_1$, and $\mathbf{I}_2$. From Fig. 17.30*a*, we have two mesh equations:

$$\mathbf{V}_1 = 12\mathbf{I}_1 + 10\mathbf{I}_2 \qquad [47]$$

$$\mathbf{V}_2 = 10\mathbf{I}_1 + 14\mathbf{I}_2 \qquad [48]$$

Solving Eq. [48] for $\mathbf{I}_1$, we get

$$\mathbf{I}_1 = 0.1\mathbf{V}_2 - 1.4\mathbf{I}_2$$

so that $\mathbf{t}_{21} = 0.1$ S and $\mathbf{t}_{22} = 1.4$. Substituting the expression for $\mathbf{I}_1$ in Eq. [47], we find

$$\mathbf{V}_1 = 12(0.1\mathbf{V}_2 - 1.4\mathbf{I}_2) + 10\mathbf{I}_2 = 1.2\mathbf{V}_2 - 6.8\mathbf{I}_2$$

and $\mathbf{t}_{11} = 1.2$ and $\mathbf{t}_{12} = 6.8\ \Omega$, once again.

For reciprocal networks, the determinant of the **t** matrix is equal to unity:

$$\Delta_t = \mathbf{t}_{11}\mathbf{t}_{22} - \mathbf{t}_{12}\mathbf{t}_{21} = 1$$

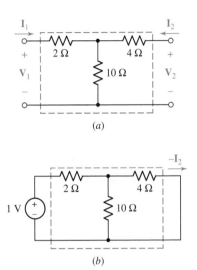

■ **FIGURE 17.30** (*a*) A two-port resistive network for which the **t** parameters are to be found. (*b*) To find $\mathbf{t}_{12}$, set $\mathbf{V}_1 = 1$ V with $\mathbf{V}_2 = 0$; then $\mathbf{t}_{12} = 1/(-\mathbf{I}_2) = 6.8\ \Omega$.

In the resistive example of Fig. 17.30, $\Delta_t = 1.2 \times 1.4 - 6.8 \times 0.1 = 1$. Good!

We conclude our two-port discussion by connecting two two-ports in cascade, as illustrated for two networks in Fig. 17.31. Terminal voltages and currents are indicated for each two-port, and the corresponding **t** parameter relationships are, for network A,

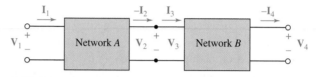

■ **FIGURE 17.31** When two-port networks A and B are cascaded, the **t** parameter matrix for the combined network is given by the matrix product $\mathbf{t} = \mathbf{t}_A \mathbf{t}_B$.

$$\begin{bmatrix} \mathbf{V}_1 \\ \mathbf{I}_1 \end{bmatrix} = \mathbf{t}_A \begin{bmatrix} \mathbf{V}_2 \\ -\mathbf{I}_2 \end{bmatrix} = \mathbf{t}_A \begin{bmatrix} \mathbf{V}_3 \\ \mathbf{I}_3 \end{bmatrix}$$

and for network B,

$$\begin{bmatrix} \mathbf{V}_3 \\ \mathbf{I}_3 \end{bmatrix} = \mathbf{t}_B \begin{bmatrix} \mathbf{V}_4 \\ -\mathbf{I}_4 \end{bmatrix}$$

Combining these results, we have

$$\begin{bmatrix} \mathbf{V}_1 \\ \mathbf{I}_1 \end{bmatrix} = \mathbf{t}_A \mathbf{t}_B \begin{bmatrix} \mathbf{V}_4 \\ -\mathbf{I}_4 \end{bmatrix}$$

Therefore, the **t** parameters for the cascaded networks are found by the matrix product,

$$\mathbf{t} = \mathbf{t}_A \mathbf{t}_B$$

This product is *not* obtained by multiplying corresponding elements in the two matrices. If necessary, review the correct procedure for matrix multiplication in Appendix 2.

EXAMPLE 17.10

Find the t parameters for the cascaded networks shown in Fig. 17.32.

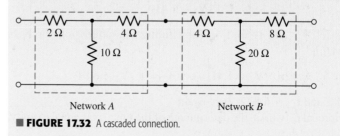

Network A Network B

■ **FIGURE 17.32** A cascaded connection.

Network A is the two-port of Fig. 17.32, and, therefore

$$\mathbf{t}_A = \begin{bmatrix} 1.2 & 6.8\ \Omega \\ 0.1\ \text{S} & 1.4 \end{bmatrix}$$

while network B has resistance values twice as large, so that

$$\mathbf{t}_B = \begin{bmatrix} 1.2 & 13.6\ \Omega \\ 0.05\ \text{S} & 1.4 \end{bmatrix}$$

For the combined network,

$$\mathbf{t} = \mathbf{t}_A \mathbf{t}_B = \begin{bmatrix} 1.2 & 6.8 \\ 0.1 & 1.4 \end{bmatrix} \begin{bmatrix} 1.2 & 13.6 \\ 0.05 & 1.4 \end{bmatrix}$$

$$= \begin{bmatrix} 1.2 \times 1.2 + 6.8 \times 0.05 & 1.2 \times 13.6 + 6.8 \times 1.4 \\ 0.1 \times 1.2 + 1.4 \times 0.05 & 0.1 \times 13.6 + 1.4 \times 1.4 \end{bmatrix}$$

and

$$\mathbf{t} = \begin{bmatrix} 1.78 & 25.84\ \Omega \\ 0.19\ \text{S} & 3.32 \end{bmatrix}$$

PRACTICE

17.12 Given $\mathbf{t} = \begin{bmatrix} 3.2 & 8\ \Omega \\ 0.2\ \text{S} & 4 \end{bmatrix}$, find (a) $\mathbf{z}$; (b) $\mathbf{t}$ for two identical networks in cascade; (c) $\mathbf{z}$ for two identical networks in cascade.

Ans: $\begin{bmatrix} 16 & 56 \\ 5 & 20 \end{bmatrix} (\Omega)$; $\begin{bmatrix} 11.84 & 57.6\ \Omega \\ 1.44\ \text{S} & 17.6 \end{bmatrix}$; $\begin{bmatrix} 8.22 & 87.1 \\ 0.694 & 12.22 \end{bmatrix} (\Omega)$.

COMPUTER-AIDED ANALYSIS

The characterization of two-port networks using $\mathbf{t}$ parameters creates the opportunity for vastly simplified analysis of cascaded two-port network circuits. As seen in this section, where, for example,

$$\mathbf{t}_A = \begin{bmatrix} 1.2 & 6.8\ \Omega \\ 0.1\ \text{S} & 1.4 \end{bmatrix}$$

and

$$\mathbf{t}_B = \begin{bmatrix} 1.2 & 13.6\ \Omega \\ 0.05\ \text{S} & 1.4 \end{bmatrix}$$

we found that the $\mathbf{t}$ parameters characterizing the cascaded network can be found by simply multiplying $\mathbf{t}_A$ and $\mathbf{t}_B$:

$$\mathbf{t} = \mathbf{t}_A \cdot \mathbf{t}_B$$

Such matrix operations are easily carried out using scientific calculators or software packages such as MATLAB. The MATLAB script,

(Continued on next page)

for example, would be

EDU» tA = [1.2 6.8; 0.1 1.4];
EDU» tB = [1.2 13.6; 0.05 1.4];
EDU» t = tA*tB

t =

1.7800	25.8400
0.1900	3.3200

as we found in Example 17.10.

In terms of entering matrices in MATLAB, each has a case-sensitive variable name (tA, tB, and t in this example). Matrix elements are entered a row at a time, beginning with the top row; rows are separated by a semicolon. Again, the reader should always be careful to remember that the order to operations is critical when performing matrix algebra. For example, tB*tA results in a totally different matrix than the one we sought:

$$\mathbf{t}_B \cdot \mathbf{t}_A = \begin{bmatrix} 2.8 & 27.2 \\ 0.2 & 2.3 \end{bmatrix}$$

For simple matrices such as seen in this example, a scientific calculator is just as handy (if not more so). However, larger cascaded networks are more easily handled on a computer, where it is more convenient to see all arrays on the screen simultaneously.

SUMMARY AND REVIEW

In this chapter we encountered a somewhat abstract way to represent networks. This new approach is especially useful if the network is *passive,* and will either be connected somehow to other networks at some point, or perhaps component values will frequently be changed. We introduced the concept through the idea of a *one-port network,* where all we really did was determine the Thévenin equivalent resistance (or impedance, more generally speaking). Our first exposure to the idea of a two-port network (perhaps one port is an input, the other an output?) was through admittance parameters, also called **y** parameters. The result is a matrix which, when multiplied by the vector containing the terminal voltages, yields a vector with the currents into each port. A little manipulation yielded what we called Δ-Y equivalents in Chap. 5. The direct counterpart to **y** parameters are **z** parameters, where each matrix element is the ratio of a voltage to a current. Occasionally **y** and **z** parameters are not particularly convenient, so we also introduced "hybrid" or **h** parameters, as well as "transmission" or **t** parameters, also referred to as *ABCD* parameters.

Table 17.1 summarizes the conversion process between **y**, **z**, **h**, and **t** parameters; having one set of parameters which completely describes a network is sufficent regardless of what type of matrix we prefer for a particular analysis.

As a convenience to the reader, we will now proceed directly to a list of key concepts in the chapter, along with correponding examples.

❏ In order to employ the analysis methods described in this chapter, it is critical to remember that each port can only be connected to either a one-port network or a port of another multiport network.

❏ The input impedance of a one-port (*passive*) linear network can be obtained using either nodal or mesh analysis; in some instances the set of coefficients can be written directly by inspection. (Examples 17.1, 17.2, 17.3)

❏ The defining equations for analyzing a two-port network in terms of its admittance (**y**) parameters are:

$$\mathbf{I}_1 = \mathbf{y}_{11}\mathbf{V}_1 + \mathbf{y}_{12}\mathbf{V}_2 \qquad \text{and} \qquad \mathbf{I}_2 = \mathbf{y}_{21}\mathbf{V}_1 + \mathbf{y}_{22}\mathbf{V}_2$$

where

$$\mathbf{y}_{11} = \left.\frac{\mathbf{I}_1}{\mathbf{V}_1}\right|_{\mathbf{V}_2=0} \qquad\qquad \mathbf{y}_{12} = \left.\frac{\mathbf{I}_1}{\mathbf{V}_2}\right|_{\mathbf{V}_1=0}$$

$$\mathbf{y}_{21} = \left.\frac{\mathbf{I}_2}{\mathbf{V}_1}\right|_{\mathbf{V}_2=0} \qquad \text{and} \qquad \mathbf{y}_{22} = \left.\frac{\mathbf{I}_2}{\mathbf{V}_2}\right|_{\mathbf{V}_1=0}$$

(Examples 17.4, 17.5, 17.7)

❏ The defining equations for analyzing a two-port network in terms of its impedance (**z**) parameters are:

$$\mathbf{V}_1 = \mathbf{z}_{11}\mathbf{I}_1 + \mathbf{z}_{12}\mathbf{I}_2 \qquad \text{and} \qquad \mathbf{V}_2 = \mathbf{z}_{21}\mathbf{I}_1 + \mathbf{z}_{22}\mathbf{I}_2$$

(Example 17.8)

❏ The defining equations for analyzing a two-port network in terms of its hybrid (**h**) parameters are:

$$\mathbf{V}_1 = \mathbf{h}_{11}\mathbf{I}_1 + \mathbf{h}_{12}\mathbf{V}_2 \qquad \text{and} \qquad \mathbf{I}_2 = \mathbf{h}_{21}\mathbf{I}_1 + \mathbf{h}_{22}\mathbf{V}_2$$

(Example 17.9)

❏ The defining equations for analyzing a two-port network in terms of its transmission (**t**) parameters (also called the **ABCD** parameters) are:

$$\mathbf{V}_1 = \mathbf{t}_{11}\mathbf{V}_2 - \mathbf{t}_{12}\mathbf{I}_2 \qquad \text{and} \qquad \mathbf{I}_1 = \mathbf{t}_{21}\mathbf{V}_2 - \mathbf{t}_{22}\mathbf{I}_2$$

(Example 17.10)

❏ It is straightforward to convert between **h, z, t**, and **y** parameters, depending on circuit analysis needs; the transformations are summarized in Table 17.1. (Example 17.6)

READING FURTHER

Further details of matrix methods for circuit analysis can be found in:

R. A. DeCarlo and P. M. Lin, *Linear Circuit Analysis,* 2nd ed. New York: Oxford University Press, 2001.

Analysis of transistor circuits using network parameters is described in:

W. H. Hayt, Jr., and G. W. Neudeck, *Electronic Circuit Analysis and Design,* 2nd ed. New York: Wiley, 1995.

EXERCISES

17.1 One-Port Networks

1. Consider the following system of equations:

$$-2\mathbf{I}_1 + 4\mathbf{I}_2 \qquad = \quad 3$$
$$5\mathbf{I}_1 + \mathbf{I}_2 - 9\mathbf{I}_3 = \quad 0$$
$$2\mathbf{I}_1 - 5\mathbf{I}_2 + 4\mathbf{I}_3 = -1$$

(a) Write the set of equations in matrix form. (b) Determine $\Delta_{\mathbf{Z}}$ and Δ_{11}. (c) Calculate $\mathbf{I}_1$.

2. For the following system of equations,

$$100\mathbf{V}_1 - \quad 45\mathbf{V}_2 + 30\mathbf{V}_3 = \quad 0.2$$
$$75\mathbf{V}_1 \qquad\qquad + 80\mathbf{V}_3 = -0.1$$
$$48\mathbf{V}_1 + 200\mathbf{V}_2 + 42\mathbf{V}_3 = \quad 0.5$$

(a) Write the set of equations in matrix form. (b) Use $\Delta_{\mathbf{Y}}$ to calculate $\mathbf{V}_2$ only.

3. With regard to the passive network depicted in Fig. 17.33, (a) obtain the four mesh equations; (b) compute $\Delta_{\mathbf{Z}}$; and (c) calculate the input impedance.

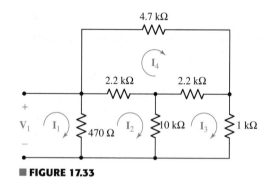

■ **FIGURE 17.33**

4. Determine the input impedance of the network shown in Fig. 17.34 after first calculating $\Delta_{\mathbf{Z}}$.

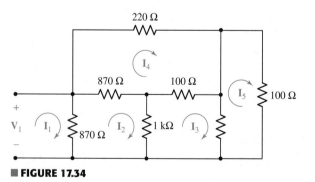

■ **FIGURE 17.34**

5. For the one-port network represented schematically in Fig. 17.35, choose the bottom node as the reference; name the junction between the 3, 10, and 20 S conductances $\mathbf{V}_2$ and the remaining node $\mathbf{V}_3$. (a) Write the three nodal equations. (b) Compute $\Delta_{\mathbf{Y}}$. (c) Calculate the input admittance.

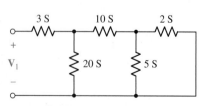

■ **FIGURE 17.35**

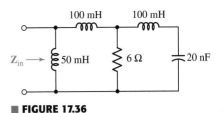

6. Calculate $\Delta_{\mathbf{Z}}$ and $\mathbf{Z}_{\text{in}}$ for the network of Fig. 17.36 if ω is equal to (a) 1 rad/s; (b) 320 krad/s.

7. Set $\omega = 100\pi$ rad/s in the one-port of Fig. 17.36. (a) Calculate $\Delta_{\mathbf{Y}}$ and the input admittance at ω, $\mathbf{Y}_{\text{in}}(\omega)$. (b) A sinusoidal current source having magnitude 100 A, frequency 100π rad/s, and $0°$ phase is connected to the network. Calculate the voltage across the current source (express answer as a phasor).

8. With reference to the one-port of Fig. 17.37, which contains a dependent current source controlled by a resistor voltage, (a) calculate $\Delta_{\mathbf{Z}}$; (b) compute $\mathbf{Z}_{\text{in}}$.

■ FIGURE 17.36

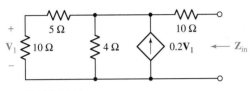

■ FIGURE 17.37

9. For the ideal op amp circuit represented in Fig. 17.38, the input resistance is defined by looking between the positive input terminal of the op amp and ground. (a) Write the appropriate nodal equations for the one-port. (b) Obtain an expression for R_{in}. Is your answer somewhat unexpected? Explain.

10. (a) If both the op amps shown in the circuit of Fig. 17.39 are assumed to be ideal ($R_i = \infty$, $R_o = 0$, and $A = \infty$), find $\mathbf{Z}_{\text{in}}$. (b) $R_1 = 4$ kΩ, $R_2 = 10$ kΩ, $R_3 = 10$ kΩ, $R_4 = 1$ kΩ, and $C = 200$ pF; show that $\mathbf{Z}_{\text{in}} = j\omega L_{\text{in}}$, where $L_{\text{in}} = 0.8$ mH.

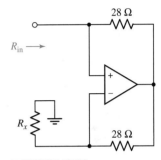

■ FIGURE 17.38

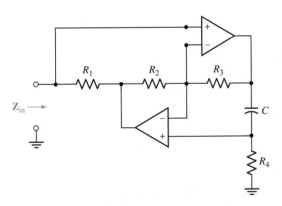

■ FIGURE 17.39

17.2 Admittance Parameters

11. Obtain a complete set of **y** parameters which describe the two-port shown in Fig. 17.40.

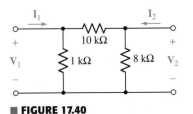

■ FIGURE 17.40

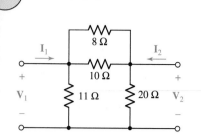

■ **FIGURE 17.41**

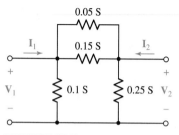

■ **FIGURE 17.42**

12. (*a*) Determine the short-circuit admittance parameters which completely describe the two-port network of Fig. 17.41. (*b*) If $V_1 = 3$ V and $V_2 = -2$ V, use your answer in part (*a*) to compute I_1 and I_2.

13. (*a*) Determine the **y** parameters for the two-port of Fig. 17.42. (*b*) Define the bottom node of Fig. 17.42 as the reference node, and apply nodal analysis to obtain expressions for I_1 and I_2 in terms of V_1 and V_2. Use these expressions to write down the admittance matrix. (*c*) If $V_1 = 2V_2 = 10$ V, calculate the power dissipated in the 100 mS conductance.

14. Obtain an complete set of **y** parameters to describe the two-port network depicted in Fig. 17.43.

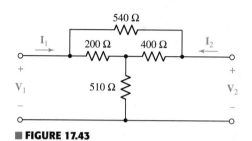

■ **FIGURE 17.43**

15. The circuit of Fig. 17.44 is simply the two-port of Fig. 17.40 terminated by a passive one-port and a separate one-port consisting of a voltage source in series with a resistor. (*a*) Determine the complete set of admittance parameters which describe the two-port network. (*Hint:* draw the two-port by itself, properly labeled with a voltage and current at each port.) (*b*) Calculate the power dissipated in the passive one-port, using your answer to part (*a*).

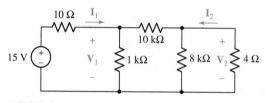

■ **FIGURE 17.44**

16. Replace the 10 Ω resistor of Fig. 17.44 with a 1 kΩ resistor, the 15 V source with a 9 V source, and the 4 Ω resistor with a 4 kΩ resistor. (*a*) Determine the complete set of admittance parameters which describe the two-port network consisting of the 1 kΩ, 10 kΩ, and 8 kΩ resistors. (*Hint:* draw the two port by itself, properly labeled with a voltage and current at each port.) (*b*) Calculate the power dissipated in the passive one-port, using your answer to part (*a*).

17. Determine the admittance parameters which describe the two-port shown in Fig. 17.45.

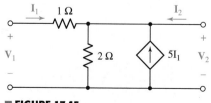

■ **FIGURE 17.45**

18. Obtain the **y** parameter for the network shown in Fig. 17.46 and use it to determine I_1 and I_2 if (a) $V_1 = 0$, $V_2 = 1$ V; (b) $V_1 = -8$ V, $V_2 = 3$ V; (c) $V_1 = V_2 = 5$ V.

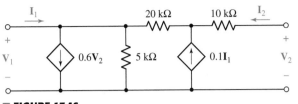

■ **FIGURE 17.46**

19. Employ an appropriate method to obtain **y** for the network of Fig. 17.47.

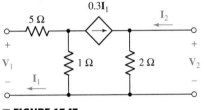

■ **FIGURE 17.47**

20. The metal-oxide-semiconductor field effect transistor (MOSFET), a three-terminal nonlinear element used in many electronics applications, is often specified in terms of its **y** parameters. The ac parameters are strongly dependent on the measurement conditions, and commonly named y_{is}, y_{rs}, y_{fs}, and y_{os}, as in

$$I_g = y_{is}V_{gs} + y_{rs}V_{ds} \qquad [49]$$

$$I_d = y_{fs}V_{gs} + y_{os}V_{ds} \qquad [50]$$

where I_g is the transistor gate current, I_d is the transistor drain current, and the third terminal (the source) is common to the input and output during the measurement. Thus, V_{gs} is the voltage between the gate and the source, and V_{ds} is the voltage between the drain and the source. The typical high-frequency model used to approximate the behavior of a MOSFET is shown in Fig. 17.48.

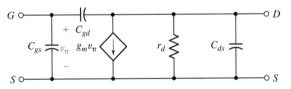

■ **FIGURE 17.48**

(a) For the configuration stated above, which transistor terminal is used as the input, and which terminal is used as the output? (b) Derive expressions for the parameters y_{is}, y_{rs}, y_{fs}, and y_{os} defined in Eqs. [49] and [50], in terms of the model parameters C_{gs}, C_{gd}, g_m, r_d, and C_{ds} of Fig. 17.48. (c) Compute y_{is}, y_{rs}, y_{fs}, and y_{os} if $g_m = 4.7$ mS, $C_{gs} = 3.4$ pF, $C_{gd} = 1.4$ pF, $C_{ds} = 0.4$ pF, and $r_d = 10$ kΩ.

17.3 Some Equivalent Networks

21. For the two-port displayed in Fig. 17.49, (a) determine the input resistance; (b) compute the power dissipated by the network if connected in parallel with a 2 A current source; (c) compute the power dissipated by the network if connected in parallel with a 9 V voltage source.

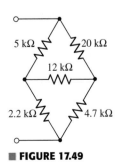

■ **FIGURE 17.49**

22. With reference to the two networks in Fig. 17.50, convert the Δ-connected network to a Y-connected network, and vice versa.

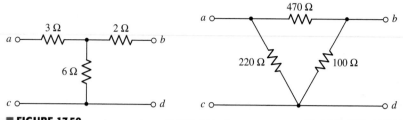

■ **FIGURE 17.50**

23. Determine the input impedance $\mathbf{Z}_{in}$ of the one-port shown in Fig. 17.51 if ω is equal to (a) 50 rad/s; (b) 1000 rad/s.

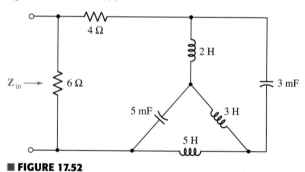

■ **FIGURE 17.51**

24. Determine the input impedance $\mathbf{Z}_{in}$ of the one-port shown in Fig. 17.52 if ω is equal to (a) 50 rad/s; (b) 1000 rad/s.

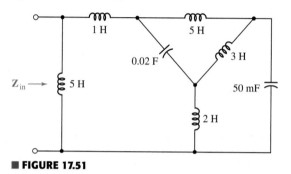

■ **FIGURE 17.52**

25. Employ Δ-Y conversion techniques as appropriate to determine the input resistance R_{in} of the one-port shown in Fig. 17.53.

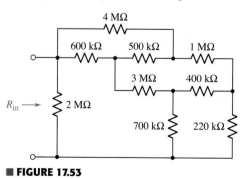

■ **FIGURE 17.53**

26. Employ appropriate techniques to find a value for the input resistance of the one-port network represented by the schematic of Fig. 17.54.

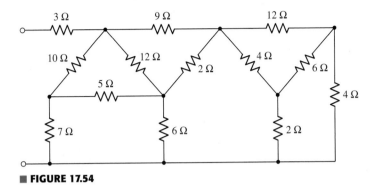

■ **FIGURE 17.54**

27. (a) Determine the parameter values required to model the network of Fig. 17.43 with the alternative network shown in Fig. 17.13a. (b) Verify that the two networks are in fact equivalent by computing the power dissipated in a 2 Ω resistor connected to the right of each network and connecting a 1 A current source to the left-hand terminals.

28. (a) The network of Fig. 17.13b is equivalent to the network of Fig. 17.43 assuming the appropriate parameter values are chosen. (a) Compute the necessary parameter values. (b) Verify the equivalence of the two networks by terminating each with a 1 Ω resistor (across their V_2 terminals), connecting a 10 mA source to the other terminals, and showing that I_1, V_1, I_2, and V_2 are equal for both networks.

29. Compute the three parameter values necessary to construct an equivalent network for Fig. 17.43 modeled after the network of Fig. 17.13c. Verify their equivalence with an appropriate PSpice simulation. (*Hint:* connect some type of source(s) and load(s).)

30. It is possible to construct an alternative two-port to the one shown in Fig. 17.47, by selecting appropriate parameter values as labeled on the diagram in Fig. 17.13. (a) Construct such an equivalent network. (b) Verify their equivalence with an appropriate PSpice simulation. (*Hint:* connect some type of source(s) and load(s).)

31. Let $\mathbf{y} = \begin{bmatrix} 0.1 & -0.05 \\ -0.5 & 0.2 \end{bmatrix}$ (S) for the two-port of Fig. 17.55. Find (a) $\mathbf{G}_V$;

(b) $\mathbf{G}_I$; (c) G_P; (d) $\mathbf{Z}_{in}$; (e) $\mathbf{Z}_{out}$. (f) If the reverse voltage gain $\mathbf{G}_{V,rev}$ is defined as V_1/V_2 with $V_s = 0$ and R_L removed, calculate $\mathbf{G}_{V,rev}$. (g) If the insertion power gain G_{ins} is defined as the ratio of $P_{5\Omega}$ with the two-port in place to $P_{5\Omega}$ with the two-port replaced by jumpers connecting each input terminal to the corresponding output terminal, calculate G_{ins}.

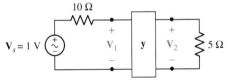

■ **FIGURE 17.55**

17.4 Impedance Parameters

32. Convert the following **z** parameters to **y** parameters, or vice versa, as appropriate:

$$\mathbf{z} = \begin{bmatrix} 2 & 3 \\ 5 & 1 \end{bmatrix} \Omega \qquad \mathbf{z} = \begin{bmatrix} 1000 & 470 \\ 2500 & 900 \end{bmatrix} \Omega$$

$$\mathbf{y} = \begin{bmatrix} 0.001 & 0.005 \\ 0.006 & 0.03 \end{bmatrix} S \qquad \mathbf{y} = \begin{bmatrix} 1 & 2 \\ -1 & 3 \end{bmatrix} S$$

33. By employing Eqs. [32] to [35], obtain a complete set of **z** parameters for the network given in Fig. 17.56.

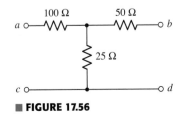

■ **FIGURE 17.56**

34. The network of Fig. 17.56 is terminated with a 10 Ω resistor across terminals b and d, and a 6 mA sinusoidal current source operating at 100 Hz in parallel with a 50 Ω resistor is connected across terminals a and c. Calculate the voltage, current, and power gains, respectively, as well as the input and output impedance.

35. The two-port networks of Fig. 17.50 are connected in series. (a) Determine the impedance parameters for the series connection by first finding the **z** parameters of the individual networks. (b) If the two networks are instead connected in parallel, determine the admittance parameters of the combination by first finding the **y** parameters of the individual networks. (c) Verify your answer to part (b) by using Table 17.1 in conjunction with your answer to part (a).

36. (a) Use an appropriate method to obtain the impedance parameters which describe the network illustrated in Fig. 17.57. (b) If a 1 V source in series with a 1 kΩ resistor is connected to the left-hand port such that the negative reference terminal of the source is connected to the common terminal of the network, and a 5 kΩ load is connected across the right-hand terminals, compute the current, voltage, and power gain.

37. Determine the impedance parameters for the two-port exhibited in Fig. 17.58.

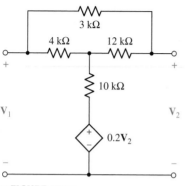

■ **FIGURE 17.57**

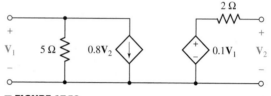

■ **FIGURE 17.58**

38. Obtain both the impedance and admittance parameters for the two-port network of Fig. 17.59.

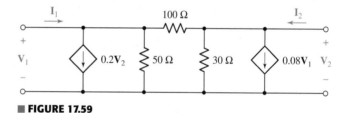

■ **FIGURE 17.59**

39. Find the four **z** parameters at $\omega = 10^8$ rad/s for the transistor high-frequency equivalent circuit shown in Fig. 17.60.

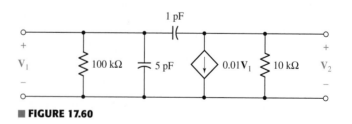

■ **FIGURE 17.60**

17.5 Hybrid Parameters

40. Determine the **h** parameters which describe the purely resistive network shown in Fig. 17.56 by connecting appropriate 1 V, 1 A, and short circuits to terminals as required.

41. Obtain the **h** parameters of the two-ports of Fig. 17.61.

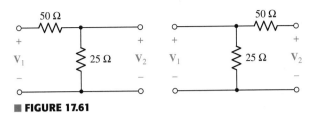

■ **FIGURE 17.61**

42. If **h** for some particular two-port is given by $\mathbf{h} = \begin{bmatrix} 2\ k\Omega & -3 \\ 5 & 0.01\ S \end{bmatrix}$, calculate (a) **z**; (b) **y**.

43. A certain two-port network is described by hybrid parameters $\mathbf{h} = \begin{bmatrix} 100\ \Omega & -2 \\ 5 & 0.1\ S \end{bmatrix}$. Determine the new **h** parameters if a 25 Ω resistor is connected in parallel with (a) the input; (b) the output.

44. A bipolar junction transistor is connected in common-emitter configuration, and found to have **h** parameters $h_{11} = 5\ k\Omega$, $h_{12} = 0.55 \times 10^{-4}$, $h_{21} = 300$, and $h_{22} = 39\ \mu S$. (a) Write **h** in matrix form. (b) Determine the small-signal current gain. (c) Determine the output resistance in kΩ. (d) If a sinusoidal voltage source having frequency 100 rad/s and amplitude 5 mV in series with a 100 Ω resistor is connected to the input terminals, calculate the peak voltage which appears across the output terminals.

45. The two-port which plays a central role in the circuit of Fig. 17.62 can be characterized by hybrid parameters $\mathbf{h} = \begin{bmatrix} 1\ \Omega & -1 \\ 2 & 0.5\ S \end{bmatrix}$. Determine $\mathbf{I}_1$, $\mathbf{I}_2$, $\mathbf{V}_1$, and $\mathbf{V}_2$.

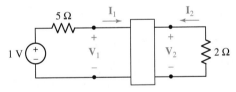

■ **FIGURE 17.62**

46. The two networks of Fig. 17.61 are connected in series by connecting the terminals as illustrated in Fig. 17.22 (assume the left-hand network of Fig. 17.61 is network A). Determine the new set of **h** parameters which describe the series connection.

47. The two networks of Fig. 17.61 are connected in parallel by tying the corresponding input terminals together, and then tying the corresponding output terminals together. Determine the new set of **h** parameters which describe the parallel connection.

48. Find **y**, **z**, and **h** for both of the two-ports shown in Fig. 17.63. If any parameter is infinite, skip that parameter set.

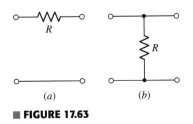

■ **FIGURE 17.63**

49. (*a*) Find **h** for the two-port of Fig. 17.64. (*b*) Find $\mathbf{Z}_{out}$ if the input contains $\mathbf{V}_s$ in series with $R_s = 200\ \Omega$.

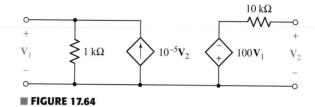

■ **FIGURE 17.64**

17.6 Transmission Parameters

50. (*a*) With the assistance of appropriate mesh equations, determine the *ABCD* matrix which represents the two-port shown in Fig. 17.9. (*b*) Convert your answer to **h**.

51. (*a*) Employ suitably written mesh equations to obtain the **t** parameters which characterize the network of Fig. 17.57. (*b*) If currents $\mathbf{I}_1$ and $\mathbf{I}_2$ are defined as flowing into the (+) reference terminals of $\mathbf{V}_1$ and $\mathbf{V}_2$, respectively, compute the voltages if $\mathbf{I}_1 = 2\mathbf{I}_2 = 3$ mA.

52. Consider the following matrices:

$$\mathbf{a} = \begin{bmatrix} 5 & 2 \\ 4 & 1 \end{bmatrix} \qquad \mathbf{b} = \begin{bmatrix} 1.5 & 1 \\ 1 & 0.5 \end{bmatrix} \qquad \mathbf{c} = \begin{bmatrix} -4 \\ 2 \end{bmatrix}$$

Calculate (*a*) $\mathbf{a} \cdot \mathbf{b}$; (*b*) $\mathbf{b} \cdot \mathbf{a}$; (*c*) $\mathbf{a} \cdot \mathbf{c}$; (*d*) $\mathbf{b} \cdot \mathbf{c}$; (*e*) $\mathbf{b} \cdot \mathbf{a} \cdot \mathbf{c}$; (*f*) $\mathbf{a} \cdot \mathbf{a}$.

53. Two networks are represented by the following impedance matrices:

$$\mathbf{z}_1 = \begin{bmatrix} 4.7 & 0.5 \\ 0.87 & 1.8 \end{bmatrix} k\Omega \text{ and } \mathbf{z}_2 = \begin{bmatrix} 1.1 & 2.2 \\ 0.89 & 1.8 \end{bmatrix} k\Omega, \text{ respectively.}$$

(*a*) Determine the **t** matrix which characterizes the cascaded network resulting from connecting network 2 to the output of network 1. (*b*) Reverse the order of the networks and compute the new **t** matrix which results.

54. The two-port of Fig. 17.65 can be viewed as three separate cascaded two-ports *A*, *B*, and *C*. (*a*) Compute **t** for each network. (*b*) Obtain **t** for the cascaded network. (*c*) Verify your answer by naming the two middle nodes V_x and V_y, respectively, writing nodal equations, obtaining the admittance parameters from your nodal equations, and converting to **t** parameters using Table 17.1.

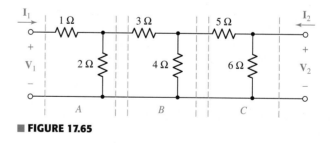

■ **FIGURE 17.65**

55. Consider the two separate two-ports of Fig. 17.61. Determine the *ABCD* matrix which characterizes the cascaded network resulting from connecting (*a*) the output of the left-hand network to the input of the right-hand network; (*b*) the output of the right-hand network to the input of the left-hand network.

56. (*a*) Determine the **t** parameters which describe the two-port of Fig. 17.58. (*b*) Compute $\mathbf{Z}_{out}$ if a practical voltage source having a 100 Ω series resistance is connected to the input terminals of the network.

57. Three identical networks as the one depicted in Fig. 17.56 are cascaded together. Determine the **t** parameters which fully represent the result.

58. (a) Find $\mathbf{t}_a$, $\mathbf{t}_b$, and $\mathbf{t}_c$ for the networks shown in Fig. 17.66a, b, and c. (b) By using the rules for interconnecting two-ports in cascade, find **t** for the network of Fig. 17.66d.

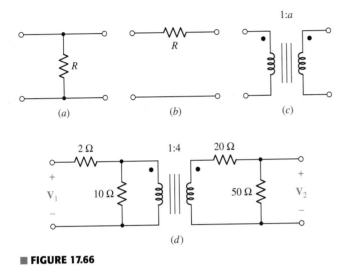

(a)   (b)   (c)

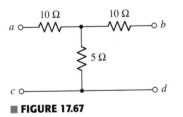

(d)

■ **FIGURE 17.66**

Chapter-Integrating Exercises

59. (a) Obtain **y**, **z**, **h**, and **t** parameters for the network shown in Fig. 17.67 using either the defining equations or mesh/nodal equations. (b) Verify your answers, using the relationships in Table 17.1.

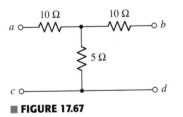

■ **FIGURE 17.67**

60. Four networks, each identical to the one depicted in Fig. 17.67, are connected in parallel such that all terminals labeled a are tied together, all terminals designated b are tied together, and all terminals labeled c and d are connected. Obtain the **y**, **z**, **h**, and **t** parameters which describe the parallel-connected network.

61. A cascaded 12-element network is formed using four two-ports identical to the one shown in Fig. 17.67. Determine the **y**, **z**, **h**, and **t** parameters which describe the result.

62. The concept of *ABCD* matrices extends to systems beyond electrical circuits. For example, they are commonly employed for ray-tracing calculations in optical systems. In that case, we envision parallel input and output planes in *xy*, skewered by an optical axis *z*. An inbound ray crosses the input plane a distance $x = r_{\text{in}}$ from the optical axis, making an angle θ_{in}. The corresponding

parameters r_{out}, θ_{out} for the outbound ray crossing the output plane are then given by the *ABCD* matrix such that

$$\begin{bmatrix} r_{out} \\ \theta_{out} \end{bmatrix} = \begin{bmatrix} A & B \\ C & D \end{bmatrix} \begin{bmatrix} r_{in} \\ \theta_{in} \end{bmatrix}$$

Each type of optical element (e.g., mirror, lens, or even propagation through free space) has its own *ABCD* matrix. If the ray passes through several elements, the net effect can be predicted by simply cascading the individual *ABCD* matrices (in the proper order).

(*a*) Obtain expressions for *A*, *B*, *C*, and *D* similar to Eqs. [32] to [35].
(*b*) If the *ABCD* matrix for a perfectly reflecting flat mirror is given by $\begin{bmatrix} 1 & 0 \\ 0 & 1 \end{bmatrix}$, sketch the system along with the inbound and outbound rays, taking care to note the orientation of the mirror.

63. Continuing from Exercise 62, the behavior of a ray propating through free space a distance *d* can be modeled with the *ABCD* matrix $\begin{bmatrix} 1 & d \\ 0 & 1 \end{bmatrix}$. (*a*) Show that the same result is obtained (r_{out}, θ_{out}) whether a single *ABCD* matrix is used with *d*, or two cascaded matrices are used, each with $d/2$. (*b*) What are the units of *A*, *B*, *C*, and *D*, respectively?

Fourier Circuit Analysis

KEY CONCEPTS

Representing Periodic
Functions as a Sum of Sines
and Cosines

Harmonic Frequencies

Even and Odd Symmetry

Half-Wave Symmetry

Complex Form of the Fourier
Series

Discrete Line Spectra

Fourier Transform

Using Fourier Series
and Fourier Transform
Techniques in Circuit Analysis

System Response and
Convolution in the
Frequency Domain

INTRODUCTION

In this chapter we continue our introduction to circuit analysis by
studying periodic functions in both the time and frequency domains.
Specifically, we will consider forcing functions which are *periodic* and
have functional natures which satisfy certain mathematical restrictions
that are characteristic of any function which we can generate in the
laboratory. Such functions may be represented as the sum of an infinite
number of sine and cosine functions which are harmonically related.
Therefore, since the forced response to each sinusoidal component can
be determined easily by sinusoidal steady-state analysis, the response
of the linear network to the general periodic forcing function may be
obtained by superposing the partial responses.

The topic of Fourier series is of vital importance in a number of
fields, particularly communications. The use of Fourier-based techniques to assist in circuit analysis, however, had been slowly falling
out of fashion for a number of years. Now as we face an increasingly larger fraction of global power usage coming from equipment
employing pulse-modulated power supplies (e.g., computers), the
subject of harmonics in power systems and power electronics is
rapidly becoming a serious problem in even large-scale generation
plants. It is only with Fourier-based analysis that the underlying
problems and possible solutions can be understood.

18.1 TRIGONOMETRIC FORM OF THE FOURIER SERIES

We know that the complete response of a linear circuit to an arbitrary forcing function is composed of the sum of a *forced response*
and a *natural response*. The natural response has been considered

both in the time domain (Chaps. 7, 8, and 9) and in the frequency domain (Chaps. 14 and 15). The forced response has also been considered from several perspectives, including the phasor-based techniques of Chap. 10. As we have discovered, in some cases we need *both* components of the total response of a particular circuit, while in others we need only the natural or the forced response. In this section, we refocus our attention on forcing functions that are *sinusoidal* in nature, and discover how to write a general periodic function as a *sum* of such functions—leading us into a discussion of a new set of circuit analysis procedures.

Harmonics

Some feeling for the validity of representing a general *periodic* function by an infinite sum of sine and cosine functions may be gained by considering a simple example. Let us first assume a cosine function of radian frequency ω_0,

$$v_1(t) = 2 \cos \omega_0 t$$

where

$$\omega_0 = 2\pi f_0$$

and the period T is

$$T = \frac{1}{f_0} = \frac{2\pi}{\omega_0}$$

Although T does not usually carry a zero subscript, it is the period of the fundamental frequency. The **harmonics** of this sinusoid have frequencies $n\omega_0$, where ω_0 is the fundamental frequency and $n = 1, 2, 3, \ldots$. The frequency of the first harmonic is the **fundamental frequency.**

Next let us select a third-harmonic voltage

$$v_{3a}(t) = \cos 3\omega_0 t$$

The fundamental $v_1(t)$, the third harmonic $v_{3a}(t)$, and the sum of these two waves are shown as functions of time in Fig. 18.1*a*. Note that the sum is also periodic, with period $T = 2\pi/\omega_0$.

The form of the resultant periodic function changes as the phase and amplitude of the third-harmonic component change. Thus, Fig. 18.1*b* shows the effect of combining $v_1(t)$ and a third harmonic of slightly larger amplitude,

$$v_{3b}(t) = 1.5 \cos 3\omega_0 t$$

By shifting the phase of the third harmonic by 90 degrees to give

$$v_{3c}(t) = \sin 3\omega_0 t$$

the sum, shown in Fig. 18.1*c*, takes on a still different character. In all cases, the period of the resultant waveform is the same as the period of the fundamental waveform. The nature of the waveform depends on the amplitude and phase of every possible harmonic component, and we will find that we are able to generate waveforms which have extremely nonsinusoidal characteristics by an appropriate combination of sinusoidal functions.

After we have become familiar with the use of the sum of an infinite number of sine and cosine functions to represent a periodic waveform, we will consider the frequency-domain representation of a general nonperiodic waveform in a manner similar to the Laplace transform.

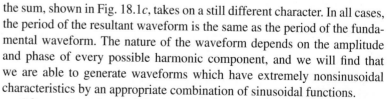

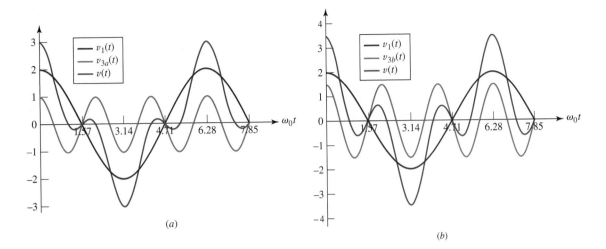

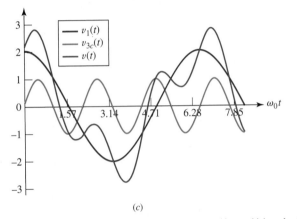

■ FIGURE 18.1 Several of the infinite number of different waveforms which may be obtained by combining a fundamental and a third harmonic. The fundamental is $v_1 = 2\cos\omega_0 t$, and the third harmonic is (a) $v_{3a} = \cos 3\omega_0 t$; (b) $v_{3b} = 1.5\cos 3\omega_0 t$; (c) $v_{3c} = \sin 3\omega_0 t$.

PRACTICE

18.1 Let a third-harmonic voltage be added to the fundamental to yield $v = 2\cos\omega_0 t + V_{m3}\sin 3\omega_0 t$, the waveform shown in Fig. 18.1c for $V_{m3} = 1$. (a) Find the value of V_{m3} so that $v(t)$ will have zero slope at $\omega_0 t = 2\pi/3$. (b) Evaluate $v(t)$ at $\omega_0 t = 2\pi/3$.

Ans: 0.577; −1.000.

The Fourier Series

We first consider a *periodic* function $f(t)$, defined in Sec. 11.2 by the functional relationship

$$f(t) = f(t + T)$$

where T is the period. We further assume that the function $f(t)$ satisfies the following properties:

1. $f(t)$ is single-valued everywhere; that is, $f(t)$ satisfies the mathematical definition of a function.
2. The integral $\int_{t_0}^{t_0+T} |f(t)|\, dt$ exists (i.e., is not infinite) for any choice of t_0.
3. $f(t)$ has a finite number of discontinuities in any one period.
4. $f(t)$ has a finite number of maxima and minima in any one period.

Given such a periodic function $f(t)$, the Fourier theorem states that $f(t)$ may be represented by the infinite series

$$f(t) = a_0 + a_1 \cos \omega_0 t + a_2 \cos 2\omega_0 t + \cdots$$
$$+ b_1 \sin \omega_0 t + b_2 \sin 2\omega_0 t + \cdots$$
$$= a_0 + \sum_{n=1}^{\infty} (a_n \cos n\omega_0 t + b_n \sin n\omega_0 t) \qquad [1]$$

where the fundamental frequency ω_0 is related to the period T by

$$\omega_0 = \frac{2\pi}{T}$$

and where a_0, a_n, and b_n are constants that depend upon n and $f(t)$. Equation [1] is the trigonometric form of the **Fourier series** *for $f(t)$*, and the process of determining the values of the constants a_0, a_n, and b_n is called *Fourier analysis*. Our object is not the proof of this theorem, but only a simple development of the procedures of Fourier analysis and a feeling that the theorem is plausible.

Some Useful Trigonometric Integrals

Before we discuss the evaluation of the constants appearing in the Fourier series, let us collect a set of useful trigonometric integrals. We let both n and k represent any element of the set of integers 1, 2, 3, ... In the following integrals, 0 and T are used as the integration limits, but it is understood that any interval of one period is equally correct.

$$\int_0^T \sin n\omega_0 t \, dt = 0 \qquad [2]$$

$$\int_0^T \cos n\omega_0 t \, dt = 0 \qquad [3]$$

$$\int_0^T \sin k\omega_0 t \cos n\omega_0 t \, dt = 0 \qquad [4]$$

As can be seen, as more terms are included, the more the plot resembles that of Fig. 18.2.

PRACTICE

18.2 A periodic waveform $f(t)$ is described as follows: $f(t) = -4$, $0 < t < 0.3$; $f(t) = 6, 0.3 < t < 0.4$; $f(t) = 0, 0.4 < t < 0.5$; $T = 0.5$. Evaluate (a) a_0; (b) a_3; (c) b_1.

18.3 Write the Fourier series for the three voltage waveforms shown in Fig. 18.4.

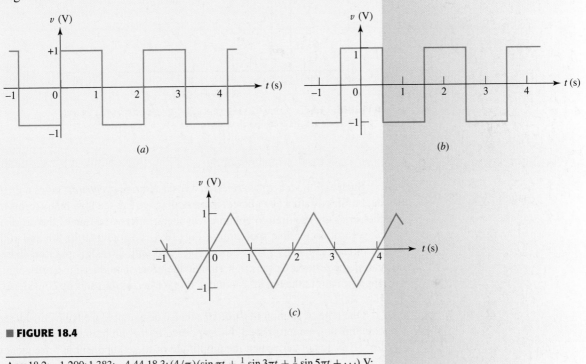

(a)

(b)

(c)

■ FIGURE 18.4

Ans: 18.2: $-1.200; 1.383; -4.44$. 18.3: $(4/\pi)(\sin \pi t + \frac{1}{3} \sin 3\pi t + \frac{1}{5} \sin 5\pi t + \cdots)$ V; $(4/\pi)(\cos \pi t - \frac{1}{3} \cos 3\pi t + \frac{1}{5} \cos 5\pi t - \cdots)$ V; $(8/\pi^2)(\sin \pi t - \frac{1}{9} \sin 3\pi t + \frac{1}{25} \sin 5\pi t - \cdots)$.

Line and Phase Spectra

We depicted the function $v(t)$ of Example 18.1 graphically in Fig. 18.2, and analytically in Eq. [12]—both representations being in the time domain. The Fourier series representation of $v(t)$ given in Eq. [16] is also a time-domain expression, but may be transformed into a *frequency-domain* representation as well. For example, Fig. 18.5 shows the amplitude of each frequency component of $v(t)$, a type of plot known as a **line spectrum.** Here, the magnitude of each frequency component (i.e., $|a_0|$, $|a_1|$, etc.) is indicated by the length of the vertical line at the corresponding frequency (f_0, f_1, etc.); for the sake of convenience, we have taken $V_m = 1$. Given a different value of V_m, we simply scale the y axis values by the new value.

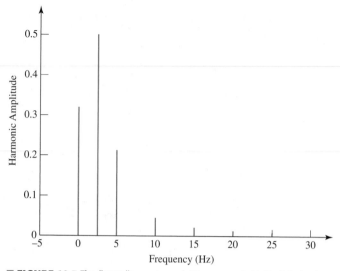

■ **FIGURE 18.5** The discrete line spectrum of $v(t)$ as represented in Eq. [16], showing the first seven frequency components. A magnitude of $V_m = 1$ has been chosen for convenience.

Such a plot, sometimes referred to as a ***discrete spectrum,*** gives a great deal of information at a glance. In particular, we can see how many terms of the series are required to obtain a reasonable approximation of the original waveform. In the line spectrum of Fig. 18.5, we note that the 8th and 10th harmonics (20 and 25 Hz, respectively) add only a small correction. Truncating the series after the 6th harmonic therefore should lead to a reasonable approximation; the reader can judge this for herself/himself by considering Fig. 18.3.

One note of caution must be injected. The example we have considered contains no sine terms, and the amplitude of the nth harmonic is therefore $|a_n|$. If b_n is not zero, then the amplitude of the component at a frequency $n\omega_0$ must be $\sqrt{a_n^2 + b_n^2}$. This is the general quantity which we must show in a line spectrum. When we discuss the complex form of the Fourier series, we will see that this amplitude is obtained more directly.

In addition to the amplitude spectrum, we may construct a discrete ***phase spectrum.*** At any frequency $n\omega_0$, we combine the cosine and sine terms to determine the phase angle ϕ_n:

$$a_n \cos n\omega_0 t + b_n \sin n\omega_0 t = \sqrt{a_n^2 + b_n^2} \cos\left(n\omega_0 t + \tan^{-1}\frac{-b_n}{a_n}\right)$$

$$= \sqrt{a_n^2 + b_n^2} \cos(n\omega_0 t + \phi_n)$$

or

$$\phi_n = \tan^{-1}\frac{-b_n}{a_n}$$

In Eq. [16], $\phi_n = 0°$ or $180°$ for every n.

The Fourier series obtained for this example includes no sine terms and no odd harmonics (except the fundamental) among the cosine terms. It is possible to anticipate the absence of certain terms in a Fourier series, before any integrations are performed, by an inspection of the symmetry of the given time function. We will investigate the use of symmetry in the following section.

18.2 THE USE OF SYMMETRY

Even and Odd Symmetry

The two types of symmetry which are most readily recognized are *even-function symmetry* and *odd-function symmetry,* or simply *even symmetry* and *odd symmetry*. We say that $f(t)$ possesses the property of even symmetry if

$$f(t) = f(-t) \qquad [17]$$

Such functions as t^2, $\cos 3t$, $\ln(\cos t)$, $\sin^2 7t$, and a constant C all possess even symmetry; the replacement of t by $(-t)$ does not change the value of any of these functions. This type of symmetry may also be recognized graphically, for if $f(t) = f(-t)$, then mirror symmetry exists about the $f(t)$ axis. The function shown in Fig. 18.6a possesses even symmetry; if the figure were to be folded along the $f(t)$ axis, then the portions of the graph for positive and negative time would fit exactly, one on top of the other.

We define odd symmetry by stating that if odd symmetry is a property of $f(t)$, then

$$f(t) = -f(-t) \qquad [18]$$

In other words, if t is replaced by $(-t)$, then the negative of the given function is obtained; for example, t, $\sin t$, $t \cos 70t$, $t\sqrt{1+t^2}$, and the function sketched in Fig. 18.6b are all odd functions and possess odd symmetry. The graphical characteristics of odd symmetry are apparent if the portion of $f(t)$ for $t > 0$ is rotated about the positive t axis and the resultant figure is then rotated about the $f(t)$ axis; the two curves will fit exactly, one on top of the other. That is, we now have symmetry about the origin, rather than about the $f(t)$ axis as we did for even functions.

Having definitions for even and odd symmetry, we should note that the product of two functions with even symmetry, or of two functions with odd symmetry, yields a function with even symmetry. Furthermore, the product of an even and an odd function gives a function with odd symmetry.

Symmetry and Fourier Series Terms

Now let us investigate the effect that even symmetry produces in a Fourier series. If we think of the expression which equates an even function $f(t)$ and the sum of an infinite number of sine and cosine functions, then it is apparent that the sum must also be an even function. A sine wave, however, is an odd function, and *no sum of sine waves can produce any even function other than zero* (which is both even and odd). It is thus plausible that the Fourier series of any even function is composed of only a constant and cosine functions. Let us now show carefully that $b_n = 0$. We have

$$b_n = \frac{2}{T} \int_{-T/2}^{T/2} f(t) \sin n\omega_0 t \, dt$$

$$= \frac{2}{T} \left[\int_{-T/2}^{0} f(t) \sin n\omega_0 t \, dt + \int_{0}^{T/2} f(t) \sin n\omega_0 t \, dt \right]$$

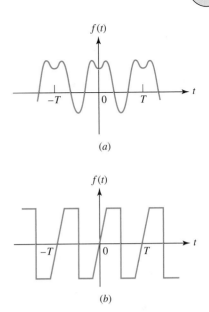

FIGURE 18.6 (*a*) A waveform showing even symmetry. (*b*) A waveform showing odd symmetry.

We replace the variable t in the first integral by $-\tau$, or $\tau = -t$, and make use of the fact that $f(t) = f(-t) = f(\tau)$:

$$b_n = \frac{2}{T}\left[\int_{T/2}^{0} f(-\tau)\sin(-n\omega_0\tau)(-d\tau) + \int_{0}^{T/2} f(t)\sin n\omega_0 t\, dt\right]$$

$$= \frac{2}{T}\left[-\int_{0}^{T/2} f(\tau)\sin n\omega_0\tau\, d\tau + \int_{0}^{T/2} f(t)\sin n\omega_0 t\, dt\right]$$

but the symbol we use to identify the variable of integration cannot affect the value of the integral. Thus,

$$\int_{0}^{T/2} f(\tau)\sin n\omega_0\tau\, d\tau = \int_{0}^{T/2} f(t)\sin n\omega_0 t\, dt$$

and

$$b_n = 0 \qquad \text{(even sym.)} \tag{19}$$

No sine terms are present. Therefore, if $f(t)$ shows even symmetry, then $b_n = 0$; conversely, if $b_n = 0$, then $f(t)$ must have even symmetry.

A similar examination of the expression for a_n leads to an integral over the *half period* extending from $t = 0$ to $t = \frac{1}{2}T$:

$$a_n = \frac{4}{T}\int_{0}^{T/2} f(t)\cos n\omega_0 t\, dt \qquad \text{(even sym.)} \tag{20}$$

The fact that a_n may be obtained for an even function by taking "twice the integral over half the range" should seem logical.

A function having odd symmetry can contain no constant term or cosine terms in its Fourier expansion. Let us prove the second part of this statement. We have

$$a_n = \frac{2}{T}\int_{-T/2}^{T/2} f(t)\cos n\omega_0 t\, dt$$

$$= \frac{2}{T}\left[\int_{-T/2}^{0} f(t)\cos n\omega_0 t\, dt + \int_{0}^{T/2} f(t)\cos n\omega_0 t\, dt\right]$$

and we now let $t = -\tau$ in the first integral:

$$a_n = \frac{2}{T}\left[\int_{T/2}^{0} f(-\tau)\cos(-n\omega_0\tau)(-d\tau) + \int_{0}^{T/2} f(t)\cos n\omega_0 t\, dt\right]$$

$$= \frac{2}{T}\left[\int_{0}^{T/2} f(-\tau)\cos n\omega_0\tau\, d\tau + \int_{0}^{T/2} f(t)\cos n\omega_0 t\, dt\right]$$

But $f(-\tau) = -f(\tau)$, and therefore

$$a_n = 0 \qquad \text{(odd sym.)} \tag{21}$$

A similar, but simpler, proof shows that

$$a_0 = 0 \qquad \text{(odd sym.)}$$

With odd symmetry, therefore, $a_n = 0$ and $a_0 = 0$; conversely, if $a_n = 0$ and $a_0 = 0$, odd symmetry is present.

The values of b_n may again be obtained by integrating over half the range:

$$b_n = \frac{4}{T} \int_0^{T/2} f(t) \sin n\omega_0 t \, dt \qquad \text{(odd sym.)} \qquad [22]$$

Half-Wave Symmetry

The Fourier series for both of these square waves have one other interesting characteristic: neither contains any even *harmonics*.[1] That is, the only frequency components present in the series have frequencies which are odd multiples of the fundamental frequency; a_n and b_n are zero for even values of n. This result is caused by another type of symmetry, called half-wave symmetry. We will say that $f(t)$ possesses *half-wave symmetry* if

$$f(t) = -f\left(t - \tfrac{1}{2}T\right)$$

or the equivalent expression,

$$f(t) = -f\left(t + \tfrac{1}{2}T\right)$$

Except for a change of sign, each half cycle is like the adjacent half cycles. Half-wave symmetry, unlike even and odd symmetry, is not a function of the choice of the point $t = 0$. Thus, we can state that the square wave (Fig. 18.4a or b) shows half-wave symmetry. Neither waveform shown in Fig. 18.6 has half-wave symmetry, but the two somewhat similar functions plotted in Fig. 18.7 do possess half-wave symmetry.

It may be shown that the Fourier series of any function which has half-wave symmetry contains only odd harmonics. Let us consider the coefficients a_n. We have again

$$a_n = \frac{2}{T} \int_{-T/2}^{T/2} f(t) \cos n\omega_0 t \, dt$$

$$= \frac{2}{T} \left[\int_{-T/2}^{0} f(t) \cos n\omega_0 t \, dt + \int_0^{T/2} f(t) \cos n\omega_0 t \, dt \right]$$

which we may represent as

$$a_n = \frac{2}{T}(I_1 + I_2)$$

Now we substitute the new variable $\tau = t + \tfrac{1}{2}T$ in the integral I_1:

$$I_1 = \int_0^{T/2} f\left(\tau - \frac{1}{2}T\right) \cos n\omega_0\left(\tau - \frac{1}{2}T\right) d\tau$$

$$= \int_0^{T/2} -f(\tau)\left(\cos n\omega_0 \tau \cos \frac{n\omega_0 T}{2} + \sin n\omega_0 \tau \sin \frac{n\omega_0 T}{2}\right) d\tau$$

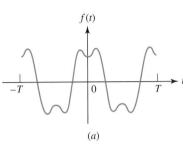

$f(t)$

(a)

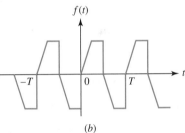

$f(t)$

(b)

■ **FIGURE 18.7** (a) A waveform somewhat similar to the one shown in Fig. 18.6a but possessing half-wave symmetry. (b) A waveform somewhat similar to the one shown in Fig. 18.6b but possessing half-wave symmetry.

(1) Constant vigilance is required to avoid confusion between an even function and an even harmonic, or between an odd function and an odd harmonic. For example, b_{10} is the coefficient of an even harmonic, and it is zero if $f(t)$ is an even function.

But $\omega_0 T$ is 2π, and thus

$$\sin \frac{n\omega_0 T}{2} = \sin n\pi = 0$$

Hence

$$I_1 = -\cos n\pi \int_0^{T/2} f(\tau) \cos n\omega_0 \tau \, d\tau$$

After noting the form of I_2, we therefore may write

$$a_n = \frac{2}{T}(1 - \cos n\pi) \int_0^{T/2} f(t) \cos n\omega_0 t \, dt$$

The factor $(1 - \cos n\pi)$ indicates that a_n is zero if n is even. Thus,

$$a_n = \begin{cases} \dfrac{4}{T} \displaystyle\int_0^{T/2} f(t) \cos n\omega_0 t \, dt & n \text{ odd} \\[4mm] 0 & n \text{ even} \end{cases} \qquad \left(\tfrac{1}{2}\text{-wave sym.}\right) \quad [23]$$

A similar investigation shows that b_n is also zero for all even n, and therefore

$$b_n = \begin{cases} \dfrac{4}{T} \displaystyle\int_0^{T/2} f(t) \sin n\omega_0 t \, dt & n \text{ odd} \\[4mm] 0 & n \text{ even} \end{cases} \qquad \left(\tfrac{1}{2}\text{-wave sym.}\right) \quad [24]$$

It should be noted that half-wave symmetry may be present in a waveform which also shows odd symmetry or even symmetry. The waveform sketched in Fig. 18.7a, for example, possesses both even symmetry and half-wave symmetry. When a waveform possesses half-wave symmetry and either even or odd symmetry, then it is possible to reconstruct the waveform if the function is known over any quarter-period interval. The value of a_n or b_n may also be found by integrating over any quarter period. Thus,

It is *always* worthwhile to spend a few moments investigating the symmetry of a function for which a Fourier series is to be determined.

$$\left. \begin{aligned} a_n &= \frac{8}{T} \int_0^{T/4} f(t) \cos n\omega_0 t \, dt & n \text{ odd} \\ a_n &= 0 & n \text{ even} \\ b_n &= 0 & \text{all } n \end{aligned} \right\} \qquad \left(\tfrac{1}{2}\text{-wave and even sym.}\right)$$

$$[25]$$

$$\left. \begin{aligned} a_n &= 0 & \text{all } n \\ b_n &= \frac{8}{T} \int_0^{T/4} f(t) \sin \omega_0 t \, dt & n \text{ odd} \\ b_n &= 0 & n \text{ even} \end{aligned} \right\} \qquad \left(\tfrac{1}{2}\text{-wave and odd sym.}\right)$$

$$[26]$$

Table 18.1 provides a short summary of the simplifications arising from the various types of symmetry discussed.

TABLE 18.1 Summary of Symmetry-Based Simplifications in Fourier Series

Symmetry Type	Characteristic	Simplification
Even	$f(t) = -f(t)$	$b_n = 0$
Odd	$f(t) = -f(-t)$	$a_n = 0$
Half-Wave	$f(t) = -f\left(t - \dfrac{T}{2}\right)$ or $f(t) = -f\left(t + \dfrac{T}{2}\right)$	$a_n = \begin{cases} \dfrac{4}{T}\displaystyle\int_0^{T/2} f(t)\cos n\omega_0 t\, dt & n \text{ odd} \\[2mm] 0 & n \text{ even} \end{cases}$ $b_n = \begin{cases} \dfrac{4}{T}\displaystyle\int_0^{T/2} f(t)\sin n\omega_0 t\, dt & n \text{ odd} \\[2mm] 0 & n \text{ even} \end{cases}$
Half-Wave and Even	$f(t) = -f\left(t - \dfrac{T}{2}\right)$ and $f(t) = -f(t)$ or $f(t) = -f\left(t + \dfrac{T}{2}\right)$ and $f(t) = -f(t)$	$a_n = \begin{cases} \dfrac{8}{T}\displaystyle\int_0^{T/4} f(t_1)\cos n\omega_0 t\, dt & n \text{ odd} \\[2mm] 0 & n \text{ even} \end{cases}$ $b_n = 0 \qquad\qquad \text{all } n$
Half-Wave and Odd	$f(t) = -f\left(t - \dfrac{T}{2}\right)$ and $f(t) = -f(-t)$ or $f(t) = -f\left(t + \dfrac{T}{2}\right)$ and $f(t) = -f(-t)$	$a_n = 0 \qquad\qquad \text{all } n$ $b_n = \begin{cases} \dfrac{8}{T}\displaystyle\int_0^{T/4} f(t)\sin n\omega_0 t\, dt & n \text{ odd} \\[2mm] 0 & n \text{ even} \end{cases}$

PRACTICE

18.4 Sketch each of the functions described; state whether or not even symmetry, odd symmetry, and half-wave symmetry are present; and give the period: (a) $v = 0$, $-2 < t < 0$ and $2 < t < 4$; $v = 5$, $0 < t < 2$; $v = -5$, $4 < t < 6$; repeats; (b) $v = 10$, $1 < t < 3$; $v = 0$, $3 < t < 7$; $v = -10$, $7 < t < 9$; repeats; (c) $v = 8t$, $-1 < t < 1$; $v = 0$, $1 < t < 3$; repeats.

18.5 Determine the Fourier series for the waveforms of Practice Problem 18.4a and b.

Ans: 18.4: No, no, yes, 8; no, no, no, 8; no, yes, no, 4.

18.5: $\displaystyle\sum_{n=1(\text{odd})}^{\infty} \frac{10}{n\pi}\left(\sin\frac{n\pi}{2}\cos\frac{n\pi t}{4} + \sin\frac{n\pi t}{4}\right)$;

$\displaystyle\sum_{n=1}^{\infty} \frac{10}{n\pi}\left[\left(\sin\frac{3n\pi}{4} - 3\sin\frac{n\pi}{4}\right)\cos\frac{n\pi t}{4} + \left(\cos\frac{n\pi}{4} - \cos\frac{3n\pi}{4}\right)\sin\frac{n\pi t}{4}\right]$.

18.3 • COMPLETE RESPONSE TO PERIODIC FORCING FUNCTIONS

Through the use of the Fourier series, we may now express an arbitrary periodic forcing function as the sum of an infinite number of sinusoidal forcing functions. The forced response to each of these functions may be determined by conventional steady-state analysis, and the form of the natural response may be determined from the poles of an appropriate network transfer function. The initial conditions existing throughout the network, including the initial value of the forced response, enable the amplitude of the natural response to be selected; the complete response is then obtained as the sum of the forced and natural responses.

EXAMPLE 18.2

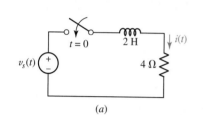

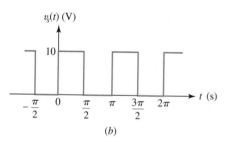

■ **FIGURE 18.8** (*a*) A simple series *RL* circuit subjected to a periodic forcing function $v_s(t)$. (*b*) The form of the forcing function.

Recall that $V_m \sin \omega t$ is equal to $V_m \cos(\omega t - 90°)$, corresponding to $V_m \underline{/-90°} = -jV_m$.

For the circuit of Fig. 18.8*a*, determine the periodic response $i(t)$ corresponding to the forcing function shown in Fig. 18.8*b* if $i(0) = 0$.

The forcing function has a fundamental frequency $\omega_0 = 2$ rad/s, and its Fourier series may be written down by comparison with the Fourier series developed for the waveform of Fig. 18.4*b* in the solution of Practice Problem 18.3,

$$v_s(t) = 5 + \frac{20}{\pi} \sum_{n=1(\text{odd})}^{\infty} \frac{\sin 2nt}{n}$$

We will find the forced response for the nth harmonic by working in the frequency domain. Thus,

$$v_{sn}(t) = \frac{20}{n\pi} \sin 2nt$$

and

$$\mathbf{V}_{sn} = \frac{20}{n\pi} \underline{/-90°} = -j\frac{20}{n\pi}$$

The impedance offered by the *RL* circuit at this frequency is

$$\mathbf{Z}_n = 4 + j(2n)2 = 4 + j4n$$

and thus the component of the forced response at this frequency is

$$\mathbf{I}_{fn} = \frac{\mathbf{V}_{sn}}{\mathbf{Z}_n} = \frac{-j5}{n\pi(1 + jn)}$$

Transforming to the time domain, we have

$$i_{fn} = \frac{5}{n\pi} \frac{1}{\sqrt{1 + n^2}} \cos(2nt - 90° - \tan^{-1} n)$$

$$= \frac{5}{\pi(1 + n^2)} \left(\frac{\sin 2nt}{n} - \cos 2nt \right)$$

Since the response to the dc component is simply 5 V/4 Ω = 1.25 A, the forced response may be expressed as the summation

$$i_f(t) = 1.25 + \frac{5}{\pi} \sum_{n=1(\text{odd})}^{\infty} \left[\frac{\sin 2nt}{n(1+n^2)} - \frac{\cos 2nt}{1+n^2} \right]$$

The familiar natural response of this simple circuit is the single exponential term [characterizing the single pole of the transfer function, $\mathbf{I}_f/\mathbf{V}_s = 1/(4+2\mathbf{s})$]

$$i_n(t) = Ae^{-2t}$$

The *complete* response is therefore the sum

$$i(t) = i_f(t) + i_n(t)$$

Letting $t = 0$, we find A using $i(0) = 0$:

$$A = -1.25 + \frac{5}{\pi} \sum_{n=1(\text{odd})}^{\infty} \frac{1}{1+n^2}$$

Although correct, it is more convenient to use the numerical value of the summation. The sum of the first 5 terms of $\Sigma \, 1/(1+n^2)$ is 0.671, the sum of the first 10 terms is 0.695, the sum of the first 20 terms is 0.708, and the exact sum is 0.720 to three significant figures. Thus

$$A = -1.25 + \frac{5}{\pi}(0.720) = -0.104$$

and

$$i(t) = -0.104e^{-2t} + 1.25$$

$$+ \frac{5}{\pi} \sum_{n=1(\text{odd})}^{\infty} \left[\frac{\sin 2nt}{n(1+n^2)} - \frac{\cos 2nt}{1+n^2} \right] \quad \text{amperes}$$

In obtaining this solution, we have had to use many of the most general concepts introduced in this and the preceding 17 chapters. Some we did not have to use because of the simple nature of this particular circuit, but their places in the general analysis were indicated. In this sense, we may look upon the solution of this problem as a significant achievement in our introductory study of circuit analysis. In spite of this glorious feeling of accomplishment, however, it must be pointed out that the complete response, as obtained in Example 18.2 in analytical form, is not of much value as it stands; it furnishes no clear picture of the nature of the response. What we really need is a sketch of $i(t)$ as a function of time. This may be obtained by a laborious calculation at a sufficient number of instants of time; a desktop computer or a programmable calculator can be of great assistance here. The sketch may be approximated by the graphical addition of the natural response, the dc term, and the first few harmonics; this is an unrewarding task.

When all is said and done, the most informative solution of this problem is probably obtained by making a repeated transient analysis. That is, the form of the response can certainly be calculated in the interval from $t = 0$ to $t = \pi/2$ s; it is an exponential rising toward 2.5 A. After determining the value at the end of this first interval, we have an initial condition for the next $(\pi/2)$-second interval. The process is repeated until the response assumes a generally periodic nature. The method is eminently suitable to this example, for there is negligible change in the current waveform in the successive periods $\pi/2 < t < 3\pi/2$ and $3\pi/2 < t < 5\pi/2$. The complete current response is sketched in Fig. 18.9.

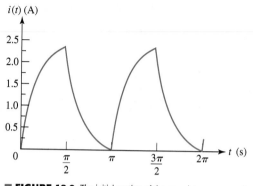

■ **FIGURE 18.9** The initial portion of the complete response of the circuit of Fig. 18.8*a* to the forcing function of Fig. 18.8*b*.

PRACTICE
●

18.6 Use the methods of Chap. 8 to determine the value of the current sketched in Fig. 18.9 at *t* equal to (*a*) $\pi/2$; (*b*) π; (*c*) $3\pi/2$.

Ans: 2.392 A; 0.1034 A; 2.396 A.

18.4 • COMPLEX FORM OF THE FOURIER SERIES

In obtaining a frequency spectrum, we have seen that the amplitude of each frequency component depends on both a_n and b_n; that is, the sine term and the cosine term both contribute to the amplitude. The exact expression for this amplitude is $\sqrt{a_n^2 + b_n^2}$. It is also possible to obtain the amplitude directly by using a form of Fourier series in which each term is a cosine function with a phase angle; the amplitude and phase angle are functions of $f(t)$ and n. An even more convenient and concise form of the Fourier series is obtained if the sines and cosines are expressed as exponential functions with complex multiplying constants.

Let us first take the trigonometric form of the Fourier series:

$$f(t) = a_0 + \sum_{n=1}^{\infty} (a_n \cos n\omega_0 t + b_n \sin n\omega_0 t)$$

and then substitute the exponential forms for the sine and cosine. After rearranging,

$$f(t) = a_0 + \sum_{n=1}^{\infty} \left(e^{jn\omega_0 t} \frac{a_n - jb_n}{2} + e^{-jn\omega_0 t} \frac{a_n + jb_n}{2} \right)$$

The reader may recall the identities

$$\sin \alpha = \frac{e^{j\alpha} - e^{-j\alpha}}{j2}$$

and

$$\cos \alpha = \frac{e^{j\alpha} + e^{-j\alpha}}{2}$$

Equations [44] and [45] are collectively called the *Fourier transform pair*. The function $\mathbf{F}(j\omega)$ is the *Fourier transform* of $f(t)$, and $f(t)$ is the *inverse Fourier transform* of $\mathbf{F}(j\omega)$.

This transform-pair relationship is very important! We should memorize it, draw arrows pointing to it, and mentally keep it on the conscious level. We emphasize the importance of these relations by repeating them in boxed form:

$$\mathbf{F}(j\omega) = \int_{-\infty}^{\infty} e^{-j\omega t} f(t)\, dt \qquad [46a]$$

$$f(t) = \frac{1}{2\pi} \int_{-\infty}^{\infty} e^{j\omega t} \mathbf{F}(j\omega)\, d\omega \qquad [46b]$$

The exponential terms in these two equations carry opposite signs for the exponents. To keep them straight, it may help to note that the positive sign is associated with the expression for $f(t)$, as it is with the complex Fourier series, Eq. [39].

It is appropriate to raise one question at this time. For the Fourier transform relationships of Eq. [46], can we obtain the Fourier transform of *any* arbitrarily chosen $f(t)$? It turns out that the answer is affirmative for almost any voltage or current that we can actually produce. A sufficient condition for the existence of $\mathbf{F}(j\omega)$ is that

$$\int_{-\infty}^{\infty} |f(t)|\, dt < \infty$$

This condition is not *necessary,* however, because some functions that do not meet it still have a Fourier transform; the step function is one such example. Furthermore, we will see later that $f(t)$ does not even need to be nonperiodic in order to have a Fourier transform; the Fourier series representation for a periodic time function is just a special case of the more general Fourier transform representation.

As we indicated earlier, the Fourier transform-pair relationship is unique. For a given $f(t)$ there is one specific $\mathbf{F}(j\omega)$; and for a given $\mathbf{F}(j\omega)$ there is one specific $f(t)$.

The reader may have already noticed a few similarities between the Fourier transform and the Laplace transform. Key differences between the two include the fact that initial energy storage is not easily incorporated in circuit analysis using Fourier transforms while it is very easily incorporated in the case of Laplace transforms. Also, there are several time functions (e.g., the *increasing* exponential) for which a Fourier transform does not exist. However, if it is spectral information as opposed to transient response in which we are primarily concerned, the Fourier transform is the ticket.

EXAMPLE 18.5

Use the Fourier transform to obtain the continuous spectrum of the single rectangular pulse Fig. 18.13a.

The pulse is a truncated version of the sequence considered previously in Fig. 18.11, and is described by

$$f(t) = \begin{cases} V_0 & t_0 < t < t_0 + \tau \\ 0 & t < t_0 \text{ and } t > t_0 + \tau \end{cases}$$

(Continued on next page)

The Fourier transform of $f(t)$ is found from Eq. [46a]:

$$\mathbf{F}(j\omega) = \int_{t_0}^{t_0+\tau} V_0 e^{-j\omega t} dt$$

and this may be easily integrated and simplified:

$$\mathbf{F}(j\omega) = V_0\tau \frac{\sin \frac{1}{2}\omega\tau}{\frac{1}{2}\omega\tau} e^{-j\omega(t_0+\tau/2)}$$

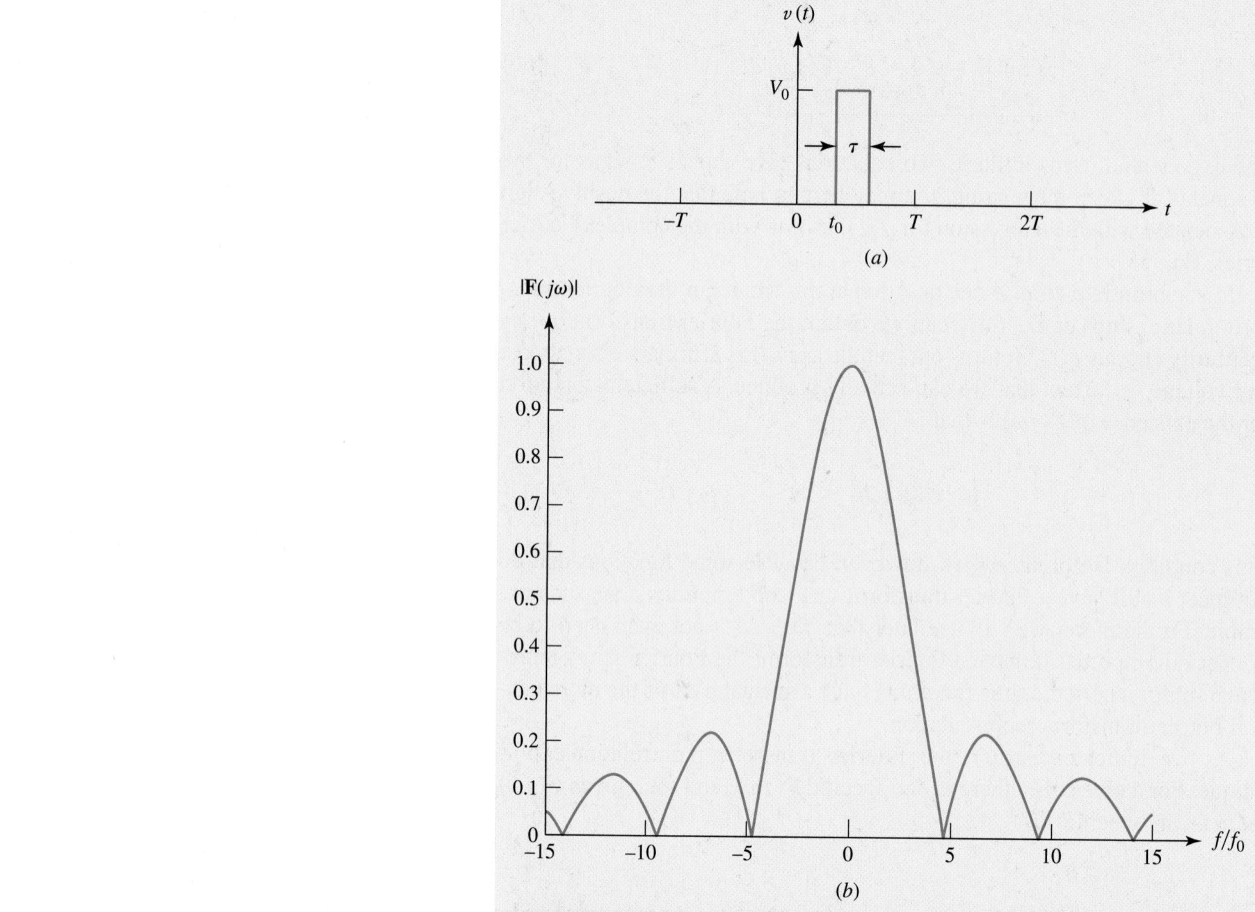

■ **FIGURE 18.13** (a) A single rectangular pulse identical to those of the sequence in Fig. 18.11. (b) A plot of $|\mathbf{F}(j\omega)|$ corresponding to the pulse, with $V_0 = 1$, $\tau = 1$, and $t_0 = 0$. The frequency axis has been normalized to the value of $f_0 = 1/1.5\pi$ corresponding to Fig. 18.12a to allow comparison; note that f_0 has no meaning or relevance in the context of $\mathbf{F}(j\omega)$.

The magnitude of $\mathbf{F}(j\omega)$ yields the continuous frequency spectrum, and it is of the form of the sampling function. The value of $\mathbf{F}(0)$ is $V_0\tau$. The shape of the spectrum is identical with the envelope in Fig. 18.12b. A plot of $|\mathbf{F}(j\omega)|$ as a function of ω does *not* indicate the magnitude of the voltage present at any given frequency. What is it, then? Examination of

Eq. [45] shows that, if $f(t)$ is a voltage waveform, then $\mathbf{F}(j\omega)$ is dimensionally "volts per unit frequency," a concept that was introduced in Sec. 15.1.

PRACTICE

18.8 If $f(t) = -10$ V, $-0.2 < t < -0.1$ s, $f(t) = 10$ V, $0.1 < t < 0.2$ s, and $f(t) = 0$ for all other t, evaluate $\mathbf{F}(j\omega)$ for ω equal to (a) 0; (b) 10π rad/s; (c) -10π rad/s; (d) 15π rad/s; (e) -20π rad/s.

18.9 If $\mathbf{F}(j\omega) = -10$ V/(rad/s) for $-4 < \omega < -2$ rad/s, $+10$ V/(rad/s) for $2 < \omega < 4$ rad/s, and 0 for all other ω, find the numerical value of $f(t)$ at t equal to (a) 10^{-4} s; (b) 10^{-2} s; (c) $\pi/4$ s; (d) $\pi/2$ s; (e) π s.

Ans: 18.8: 0; $j1.273$ V/(rad/s); $-j1.273$ V/(rad/s); $-j0.424$ V/(rad/s); 0.
18.9: $j1.9099 \times 10^{-3}$ V; $j0.1910$ V; $j4.05$ V; $-j4.05$ V; 0.

18.6 SOME PROPERTIES OF THE FOURIER TRANSFORM

Our object in this section is to establish several of the mathematical properties of the Fourier transform and, even more important, to understand its physical significance. We begin by using Euler's identity to replace $e^{-j\omega t}$ in Eq. [46a]:

$$\mathbf{F}(j\omega) = \int_{-\infty}^{\infty} f(t) \cos \omega t \, dt - j \int_{-\infty}^{\infty} f(t) \sin \omega t \, dt \qquad [47]$$

Since $f(t)$, $\cos \omega t$, and $\sin \omega t$ are all real functions of time, both the integrals in Eq. [47] are real functions of ω. Thus, by letting

$$\mathbf{F}(j\omega) = A(\omega) + jB(\omega) = |\mathbf{F}(j\omega)|e^{j\phi(\omega)} \qquad [48]$$

we have

$$A(\omega) = \int_{-\infty}^{\infty} f(t) \cos \omega t \, dt \qquad [49]$$

$$B(\omega) = -\int_{-\infty}^{\infty} f(t) \sin \omega t \, dt \qquad [50]$$

$$|\mathbf{F}(j\omega)| = \sqrt{A^2(\omega) + B^2(\omega)} \qquad [51]$$

and

$$\phi(\omega) = \tan^{-1} \frac{B(\omega)}{A(\omega)} \qquad [52]$$

Replacing ω by $-\omega$ shows that $A(\omega)$ and $|\mathbf{F}(j\omega)|$ are both even functions of ω, while $B(\omega)$ and $\phi(\omega)$ are both odd functions of ω.

Now, if $f(t)$ is an even function of t, then the integrand of Eq. [50] is an odd function of t, and the symmetrical limits force $B(\omega)$ to be zero; thus, if $f(t)$ is even, its Fourier transform $\mathbf{F}(j\omega)$ is a real, even function of ω, and the phase function $\phi(\omega)$ is zero or π for all ω. However, if $f(t)$ is an odd function of t, then $A(\omega) = 0$ and $\mathbf{F}(j\omega)$ is both odd and a pure imaginary function of ω; $\phi(\omega)$ is $\pm\pi/2$. In general, however, $\mathbf{F}(j\omega)$ is a complex function of ω.

Finally, we note that the replacement of ω by $-\omega$ in Eq. [47] forms the *conjugate* of $\mathbf{F}(j\omega)$. Thus,

$$\mathbf{F}(-j\omega) = A(\omega) - jB(\omega) = \mathbf{F}^*(j\omega)$$

and we have

$$\mathbf{F}(j\omega)\mathbf{F}(-j\omega) = \mathbf{F}(j\omega)\mathbf{F}^*(j\omega) = A^2(\omega) + B^2(\omega) = |\mathbf{F}(j\omega)|^2$$

Physical Significance of the Fourier Transform

With these basic mathematical properties of the Fourier transform in mind, we are now ready to consider its physical significance. Let us suppose that $f(t)$ is either the voltage across or the current through a 1 Ω resistor, so that $f^2(t)$ is the instantaneous power delivered to the 1 Ω resistor by $f(t)$. Integrating this power over all time, we obtain the total energy delivered by $f(t)$ to the 1 Ω resistor,

$$W_{1\Omega} = \int_{-\infty}^{\infty} f^2(t)\, dt \qquad [53]$$

Now let us resort to a little trickery. Thinking of the integrand in Eq. [53] as $f(t)$ times itself, we replace one of those functions with Eq. [46*b*]:

$$W_{1\Omega} = \int_{-\infty}^{\infty} f(t) \left[\frac{1}{2\pi} \int_{-\infty}^{\infty} e^{j\omega t} \mathbf{F}(j\omega)\, d\omega \right] dt$$

Since $f(t)$ is not a function of the variable of integration ω, we may move it inside the bracketed integral and then interchange the order of integration:

$$W_{1\Omega} = \frac{1}{2\pi} \int_{-\infty}^{\infty} \left[\int_{-\infty}^{\infty} \mathbf{F}(j\omega) e^{j\omega t} f(t)\, dt \right] d\omega$$

Next we shift $\mathbf{F}(j\omega)$ outside the inner integral, causing that integral to become $\mathbf{F}(-j\omega)$:

$$W_{1\Omega} = \frac{1}{2\pi} \int_{-\infty}^{\infty} \mathbf{F}(j\omega)\mathbf{F}(-j\omega)\, d\omega = \frac{1}{2\pi} \int_{-\infty}^{\infty} |\mathbf{F}(j\omega)|^2\, d\omega$$

Collecting these results,

$$\int_{-\infty}^{\infty} f^2(t)\, dt = \frac{1}{2\pi} \int_{-\infty}^{\infty} |\mathbf{F}(j\omega)|^2\, d\omega \qquad [54]$$

Marc Antoine Parseval-Deschenes was a rather obscure French mathematician, geographer, and occasional poet who published these results in 1805, seventeen years before Fourier published his theorem.

Equation [54] is a very useful expression known as Parseval's theorem. This theorem, along with Eq. [53], tells us that the energy associated with $f(t)$ can be obtained either from an integration over all time in the time domain or by $1/(2\pi)$ times an integration over all (radian) frequency in the frequency domain.

Parseval's theorem also leads us to a greater understanding and interpretation of the meaning of the Fourier transform. Consider a voltage $v(t)$ with Fourier transform $\mathbf{F}_v(j\omega)$ and 1 Ω energy $W_{1\Omega}$:

$$W_{1\Omega} = \frac{1}{2\pi} \int_{-\infty}^{\infty} |\mathbf{F}_v(j\omega)|^2\, d\omega = \frac{1}{\pi} \int_{0}^{\infty} |\mathbf{F}_v(j\omega)|^2\, d\omega$$

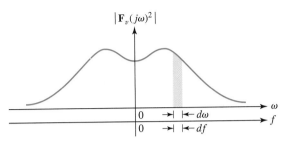

■ **FIGURE 18.14** The area of the slice $|F_v(j\omega)|^2$ is the 1 Ω energy associated with $v(t)$ lying in the bandwidth df.

where the rightmost equality follows from the fact that $|\mathbf{F}_v(j\omega)|^2$ is an even function of ω. Then, since $\omega = 2\pi f$, we can write

$$W_{1\Omega} = \int_{-\infty}^{\infty} |\mathbf{F}_v(j\omega)|^2\, df = 2\int_{0}^{\infty} |\mathbf{F}_v(j\omega)|^2\, df \qquad [55]$$

Figure 18.14 illustrates a typical plot of $|\mathbf{F}_v(j\omega)|^2$ as a function of both ω and f. If we divide the frequency scale up into vanishingly small increments df, Eq. [55] shows us that the area of a differential slice under the $|\mathbf{F}_v(j\omega)|^2$ curve, having a width df, is $|\mathbf{F}_v(j\omega)|^2\, df$. This area is shown shaded. The sum of all such areas, as f ranges from minus to plus infinity, is the total 1 Ω energy contained in $v(t)$. Thus, $|\mathbf{F}_v(j\omega)|^2$ is the (1 Ω) **energy density** or energy per unit bandwidth (J/Hz) of $v(t)$, and this energy density is always a real, even, nonnegative function of ω. By integrating $|\mathbf{F}_v(j\omega)|^2$ over an appropriate frequency interval, we are able to calculate that portion of the total energy lying within the chosen interval. Note that the energy density is not a function of the phase of $\mathbf{F}_v(j\omega)$, and thus there are an infinite number of time functions and Fourier transforms that possess identical energy-density functions.

EXAMPLE **18.6**

The one-sided [i.e., $v(t) = 0$ for $t < 0$] exponential pulse

$$v(t) = 4e^{-3t}u(t) \qquad \text{V}$$

is applied to the input of an ideal bandpass filter. If the filter passband is defined by $1 < |f| < 2$ Hz, calculate the total output energy.

We call the filter output voltage $v_o(t)$. The energy in $v_o(t)$ will therefore be equal to the energy of that part of $v(t)$ having frequency components in the intervals $1 < f < 2$ and $-2 < f < -1$. We determine the Fourier transform of $v(t)$,

$$\mathbf{F}_v(j\omega) = 4\int_{-\infty}^{\infty} e^{-j\omega t} e^{-3t} u(t)\, dt$$

$$= 4\int_{0}^{\infty} e^{-(3+j\omega)t}\, dt = \frac{4}{3+j\omega}$$

(Continued on next page)

and then we may calculate the total 1 Ω energy in the input signal by either

$$W_{1\Omega} = \frac{1}{2\pi} \int_{-\infty}^{\infty} |\mathbf{F}_v(j\omega)|^2 \, d\omega$$

$$= \frac{8}{\pi} \int_{-\infty}^{\infty} \frac{d\omega}{9+\omega^2} = \frac{16}{\pi} \int_{0}^{\infty} \frac{d\omega}{9+\omega^2} = \frac{8}{3} \text{ J}$$

or

$$W_{1\Omega} = \int_{-\infty}^{\infty} v^2(t) \, dt = 16 \int_{0}^{\infty} e^{-6t} \, dt = \frac{8}{3} \text{ J}$$

The total energy in $v_o(t)$, however, is smaller:

$$W_{o1} = \frac{1}{2\pi} \int_{-4\pi}^{-2\pi} \frac{16\,d\omega}{9+\omega^2} + \frac{1}{2\pi} \int_{2\pi}^{4\pi} \frac{16\,d\omega}{9+\omega^2}$$

$$= \frac{16}{\pi} \int_{2\pi}^{4\pi} \frac{d\omega}{9+\omega^2} = \frac{16}{3\pi} \left(\tan^{-1} \frac{4\pi}{3} - \tan^{-1} \frac{2\pi}{3} \right) = 358 \text{ mJ}$$

In general, we see that an ideal bandpass filter enables us to remove energy from prescribed frequency ranges while still retaining the energy contained in other frequency ranges. The Fourier transform helps us to describe the filtering action quantitatively without actually evaluating $v_o(t)$, although we will see later that the Fourier transform can also be used to obtain the expression for $v_o(t)$ if we wish to do so.

PRACTICE

18.10 If $i(t) = 10e^{20t}[u(t+0.1) - u(t-0.1)]$ A, find (a) $\mathbf{F}_i(j0)$; (b) $\mathbf{F}_i(j10)$; (c) $A_i(10)$; (d) $B_i(10)$; (e) $\phi_i(10)$.
18.11 Find the 1 Ω energy associated with the current $i(t) = 20e^{-10t}u(t)$ A in the interval (a) $-0.1 < t < 0.1$ s; (b) $-10 < \omega < 10$ rad/s; (c) $10 < \omega < \infty$ rad/s.

Ans: 18.10: 3.63 A/(rad/s); 3.33$\underline{/-31.7°}$ A/(rad/s); 2.83 A/(rad/s); −1.749 A/(rad/s); −31.7°. 18.11: 17.29 J; 10 J; 5 J.

18.7 FOURIER TRANSFORM PAIRS FOR SOME SIMPLE TIME FUNCTIONS

The Unit-Impulse Function

We now seek the Fourier transform of the unit impulse $\delta(t - t_0)$, a function we introduced in Sec. 14.4. That is, we are interested in the spectral properties or frequency-domain description of this singularity function. If we use the notation $\mathcal{F}\{\,\}$ to symbolize "Fourier transform of $\{\,\}$," then

$$\mathcal{F}\{\delta(t - t_0)\} = \int_{-\infty}^{\infty} e^{-j\omega t} \delta(t - t_0) \, dt$$

From our earlier discussion of this type of integral, we have

$$\mathcal{F}\{\delta(t - t_0)\} = e^{-j\omega t_0} = \cos \omega t_0 - j \sin \omega t_0$$

This complex function of ω leads to the 1 Ω energy-density function,

$$|\mathcal{F}\{\delta(t - t_0)\}|^2 = \cos^2 \omega t_0 + \sin^2 \omega t_0 = 1$$

This remarkable result says that the (1 Ω) energy per unit bandwidth is unity *at all frequencies,* and that the total energy in the unit impulse is infinitely large. No wonder, then, that we must conclude that the unit impulse is "impractical" in the sense that it cannot be generated in the laboratory. Moreover, even if one were available to us, it must appear distorted after being subjected to the finite bandwidth of any practical laboratory instrument.

Since there is a unique one-to-one correspondence between a time function and its Fourier transform, we can say that the inverse Fourier transform of $e^{-j\omega t_0}$ is $\delta(t - t_0)$. Utilizing the symbol $\mathcal{F}^{-1}\{\,\}$ for the inverse transform, we have

$$\mathcal{F}^{-1}\{e^{-j\omega t_0}\} = \delta(t - t_0)$$

Thus, we now know that

$$\frac{1}{2\pi} \int_{-\infty}^{\infty} e^{j\omega t} e^{-j\omega t_0} \, d\omega = \delta(t - t_0)$$

even though we would fail in an attempt at the direct evaluation of this improper integral. Symbolically, we may write

$$\delta(t - t_0) \Leftrightarrow e^{-j\omega t_0} \qquad [56]$$

where $\Leftrightarrow$ indicates that the two functions constitute a Fourier transform pair.

Continuing with our consideration of the unit-impulse function, let us consider a Fourier transform in that form,

$$\mathbf{F}(j\omega) = \delta(\omega - \omega_0)$$

which is a unit impulse *in the frequency domain* located at $\omega = \omega_0$. Then $f(t)$ must be

$$f(t) = \mathcal{F}^{-1}\{\mathbf{F}(j\omega)\} = \frac{1}{2\pi} \int_{-\infty}^{\infty} e^{j\omega t} \delta(\omega - \omega_0) \, d\omega = \frac{1}{2\pi} e^{j\omega_0 t}$$

where we have used the sifting property of the unit impulse. Thus we may now write

$$\frac{1}{2\pi} e^{j\omega_0 t} \Leftrightarrow \delta(\omega - \omega_0)$$

or

$$e^{j\omega_0 t} \Leftrightarrow 2\pi\delta(\omega - \omega_0) \qquad [57]$$

Also, by a simple sign change we obtain

$$e^{-j\omega_0 t} \Leftrightarrow 2\pi\delta(\omega + \omega_0) \qquad [58]$$

Clearly, the time function is complex in both expressions [57] and [58], and does not exist in the real world of the laboratory.

However, we know that

$$\cos \omega_0 t = \tfrac{1}{2} e^{j\omega_0 t} + \tfrac{1}{2} e^{-j\omega_0 t}$$

and it is easily seen from the definition of the Fourier transform that

$$\mathcal{F}\{f_1(t)\} + \mathcal{F}\{f_2(t)\} = \mathcal{F}\{f_1(t) + f_2(t)\} \qquad [59]$$

Therefore,

$$\mathcal{F}\{\cos \omega_0 t\} = \mathcal{F}\left\{\tfrac{1}{2}e^{j\omega_0 t}\right\} + \mathcal{F}\left\{\tfrac{1}{2}e^{-j\omega_0 t}\right\}$$
$$= \pi \delta(\omega - \omega_0) + \pi \delta(\omega + \omega_0)$$

which indicates that the frequency-domain description of $\cos \omega_0 t$ shows a *pair* of impulses, located at $\omega = \pm \omega_0$. This should not be a great surprise, for in our first discussion of complex frequency in Chap. 14, we noted that a sinusoidal function of time was always represented by a pair of imaginary frequencies located at $\mathbf{s} = \pm j\omega_0$. We have, therefore,

$$\cos \omega_0 t \Leftrightarrow \pi[\delta(\omega + \omega_0) + \delta(\omega - \omega_0)] \qquad [60]$$

The Constant Forcing Function

To find the Fourier transform of a constant function of time, $f(t) = K$, our first inclination might be to substitute this constant in the defining equation for the Fourier transform and evaluate the resulting integral. If we did, we would find ourselves with an indeterminate expression on our hands. Fortunately, however, we have already solved this problem, for from expression [58],

$$e^{-j\omega_0 t} \Leftrightarrow 2\pi \delta(\omega + \omega_0)$$

We see that if we simply let $\omega_0 = 0$, then the resulting transform pair is

$$1 \Leftrightarrow 2\pi \delta(\omega) \qquad [61]$$

from which it follows that

$$K \Leftrightarrow 2\pi K \delta(\omega) \qquad [62]$$

and our problem is solved. The frequency spectrum of a constant function of time consists only of a component at $\omega = 0$, which we knew all along.

The Signum Function

As another example, let us obtain the Fourier transform of a singularity function known as the ***signum function,*** $\mathrm{sgn}(t)$, defined by

$$\mathrm{sgn}(t) = \begin{cases} -1 & t < 0 \\ 1 & t > 0 \end{cases} \qquad [63]$$

or

$$\mathrm{sgn}(t) = u(t) - u(-t)$$

Again, if we should try to substitute this time function in the defining equation for the Fourier transform, we would face an indeterminate expression upon substitution of the limits of integration. This same problem will arise every time we attempt to obtain the Fourier transform of a time function that does not approach zero as $|t|$ approaches infinity. Fortunately, we can avoid this situation by using the *Laplace transform,* as it contains a built-in convergence factor that cures many of the inconvenient ills associated with the evaluation of certain Fourier transforms.

Along those lines, the signum function under consideration can be written as

$$\text{sgn}(t) = \lim_{a \to 0}[e^{-at}u(t) - e^{at}u(-t)]$$

Notice that the expression within the brackets *does* approach zero as $|t|$ gets very large. Using the definition of the Fourier transform, we obtain

$$\mathcal{F}\{\text{sgn}(t)\} = \lim_{a \to 0}\left[\int_0^\infty e^{-j\omega t}e^{-at}\,dt - \int_{-\infty}^0 e^{-j\omega t}e^{at}\,dt\right]$$

$$= \lim_{a \to 0}\frac{-j2\omega}{\omega^2 + a^2} = \frac{2}{j\omega}$$

The real component is zero, since $\text{sgn}(t)$ is an odd function of t. Thus,

$$\text{sgn}(t) \Leftrightarrow \frac{2}{j\omega} \qquad\qquad [64]$$

The Unit-Step Function

As a final example in this section, let us look at the familiar unit-step function, $u(t)$. Making use of our work on the signum function in the preceding paragraphs, we represent the unit step by

$$u(t) = \tfrac{1}{2} + \tfrac{1}{2}\text{sgn}(t)$$

and obtain the Fourier transform pair

$$u(t) \Leftrightarrow \left[\pi\delta(\omega) + \frac{1}{j\omega}\right] \qquad\qquad [65]$$

Table 18.2 presents the conclusions drawn from the examples discussed in this section, along with a few others that have not been detailed here.

EXAMPLE 18.7

Use Table 18.2 to find the Fourier transform of the time function $3e^{-t}\cos 4t\, u(t)$.

From the next to the last entry in the table, we have

$$e^{-\alpha t}\cos\omega_d t\, u(t) \Leftrightarrow \frac{\alpha + j\omega}{(\alpha + j\omega)^2 + \omega_d^2}$$

We therefore identify α as 1 and ω_d as 4, and have

$$\mathbf{F}(j\omega) = 3\frac{1 + j\omega}{(1 + j\omega)^2 + 16}$$

PRACTICE

18.12 Evaluate the Fourier transform at $\omega = 12$ for the time function
(a) $4u(t) - 10\delta(t)$; (b) $5e^{-8t}u(t)$; (c) $4\cos 8tu(t)$; (d) $-4\,\text{sgn}(t)$.
18.13 Find $f(t)$ at $t = 2$ if $\mathbf{F}(j\omega)$ is equal to (a) $5e^{-j3\omega} - j(4/\omega)$;
(b) $8[\delta(\omega - 3) + \delta(\omega + 3)]$; (c) $(8/\omega)\sin 5\omega$.

Ans: 18.12: $10.01\underline{/-178.1°}; 0.347\underline{/-56.3°}; -j0.6; j0.667$. 18.13: $2.00; 2.45; 4.00$.

TABLE 18.2 A Summary of Some Fourier Transform Pairs

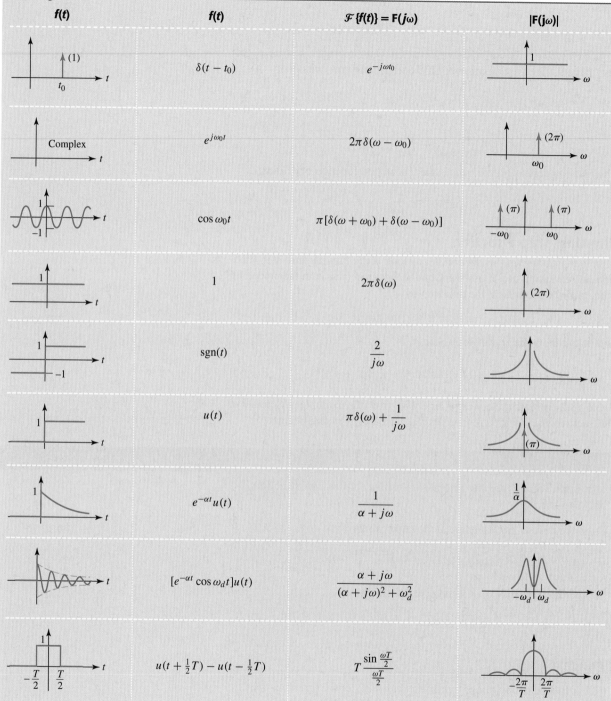

$f(t)$	$f(t)$	$\mathcal{F}\{f(t)\} = \mathbf{F}(j\omega)$	$|\mathbf{F}(j\omega)|$
	$\delta(t - t_0)$	$e^{-j\omega t_0}$	
Complex	$e^{j\omega_0 t}$	$2\pi\delta(\omega - \omega_0)$	
	$\cos\omega_0 t$	$\pi[\delta(\omega + \omega_0) + \delta(\omega - \omega_0)]$	
	1	$2\pi\delta(\omega)$	
	$\operatorname{sgn}(t)$	$\dfrac{2}{j\omega}$	
	$u(t)$	$\pi\delta(\omega) + \dfrac{1}{j\omega}$	
	$e^{-\alpha t}u(t)$	$\dfrac{1}{\alpha + j\omega}$	
	$[e^{-\alpha t}\cos\omega_d t]u(t)$	$\dfrac{\alpha + j\omega}{(\alpha + j\omega)^2 + \omega_d^2}$	
	$u(t + \tfrac{1}{2}T) - u(t - \tfrac{1}{2}T)$	$T\dfrac{\sin\frac{\omega T}{2}}{\frac{\omega T}{2}}$	

18.8 THE FOURIER TRANSFORM OF A GENERAL PERIODIC TIME FUNCTION

In Sec. 18.5 we remarked that we would be able to show that periodic time functions, as well as nonperiodic functions, possess Fourier transforms. Let us now establish this fact on a rigorous basis. Consider a periodic time function $f(t)$ with period T and Fourier series expansion, as outlined by Eqs. [39], [40], and [41], repeated here for convenience:

$$f(t) = \sum_{n=-\infty}^{\infty} \mathbf{c}_n e^{jn\omega_0 t} \tag{39}$$

$$\mathbf{c}_n = \frac{1}{T} \int_{-T/2}^{T/2} f(t) e^{-jn\omega_0 t} \, dt \tag{40}$$

and

$$\omega_0 = \frac{2\pi}{T} \tag{41}$$

Bearing in mind that the Fourier transform of a sum is just the sum of the transforms of the terms in the sum, and that $\mathbf{c}_n$ is not a function of time, we can write

$$\mathcal{F}\{f(t)\} = \mathcal{F}\left\{ \sum_{n=-\infty}^{\infty} \mathbf{c}_n e^{jn\omega_0 t} \right\} = \sum_{n=-\infty}^{\infty} \mathbf{c}_n \mathcal{F}\{e^{jn\omega_0 t}\}$$

After obtaining the transform of $e^{jn\omega_0 t}$ from expression [57], we have

$$f(t) \Leftrightarrow 2\pi \sum_{n=-\infty}^{\infty} \mathbf{c}_n \delta(\omega - n\omega_0) \tag{66}$$

This shows that $f(t)$ has a discrete spectrum consisting of impulses located at points on the ω axis given by $\omega = n\omega_0$, $n = \ldots, -2, -1, 0, 1, \ldots$. The strength of each impulse is 2π times the value of the corresponding Fourier coefficient appearing in the complex form of the Fourier series expansion for $f(t)$.

As a check on our work, let us see whether the inverse Fourier transform of the right side of expression [66] is once again $f(t)$. This inverse transform can be written as

$$\mathcal{F}^{-1}\{\mathbf{F}(j\omega)\} = \frac{1}{2\pi} \int_{-\infty}^{\infty} e^{j\omega t} \left[2\pi \sum_{n=-\infty}^{\infty} \mathbf{c}_n \delta(\omega - n\omega_0) \right] d\omega \stackrel{?}{=} f(t)$$

Since the exponential term does not contain the index of summation n, we can interchange the order of the integration and summation operations:

$$\mathcal{F}^{-1}\{\mathbf{F}(j\omega)\} = \sum_{n=-\infty}^{\infty} \int_{-\infty}^{\infty} \mathbf{c}_n e^{j\omega t} \delta(\omega - n\omega_0) \, d\omega \stackrel{?}{=} f(t)$$

Because it is not a function of the variable of integration, $\mathbf{c}_n$ can be treated as a constant. Then, using the sifting property of the impulse, we obtain

$$\mathcal{F}^{-1}\{\mathbf{F}(j\omega)\} = \sum_{n=-\infty}^{\infty} \mathbf{c}_n e^{jn\omega_0 t} \stackrel{?}{=} f(t)$$

which is exactly the same as Eq. [39], the complex Fourier series expansion for $f(t)$. The question marks in the preceding equations can now be removed, and the existence of the Fourier transform for a periodic time function is established. This should come as no great surprise, however. In the last section we evaluated the Fourier transform of a cosine function, which is certainly periodic, although we made no direct reference to its periodicity. However, we did use a backhanded approach in getting the transform. But now we have a mathematical tool by which the transform can be obtained more directly. To demonstrate this procedure, consider $f(t) = \cos \omega_0 t$ once more. First we evaluate the Fourier coefficients $\mathbf{c}_n$:

$$\mathbf{c}_n = \frac{1}{T} \int_{-T/2}^{T/2} \cos \omega_0 t \, e^{-jn\omega_0 t} dt = \begin{cases} \frac{1}{2} & n = \pm 1 \\ 0 & \text{otherwise} \end{cases}$$

Then

$$\mathcal{F}\{f(t)\} = 2\pi \sum_{n=-\infty}^{\infty} \mathbf{c}_n \delta(\omega - n\omega_0)$$

This expression has values that are nonzero only when $n = \pm 1$, and it follows, therefore, that the entire summation reduces to

$$\mathcal{F}\{\cos \omega_0 t\} = \pi [\delta(\omega - \omega_0) + \delta(\omega + \omega_0)]$$

which is precisely the expression that we obtained before. What a relief!

PRACTICE

18.14 Find (a) $\mathcal{F}\{5 \sin^2 3t\}$; (b) $\mathcal{F}\{A \sin \omega_0 t\}$; (c) $\mathcal{F}\{6 \cos(8t + 0.1\pi)\}$.

Ans: $2.5\pi[2\delta(\omega) - \delta(\omega + 6) - \delta(\omega - 6)]$; $j\pi A[\delta(\omega + \omega_0) - \delta(\omega - \omega_0)]$; $[18.85\underline{/18°}]\delta(\omega - 8) + [18.85\underline{/-18°}]\delta(\omega + 8)$.

18.9 THE SYSTEM FUNCTION AND RESPONSE IN THE FREQUENCY DOMAIN

In Sec. 15.5, the problem of determining the output of a physical system in terms of the input and the impulse response was solved by using the convolution integral and initially working in the time domain. The input, the output, and the impulse response are all time functions. Subsequently, we found that it was often more convenient to perform such operations in the frequency domain, as the Laplace transform of the convolution of two functions is simply the product of each function in the frequency domain. Along the same lines, we find the same is true when working with Fourier transforms.

To do this we examine the Fourier transform of the system output. Assuming arbitrarily that the input and output are voltages, we apply the basic definition of the Fourier transform and express the output by the convolution integral:

$$\mathcal{F}\{v_0(t)\} = \mathbf{F}_0(j\omega) = \int_{-\infty}^{\infty} e^{-j\omega t} \left[\int_{-\infty}^{\infty} v_i(t-z)h(z)\,dz \right] dt$$

where we again assume no initial energy storage. At first glance this expression may seem rather formidable, but it can be reduced to a result that is surprisingly simple. We may move the exponential term inside the inner integral because it does not contain the variable of integration z. Next we reverse the order of integration, obtaining

$$\mathbf{F}_0(j\omega) = \int_{-\infty}^{\infty} \left[\int_{-\infty}^{\infty} e^{-j\omega t} v_i(t-z)h(z)\,dt \right] dz$$

Since it is not a function of t, we can extract $h(z)$ from the inner integral and simplify the integration with respect to t by a change of variable, $t - z = x$:

$$\mathbf{F}_0(j\omega) = \int_{-\infty}^{\infty} h(z) \left[\int_{-\infty}^{\infty} e^{-j\omega(x+z)} v_i(x)\,dx \right] dz$$

$$= \int_{-\infty}^{\infty} e^{-j\omega z} h(z) \left[\int_{-\infty}^{\infty} e^{-j\omega x} v_i(x)\,dx \right] dz$$

But now the sun is starting to break through, for the inner integral is merely the Fourier transform of $v_i(t)$. Furthermore, it contains no z terms and can be treated as a constant in any integration involving z. Thus, we can move this transform, $\mathbf{F}_i(j\omega)$, completely outside all the integral signs:

$$\mathbf{F}_0(j\omega) = \mathbf{F}_i(j\omega) \int_{-\infty}^{\infty} e^{-j\omega z} h(z)\,dz$$

Finally, the remaining integral exhibits our old friend once more, another Fourier transform! This one is the Fourier transform of the impulse response, which we will designate by the notation $\mathbf{H}(j\omega)$. Therefore, all our work has boiled down to the simple result:

$$\mathbf{F}_0(j\omega) = \mathbf{F}_i(j\omega)\mathbf{H}(j\omega) = \mathbf{F}_i(j\omega)\mathcal{F}\{h(t)\}$$

This is another important result: it defines the *system function* $\mathbf{H}(j\omega)$ as the ratio of the Fourier transform of the response function to the Fourier transform of the forcing function. Moreover, the system function and the impulse response constitute a Fourier transform pair:

$$h(t) \Leftrightarrow \mathbf{H}(j\omega) \qquad [67]$$

The development in the preceding paragraph also serves to prove the general statement that the Fourier transform of the convolution of two time functions is the product of their Fourier transforms,

$$\boxed{\mathcal{F}\{f(t) * g(t)\} = \mathbf{F}_f(j\omega)\mathbf{F}_g(j\omega)} \qquad [68]$$

To recapitulate, if we know the Fourier transforms of the forcing function and the impulse response, then the Fourier transform of the response function can be obtained as their product. The result is a description of the response function in the frequency domain; the time-domain description of the response function is obtained by simply taking the inverse Fourier transform. Thus we see that the process of convolution in the time domain is equivalent to the relatively simple operation of multiplication in the frequency domain.

The foregoing comments might make us wonder once again why we would ever choose to work in the time domain at all, but we must always remember that we seldom get something for nothing. A poet once said, *"Our sincerest laughter/with some pain is fraught."*[2] The pain herein is the occasional difficulty in obtaining the inverse Fourier transform of a response function, for reasons of mathematical complexity. On the other hand, a simple desktop computer can convolve two time functions with magnificent celerity. For that matter, it can also obtain an FFT (fast Fourier transform) quite rapidly. Consequently there is no clear-cut advantage between working in the time domain and in the frequency domain. A decision must be made each time a new problem arises; it should be based on the information available and on the computational facilities at hand.

Consider a forcing function of the form

$$v_i(t) = u(t) - u(t-1)$$

and a unit-impulse response defined by

$$h(t) = 2e^{-t}u(t)$$

We first obtain the corresponding Fourier transforms. The forcing function is the difference between two unit-step functions. These two functions are identical, except that one is initiated 1 s after the other. We will evaluate the response due to $u(t)$; the response due to $u(t-1)$ is the same, but delayed in time by 1 s. The difference between these two partial responses will be the total response due to $v_i(t)$.

The Fourier transform of $u(t)$ was obtained in Sec. 18.7:

$$\mathcal{F}\{u(t)\} = \pi\delta(\omega) + \frac{1}{j\omega}$$

The system function is obtained by taking the Fourier transform of $h(t)$, listed in Table 18.2,

$$\mathcal{F}\{h(t)\} = \mathbf{H}(j\omega) = \mathcal{F}\{2e^{-t}u(t)\} = \frac{2}{1+j\omega}$$

The inverse transform of the product of these two functions yields that component of $v_o(t)$ caused by $u(t)$,

$$v_{o1}(t) = \mathcal{F}^{-1}\left\{\frac{2\pi\delta(\omega)}{1+j\omega} + \frac{2}{j\omega(1+j\omega)}\right\}$$

Using the sifting property of the unit impulse, the inverse transform of the first term is just a constant equal to unity. Thus,

$$v_{o1}(t) = 1 + \mathcal{F}^{-1}\left\{\frac{2}{j\omega(1+j\omega)}\right\}$$

The second term contains a product of terms in the denominator, each of the form $(\alpha + j\omega)$, and its inverse transform is found most easily by making use of the partial-fraction expansion that we developed in Sec. 14.5. Let us

(2) P. B. Shelley, "To a Skylark," 1821.

select a technique for obtaining a partial-fraction expansion that has one big advantage—it always works, although faster methods are usually available for most situations. We assign an unknown quantity in the numerator of each fraction, here two in number,

$$\frac{2}{j\omega(1+j\omega)} = \frac{A}{j\omega} + \frac{B}{1+j\omega}$$

and then substitute a corresponding number of simple values for $j\omega$. Here we let $j\omega = 1$:

$$1 = A + \frac{B}{2}$$

and then let $j\omega = -2$:

$$1 = -\frac{A}{2} - B$$

This leads to $A = 2$ and $B = -2$. Thus,

$$\mathcal{F}^{-1}\left\{\frac{2}{j\omega(1+j\omega)}\right\} = \mathcal{F}^{-1}\left\{\frac{2}{j\omega} - \frac{2}{1+j\omega}\right\} = \text{sgn}(t) - 2e^{-t}u(t)$$

so that

$$v_{o1}(t) = 1 + \text{sgn}(t) - 2e^{-t}u(t)$$
$$= 2u(t) - 2e^{-t}u(t)$$
$$= 2(1 - e^{-t})u(t)$$

It follows that $v_{o2}(t)$, the component of $v_o(t)$ produced by $u(t-1)$, is

$$v_{o2}(t) = 2(1 - e^{-(t-1)})u(t-1)$$

Therefore,

$$v_o(t) = v_{o1}(t) - v_{o2}(t)$$
$$= 2(1 - e^{-t})u(t) - 2(1 - e^{-t+1})u(t-1)$$

The discontinuities at $t = 0$ and $t = 1$ dictate a separation into three time intervals:

$$v_o(t) = \begin{cases} 0 & t < 0 \\ 2(1 - e^{-t}) & 0 < t < 1 \\ 2(e - 1)e^{-t} & t > 1 \end{cases}$$

PRACTICE

18.15 The impulse response of a certain linear network is $h(t) = 6e^{-20t}u(t)$. The input signal is $3e^{-6t}u(t)$ V. Find (a) $\mathbf{H}(j\omega)$; (b) $\mathbf{V}_i(j\omega)$; (c) $V_o(j\omega)$; (d) $v_o(0.1)$; (e) $v_o(0.3)$; (f) $v_{o,\text{max}}$.

Ans: $6/(20 + j\omega)$; $3/(6 + j\omega)$; $18/[(20 + j\omega)(6 + j\omega)]$; 0.532 V; 0.209 V; 0.5372.

COMPUTER-AIDED ANALYSIS

The material presented in this chapter forms the foundation for many advanced fields of study, including signal processing, communications, and controls. We are only able to introduce some of the more fundamental concepts within the context of an introductory circuits text, but even at this point some of the power of Fourier-based analysis can be brought to bear. As a first example, consider the op amp circuit of Fig. 18.15, constructed in PSpice using a μA741 operational amplifier.

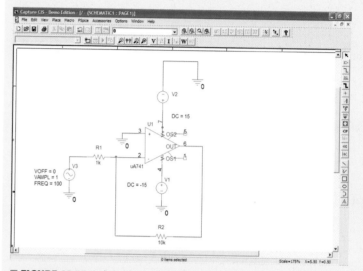

■ **FIGURE 18.15** An inverting amplifier circuit with a voltage gain of -10, driven by a sinusoidal input operating at 100 Hz.

The circuit has a voltage gain of -10, and so we would expect a sinusoidal output of 10 V amplitude. This is indeed what we obtain from a transient analysis of the circuit, as shown in Fig. 18.16.

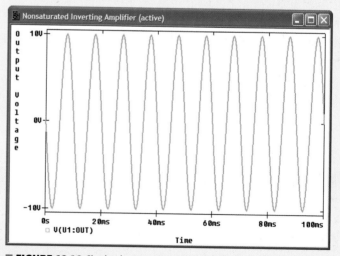

■ **FIGURE 18.16** Simulated output voltage of the amplifier circuit shown in Fig. 18.15.

PSpice allows us to determine the frequency spectrum of the output voltage through what is known as a fast Fourier transform (FFT), a discrete-time approximation to the exact Fourier transform of the signal. From within Probe, we select **Fourier** under the **Trace** menu; the result is the plot shown in Fig. 18.17. As expected, the line spectrum for the output voltage of this amplifier circuit consists of a single feature at a frequency of 100 Hz.

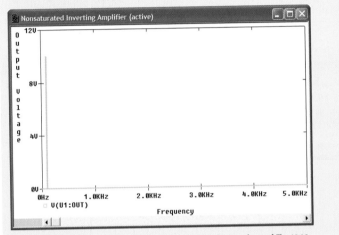

■ **FIGURE 18.17** Discrete approximation to the Fourier transform of Fig. 18.16.

As the input voltage magnitude is increased, the output of the amplifier approaches the saturation condition determined by the positive and negative dc supply voltages (±15 V in this example). This behavior is evident in the simulation result of Fig. 18.18, which corresponds to an input voltage magnitude of 1.8 V. A key feature of interest is that the output voltage waveform is no longer a pure sinusoid. As a result, we expect nonzero values at harmonic frequencies to appear in the

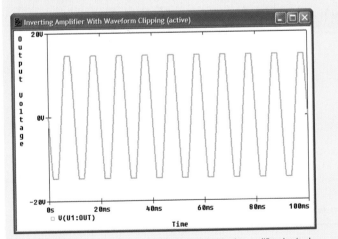

■ **FIGURE 18.18** Transient analysis simulation results for the amplifier circuit when the input voltage magnitude is increased to 1.8 V. Saturation effects manifest themselves in the plot as clipped waveform extrema.

(Continued on next page)

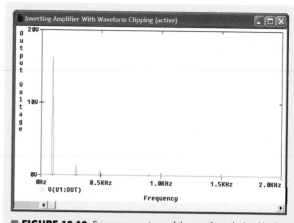

■ **FIGURE 18.19** Frequency spectrum of the waveform depicted in Fig. 18.18, showing the presence of several harmonic components in addition to the fundamental frequency. The finite width of the features is an artifact of the numerical discretization (a set of discrete time values was used).

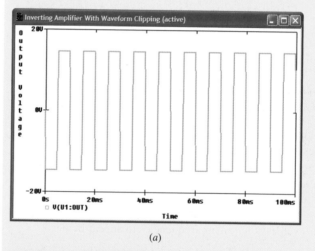

(a)

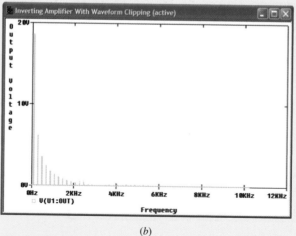

(b)

■ **FIGURE 18.20** (a) Severe effects of amplifier saturation are observed in the simulated response to a 15 V sinusoidal input. (b) An FFT of the waveform shows a significant increase in the fraction of energy present in harmonics as opposed to the fundamental frequency of 100 Hz.

frequency spectrum of the function, as is the case in Fig. 18.19. The effect of reaching saturation in the amplifier circuit is a distortion of the signal; if connected to a speaker, we do not hear a "clean" 100 Hz waveform. Instead, we now hear a superposition of waveforms which include not only the 100 Hz fundamental frequency, but significant harmonic components at 300 and 500 Hz as well. Further distortion of the waveform would increase the amount of energy in harmonic frequencies, so that contributions from higher-frequency harmonics would become more significant. This is evident in the simulation results of Fig. 18.20*a* and *b*, which show the output voltage in the time and frequency domains, respectively.

18.10 THE PHYSICAL SIGNIFICANCE OF THE SYSTEM FUNCTION

In this section we will try to connect several aspects of the Fourier transform with work we completed in earlier chapters.

Given a general linear two-port network N without any initial energy storage, we assume sinusoidal forcing and response functions, arbitrarily taken to be voltages, as shown in Fig. 18.21. We let the input voltage be simply $A \cos(\omega_x t + \theta)$, and the output can be described in general terms as $v_o(t) = B \cos(\omega_x t + \phi)$, where the amplitude B and phase angle ϕ are functions of ω_x. In phasor form, we can write the forcing and response functions as $\mathbf{V}_i = Ae^{j\theta}$ and $\mathbf{V}_o = Be^{j\phi}$. The ratio of the phasor response to the phasor forcing function is a complex number that is a function of ω_x:

$$\frac{\mathbf{V}_o}{\mathbf{V}_i} = \mathbf{G}(\omega_x) = \frac{B}{A}e^{j(\phi - \theta)}$$

where B/A is the amplitude of $\mathbf{G}$ and $\phi - \theta$ is its phase angle. This transfer function $\mathbf{G}(\omega_x)$ could be obtained in the laboratory by varying ω_x over a large range of values and measuring the amplitude B/A and phase $\phi - \theta$ for each value of ω_x. If we then plotted each of these parameters as a function of frequency, the resultant pair of curves would completely describe the transfer function.

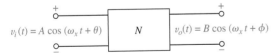

$v_i(t) = A \cos(\omega_x t + \theta)$ N $v_o(t) = B \cos(\omega_x t + \phi)$

■ **FIGURE 18.21** Sinusoidal analysis can be used to determine the transfer function $\mathsf{H}(j\omega_x) = (B/A)e^{j(\phi - \theta)}$, where B and ϕ are functions of ω_x.

Now let us hold these comments in the backs of our minds for a moment as we consider a slightly different aspect of the same analysis problem.

For the circuit with sinusoidal input and output shown in Fig. 18.21, what is the system function $\mathbf{H}(j\omega)$? To answer this question, we begin with the definition of $\mathbf{H}(j\omega)$ as the ratio of the Fourier transforms of the output and the input. Both of these time functions involve the functional form

$\cos(\omega_x t + \beta)$, whose Fourier transform we have not evaluated as yet, although we can handle $\cos \omega_x t$. The transform we need is

$$\mathcal{F}\{\cos(\omega_x t + \beta)\} = \int_{-\infty}^{\infty} e^{-j\omega t} \cos(\omega_x t + \beta)\, dt$$

If we make the substitution $\omega_x t + \beta = \omega_x \tau$, then

$$\mathcal{F}\{\cos(\omega_x t + \beta)\} = \int_{-\infty}^{\infty} e^{-j\omega\tau + j\omega\beta/\omega_x} \cos \omega_x \tau\, d\tau$$

$$= e^{j\omega\beta/\omega_x} \mathcal{F}\{\cos \omega_x t\}$$

$$= \pi e^{j\omega\beta/\omega_x}[\delta(\omega - \omega_x) + \delta(\omega + \omega_x)]$$

This is a new Fourier transform pair,

$$\cos(\omega_x t + \beta) \Leftrightarrow \pi e^{j\omega\beta/\omega_x}[\delta(\omega - \omega_x) + \delta(\omega + \omega_x)] \qquad [69]$$

which we can now use to evaluate the desired system function,

$$\mathbf{H}(j\omega) = \frac{\mathcal{F}\{B \cos(\omega_x t + \phi)\}}{\mathcal{F}\{A \cos(\omega_x t + \theta)\}}$$

$$= \frac{\pi B e^{j\omega\phi/\omega_x}[\delta(\omega - \omega_x) + \delta(\omega + \omega_x)]}{\pi A e^{j\omega\theta/\omega_x}[\delta(\omega - \omega_x) + \delta(\omega + \omega_x)]}$$

$$= \frac{B}{A} e^{j\omega(\phi-\theta)/\omega_x}$$

Now we recall the expression for $\mathbf{G}(\omega_x)$,

$$\mathbf{G}(\omega_x) = \frac{B}{A} e^{j(\phi-\theta)}$$

where B and ϕ were evaluated at $\omega = \omega_x$, and we see that evaluating $\mathbf{H}(j\omega)$ at $\omega = \omega_x$ gives

$$\mathbf{H}(\omega_x) = \mathbf{G}(\omega_x) = \frac{B}{A} e^{j(\phi-\theta)}$$

Since there is nothing special about the x subscript, we conclude that the system function and the transfer function are identical:

$$\mathbf{H}(j\omega) = \mathbf{G}(\omega) \qquad [70]$$

The fact that one argument is ω while the other is indicated by $j\omega$ is immaterial and arbitrary; the j merely makes possible a more direct comparison between the Fourier and Laplace transforms.

Equation [70] represents a direct connection between Fourier transform techniques and sinusoidal steady-state analysis. Our previous work on steady-state sinusoidal analysis using phasors was but a special case of the more general techniques of Fourier transform analysis. It was "special" in the sense that the inputs and outputs were sinusoids, whereas the use of Fourier transforms and system functions enables us to handle nonsinusoidal forcing functions and responses.

Thus, to find the system function $\mathbf{H}(j\omega)$ for a network, all we need to do is to determine the corresponding sinusoidal transfer function as a function of ω (or $j\omega$).

EXAMPLE **18.8**

Find the voltage across the inductor of the circuit shown in Fig. 18.22a when the input voltage is a simple exponentially decaying pulse, as indicated.

We need the system function; but it is not necessary to apply an impulse, find the impulse response, and then determine its inverse transform. Instead we use Eq. [70] to obtain the system function $\mathbf{H}(j\omega)$ by assuming that the input and output voltages are both sinusoids described by their corresponding phasors, as shown in Fig. 18.22b. Using voltage division, we have

$$\mathbf{H}(j\omega) = \frac{\mathbf{V}_o}{\mathbf{V}_i} = \frac{j2\omega}{4 + j2\omega}$$

The transform of the forcing function is

$$\mathcal{F}\{v_i(t)\} = \frac{5}{3 + j\omega}$$

and thus the transform of $v_o(t)$ is given as

$$\mathcal{F}\{v_o(t)\} = \mathbf{H}(j\omega)\mathcal{F}\{v_i(t)\}$$
$$= \frac{j2\omega}{4 + j2\omega} \frac{5}{3 + j\omega}$$
$$= \frac{15}{3 + j\omega} - \frac{10}{2 + j\omega}$$

where the partial fractions appearing in the last step help to determine the inverse Fourier transform

$$v_o(t) = \mathcal{F}^{-1}\left\{\frac{15}{3 + j\omega} - \frac{10}{2 + j\omega}\right\}$$
$$= 15e^{-3t}u(t) - 10e^{-2t}u(t)$$
$$= 5(3e^{-3t} - 2e^{-2t})u(t)$$

Our problem is completed without fuss, convolution, or differential equations.

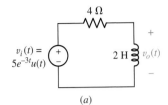

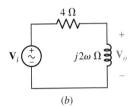

■ **FIGURE 18.22** (a) The response $v_o(t)$ caused by $v_i(t)$ is desired. (b) The system function $\mathbf{H}(j\omega)$ may be determined by sinusoidal steady-state analysis: $\mathbf{H}(j\omega) = \mathbf{V}_o/\mathbf{V}_i$.

PRACTICE

18.16 Use Fourier transform techniques on the circuit of Fig. 18.23 to find $i_1(t)$ at $t = 1.5$ ms if i_s equals (a) $\delta(t)$ A; (b) $u(t)$ A; (c) $\cos 500t$ A.

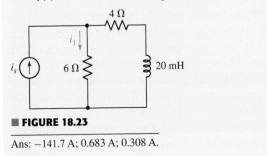

■ **FIGURE 18.23**

Ans: -141.7 A; 0.683 A; 0.308 A.

Image Processing

Although a great deal of progress has been made toward developing a complete understanding of the function of muscle, there remain many open questions. A great deal of research in this field has been carried out using vertebrate skeletal muscle, in particular the *sartorius* or leg muscle of the frog (Fig. 18.24).

■ **FIGURE 18.24** Close-up of a frog against an orange background.
© IT Stock/PunchStock/RF.

Of the many analytical techniques scientists use, one of the most common is electron microscopy. Figure 18.25

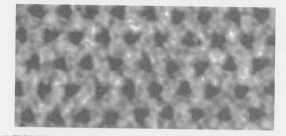

■ **FIGURE 18.25** Electron micrograph of a region of frog sartorius muscle tissue. False color has been employed for clarity.
Courtesy Professor John M. Squire, Imperial College, London.

shows an electron micrograph of frog sartorius muscle tissue, sectioned in such a fashion as to highlight the regular arrangement of *myosin,* a filamentary type of contractile protein. Of interest to structural biologists are the periodicity and disorder of these proteins over a large area of muscle tissue. In order to develop a model for these characteristics, a numerical approach is preferable, where the analysis of such images can be automated. As can be seen in the figure, however, the image produced by the electron microscope can be contaminated by a high level of background noise, making automated identification of the myosin filaments prone to error.

Introduced with the intent of ultimately assisting us in the analysis of time-varying linear circuits, the Fourier-based techniques of this chapter are in fact very powerful general methods which find application in many other situations. Among these, the field of *image processing* makes frequent use of Fourier techniques, especially through the fast Fourier transform and related numerical methods. The image of Fig. 18.25 can be described by a spatial function $f(x, y)$ where $f(x, y) = 0$ corresponds to white, $f(x, y) = 1$ corresponds to red, and (x, y) denotes a pixel location in the image. Defining a filter function $h(x, y)$ that has the appearance of Fig. 18.26a, the convolution operation

$$g(x, y) = f(x, y) * h(x, y)$$

results in the image of Fig. 18.26b in which the myosin filaments (viewed on end) are more clearly identifiable.

In practice, this image processing is performed in the frequency domain, where the FFT of both f and h are calculated, and the resulting matrices multiplied together.

Epilogue

Returning again to Eq. [70], the identity between the system function $\mathbf{H}(j\omega)$ and the sinusoidal steady-state transfer function $\mathbf{G}(\omega)$, we may now consider the system function as the ratio of the output phasor to the input phasor. Suppose that we hold the input-phasor amplitude at unity and the phase angle at zero. Then the output phasor is $\mathbf{H}(j\omega)$. Under these conditions, if we record the output amplitude and phase as functions of ω, for all ω, we have recorded the system function $\mathbf{H}(j\omega)$ as a function of ω, for all ω. We thus have examined the system response under the condition that an infinite number of sinusoids, all with unity amplitude and zero phase, were successively applied at the input. Now suppose that our input is a single unit impulse, and look at the

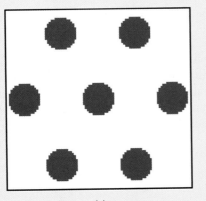

(a)

(b)

■ **FIGURE 18.26** (*a*) Spatial filter having hexagonal symmetry. (*b*) Image after convolution and inverse discrete Fourier transform are performed, showing a reduction in background noise. *Courtesy Professor John M. Squire, Imperial College, London.*

An inverse FFT operation then produces the filtered image of Fig. 18.26*b*. Why does this convolution equate to a filtering operation? The myosin filament arrangement possesses hexagonal symmetry, as does the filter function $h(x, y)$—in a sense, both the myosin filament arrangement and the filter function possess the same

spatial frequencies. The convolution of f with h results in a reinforcement of the hexagonal pattern within the original image, and the removal of noise pixels (which do not possess hexagonal symmetry). This can be understood qualitatively if we model a horizontal row of Fig. 18.25 as a sinusoidal function $f(x) = \cos \omega_0 t$, which has the Fourier transform shown in Fig. 18.27*a*—a matched pair of impulse functions separated by $2\omega_0$. If we convolve this function with a filter function $h(x) = \cos \omega_1 t$, the Fourier transform of which is depicted in Fig. 18.27*b*, we get zero if $\omega_1 \neq \omega_0$; the frequencies (periodicities) of the two functions do not match. If, instead, we choose a filter function with the same frequency as $f(x)$, the convolution has a nonzero value at $\omega = \pm\omega_0$.

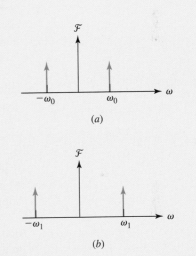

(a)

(b)

■ **FIGURE 18.27** (*a*) Fourier transform of $f(x) = \cos \omega_0 t$. (*b*) Fourier transform of $h(x) = \cos \omega_1 t$.

impulse response $h(t)$. Is the information we examine really any different from what we just obtained? The Fourier transform of the unit impulse is a constant equal to unity, indicating that all frequency components are present, all with the same magnitude, and all with zero phase. Our system response is the sum of the responses to all these components. The result might be viewed at the output on a cathode-ray oscilloscope. It is evident that the system function and the impulse-response function contain equivalent information regarding the response of the system.

We therefore have two different methods of describing the response of a system to a general forcing function; one is a time-domain description, and the other a frequency-domain description. Working in the time domain, we convolve the forcing function with the impulse response of the system to obtain the response function. As we saw when we first considered convolution, this procedure may be interpreted by thinking of the input as a continuum of impulses of different strengths and times of application; the output which results is a continuum of impulse responses.

In the frequency domain, however, we determine the response by multiplying the Fourier transform of the forcing function by the system function. In this case we interpret the transform of the forcing function as a frequency spectrum, or a continuum of sinusoids. Multiplying this by the system function, we obtain the response function, also as a continuum of sinusoids.

SUMMARY AND REVIEW

Whether we choose to think of the output as a continuum of impulse responses or as a continuum of sinusoidal responses, the linearity of the network and the superposition principle enable us to determine the total output as a time function by summing over all frequencies (the inverse Fourier transform), or as a frequency function by summing over all time (the Fourier transform).

Unfortunately, both of these techniques have some difficulties or limitations associated with their use. In using convolution, the integral itself can often be rather difficult to evaluate when complicated forcing functions or impulse response functions are present. Furthermore, from the experimental point of view, we cannot really measure the impulse response of a system because we cannot actually generate an impulse. Even if we approximated the impulse by a narrow high-amplitude pulse, we would probably drive our system into saturation and out of its linear operating range.

With regard to the frequency domain, we encounter one absolute limitation in that we may easily hypothesize forcing functions that we would like to apply theoretically that do not possess Fourier transforms. Moreover, if we wish to find the time-domain description of the response function, we must evaluate an inverse Fourier transform, and some of these inversions can be extremely difficult.

Finally, neither of these techniques offers a very convenient method of handling initial conditions. For this, the Laplace transform is clearly superior.

The greatest benefits derived from the use of the Fourier transform arise through the abundance of useful information it provides about the spectral properties of a signal, particularly the energy or power per unit bandwidth. Some of this information is also easily obtained through the Laplace transform; we must leave a detailed discussion of the relative merits of each to more advanced signals and systems courses.

So, why has this all been withheld until now? The best answer is probably that these powerful techniques can overcomplicate the solution of simple problems and tend to obscure the physical interpretation of the performance of the simpler networks. For example, if we are interested only in the forced response, then there is little point in using the Laplace transform and obtaining both the forced and natural response after laboring through a difficult inverse transform operation.

Well, we could go on, but all good things must come to an end. Best of luck to you in your future studies.

❑ The harmonic frequencies of a sinusoid having the fundamental frequency ω_0 are $n\omega_0$, where n is an integer. (Examples 18.1, 18.2)

❑ The Fourier theorem states that provided a function $f(t)$ satisfies certain key properties, it may be represented by the infinite series $a_0 + \sum_{n=1}^{\infty}(a_n \cos n\omega_0 t + b_n \sin n\omega_0 t)$, where $a_0 = (1/T)\int_0^T f(t)\,dt$, $a_n = (2/T)\int_0^T f(t)\cos n\omega_0 t\,dt$, and $b_n = (2/T)\int_0^T f(t)\sin n\omega_0 t\,dt$. (Example 18.1)

❑ A function $f(t)$ possesses *even* symmetry if $f(t) = f(-t)$.

❑ A function $f(t)$ possesses *odd* symmetry if $f(t) = -f(-t)$.

❑ A function $f(t)$ possesses *half-wave* symmetry if $f(t) = -f(t - \frac{1}{2}T)$.

❑ The Fourier series of an even function is composed of only a constant and cosine functions.

❑ The Fourier series of an odd function is composed of only sine functions.

❑ The Fourier series of any function possessing half-wave symmetry contains only odd harmonics.

❑ The Fourier series of a function may also be expressed in complex or exponential form, where $f(t) = \sum_{n=-\infty}^{\infty} \mathbf{c}_n e^{jn\omega_0 t}$ and $\mathbf{c}_n = (1/T)\int_{-T/2}^{T/2} f(t)e^{-jn\omega_0 t}\,dt$. (Examples 18.3, 18.4)

❑ The Fourier transform allows us to represent time-varying functions in the frequency domain, in a manner similar to that of the Laplace transform. The defining equations are $\mathbf{F}(j\omega) = \int_{-\infty}^{\infty} e^{-j\omega t} f(t)\,dt$ and $f(t) = (1/2\pi)\int_{-\infty}^{\infty} e^{j\omega t}\mathbf{F}(j\omega)\,d\omega$. (Examples 18.5, 18.6, 18.7)

❑ Fourier transform analysis can be implemented to analyze circuits containing resistors, inductors, and/or capacitors in a manner similar to what is done using Laplace transforms. (Example 18.8)

READING FURTHER

A very readable treatment of Fourier analysis can be found in:

A. Pinkus and S. Zafrany, *Fourier Series and Integral Transforms.* Cambridge: Cambridge University Press, 1997.

Finally, for those interested in learning more about muscle research, including electron microscopy of tissue, an excellent treatment can be found in:

J. Squire, *The Structural Basis of Muscular Contraction.* New York: Plenum Press, 1981.

EXERCISES

18.1 Trigonometric Form of the Fourier Series

1. Determine the fundamental frequency, fundamental radian frequency, and period of the following: (*a*) $5\sin 9t$; (*b*) $200\cos 70t$; (*c*) $4\sin(4t - 10°)$; (*d*) $4\sin(4t + 10°)$.

2. Plot multiple periods of the first, third, and fifth harmonics on the same graph of each of the following periodic waveforms (three separate graphs in total are desired): (*a*) $3\sin t$; (*b*) $40\cos 100t$; (*c*) $2\cos(10t - 90°)$.

3. Calculate a_0 for the following: (a) $4 \sin 4t$; (b) $4 \cos 4t$; (c) $4 + \cos 4t$; (d) $4 \cos(4t + 40°)$.

4. Compute a_0, a_1, and b_1 for the following functions: (a) $2 \cos 3t$; (b) $3 - \cos 3t$; (c) $4 \sin(4t - 35°)$.

5. (a) Calculate the Fourier coefficients a_0, a_1, a_2, a_3, b_1, b_2, and b_3 for the periodic function $f(t) = 2u(t) - 2u(t+1) + 2u(t+2) - 2u(t+3) + \cdots$. (b) Sketch $f(t)$ and the Fourier series truncated after $n = 3$ over 3 periods.

6. (a) Compute the Fourier coefficients a_0, a_1, a_2, a_3, a_4, b_1, b_2, b_3, and b_4 for the periodic function $g(t)$ partially sketched in Fig. 18.28. (b) Plot $g(t)$ along with the Fourier series representation truncated after $n = 4$.

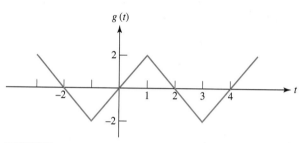

■ FIGURE 18.28

7. For the periodic waveform $f(t)$ represented in Fig. 18.29, calculate a_1, a_2, a_3 and b_1, b_2, b_3.

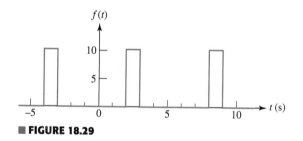

■ FIGURE 18.29

8. With respect to the periodic waveform sketched in Fig. 18.29, let $g_n(t)$ represent the Fourier series representation of $f(t)$ truncated at n. [For example, if $n = 1$, $g_1(t)$ has three terms, defined through a_0, a_1 and b_1.] (a) Sketch $g_2(t)$, $g_3(t)$, and $g_5(t)$, along with $f(t)$. (b) Calculate $f(2.5)$, $g_2(2.5)$, $g_3(2.5)$, and $g_5(2.5)$.

9. With respect to the periodic waveform $g(t)$ sketched in Fig. 18.28, define $y_n(t)$ which represents the Fourier series representation truncated at n. (For example, $y_2(t)$ has five terms, defined through a_0, a_1, a_2, b_1, and b_2.) (a) Sketch $y_3(t)$ and $y_5(t)$ along with $g(t)$. (b) Compute $y_1(0.5)$, $y_2(0.5)$, $y_3(0.5)$, and $g(0.5)$.

10. Determine expressions for a_n and b_n for $g(t - 1)$ if the periodic waveform $g(t)$ is defined as sketched in Fig. 18.28.

11. Plot the line spectrum (limited to the six largest terms) for the waveform shown in Fig. 18.4a.

12. Plot the line spectrum (limited to the five largest terms) for the waveform of Fig. 18.4b.

13. Plot the line spectrum (limited to the five largest terms) for the waveform represented by the graph of Fig. 18.4c.

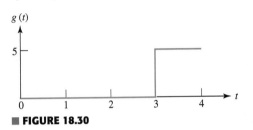

18.2 The Use of Symmetry

14. State whether the following exhibit odd symmetry, even symmetry, and/or half-wave symmetry: (*a*) $4 \sin 100t$; (*b*) $4 \cos 100t$; (*c*) $4 \cos(4t + 70°)$; (*d*) $4 \cos 100t + 4$; (*e*) each waveform in Fig. 18.4.

15. Determine whether the following exhibit odd symmetry, even symmetry, and/or half-wave symmetry: (*a*) the waveform in Fig. 18.28; (*b*) $g(t - 1)$, if $g(t)$ is represented in Fig. 18.28; (*c*) $g(t + 1)$, if $g(t)$ is represented in Fig. 18.28; (*d*) the waveform of Fig. 18.29.

16. The nonperiodic waveform $g(t)$ is defined in Fig. 18.30. Use it to create a new function $y(t)$ such that $y(t)$ is identical to $g(t)$ over the range of $0 < t < 4$ and also is characterized by a period $T = 8$ and has (*a*) odd symmetry; (*b*) even symmetry; (*c*) both even and half-wave symmetry; (*d*) both odd and half-wave symmetry.

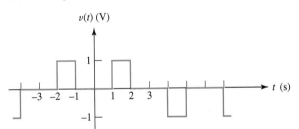

■ FIGURE 18.30

17. Calculate a_0, a_1, a_2, a_3 and b_1, b_2, b_3 for the periodic waveform $v(t)$ represented in Fig. 18.31.

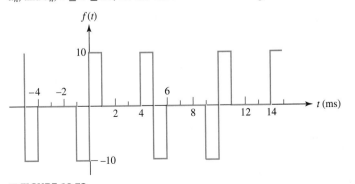

■ FIGURE 18.31

18. The waveform of Fig. 18.31 is shifted to create a new waveform such that $v_{\text{new}}(t) = v(t + 1)$. Calculate a_0, a_1, a_2, a_3 and b_1, b_2, b_3.

19. Design a triangular waveform having a peak magnitude of 3, a period of 2 s, and characterized by (*a*) half-wave and even symmetry; (*b*) half-wave and odd symmetry.

20. Make use of symmetry as much as possible to obtain numerical values for a_0, a_n, and b_n, $1 \le n \le 10$, for the waveform shown in Fig. 18.32.

■ FIGURE 18.32

18.3 Complete Response to Periodic Forcing Functions

21. For the circuit of Fig. 18.33a, calculate $v(t)$ if $i_s(t)$ is given by Fig. 18.33b and $v(0) = 0$.

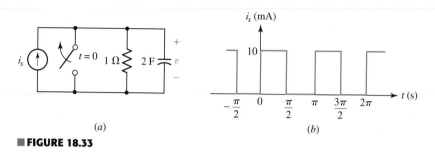

(a)   (b)

■ FIGURE 18.33

22. If the waveform shown in Fig. 18.34 is applied to the circuit of Fig. 18.8a, calculate $i(t)$.

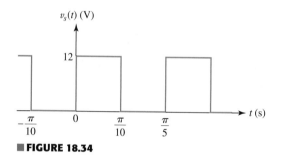

■ FIGURE 18.34

23. The circuit of Fig. 18.35a is subjected to the waveform depicted in Fig. 18.35b. Determine the steady-state voltage $v(t)$.

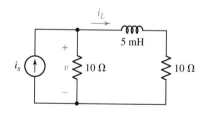

(a)

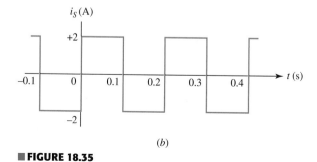

(b)

■ FIGURE 18.35

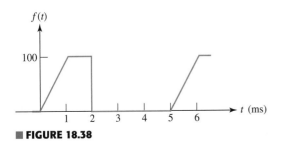

24. Apply the waveform of Fig. 18.36 to the circuit of Fig. 18.35b, and calculate the steady-state current $i_L(t)$.

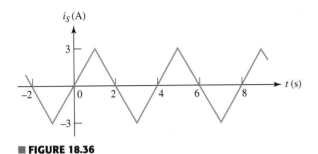

■ FIGURE 18.36

25. If the current waveform of Fig. 18.36 is applied to the circuit of Fig. 18.33a, calculate the steady-state voltage $v(t)$.

18.4 Complex Form of the Fourier Series

26. Let the function $v(t)$ be defined as indicated in Fig. 18.10. Determine $\mathbf{c}_n$ for (a) $v(t + 0.5)$; (b) $v(t-0.5)$.

27. Calculate $\mathbf{c}_0$, $\mathbf{c}_{\pm 1}$, and $\mathbf{c}_{\pm 2}$ for the waveform of Fig. 18.36.

28. Determine the first five terms of the exponential Fourier series representation of the waveform graphed in Fig. 18.33b.

29. For the periodic waveform shown in Fig. 18.37, determine (a) the period T; (b) $\mathbf{c}_0$, $\mathbf{c}_{\pm 1}$, $\mathbf{c}_{\pm 2}$, and $\mathbf{c}_{\pm 3}$.

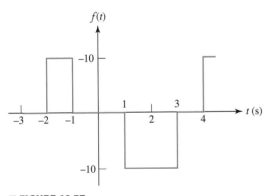

■ FIGURE 18.37

30. For the periodic waveform represented in Fig. 18.38, calculate (a) the period T; (b) $\mathbf{c}_0$, $\mathbf{c}_{\pm 1}$, $\mathbf{c}_{\pm 2}$, and $\mathbf{c}_{\pm 3}$.

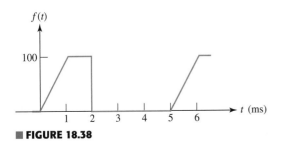

■ FIGURE 18.38

31. A pulse sequence has a period of 5 μs, an amplitude of unity for $-0.6 < t < -0.4$ μs and for $0.4 < t < 0.6$ μs, and zero amplitude elsewhere in the period interval. This series of pulses might represent the decimal number 3 being transmitted in binary form by a digital computer. (a) Find c_n. (b) Evaluate c_4. (c) Evaluate c_0. (d) Find $|c_n|_{max}$. (e) Find N so that $|c_n| \le 0.1|c_n|_{max}$ for all $n > N$. (f) What bandwidth is required to transmit this portion of the spectrum?

32. Let a periodic voltage $v_s(t)$ is equal to 40 V for $0 < t < \frac{1}{96}$ s, and to 0 for $\frac{1}{96} < t < \frac{1}{16}$ s. If $T = \frac{1}{16}$ s, find (a) c_3; (b) the power delivered to the load in the circuit of Fig. 18.39.

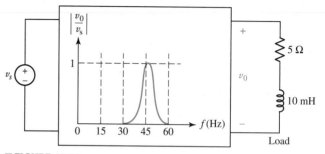

■ **FIGURE 18.39**

18.5 Definition of the Fourier Transform

33. Given

$$g(t) = \begin{cases} 5 & -1 < t < 1 \\ 0 & \text{elsewhere} \end{cases}$$

sketch (a) $g(t)$; (b) $\mathbf{G}(j\omega)$.

34. For the function $v(t) = 2u(t) - 2u(t+2) + 2u(t+4) - 2u(t+6)$ V, sketch (a) $v(t)$; (b) $\mathbf{V}(j\omega)$.

35. Employ Eq. [46a] to calculate $\mathbf{G}(j\omega)$ if $g(t)$ is (a) $5e^{-t}u(t)$; (b) $5te^{-t}u(t)$.

36. Obtain the Fourier transform $\mathbf{F}(j\omega)$ of the single triangle pulse plotted in Fig. 18.40.

37. Determine the Fourier transform $\mathbf{F}(j\omega)$ of the single sinusoidal pulse waveform shown in Fig. 18.41.

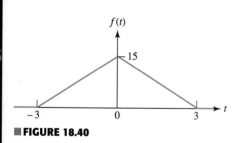

■ **FIGURE 18.40**

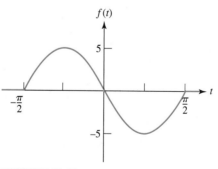

■ **FIGURE 18.41**

18.6 Some Properties of the Fourier Transform

38. For $g(t) = 3e^{-t}u(t)$, calculate (a) $\mathbf{G}(j\omega)$; (b) $\mathbf{A}_g(1)$; (c) $\mathbf{B}_g(1)$; (d) $\phi(\omega)$.

39. The voltage pulse $2e^{-t}u(t)$ V is applied to the input of an ideal bandpass filter. The passband of the filter is defined by $100 < |f| < 500$ Hz. Calculate the total output energy.

40. Given that $v(t) = 4e^{-|t|}$ V, calculate the frequency range in which 85% of the 1 Ω energy lies.

41. Calculate the 1 Ω energy associated with the function $f(t) = 4te^{-3t}u(t)$.

42. Use the definition of the Fourier transform to prove the following results, where $\mathcal{F}\{f(t)\} = \mathbf{F}(j\omega)$: (a) $\mathcal{F}\{f(t - t_0)\} = e^{-j\omega t_0}\mathcal{F}\{f(t)\}$; (b) $\mathcal{F}\{df(t)/dt\} = j\omega\mathcal{F}\{f(t)\}$; (c) $\mathcal{F}\{f(kt)\} = (1/|k|)\mathbf{F}(j\omega/k)$; (d) $\mathcal{F}\{f(-t)\} = \mathbf{F}(-j\omega)$; (e) $\mathcal{F}\{tf(t)\} = j\, d[\mathbf{F}(j\omega)]/d\omega$.

18.7 Fourier Transform Pairs for Some Simple Time Functions

43. Determine the Fourier transform of the following: (a) $5u(t) - 2\,\text{sgn}(t)$; (b) $2\cos 3t - 2$; (c) $4e^{-j3t} + 4e^{j3t} + 5u(t)$.

44. Find the Fourier transform of each of the following: (a) $85u(t + 2) - 50u(t - 2)$; (b) $5\delta(t) - 2\cos 4t$.

45. Sketch $f(t)$ and $|\mathbf{F}(j\omega)|$ if $f(t)$ is given by (a) $2\cos 10t$; (b) $e^{-4t}u(t)$; (c) $5\,\text{sgn}(t)$.

46. Determine $f(t)$ if $\mathbf{F}(j\omega)$ is given by (a) $4\delta(\omega)$; (b) $2/(5000 + j\omega)$; (c) $e^{-j120\omega}$.

47. Obtain an expression for $f(t)$ if $\mathbf{F}(j\omega)$ is given by

$(a) -j\dfrac{231}{\omega}$; $(b)\ \dfrac{1 + j2}{1 + j4}$; $(c)\ 5\delta(\omega) + \dfrac{1}{2 + j10}$.

18.8 The Fourier Transform of a General Periodic Time Function

48. Calculate the Fourier transform of the following functions: (a) $2\cos^2 5t$; (b) $7\sin 4t \cos 3t$; (c) $3\sin(4t - 40°)$.

49. Determine the Fourier transform of the periodic function $g(t)$, which is defined over the range $0 < t < 10$ s by $g(t) = 2u(t) - 3u(t - 4) + 2u(t - 8)$.

50. If $\mathbf{F}(j\omega) = 20\sum_{n=1}^{\infty}[1/(|n|! + 1)]\delta(\omega - 20n)$, find the value of $f(0.05)$.

51. Given the periodic waveform shown in Fig. 18.42, determine its Fourier transform.

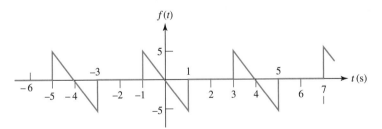

■ **FIGURE 18.42**

18.9 The System Function and Response in the Frequency Domain

52. If a system is described by transfer function $h(t) = 2u(t) + 2u(t - 1)$, use convolution to calculate the output (time domain) if the input is (a) $2u(t)$; (b) $2te^{-2t}u(t)$.

53. Given the input function $x(t) = 5e^{-5t}u(t)$, employ convolution to obtain a time-domain output if the system transfer function $h(t)$ is given by (a) $3u(t + 1)$; (b) $10te^{-t}u(t)$.

54. (a) Design a noninverting amplifier having a gain of 10. If the circuit is constructed using a μA741 op amp powered by ±15 V supplies, determine the FFT of the output through appropriate simulations if the input voltage operates at 1 kHz and has magnitude (b) 10 mV; (c) 1 V; (d) 2 V.

55. (a) Design an inverting amplifier having a gain of 5. If the circuit is constructed using a μA741 op amp powered by ± 10 V supplies, perform appropriate simulations to determine the FFT of the output voltage if the input voltage has a frequency of 10 kHz and magnitude (b) 500 mV; (c) 1.8 V; (d) 3 V.

18.10 The Physical Significance of the System Function

56. With respect to the circuit of Fig. 18.43, calculate $v_o(t)$ using Fourier techniques if $v_i(t) = 2te^{-t}u(t)$ V.

57. After the inductor of Fig. 18.43 is surreptitiously replaced with a 2 F capacitor, calculate $v_o(t)$ using Fourier techniques if $v_i(t)$ is equal to (a) $5u(t)$ V; (b) $3e^{-4t}u(t)$ V.

58. Employ Fourier-based techniques to calculate $v_C(t)$ as labeled in Fig. 18.44 if $v_i(t)$ is equal to (a) $2u(t)$ V; (b) $2\delta(t)$ V.

59. Employ Fourier-based techniques to calculate $v_o(t)$ as labeled in Fig. 18.45 if $v_i(t)$ is equal to (a) $5u(t)$ V; (b) $3\delta(t)$ V.

60. Employ Fourier-based techniques to calculate $v_o(t)$ as labeled in Fig. 18.45 if $v_i(t)$ is equal to (a) $5u(t-1)$ V; (b) $2 + 8e^{-t}u(t)$ V.

Chapter-Integrating Exercises

61. Apply the pulse waveform of Fig. 18.46a as the voltage input $v_i(t)$ to the circuit shown in Fig. 18.44, and calculate $v_C(t)$.

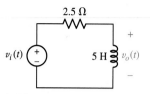

■ **FIGURE 18.43**

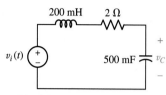

■ **FIGURE 18.44**

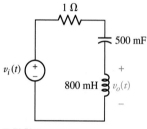

■ **FIGURE 18.45**

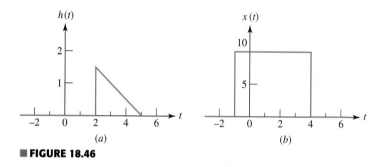

(a)   (b)

■ **FIGURE 18.46**

62. Apply the pulse waveform of Fig. 18.46b as the voltage input $v_i(t)$ to the circuit shown in Fig. 18.44, and calculate $v_C(t)$.

63. Apply the pulse waveform of Fig. 18.46a as the voltage input $v_i(t)$ to the circuit shown in Fig. 18.44, and calculate $i_C(t)$, defined consistent with the passive sign convention.

64. Apply the pulse waveform of Fig. 18.46b as the voltage input $v_i(t)$ to the circuit shown in Fig. 18.45, and calculate $v_o(t)$.

65. Apply the pulse waveform of Fig. 18.46b as the voltage input $v_i(t)$ to the circuit shown in Fig. 18.45, and calculate $v_o(t)$.

AN INTRODUCTION TO NETWORK TOPOLOGY

After working many circuits problems, it slowly becomes evident that many of the circuits we see have quite a bit in common, at least in terms of the arrangement of components. From this realization, it is possible to create a more abstract view of circuits which we call *network topology*, a subject we introduce in this appendix.

A1.1 TREES AND GENERAL NODAL ANALYSIS

We now plan to generalize the method of nodal analysis that we have come to know and love. Since nodal analysis is applicable to any network, we cannot promise that we will be able to solve a wider class of circuit problems. We can, however, look forward to being able to select a general nodal analysis method for any particular problem that may result in fewer equations and less work.

We must first extend our list of definitions relating to network topology. We begin by defining *topology* itself as a branch of geometry which is concerned with those properties of a geometrical figure which are unchanged when the figure is twisted, bent, folded, stretched, squeezed, or tied in knots, with the provision that no parts of the figure are to be cut apart or to be joined together. A sphere and a tetrahedron are topologically identical, as are a square and a circle. In terms of electric circuits, then, we are not now concerned with the particular types of elements appearing in the circuit, but only with the way in which branches and nodes are arranged. As a matter of fact, we usually suppress the nature of the elements and simplify the drawing of the circuit by showing the elements as lines. The resultant drawing is called a linear graph, or simply a graph. A circuit and its graph are shown in Fig. A1.1. Note that all nodes are identified by heavy dots in the graph.

Since the topological properties of the circuit or its graph are unchanged when it is distorted, the three graphs shown in Fig. A1.2 are all topologically identical with the circuit and graph of Fig. A1.1.

Topological terms that we already know and have been using correctly are

Node: A point at which two or more elements have a common connection.
Path: A set of elements that may be traversed in order without passing through the same node twice.
Branch: A single path, containing one simple element, which connects one node to any other node.

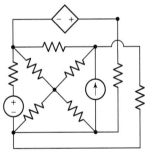

(a)

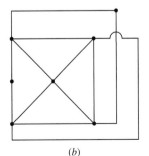

(b)

■ **FIGURE A1.1** (*a*) A given circuit. (*b*) The linear graph of this circuit.

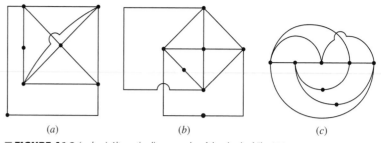

■ **FIGURE A1.2** (*a, b, c*) Alternative linear graphs of the circuit of Fig. A1.1.

Loop: A closed path.
Mesh: A loop which does not contain any other loops within it.
Planar circuit: A circuit which may be drawn on a plane surface in such a way that no branch passes over or under any other branch.
Nonplanar circuit: Any circuit which is not planar.

The graphs of Fig. A1.2 each contain 12 branches and 7 nodes.

Three new properties of a linear graph must now be defined—a *tree,* a *cotree,* and a *link.* We define a *tree* as any set of branches which does not contain any loops and yet connects every node to every other node, not necessarily directly. There are usually a number of different trees which may be drawn for a network, and the number increases rapidly as the complexity of the network increases. The simple graph shown in Fig. A1.3*a* has eight possible trees, four of which are shown by heavy lines in Fig. A1.3*b, c, d,* and *e.*

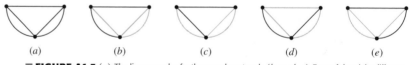

(*a*) (*b*) (*c*) (*d*) (*e*)

■ **FIGURE A1.3** (*a*) The linear graph of a three-node network. (*b, c, d, e*) Four of the eight different trees which may be drawn for this graph are shown by the black lines.

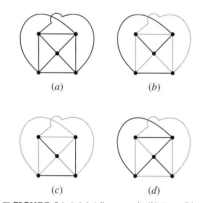

(*a*) (*b*)

(*c*) (*d*)

■ **FIGURE A1.4** (*a*) A linear graph. (*b*) A possible tree for this graph. (*c, d*) These sets of branches do not satisfy the definition of a tree.

In Fig. A1.4*a* a more complex graph is shown. Figure A1.4*b* shows one possible tree, and Fig. A1.4*c* and *d* show sets of branches which are not trees because neither set satisfies the definition.

After a tree has been specified, those branches that are not part of the tree form the *cotree,* or complement of the tree. The lightly drawn branches in Fig. A1.3*b* to *e* show the cotrees that correspond to the heavier trees.

Once we understand the construction of a tree and its cotree, the concept of the *link* is very simple, for a link is any branch belonging to the cotree. It is evident that any particular branch may or may not be a link, depending on the particular tree which is selected.

The number of links in a graph may easily be related to the number of branches and nodes. If the graph has N nodes, then exactly $(N - 1)$ branches are required to construct a tree because the first branch chosen connects two nodes and each additional branch includes one more node.

Thus, given B branches, the number of links L must be

$$L = B - (N - 1)$$

or

$$L = B - N + 1 \qquad [1]$$

There are L branches in the cotree and $(N - 1)$ branches in the tree.

In any of the graphs shown in Fig. A1.3, we note that $3 = 5 - 3 + 1$, and in the graph of Fig. A1.4b, $6 = 10 - 5 + 1$. A network may be in several disconnected parts, and Eq. [1] may be made more general by replacing $+1$ with $+S$, where S is the number of separate parts. However, it is also possible to connect two separate parts by a single conductor, thus causing two nodes to form one node; no current can flow through this single conductor. This process may be used to join any number of separate parts, and thus we will not suffer any loss of generality if we restrict our attention to circuits for which $S = 1$.

We are now ready to discuss a method by which we may write a set of nodal equations that are independent and sufficient. The method will enable us to obtain many different sets of equations for the same network, and all the sets will be valid. However, the method does not provide us with every possible set of equations. Let us first describe the procedure, illustrate it by three examples, and then point out the reason that the equations are independent and sufficient.

Given a network, we should:

1. Draw a graph and then identify a tree.
2. Place all voltage sources in the tree.
3. Place all current sources in the cotree.
4. Place all control-voltage branches for voltage-controlled dependent sources in the tree, if possible.
5. Place all control-current branches for current-controlled dependent sources in the cotree, if possible.

These last four steps effectively associate voltages with the tree and currents with the cotree.

We now assign a voltage variable (with its plus-minus pair) across each of the $(N - 1)$ branches in the tree. A branch containing a voltage source (dependent or independent) should be assigned that source voltage, and a branch containing a controlling voltage should be assigned that controlling voltage. The number of new variables that we have introduced is therefore equal to the number of branches in the tree $(N - 1)$, reduced by the number of voltage sources in the tree, and reduced also by the number of control voltages we were able to locate in the tree. In Example A1.3, we will find that the number of new variables required may be zero.

Having a set of variables, we now need to write a set of equations that are sufficient to determine these variables. The equations are obtained through the application of KCL. Voltage sources are handled in the same way that they were in our earlier attack on nodal analysis; each voltage source and the two nodes at its terminals constitute a supernode or a part of a supernode. Kirchhoff's current law is then applied at all but one of the remaining nodes and supernodes. We set the sum of the currents leaving the node in all of the branches connected to it equal to zero. Each current is expressed in terms of the voltage variables we just assigned. One node may be ignored, just as was the case earlier for the reference node. Finally, in case there are current-controlled dependent sources, we must write an equation for each control current that relates it to the voltage variables; this also is no different from the procedure used before with nodal analysis.

Let us try out this process on the circuit shown in Fig. A1.5a. It contains four nodes and five branches, and its graph is shown in Fig. A1.5b.

EXAMPLE **A1.1**

Find the value of v_x in the circuit of Fig. A1.5a.

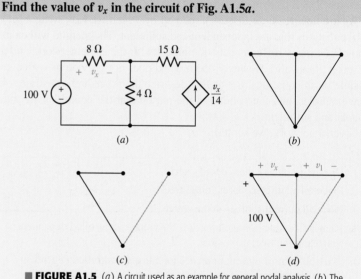

(a)

(b)

(c)

(d)

■ **FIGURE A1.5** (a) A circuit used as an example for general nodal analysis. (b) The graph of the given circuit. (c) The voltage source and the control voltage are placed in the tree, while the current source goes in the cotree. (d) The tree is completed and a voltage is assigned across each tree branch.

In accordance with steps 2 and 3 of the tree-drawing procedure, we place the voltage source in the tree and the current source in the cotree. Following step 4, we see that the v_x branch may also be placed in the tree, since it does not form any loop which would violate the definition of a tree. We have now arrived at the two tree branches and the single link shown in Fig. A1.5c, and we see that we do not yet have a tree, since the right node is not connected to the others by a path through tree branches. The only possible way to complete the tree is shown in Fig. A1.5d. The 100 V source voltage, the control voltage v_x, and a new voltage variable v_1 are next assigned to the three tree branches as shown.

We therefore have two unknowns, v_x and v_1, and we need to obtain two equations in terms of them. There are four nodes, but the presence of the voltage source causes two of them to form a single supernode. Kirchhoff's current law may be applied at any two of the three remaining nodes or supernodes. Let's attack the right node first. The current leaving to the left is $-v_1/15$, while that leaving downward is $-v_x/14$. Thus, our first equation is

$$-\frac{v_1}{15} + \frac{-v_x}{14} = 0$$

The central node at the top looks easier than the supernode, and so we set the sum of the current to the left ($-v_x/8$), the current to the right ($v_1/15$), and the downward current through the 4-Ω resistor equal to zero. This latter current is given by the voltage across the resistor divided by 4 Ω, but there is no voltage labeled on that link. However, when a tree is constructed according to the definition, there is a path through it from any node to any other node. Then, since every branch in the tree is assigned a voltage, we may express the voltage across any link in terms of the tree-branch voltages. This downward current is therefore $(-v_x + 100)/4$, and we have the second equation,

$$-\frac{v_x}{8} + \frac{v_1}{15} + \frac{-v_x + 100}{4} = 0$$

The simultaneous solution of these two nodal equations gives

$$v_1 = -60 \text{ V} \qquad v_x = 56 \text{ V}$$

EXAMPLE A1.2

Find the values of v_x and v_y in the circuit of Fig. A1.6a.

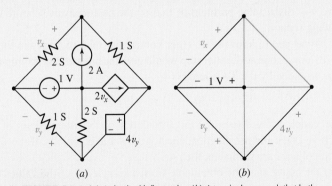

(a)   (b)

■ **FIGURE A1.6** (a) A circuit with five nodes. (b) A tree is chosen such that both voltage sources and both control voltages are tree branches.

We draw a tree so that both voltage sources and both control voltages appear as tree-branch voltages and, hence, as assigned variables. As it happens, these four branches constitute a tree, Fig. A1.6b, and tree-branch voltages v_x, 1, v_y, and $4v_y$ are chosen, as shown.

(Continued on next page)

Both voltage sources define supernodes, and we apply KCL twice, once to the top node,

$$2v_x + 1(v_x - v_y - 4v_y) = 2$$

and once to the supernode consisting of the right node, the bottom node, and the dependent voltage source,

$$1v_y + 2(v_y - 1) + 1(4v_y + v_y - v_x) = 2v_x$$

Instead of the four equations we would expect using previously studied techniques, we have only two, and we find easily that $v_x = \frac{26}{9}$ V and $v_y = \frac{4}{3}$ V.

EXAMPLE A1.3

Find the value of v_x in the circuit of Fig. A1.7a.

The two voltage sources and the control voltage establish the three-branch tree shown in Fig. A1.7b. Since the two upper nodes and the lower right node all join to form one supernode, we need write only one KCL equation. Selecting the lower left node, we have

$$-1 - \frac{v_x}{4} + 3 + \frac{-v_x + 30 + 6v_x}{5} = 0$$

and it follows that $v_x = -\frac{32}{3}$ V. In spite of the apparent complexity of this circuit, the use of general nodal analysis has led to an easy solution. Employing mesh currents or node-to-reference voltages would require more equations and more effort.

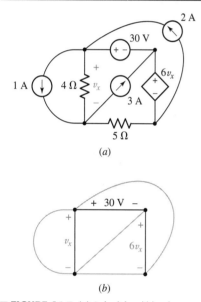

(a)

(b)

■ **FIGURE A1.7** (a) A circuit for which only one general nodal equation need be written. (b) The tree and the tree-branch voltages used.

We will discuss the problem of finding the best analysis scheme in the following section.

If we needed to know some other voltage, current, or power in the previous example, one additional step would give the answer. For example, the power provided by the 3 A source is

$$3\left(-30 - \tfrac{32}{3}\right) = -122 \text{ W}$$

Let us conclude by discussing the sufficiency of the assumed set of tree-branch voltages and the independence of the nodal equations. If these tree-branch voltages are sufficient, then the voltage of every branch in either the tree or the cotree must be obtainable from a knowledge of the values of all the tree-branch voltages. This is certainly true for those branches in the tree. For the links we know that each link extends between two nodes, and, by definition, the tree must also connect those two nodes. Hence, every link voltage may also be established in terms of the tree-branch voltages.

Once the voltage across every branch in the circuit is known, then all the currents may be found by using either the given value of the current if the branch consists of a current source, by using Ohm's law if it is a resistive branch, or by using KCL and these current values if the branch happens to be a voltage source. Thus, all the voltages and currents are determined and sufficiency is demonstrated.

EXAMPLE **A1.4**

For the example of Fig. A1.9, find the values of i_A and i_B.

We first traverse the i_A loop, proceeding clockwise from its lower left corner. The current going our way in the 1 Ω resistor is $(i_A - 7)$, in the 2 Ω element it is $(i_A + i_B)$, and in the link it is simply i_B. Thus

$$1(i_A - 7) + 2(i_A + i_B) + 3i_A = 0$$

For the i_B link, clockwise travel from the lower left corner leads to

$$-7 + 2(i_A + i_B) + 1i_B = 0$$

Traversal of the loop defined by the 7 A link is not required. Solving, we have $i_A = 0.5$ A, $i_B = 2$ A, once again. The solution has been achieved with one less equation than before!

EXAMPLE **A1.5**

Evaluate i_1 in the circuit shown in Fig. A1.10a.

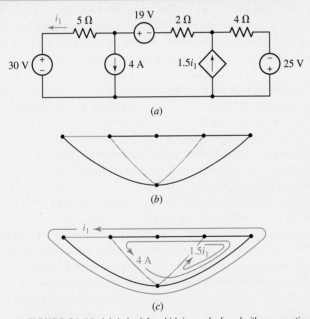

(a)

(b)

(c)

■ **FIGURE A1.10** (a) A circuit for which i_1 may be found with one equation using general loop analysis. (b) The only tree that satisfies the rules outlined in Sec. A1.1. (c) The three link currents are shown with their loops.

This circuit contains six nodes, and its tree therefore must have five branches. Since there are eight elements in the network, there are three links in the cotree. If we place the three voltage sources in the tree and the two current sources and the current control in the cotree, we are led to the tree shown in Fig. A1.10b. The source current of 4 A defines a

(Continued on next page)

loop as shown in Fig. A1.10*c*. The dependent source establishes the loop current $1.5i_1$ around the right mesh, and the control current i_1 gives us the remaining loop current about the perimeter of the circuit. Note that all three currents flow through the 4 Ω resistor.

We have only one unknown quantity, i_1, and after discarding the loops defined by the two current sources, we apply KVL around the outside of the circuit:

$$-30 + 5(-i_1) + 19 + 2(-i_1 - 4) + 4(-i_1 - 4 + 1.5i_1) - 25 = 0$$

Besides the three voltage sources, there are three resistors in this loop. The 5 Ω resistor has one loop current in it, since it is also a link; the 2 Ω resistor contains two loop currents; and the 4 Ω resistor has three. A carefully drawn set of loop currents is a necessity if errors in skipping currents, utilizing extra ones, or erring in choosing the correct direction are to be avoided. The foregoing equation is guaranteed, however, and it leads to $i_1 = -12$ A.

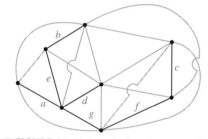

■ **FIGURE A1.11** A tree that is used as an example to illustrate the sufficiency of the link currents.

How may we demonstrate sufficiency? Let us visualize a tree. It contains no loops and therefore contains at least two nodes to each of which only one tree branch is connected. The current in each of these two branches is easily found from the known link currents by applying KCL. If there are other nodes at which only one tree branch is connected, these tree-branch currents may also be immediately obtained. In the tree shown in Fig. A1.11, we thus have found the currents in branches **a**, **b**, **c**, and **d**. Now we move along the branches of the tree, finding the currents in tree branches **e** and **f**; the process may be continued until all the branch currents are determined. The link currents are therefore sufficient to determine all branch currents. It is helpful to look at the situation where an incorrect "tree" has been drawn which contains a loop. Even if all the link currents were zero, a current might still circulate about this "tree loop." Hence, the link currents could not determine this current, and they would not represent a sufficient set. Such a "tree" is by definition impossible.

To demonstrate independence, let us satisfy ourselves by assuming the situation where the only sources in the network are independent voltage sources. As we have noticed earlier, independent current sources in the circuit result in fewer equations, while dependent sources usually necessitate a greater number of equations. If only independent voltage sources are present, there will then be precisely $(B - N + 1)$ loop equations written in terms of the $(B - N + 1)$ link currents. To show that these $(B - N + 1)$ loop equations are independent, it is necessary only to point out that each represents the application of KVL around a loop which contains one link not appearing in any other equation. We might visualize a different resistance $R_1, R_2, \ldots, R_{B-N+1}$ in each of these links, and it is then apparent that one equation can never be obtained from the others, since each contains one coefficient not appearing in any other equation.

Hence, the link currents are sufficient to enable a complete solution to be obtained, and the set of loop equations which we use to find the link currents is a set of independent equations.

Having looked at both general nodal analysis and loop analysis, we should now consider the advantages and disadvantages of each method so that an intelligent choice of a plan of attack can be made on a given analysis problem.

The nodal method in general requires $(N - 1)$ equations, but this number is reduced by 1 for each independent or dependent voltage source in a tree branch, and increased by 1 for each dependent source that is voltage-controlled by a link voltage, or current-controlled.

The loop method basically involves $(B - N + 1)$ equations. However, each independent or dependent current source in a link reduces this number by 1, while each dependent source that is current-controlled by a tree-branch current, or is voltage-controlled, increases the number by 1.

As a grand finale for this discussion, let us inspect the T-equivalent circuit model for a transistor shown in Fig. A1.12, to which is connected a sinusoidal source, 4 sin 1000t mV, and a 10 kΩ load.

EXAMPLE A1.6

Find the input (emitter) current i_e and the load voltage v_L in the circuit of Fig. A1.12, assuming typical values for the emitter resistance $r_e = 50\ \Omega$; the base resistance $r_b = 500\ \Omega$; the collector resistance $r_c = 20\ \text{k}\Omega$; and the common-base forward-current-transfer ratio $\alpha = 0.99$.

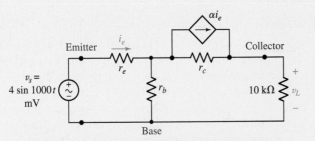

■ **FIGURE A1.12** A sinusoidal voltage source and a 10 kΩ load are connected to the T-equivalent circuit of a transistor. The common connection between the input and output is at the base terminal of the transistor, and the arrangement is called the *common-base* configuration.

Although the details are requested in the practice problems that follow, we should see readily that the analysis of this circuit might be accomplished by drawing trees requiring three general nodal equations $(N - 1 - 1 + 1)$ or two loop equations $(B - N + 1 - 1)$. We might also note that three equations are required in terms of node-to-reference voltages, as are three mesh equations.

No matter which method we choose, these results are obtained for this specific circuit:

$$i_e = 18.42 \sin 1000t \quad \mu A$$
$$v_L = 122.6 \sin 1000t \quad mV$$

(Continued on next page)

and we therefore find that this transistor circuit provides a voltage gain (v_L/v_s) of 30.6, a current gain $(v_L/10,000i_e)$ of 0.666, and a power gain equal to the product $30.6(0.666) = 20.4$. Higher gains could be secured by operating this transistor in a common-emitter configuration.

PRACTICE

A1.2 Draw a suitable tree and use general loop analysis to find i_{10} in the circuit of (a) Fig. A1.13a by writing just one equation with i_{10} as the variable; (b) Fig. A1.13b by writing just two equations with i_{10} and i_3 as the variables.

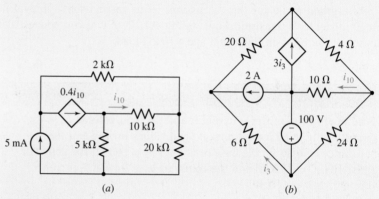

(a)   (b)

■ **FIGURE A1.13**

A1.3 For the transistor amplifier equivalent circuit shown in Fig. A1.12, let $r_e = 50\ \Omega$, $r_b = 500\ \Omega$, $r_c = 20\ \text{k}\Omega$, and $\alpha = 0.99$, and find both i_e and v_L by drawing a suitable tree and using (a) two loop equations; (b) three nodal equations with a common reference node for the voltage; (c) three nodal equations without a common reference node.

A1.4 Determine the Thévenin and Norton equivalent circuits presented to the 10 kΩ load in Fig. A1.12 by finding (a) the open-circuit value of v_L; (b) the (downward) short-circuit current; (c) the Thévenin equivalent resistance. All circuit values are given in Practice Problem A1.3.

Ans: A1.2: -4.00 mA; 4.69 A.   A1.3: $18.42 \sin 1000t\ \mu$A; $122.6 \sin 1000t$ mV.
A1.4: $147.6 \sin 1000t$ mV; $72.2 \sin 1000t\ \mu$A; 2.05 kΩ.

APPENDIX 2

SOLUTION OF SIMULTANEOUS EQUATIONS

Consider the simple system of equations

$$7v_1 - 3v_2 - 4v_3 = -11 \qquad [1]$$
$$-3v_1 + 6v_2 - 2v_3 = 3 \qquad [2]$$
$$-4v_1 - 2v_2 + 11v_3 = 25 \qquad [3]$$

This set of equations *could* be solved by a systematic elimination of the variables. Such a procedure is lengthy, however, and may never yield answers if done unsystematically for a greater number of simultaneous equations. Fortunately, there are many more options available to us, some of which we will explore in this appendix.

The Scientific Calculator

Perhaps the most straightforward approach when faced with a system of equations such as Eqs. [1] to [3], in which we have numerical coefficients and are only interested in the specific values of our unknowns (as opposed to algebraic relationships), is to employ any of the various scientific calculators presently on the market. For example, on a Texas Instruments TI-84, we can employ the Polynomial Root Finder and Simultaneous Equation Solver (you may need to install the application using TI Connect™). Pressing the **APPS** key and scrolling down, find the application named **PLYSmlt2**. Running and continuing past the welcome screen shows the Main Menu of Fig. A2.1*a*. Selecting the second menu item results in the screen shown in Fig. A2.1*b*, where we have chosen three equations in three unknowns. After

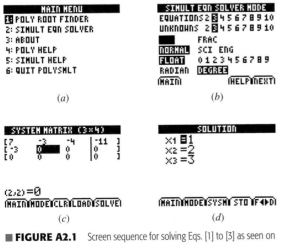

(a)

(b)

(c)

(d)

■ **FIGURE A2.1** Screen sequence for solving Eqs. [1] to [3] as seen on a TI-84 running the Simultaneous Equation Solver application.

pressing **NEXT**, we are presented with a screen similar to that shown in Fig. A2.1c. After we have finished entering all coefficients, pressing the **SOLVE** button yields the Solution screen depicted in Fig. A2.1d. Since we did not name the variables, a slight mental conversion is required to realize $X_1 = v_1$, $X_2 = v_2$, etc.

It should be noted that each calculator capable of solving simultaneous equations has its own procedure for entering the required information— therefore, it is a good idea not to throw away anything marked "Owner's Manual" or "Instructions," no matter how tempting such an action might be.

Matrices

Another powerful approach to the solution of a system of equations is based on the concept of matrices. Consider Eqs. [1], [2], and [3]. The array of the constant coefficients of the equations,

$$\mathbf{G} = \begin{bmatrix} 7 & -3 & -4 \\ -3 & 6 & -2 \\ -4 & -2 & 11 \end{bmatrix}$$

is called a *matrix;* the symbol $\mathbf{G}$ has been selected since each element of the matrix is a conductance value. A matrix itself has no "value"; it is merely an ordered array of elements. We use a letter in boldface type to represent a matrix, and we enclose the array itself by square brackets.

A matrix having m rows and n columns is called an $(m \times n)$ (pronounced "m by n") matrix. Thus,

$$\mathbf{A} = \begin{bmatrix} 2 & 0 & 5 \\ -1 & 6 & 3 \end{bmatrix}$$

is a (2×3) matrix, and the $\mathbf{G}$ matrix of our example is a (3×3) matrix. An $(n \times n)$ matrix is a *square matrix* of order n.

An $(m \times 1)$ matrix is called a *column matrix,* or a *vector.* Thus,

$$\mathbf{V} = \begin{bmatrix} \mathbf{V}_1 \\ \mathbf{V}_2 \end{bmatrix}$$

is a (2×1) column matrix of phasor voltages, and

$$\mathbf{I} = \begin{bmatrix} \mathbf{I}_1 \\ \mathbf{I}_2 \end{bmatrix}$$

is a (2×1) phasor-current vector. A $(1 \times n)$ matrix is known as a *row vector.*

Two $(m \times n)$ matrices are equal if their corresponding elements are equal. Thus, if a_{jk} is that element of $\mathbf{A}$ located in row j and column k and b_{jk} is the element at row j and column k in matrix $\mathbf{B}$, then $\mathbf{A} = \mathbf{B}$ *if and only if* $a_{jk} = b_{jk}$ for all $1 \leq j \leq m$ and $1 \leq k \leq n$. Thus, if

$$\begin{bmatrix} \mathbf{V}_1 \\ \mathbf{V}_2 \end{bmatrix} = \begin{bmatrix} \mathbf{z}_{11}\mathbf{I}_1 + \mathbf{z}_{12}\mathbf{I}_2 \\ \mathbf{z}_{21}\mathbf{I}_1 + \mathbf{z}_{22}\mathbf{I}_2 \end{bmatrix}$$

then $\mathbf{V}_1 = \mathbf{z}_{11}\mathbf{I}_1 + \mathbf{z}_{12}\mathbf{I}_2$ and $\mathbf{V}_2 = \mathbf{z}_{21}\mathbf{I}_1 + \mathbf{z}_{22}\mathbf{I}_2$.

Two $(m \times n)$ matrices may be added by adding corresponding elements. Thus,

$$\begin{bmatrix} 2 & 0 & 5 \\ -1 & 6 & 3 \end{bmatrix} + \begin{bmatrix} 1 & 2 & 3 \\ -3 & -2 & -1 \end{bmatrix} = \begin{bmatrix} 3 & 2 & 8 \\ -4 & 4 & 2 \end{bmatrix}$$

Next let us consider the matrix product $\mathbf{AB}$, where $\mathbf{A}$ is an $(m \times n)$ matrix and $\mathbf{B}$ is a $(p \times q)$ matrix. If $n = p$, the matrices are said to be *conformal,* and their product exists. That is, matrix multiplication is defined only for the case where the number of columns of the first matrix in the product is equal to the number of rows in the second matrix.

The formal definition of matrix multiplication states that the product of the $(m \times n)$ matrix $\mathbf{A}$ and the $(n \times q)$ matrix $\mathbf{B}$ is an $(m \times q)$ matrix having elements c_{jk}, $1 \le j \le m$ and $1 \le k \le q$, where

$$c_{jk} = a_{j1}b_{1k} + a_{j2}b_{2k} + \cdots + a_{jn}b_{nk}$$

That is, to find the element in the second row and third column of the product, we multiply each of the elements in the second row of $\mathbf{A}$ by the corresponding element in the third column of $\mathbf{B}$ and then add the n results. For example, given the (2×3) matrix $\mathbf{A}$ and the (3×2) matrix $\mathbf{B}$,

$$\begin{bmatrix} a_{11} & a_{12} & a_{13} \\ a_{21} & a_{22} & a_{23} \end{bmatrix} \begin{bmatrix} b_{11} & b_{12} \\ b_{21} & b_{22} \\ b_{31} & b_{32} \end{bmatrix} =$$

$$\begin{bmatrix} (a_{11}b_{11} + a_{12}b_{21} + a_{13}b_{31}) & (a_{11}b_{12} + a_{12}b_{22} + a_{13}b_{32}) \\ (a_{21}b_{11} + a_{22}b_{21} + a_{23}b_{31}) & (a_{21}b_{12} + a_{22}b_{22} + a_{23}b_{32}) \end{bmatrix}$$

The result is a (2×2) matrix.

As a numerical example of matrix multiplication, we take

$$\begin{bmatrix} 3 & 2 & 1 \\ -2 & -2 & 4 \end{bmatrix} \begin{bmatrix} 2 & 3 \\ -2 & -1 \\ 4 & -3 \end{bmatrix} = \begin{bmatrix} 6 & 4 \\ 16 & -16 \end{bmatrix}$$

where $6 = (3)(2) + (2)(-2) + (1)(4)$, $\quad 4 = (3)(3) + (2)(-1) + (1)(-3)$, and so forth.

Matrix multiplication is not commutative. For example, given the (3×2) matrix $\mathbf{C}$ and the (2×1) matrix $\mathbf{D}$, it is evident that the product $\mathbf{CD}$ may be calculated, but the product $\mathbf{DC}$ is not even defined.

As a final example, let

$$\mathbf{t}_A = \begin{bmatrix} 2 & 3 \\ -1 & 4 \end{bmatrix}$$

and

$$\mathbf{t}_B = \begin{bmatrix} 3 & 1 \\ 5 & 0 \end{bmatrix}$$

so that both $\mathbf{t}_A \mathbf{t}_B$ and $\mathbf{t}_B \mathbf{t}_A$ are defined. However,

$$\mathbf{t}_A \mathbf{t}_B = \begin{bmatrix} 21 & 2 \\ 17 & -1 \end{bmatrix}$$

while

$$\mathbf{t}_B \mathbf{t}_A = \begin{bmatrix} 5 & 13 \\ 10 & 15 \end{bmatrix}$$

PRACTICE

A2.1 Given $\mathbf{A} = \begin{bmatrix} 1 & -3 \\ 3 & 5 \end{bmatrix}$, $\mathbf{B} = \begin{bmatrix} 4 & -1 \\ -2 & 3 \end{bmatrix}$, $\mathbf{C} = \begin{bmatrix} 50 \\ 30 \end{bmatrix}$, and $\mathbf{V} = \begin{bmatrix} V_1 \\ V_2 \end{bmatrix}$, find (a) $\mathbf{A} + \mathbf{B}$; (b) $\mathbf{AB}$; (c) $\mathbf{BA}$; (d) $\mathbf{AV} + \mathbf{BC}$; (e) $\mathbf{A}^2 = \mathbf{AA}$.

Ans: $\begin{bmatrix} 5 & -4 \\ 1 & 8 \end{bmatrix}$; $\begin{bmatrix} 10 & -10 \\ 2 & 12 \end{bmatrix}$; $\begin{bmatrix} 1 & -17 \\ 7 & 21 \end{bmatrix}$; $\begin{bmatrix} V_1 - 3V_2 + 170 \\ 3V_1 + 5V_2 - 10 \end{bmatrix}$; $\begin{bmatrix} -8 & -18 \\ 18 & 16 \end{bmatrix}$.

Matrix Inversion

If we write our system of equations using matrix notation,

$$\begin{bmatrix} 7 & -3 & -4 \\ -3 & 6 & -2 \\ -4 & -2 & 11 \end{bmatrix} \begin{bmatrix} v_1 \\ v_2 \\ v_3 \end{bmatrix} = \begin{bmatrix} -11 \\ 3 \\ 25 \end{bmatrix} \qquad [4]$$

we may solve for the voltage vector by multiplying both sides of Eq. [4] by the inverse of our matrix $\mathbf{G}$:

$$\mathbf{G}^{-1} \begin{bmatrix} 7 & -3 & -4 \\ -3 & 6 & -2 \\ -4 & -2 & 11 \end{bmatrix} \begin{bmatrix} v_1 \\ v_2 \\ v_3 \end{bmatrix} = \mathbf{G}^{-1} \begin{bmatrix} -11 \\ 3 \\ 25 \end{bmatrix} \qquad [5]$$

This procedure makes use of the identity $\mathbf{G}^{-1}\mathbf{G} = \mathbf{I}$, where $\mathbf{I}$ is the identity matrix, a square matrix of the same size as $\mathbf{G}$, with zeros everywhere except on the diagonal. Each element on the diagonal of an identity matrix is unity. Thus, Eq. [5] becomes

$$\begin{bmatrix} 1 & 0 & 0 \\ 0 & 1 & 0 \\ 0 & 0 & 1 \end{bmatrix} \begin{bmatrix} v_1 \\ v_2 \\ v_3 \end{bmatrix} = \mathbf{G}^{-1} \begin{bmatrix} -11 \\ 3 \\ 25 \end{bmatrix}$$

which may be simplified to

$$\begin{bmatrix} v_1 \\ v_2 \\ v_3 \end{bmatrix} = \mathbf{G}^{-1} \begin{bmatrix} -11 \\ 3 \\ 25 \end{bmatrix}$$

since the identity matrix times any vector is simply equal to that vector (the proof is left to the reader as a 30-second exercise). The solution of our system of equations has therefore been transformed into the problem of obtaining the inverse matrix of $\mathbf{G}$. Many scientific calculators provide the means of performing matrix algebra.

Returning to the TI-84, we press **2ND** and **MATRIX** to obtain the screen shown in Fig. A2.2*a*. Scrolling horizontally to **EDIT**, we press the **ENTER** key, and select a 3×3 matrix, resulting in a screen similar to that shown in Fig. A2.2*b*. Once we have finished entering the matrix, we press **2ND** and **QUIT**. Returning to the **MATRIX** editor, we create a 3×1 vector named **B**, as shown in Fig. A2.2*c*. We are now (finally) ready to solve for the solution vector. Pressing **2ND** and **MATRIX**, under **NAMES** we select **[A]** and press **ENTER**, followed by the x^{-1} key. Next, we select **[B]** the same way (we could have pressed the multiplication key in between but it is not necessary). The result of our calculation is shown in Fig. A2.2*d*, and agrees with our earlier exercise.

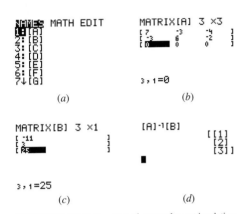

■ **FIGURE A2.2** Sequence of screens for matrix solution. (*a*) Matrix editor screen; (*b*) entering terms; (*c*) creating right-hand side vector; (*d*) solving matrix equation.

Determinants

Although a matrix *itself* has no "value," the ***determinant*** of a square matrix *does* have a value. To be precise, we should say that the determinant of a matrix is a value, but common usage enables us to speak of both the array itself and its value as the determinant. We shall symbolize a determinant by Δ, and employ a suitable subscript to denote the matrix to which the determinant refers. Thus,

$$\Delta_G = \begin{vmatrix} 7 & -3 & -4 \\ -3 & 6 & -2 \\ -4 & -2 & 11 \end{vmatrix}$$

Note that simple vertical lines are used to enclose the determinant.

The value of any determinant is obtained by expanding it in terms of its minors. To do this, we select any row j or any column k, multiply each element in that row or column by its minor and by $(-1)^{j+k}$, and then add the products. The minor of the element appearing in both row j and column k

is the determinant obtained when row j and column k are removed; it is indicated by Δ_{jk}.

As an example, let us expand the determinant Δ_G along column 3. We first multiply the (-4) at the top of this column by $(-1)^{1+3} = 1$ and then by its minor:

$$(-4)(-1)^{1+3}\begin{vmatrix} -3 & 6 \\ -4 & -2 \end{vmatrix}$$

and then repeat for the other two elements in column 3, adding the results:

$$-4\begin{vmatrix} -3 & 6 \\ -4 & -2 \end{vmatrix} + 2\begin{vmatrix} 7 & -3 \\ -4 & -2 \end{vmatrix} + 11\begin{vmatrix} 7 & -3 \\ -3 & 6 \end{vmatrix}$$

The minors contain only two rows and two columns. They are of order 2, and their values are easily determined by expanding in terms of minors again, here a trivial operation. Thus, for the first determinant, we expand along the first column by multiplying (-3) by $(-1)^{1+1}$ and its minor, which is merely the element (-2), and then multiplying (-4) by $(-1)^{2+1}$ and by 6. Thus,

$$\begin{vmatrix} -3 & 6 \\ -4 & -2 \end{vmatrix} = (-3)(-2) - 4(-6) = 30$$

It is usually easier to remember the result for a second-order determinant as "upper left times lower right minus upper right times lower left." Finally,

$$\begin{aligned} \Delta_G &= -4[(-3)(-2) - 6(-4)] \\ &\quad + 2[(7)(-2) - (-3)(-4)] \\ &\quad + 11[(7)(6) - (-3)(-3)] \\ &= -4(30) + 2(-26) + 11(33) \\ &= 191 \end{aligned}$$

For practice, let us expand this same determinant along the first row:

$$\begin{aligned} \Delta_G &= 7\begin{vmatrix} 6 & -2 \\ -2 & 11 \end{vmatrix} - (-3)\begin{vmatrix} -3 & -2 \\ -4 & 11 \end{vmatrix} + (-4)\begin{vmatrix} -3 & 6 \\ -4 & -2 \end{vmatrix} \\ &= 7(62) + 3(-41) - 4(30) \\ &= 191 \end{aligned}$$

The expansion by minors is valid for a determinant of any order.

Repeating these rules for evaluating a determinant in more general terms, we would say, given a matrix $\mathbf{a}$,

$$\mathbf{a} = \begin{bmatrix} a_{11} & a_{12} & \cdots & a_{1N} \\ a_{21} & a_{22} & \cdots & a_{2N} \\ \cdots & \cdots & \cdots & \cdots \\ a_{N1} & a_{N2} & \cdots & a_{NN} \end{bmatrix}$$

that Δ_a may be obtained by expansion in terms of minors along any row j:

$$\begin{aligned} \Delta_a &= a_{j1}(-1)^{j+1}\Delta_{j1} + a_{j2}(-1)^{j+2}\Delta_{j2} + \cdots + a_{jN}(-1)^{j+N}\Delta_{jN} \\ &= \sum_{n=1}^{N} a_{jn}(-1)^{j+n}\Delta_{jn} \end{aligned}$$

or along any column k:

$$\Delta_a = a_{1k}(-1)^{1+k}\Delta_{1k} + a_{2k}(-1)^{2+k}\Delta_{2k} + \cdots + a_{Nk}(-1)^{N+k}\Delta_{Nk}$$

$$= \sum_{n=1}^{N} a_{nk}(-1)^{n+k}\Delta_{nk}$$

The cofactor C_{jk} of the element appearing in both row j and column k is simply $(-1)^{j+k}$ times the minor Δ_{jk}. Thus, $C_{11} = \Delta_{11}$, but $C_{12} = -\Delta_{12}$. We may now write

$$\Delta_a = \sum_{n=1}^{N} a_{jn}C_{jn} = \sum_{n=1}^{N} a_{nk}C_{nk}$$

As an example, let us consider this fourth-order determinant:

$$\Delta = \begin{vmatrix} 2 & -1 & -2 & 0 \\ -1 & 4 & 2 & -3 \\ -2 & -1 & 5 & -1 \\ 0 & -3 & 3 & 2 \end{vmatrix}$$

We find

$$\Delta_{11} = \begin{vmatrix} 4 & 2 & -3 \\ -1 & 5 & -1 \\ -3 & 3 & 2 \end{vmatrix} = 4(10+3) + 1(4+9) - 3(-2+15) = 26$$

$$\Delta_{12} = \begin{vmatrix} -1 & 2 & -3 \\ -2 & 5 & -1 \\ 0 & 3 & 2 \end{vmatrix} = -1(10+3) + 2(4+9) + 0 = 13$$

and $C_{11} = 26$, whereas $C_{12} = -13$. Finding the value of Δ for practice, we have

$$\Delta = 2C_{11} + (-1)C_{12} + (-2)C_{13} + 0$$
$$= 2(26) + (-1)(-13) + (-2)(3) + 0 = 59$$

Cramer's Rule

We next consider Cramer's rule, which enables us to find the values of the unknown variables. It is also useful in solving systems of equations where numerical coefficients have not yet been specified, thus confounding our calculators. Let us again consider Eqs. [1], [2], and [3]; we define the determinant Δ_1 as that determinant which is obtained when the first column of Δ_G is replaced by the three constants on the right-hand sides of the three equations. Thus,

$$\Delta_1 = \begin{vmatrix} -11 & -3 & -4 \\ 3 & 6 & -2 \\ 25 & -2 & 11 \end{vmatrix}$$

We expand along the first column:

$$\Delta_1 = -11 \begin{vmatrix} 6 & -2 \\ -2 & 11 \end{vmatrix} - 3 \begin{vmatrix} -3 & -4 \\ -2 & 11 \end{vmatrix} + 25 \begin{vmatrix} -3 & -4 \\ 6 & -2 \end{vmatrix}$$
$$= -682 + 123 + 750 = 191$$

Cramer's rule then states that

$$v_1 = \frac{\Delta_1}{\Delta_G} = \frac{191}{191} = 1\ V$$

and

$$v_2 = \frac{\Delta_2}{\Delta_G} = \begin{vmatrix} 7 & -11 & -4 \\ -3 & 3 & -2 \\ -4 & 25 & 11 \end{vmatrix} = \frac{581 - 63 - 136}{191} = 2\ V$$

and finally,

$$v_3 = \frac{\Delta_3}{\Delta_G} = \begin{vmatrix} 7 & -3 & -11 \\ -3 & 6 & 3 \\ -4 & -2 & 25 \end{vmatrix} = \frac{1092 - 291 - 228}{191} = 3\ V$$

Cramer's rule is applicable to a system of N simultaneous linear equations in N unknowns; for the ith variable v_i:

$$v_i = \frac{\Delta_i}{\Delta_G}$$

PRACTICE

A2.2 Evaluate

(a) $\begin{vmatrix} 2 & -3 \\ -2 & 5 \end{vmatrix}$; (b) $\begin{vmatrix} 1 & -1 & 0 \\ 4 & 2 & -3 \\ 3 & -2 & 5 \end{vmatrix}$; (c) $\begin{vmatrix} 2 & -3 & 1 & 5 \\ -3 & 1 & -1 & 0 \\ 0 & 4 & 2 & -3 \\ 6 & 3 & -2 & 5 \end{vmatrix}$;

(d) Find i_2 if $5i_1 - 2i_2 - i_3 = 100$, $-2i_1 + 6i_2 - 3i_3 - i_4 = 0$, $-i_1 - 3i_2 + 4i_3 - i_4 = 0$, and $-i_2 - i_3 = 0$.

Ans: 4; 33; −411; 1.266.

A PROOF OF THÉVENIN'S THEOREM

Here we prove Thévenin's theorem in the same form in which it is stated in Sec. 5.4 of Chap. 5:

> Given any linear circuit, rearrange it in the form of two networks A and B connected by two wires. Define a voltage v_{oc} as the open-circuit voltage which appears across the terminals of A when B is disconnected. Then all currents and voltages in B will remain unchanged if all *independent* voltage and current sources in A are "killed" or "zeroed out," and an independent voltage source v_{oc} is connected, with proper polarity, in series with the dead (inactive) A network.

We will effect our proof by showing that the original A network and the Thévenin equivalent of the A network both cause the same current to flow into the terminals of the B network. If the currents are the same, then the voltages must be the same; in other words, if we apply a certain current, which we might think of as a current source, to the B network, then the current source and the B network constitute a circuit that has a specific input voltage as a response. Thus, the current determines the voltage. Alternatively we could, if we wished, show that the terminal voltage at B is unchanged, because the voltage also determines the current uniquely. If the input voltage and current to the B network are unchanged, then it follows that the currents and voltages *throughout* the B network are also unchanged.

Let us first prove the theorem for the case where the B network is inactive (no independent sources). After this step has been accomplished, we may then use the superposition principle to extend the theorem to include B networks that contain independent sources. Each network may contain dependent sources, provided that their control variables are in the same network.

The current i, flowing in the upper conductor from the A network to the B network in Fig. A3.1a, is therefore caused entirely by the independent

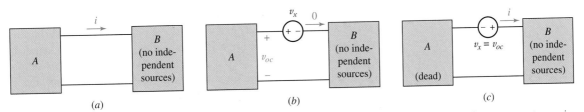

■ FIGURE A3.1 (a) A general linear network A and a network B that contains no independent sources. Controls for dependent sources must appear in the same part of the network. (b) The Thévenin source is inserted in the circuit and adjusted until $i = 0$. No voltage appears across network B and thus $v_x = v_{oc}$. The Thévenin source thus produces a current $-i$ while network A provides i. (c) The Thévenin source is reversed and network A is killed. The current is therefore i.

sources present in the A network. Suppose now that we add an additional voltage source v_x, which we shall call the Thévenin source, in the conductor in which i is measured, as shown in Fig. A3.1b, and then adjust the magnitude and time variation of v_x until the current is reduced to zero. By our definition of v_{oc}, then, the voltage across the terminals of A must be v_{oc}, since $i = 0$. Network B contains no independent sources, and no current is entering its terminals; therefore, there is no voltage across the terminals of the B network, and by Kirchhoff's voltage law the voltage of the Thévenin source is v_{oc} volts, $v_x = v_{oc}$. Moreover, since the Thévenin source and the A network jointly deliver no current to B, and since the A network by itself delivers a current i, superposition requires that the Thévenin source acting by itself must deliver a current of $-i$ to B. The source acting alone in a reversed direction, as shown in Fig. A3.1c, therefore produces a current i in the upper lead. This situation, however, is the same as the conclusion reached by Thévenin's theorem: the Thévenin source v_{oc} acting in series with the inactive A network is equivalent to the given network.

Now let us consider the case where the B network may be an active network. We now think of the current i, flowing from the A network to the B network in the upper conductor, as being composed of two parts, i_A and i_B, where i_A is the current produced by A acting alone and the current i_B is due to B acting alone. Our ability to divide the current into these two components is a direct consequence of the applicability of the superposition principle to these two *linear* networks; the complete response and the two partial responses are indicated by the diagrams of Fig. A3.2.

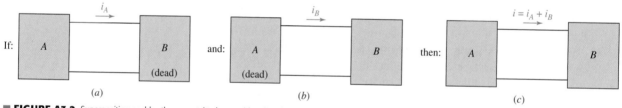

■ **FIGURE A3.2** Superposition enables the current i to be considered as the sum of two partial responses.

The partial response i_A has already been considered; if network B is inactive, we know that network A may be replaced by the Thévenin source and the inactive A network. In other words, of the three sources which we must keep in mind—those in A, those in B, and the Thévenin source—the partial response i_A occurs when A and B are dead and the Thévenin source is active. Preparing for the use of superposition, we now let A remain inactive, but turn on B and turn off the Thévenin source; by definition, the partial response i_B is obtained. Superimposing the results, the response when A is dead and both the Thévenin source and B are active is $i_A + i_B$. This sum is the original current i, and the situation wherein the Thévenin source and B are active but A is dead is the desired Thévenin equivalent circuit. Thus the active network A may be replaced by its Thévenin source, the open-circuit voltage, in series with the inactive A network, regardless of the status of the B network; it may be either active or inactive.

A PSpice® TUTORIAL

SPICE is an acronym for Simulation Program with Integrated Circuit Emphasis. A powerful program, it is an industry standard and used throughout the world for a variety of circuit analysis applications. SPICE was originally developed in the early 1970s by Donald O. Peterson and coworkers at the University of California at Berkeley. Interestingly, Peterson advocated free and unhindered distribution of knowledge created in university labs, choosing to make an impact as opposed to profiting financially. In 1984, MicroSim Corporation introduced a PC version of SPICE called PSpice®, which built intuitive graphical interfaces around the core SPICE software routines. There are now several variations of SPICE available commercially, as well as competing software products.

The goal of this appendix is to simply introduce the basics of computer-aided circuit analysis; more details are presented in the main text as well as in the references listed under Reading Further. Advanced topics covered in the references include how to determine the sensitivity of an output variable to changes in a specific component value; how to obtain plots of the output versus a source value; determining ac output as a function of source frequency; methods for performing noise and distortion analyses; nonlinear component models; and how to model temperature effects on specific types of circuits.

The acquisition of MicroSim by OrCAD, and the subsequent acquisition of OrCAD by Cadence, has led to quite a few changes in this popular circuit simulation package. At the time of this writing, OrCAD Capture CIS-Demo 16.3, is the current professional release; a scaled-back version is available for free download (www.cadence.com). This new version replaces the popular PSpice Student Release 9.1, and although slightly different, particularly in terms of the schematic editing, should seem generally familiar to users of previous PSpice releases.

The documentation which accompanies the demo version of OrCAD 16.3 lists several restrictions that do not apply to the professional (commercially available) version. The most significant is that only circuits having 60 or fewer parts may be saved and simulated; larger circuits can be drawn and viewed, however. We have chosen to work with the OrCAD Capture schematics editor, as the current version is very similar fundamentally to the PSpice A/D Schematic Capture editor. Although at present Cadence also provides PSpice A/D for download, it is no longer supported.

Getting Started

A computer-aided circuit analysis consists of three separate steps: (1) drawing the schematic, (2) simulating the circuit, and (3) extracting the desired information from the simulation output. The OrCAD Capture schematic editor is launched through the Windows programs list found under the **start** menu; upon selecting OrCAD Capture CIS-Demo, the schematics editor opens, as shown in Fig. A4.1.

■ FIGURE A4.1 Capture CIS-Demo window.

Under the **File** menu, select **New,** then **Project;** the window of Fig. A4.2*a* appears. After providing a simulation filename and a directory path, the window of Fig. A4.2*b* appears (simply select the "blank project" option). We are now presented with the main schematics editor window, as in Fig. A4.3.

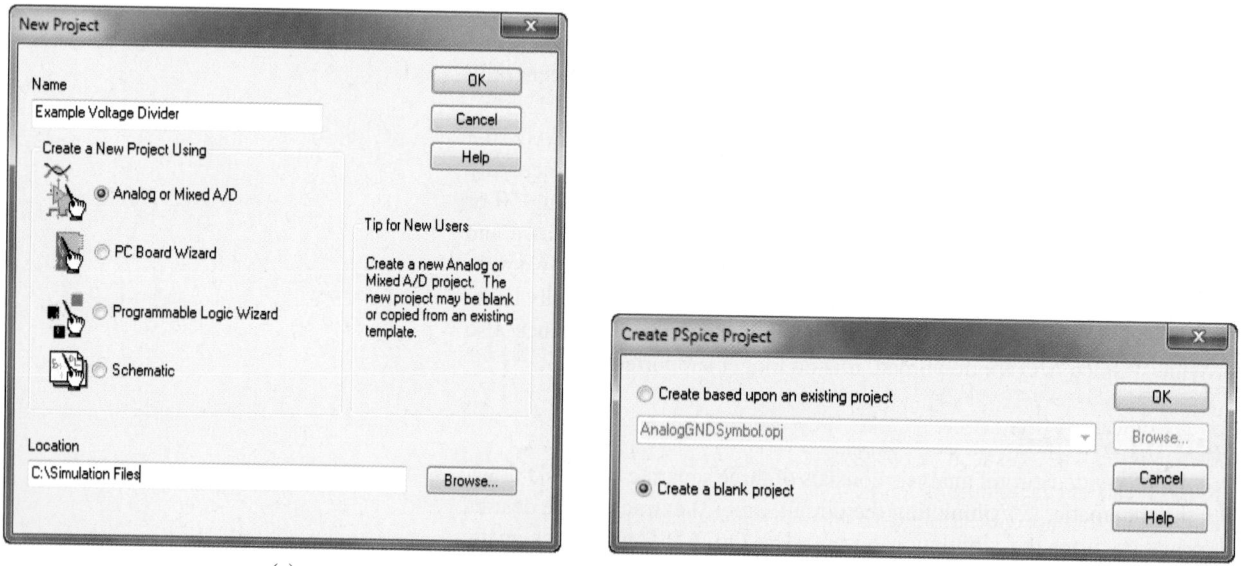

(*a*)

(*b*)

■ FIGURE A4.2 (*a*) New Project window. (*b*) Create PSpice Project window.

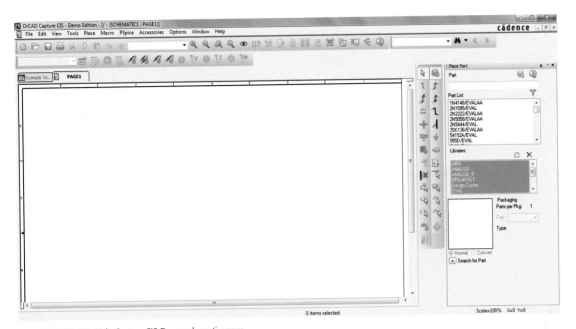

■ **FIGURE A4.3** Main Capture CIS-Demo schematics page.

At this point, we're ready to draw a circuit, so let's try a simple voltage divider for purposes of illustration. We will first place the necessary components on the grid, and then wire them together.

Pulling down the **Place** menu, we choose **Part.** Typing a lowercase "r" as shown, we click OK and are now able to move a resistor symbol around the schematic window using the mouse. A single left click places a resistor (named R1) at the mouse location; a second left click places a second resistor on our schematic (named R2). A single right click and selecting **End Mode** cancel further resistor placements. The second resistor does not have the appropriate orientation, but is easily manipulated by highlighting it with a single right click, then selecting **Rotate.** If we do not know the name of the desired part, we can browse through the library of parts provided. If 1 kΩ resistors are not desired—for example, perhaps two 500 Ω resistors were called for—we change the default values simply by double-clicking the "1k" next to the appropriate symbol.

No voltage divider circuit is complete of course without a voltage source (**vsrc**). Double-clicking **DC=**, we choose a value of 9 V for our source. One further component is required: SPICE requires a reference (or ground) node to be specified. Clicking the GND symbol to the far right of the schematic window, we choose 0/Source from the options. Our progress so far is shown in Fig. A4.4*a*; all that remains is to wire the components together. This is accomplished by selecting the **Place wire (W)** icon. The left and right mouse keys control each wire (some experimenting is called for here—afterward, select any unwanted wire segments and hit the Delete key). Our final circuit is shown in Fig. A4.4*b*. It is worth noting that the editor will allow the user to wire through a resistor (thus shorting it out), which can be difficult to see. Generally a warning symbol appears before wiring to an inappropriate location.

Prior to simulating our circuit, we save it by clicking the save icon or selecting **Save** from the **File** menu. From the **PSpice** menu, we select **New Simulation Profile,** and type Voltage Divider in the dialog box that appears.

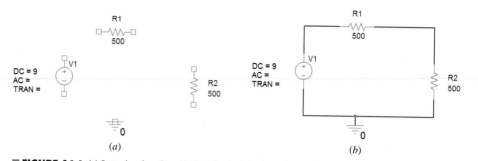

■ **FIGURE A4.4** (a) Parts placed on the grid. (b) Fully wired circuit, ready to simulate.

The Simulation Settings dialog box that appears allows us to set parameters for a variety of types of simulations; for the present example we need to select **Bias Point** under the **Analysis type** menu. Once again pulling down the **PSpice** menu, we select **Run.** The simulation results are shown in Fig. A4.5.

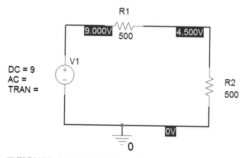
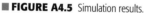

■ **FIGURE A4.5** Simulation results.

Fortunately, our simulation yields the expected result—an even split of our source voltage across the two equal-valued resistors. We can also view the simulation results by selecting **View Output File** under the **PSpice** menu. Scrolling down to the end of this file, we see the following lines:

NODE   VOLTAGE   NODE   VOLTAGE

(N00109) 9.0000   (N00116)   4.5000

where node 109 is the positive reference of our voltage source, and node 116 is the junction between the two resistors. This information is available at the top of the file.

READING FURTHER

Two very good books devoted to SPICE and PSpice simulation are:

P. W. Tuinenga, *SPICE: A Guide to Circuit Simulation and Analysis Using PSpice.* Englewood Cliffs, N.J.: Prentice-Hall, 1995.

R. W. Goody, *OrCAD PSpice for Windows Volume 1: DC and AC Circuits,* 3rd ed. Englewood Cliffs, N.J.: Prentice-Hall, 2001.

An interesting history of circuit simulators, as well as Donald Peterson's contributions to the field, can be found in:

T. Perry, "Donald O. Peterson [electronic engineering biography]," *IEEE Spectrum* **35** (1998) 22–27.

COMPLEX NUMBERS

This appendix includes sections covering the definition of a complex number, the basic arithmetic operations for complex numbers, Euler's identity, and the exponential and polar forms of the complex number. We first introduce the concept of a complex number.

A5.1 THE COMPLEX NUMBER

Our early training in mathematics dealt exclusively with real numbers, such as 4, $-\frac{2}{7}$, and π. Soon, however, we began to encounter algebraic equations, such as $x^2 = -3$, which could not be satisfied by any real number. Such an equation can be solved only through the introduction of the *imaginary unit*, or the *imaginary operator*, which we shall designate by the symbol j. By definition, $j^2 = -1$, and thus $j = \sqrt{-1}$, $j^3 = -j$, $j^4 = 1$, and so forth. The product of a real number and the imaginary operator is called an *imaginary number*, and the sum of a real number and an imaginary number is called a *complex number*. Thus, a number having the form $a + jb$, where a and b are real numbers, is a complex number.

We shall designate a complex number by means of a special single symbol; thus, $\mathbf{A} = a + jb$. The complex nature of the number is indicated by the use of boldface type; in handwritten material, a bar over the letter is customary. The complex number $\mathbf{A}$ just shown is described as having a *real component* or *real part a* and an *imaginary component* or *imaginary part b*. This is also expressed as

$$\text{Re}\{\mathbf{A}\} = a \qquad \text{Im}\{\mathbf{A}\} = b$$

The imaginary component of $\mathbf{A}$ is *not jb*. By definition, the imaginary component is a real number.

It should be noted that all real numbers may be regarded as complex numbers having imaginary parts equal to zero. The real numbers are therefore included in the system of complex numbers, and we may now consider them as a special case. When we define the fundamental arithmetic operations for complex numbers, we should therefore expect them to reduce to the corresponding definitions for real numbers if the imaginary part of every complex number is set equal to zero.

Since any complex number is completely characterized by a pair of real numbers, such as a and b in the previous example, we can obtain some visual assistance by representing a complex number graphically on a rectangular, or Cartesian, coordinate system. By providing ourselves with a real axis and an imaginary axis, as shown in Fig. A5.1, we form a *complex plane*, or *Argand diagram*, on which any complex number can be represented as a single point. The complex numbers $\mathbf{M} = 3 + j1$ and $\mathbf{N} = 2 - j2$

Mathematicians designate the imaginary operator by the symbol i, but it is customary to use j in electrical engineering in order to avoid confusion with the symbol for current.

The choice of the words *imaginary* and *complex* is unfortunate. They are used here and in the mathematical literature as technical terms to designate a class of numbers. To interpret imaginary as "not pertaining to the physical world" or complex as "complicated" is neither justified nor intended.

are indicated. It is important to understand that this complex plane is only a visual aid; it is not at all essential to the mathematical statements which follow.

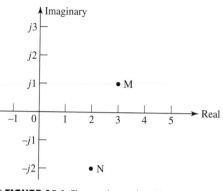

■ **FIGURE A5.1** The complex numbers $M = 3 + j1$ and $N = 2 - j2$ are shown on the complex plane.

We define two complex numbers as being equal if, and only if, their real parts are equal and their imaginary parts are equal. Graphically, then, to each point in the complex plane there corresponds only one complex number, and conversely, to each complex number there corresponds only one point in the complex plane. Thus, suppose we are given the two complex numbers

$$\mathbf{A} = a + jb \qquad \text{and} \qquad \mathbf{B} = c + jd$$

Then, if

$$\mathbf{A} = \mathbf{B}$$

it is necessary that

$$a = c \qquad \text{and} \qquad b = d$$

A complex number expressed as the sum of a real number and an imaginary number, such as $\mathbf{A} = a + jb$, is said to be in *rectangular* or *cartesian* form. Other forms for a complex number will appear shortly.

Let us now define the fundamental operations of addition, subtraction, multiplication, and division for complex numbers. The sum of two complex numbers is defined as the complex number whose real part is the sum of the real parts of the two complex numbers and whose imaginary part is the sum of the imaginary parts of the two complex numbers. Thus,

$$(a + jb) + (c + jd) = (a + c) + j(b + d)$$

For example,

$$(3 + j4) + (4 - j2) = 7 + j2$$

The difference of two complex numbers is taken in a similar manner; for example,

$$(3 + j4) - (4 - j2) = -1 + j6$$

Addition and subtraction of complex numbers may also be accomplished graphically on the complex plane. Each complex number is represented as a vector, or directed line segment, and the sum is obtained by completing the parallelogram, illustrated by Fig. A5.2a, or by connecting the vectors in a head-to-tail manner, as shown in Fig. A5.2b. A graphical sketch is often useful as a check for a more exact numerical solution.

The product of two complex numbers is defined by

$$(a + jb)(c + jd) = (ac - bd) + j(bc + ad)$$

This result may be easily obtained by a direct multiplication of the two binomial terms, using the rules of the algebra of real numbers, and then simplifying the result by letting $j^2 = -1$. For example,

$$(3 + j4)(4 - j2) = 12 - j6 + j16 - 8j^2$$
$$= 12 + j10 + 8$$
$$= 20 + j10$$

It is easier to multiply the complex numbers by this method, particularly if we immediately replace j^2 by -1, than it is to substitute in the general formula that defines the multiplication.

Before defining the operation of division for complex numbers, we should define the conjugate of a complex number. The *conjugate* of the complex number $\mathbf{A} = a + jb$ is $a - jb$ and is represented as $\mathbf{A}^*$. The conjugate of any complex number is therefore easily obtained by merely changing the sign of the imaginary part of the complex number. Thus, if

$$\mathbf{A} = 5 + j3$$

then

$$\mathbf{A}^* = 5 - j3$$

It is evident that the conjugate of any complicated complex expression may be found by replacing every complex term in the expression by its conjugate, which may be obtained by replacing every j in the expression by $-j$.

The definitions of addition, subtraction, and multiplication show that the following statements are true: the sum of a complex number and its conjugate is a real number; the difference of a complex number and its conjugate is an imaginary number; and the product of a complex number and its conjugate is a real number. It is also evident that if $\mathbf{A}^*$ is the conjugate of $\mathbf{A}$, then $\mathbf{A}$ is the conjugate of $\mathbf{A}^*$; in other words, $\mathbf{A} = (\mathbf{A}^*)^*$. A complex number and its conjugate are said to form a *conjugate complex pair* of numbers.

We now define the quotient of two complex numbers:

$$\frac{\mathbf{A}}{\mathbf{B}} = \frac{(\mathbf{A})(\mathbf{B}^*)}{(\mathbf{B})(\mathbf{B}^*)}$$

and thus

$$\frac{a + jb}{c + jd} = \frac{(ac + bd) + j(bc - ad)}{c^2 + d^2}$$

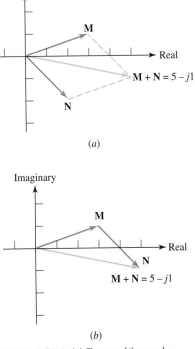

■ **FIGURE A5.2** (a) The sum of the complex numbers $\mathsf{M} = 3 + j1$ and $\mathsf{N} = 2 - j2$ is obtained by constructing a parallelogram. (b) The sum of the same two complex numbers is found by a head-to-tail combination.

Inevitably in a physical problem a complex number is somehow accompanied by its conjugate.

We multiply numerator and denominator by the conjugate of the denominator in order to obtain a denominator which is real; this process is called *rationalizing the denominator.* As a numerical example,

$$\frac{3+j4}{4-j2} = \frac{(3+j4)(4+j2)}{(4-j2)(4+j2)}$$
$$= \frac{4+j22}{16+4} = 0.2 + j1.1$$

The addition or subtraction of two complex numbers which are each expressed in rectangular form is a relatively simple operation; multiplication or division of two complex numbers in rectangular form, however, is a rather unwieldy process. These latter two operations will be found to be much simpler when the complex numbers are given in either exponential or polar form. These forms will be introduced in Secs. A5.3 and A5.4.

PRACTICE

A5.1 Let $\mathbf{A} = -4 + j5$, $\mathbf{B} = 3 - j2$, and $\mathbf{C} = -6 - j5$, and find
(a) $\mathbf{C} - \mathbf{B}$; (b) $2\mathbf{A} - 3\mathbf{B} + 5\mathbf{C}$; (c) $j^5\mathbf{C}^2(\mathbf{A} + \mathbf{B})$; (d) $\mathbf{B}\,\text{Re}\{\mathbf{A}\} + \mathbf{A}\,\text{Re}\{\mathbf{B}\}$.
A5.2 Using the same values for $\mathbf{A}$, $\mathbf{B}$, and $\mathbf{C}$ as in the previous problem, find (a) $[(\mathbf{A} - \mathbf{A}^*)(\mathbf{B} + \mathbf{B}^*)^*]^*$; (b) $(1/\mathbf{C}) - (1/\mathbf{B})^*$; (c) $(\mathbf{B} + \mathbf{C})/(2\mathbf{B}\mathbf{C})$.

Ans: A5.1: $-9 - j3$; $-47 - j9$; $27 - j191$; $-24 + j23$.   A5.2: $-j60$; $-0.329 + j0.236$; $0.0662 + j0.1179$.

A5.2 EULER'S IDENTITY

In Chap. 9 we encounter functions of time which contain complex numbers, and we are concerned with the differentiation and integration of these functions with respect to the real variable t. We differentiate and integrate such functions with respect to t by exactly the same procedures we use for real functions of time. That is, the complex constants are treated just as though they were real constants when performing the operations of differentiation or integration. If $\mathbf{f}(t)$ is a complex function of time, such as

$$\mathbf{f}(t) = a\cos ct + jb\sin ct$$

then

$$\frac{d\mathbf{f}(t)}{dt} = -ac\sin ct + jbc\cos ct$$

and

$$\int \mathbf{f}(t)\, dt = \frac{a}{c}\sin ct - j\frac{b}{c}\cos ct + \mathbf{C}$$

where the constant of integration $\mathbf{C}$ is a complex number in general.

It is sometimes necessary to differentiate or integrate a function of a complex variable with respect to that complex variable. In general, the successful accomplishment of either of these operations requires that the function which is to be differentiated or integrated satisfy certain conditions. All our functions do meet these conditions, and integration or differentiation with respect to a complex variable is achieved by using methods identical to those used for real variables.

At this time we must make use of a very important fundamental relationship known as Euler's identity (pronounced "oilers"). We shall prove this identity, for it is extremely useful in representing a complex number in a form other than rectangular form.

The proof is based on the power series expansions of $\cos\theta$, $\sin\theta$, and e^z, given toward the back of your favorite college calculus text:

$$\cos\theta = 1 - \frac{\theta^2}{2!} + \frac{\theta^4}{4!} - \frac{\theta^6}{6!} + \cdots$$

$$\sin\theta = \theta - \frac{\theta^3}{3!} + \frac{\theta^5}{5!} - \frac{\theta^7}{7!} + \cdots$$

or

$$\cos\theta + j\sin\theta = 1 + j\theta - \frac{\theta^2}{2!} - j\frac{\theta^3}{3!} + \frac{\theta^4}{4!} + j\frac{\theta^5}{5!} - \cdots$$

and

$$e^z = 1 + z + \frac{z^2}{2!} + \frac{z^3}{3!} + \frac{z^4}{4!} + \frac{z^5}{5!} + \cdots$$

so that

$$e^{j\theta} = 1 + j\theta - \frac{\theta^2}{2!} - j\frac{\theta^3}{3!} + \frac{\theta^4}{4!} + \cdots$$

We conclude that

$$e^{j\theta} = \cos\theta + j\sin\theta \qquad [1]$$

or, if we let $z = -j\theta$, we find that

$$e^{-j\theta} = \cos\theta - j\sin\theta \qquad [2]$$

By adding and subtracting Eqs. [1] and [2], we obtain the two expressions which we use without proof in our study of the underdamped natural response of the parallel and series *RLC* circuits,

$$\cos\theta = \tfrac{1}{2}(e^{j\theta} + e^{-j\theta}) \qquad [3]$$

$$\sin\theta = -j\tfrac{1}{2}(e^{j\theta} - e^{-j\theta}) \qquad [4]$$

PRACTICE

A5.3 Use Eqs. [1] through [4] to evaluate (*a*) e^{-j1}; (*b*) e^{1-j1}; (*c*) $\cos(-j1)$; (*d*) $\sin(-j1)$.

A5.4 Evaluate at $t = 0.5$: (*a*)$(d/dt)(3\cos 2t - j2\sin 2t)$; (*b*)$\int_0^t (3\cos 2t - j2\sin 2t)\, dt$; Evaluate at $\mathbf{s} = 1 + j2$: (*c*) $\int_s^\infty \mathbf{s}^{-3}\, d\mathbf{s}$; (*d*) $(d/d\mathbf{s})[3/(\mathbf{s} + 2)]$.

Ans: A5.3: $0.540 - j0.841$; $1.469 - j2.29$; 1.543; $-j1.175$. A5.4: $-5.05 - j2.16$; $1.262 - j0.460$; $-0.06 - j0.08$; $-0.0888 + j0.213$.

A5.3 THE EXPONENTIAL FORM

Let us now take Euler's identity

$$e^{j\theta} = \cos\theta + j\sin\theta$$

and multiply each side by the real positive number C:

$$Ce^{j\theta} = C\cos\theta + jC\sin\theta \qquad [5]$$

The right-hand side of Eq. [5] consists of the sum of a real number and an imaginary number and thus represents a complex number in rectangular form; let us call this complex number **A**, where $\mathbf{A} = a + jb$. By equating the real parts

$$a = C\cos\theta \qquad [6]$$

and the imaginary parts

$$b = C\sin\theta \qquad [7]$$

then squaring and adding Eqs. [6] and [7],

$$a^2 + b^2 = C^2$$

or

$$C = +\sqrt{a^2 + b^2} \qquad [8]$$

and dividing Eq. [7] by Eq. [6]:

$$\frac{b}{a} = \tan\theta$$

or

$$\theta = \tan^{-1}\frac{b}{a} \qquad [9]$$

we obtain the relationships of Eqs. [8] and [9], which enable us to determine C and θ from a knowledge of a and b. For example, if $\mathbf{A} = 4 + j2$, then we identify a as 4 and b as 2 and find C and θ:

$$C = \sqrt{4^2 + 2^2} = 4.47$$
$$\theta = \tan^{-1}\tfrac{2}{4} = 26.6°$$

We could use this new information to write **A** in the form

$$\mathbf{A} = 4.47\cos 26.6° + j4.47\sin 26.6°$$

but it is the form of the left-hand side of Eq. [5] which will prove to be the more useful:

$$\mathbf{A} = Ce^{j\theta} = 4.47e^{j26.6°}$$

A complex number expressed in this manner is said to be in *exponential form*. The real positive multiplying factor C is known as the *amplitude* or *magnitude*, and the real quantity θ appearing in the exponent is called the *argument* or *angle*. A mathematician would always express θ in radians and would write

$$\mathbf{A} = 4.47e^{j0.464}$$

but engineers customarily work in terms of degrees. The use of the degree symbol (°) in the exponent should make confusion impossible.

To recapitulate, if we have a complex number which is given in rectangular form,

$$\mathbf{A} = a + jb$$

and wish to express it in exponential form,

$$\mathbf{A} = Ce^{j\theta}$$

we may find C and θ by Eqs. [8] and [9]. If we are given the complex number in exponential form, then we may find a and b by Eqs. [6] and [7].

When $\mathbf{A}$ is expressed in terms of numerical values, the transformation between exponential (or polar) and rectangular forms is available as a built-in operation on most handheld scientific calculators.

One question will be found to arise in the determination of the angle θ by using the arctangent relationship of Eq. [9]. This function is multivalued, and an appropriate angle must be selected from various possibilities. One method by which the choice may be made is to select an angle for which the sine and cosine have the proper signs to produce the required values of a and b from Eqs. [6] and [7]. For example, let us convert

$$\mathbf{V} = 4 - j3$$

to exponential form. The amplitude is

$$C = \sqrt{4^2 + (-3)^2} = 5$$

and the angle is

$$\theta = \tan^{-1} \frac{-3}{4} \qquad [10]$$

A value of θ has to be selected which leads to a positive value for $\cos \theta$, since $4 = 5 \cos \theta$, and a negative value for $\sin \theta$, since $-3 = 5 \sin \theta$. We therefore obtain $\theta = -36.9°$, $323.1°$, $-396.9°$, and so forth. Any of these angles is correct, and we usually select that one which is the simplest, here, $-36.9°$. We should note that the alternative solution of Eq. [10], $\theta = 143.1°$, is not correct, because $\cos \theta$ is negative and $\sin \theta$ is positive.

A simpler method of selecting the correct angle is available if we represent the complex number graphically in the complex plane. Let us first select a complex number, given in rectangular form, $\mathbf{A} = a + jb$, which lies in the first quadrant of the complex plane, as illustrated in Fig. A5.3. If we draw a line from the origin to the point which represents the complex number, we shall have constructed a right triangle whose hypotenuse is evidently the amplitude of the exponential representation of the complex number. In other words, $C = \sqrt{a^2 + b^2}$. Moreover, the counterclockwise angle which the line makes with the positive real axis is seen to be the angle θ of the exponential representation, because $a = C \cos \theta$ and $b = C \sin \theta$. Now if we are given the rectangular form of a complex number which lies in another quadrant, such as $\mathbf{V} = 4 - j3$, which is depicted in Fig. A5.4, the correct angle is graphically evident, either $-36.9°$ or $323.1°$ for this example. The sketch may often be visualized and need not be drawn.

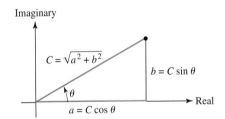

■ **FIGURE A5.3** A complex number may be represented by a point in the complex plane through choosing the correct real and imaginary parts from the rectangular form, or by selecting the magnitude and angle from the exponential form.

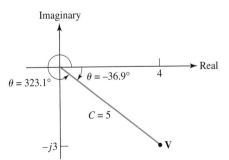

■ **FIGURE A5.4** The complex number $\mathbf{V} = 4 - j3 = 5e^{-j36.9°}$ is represented in the complex plane.

If the rectangular form of the complex number has a negative real part, it is often easier to work with the negative of the complex number, thus avoiding angles greater than $90°$ in magnitude. For example, given

$$\mathbf{I} = -5 + j2$$

we write

$$\mathbf{I} = -(5 - j2)$$

and then transform $(5 - j2)$ to exponential form:

$$\mathbf{I} = -Ce^{j\theta}$$

where

$$C = \sqrt{29} = 5.39 \qquad \text{and} \qquad \theta = \tan^{-1} \frac{-2}{5} = -21.8°$$

We therefore have

$$\mathbf{I} = -5.39e^{-j21.8°}$$

The negative sign may be removed from the complex number by increasing or decreasing the angle by $180°$, as shown by reference to a sketch in the complex plane. Thus, the result may be expressed in exponential form as

$$\mathbf{I} = 5.39e^{j158.2°} \qquad \text{or} \qquad \mathbf{I} = 5.39e^{-j201.8°}$$

Note that use of an electronic calculator in the inverse tangent mode always yields angles having magnitudes less than $90°$. Thus, both $\tan^{-1}[(-3)/4]$ and $\tan^{-1}[3/(-4)]$ come out as $-36.9°$. Calculators that provide rectangular-to-polar conversion, however, give the correct angle in all cases.

One last remark about the exponential representation of a complex number should be made. Two complex numbers, both written in exponential form, are equal if, and only if, their amplitudes are equal and their angles are equivalent. Equivalent angles are those which differ by multiples of $360°$. For example, if $\mathbf{A} = Ce^{j\theta}$ and $\mathbf{B} = De^{j\phi}$, then if $\mathbf{A} = \mathbf{B}$, it is necessary that $C = D$ and $\theta = \phi \pm (360°)n$, where $n = 0, 1, 2, 3, \ldots$.

PRACTICE

A5.5 Express each of the following complex numbers in exponential form, using an angle lying in the range $-180° < \theta \leq 180°$:
(a) $-18.5 - j26.1$; (b) $17.9 - j12.2$; (c) $-21.6 + j31.2$.
A5.6 Express each of these complex numbers in rectangular form:
(a) $61.2e^{-j111.1°}$; (b) $-36.2e^{j108°}$; (c) $5e^{-j2.5}$.

Ans: A5.5: $32.0e^{-j125.3°}$; $21.7e^{-j34.3°}$; $37.9e^{j124.7°}$.   A5.6: $-22.0 - j57.1$;
$11.19 - j34.4$; $-4.01 - j2.99$.

A5.4 •THE POLAR FORM

The third (and last) form in which we may represent a complex number is essentially the same as the exponential form, except for a slight difference in symbolism. We use an angle sign ($\underline{/}$) to replace the combination e^j. Thus, the exponential representation of a complex number **A**,

$$\mathbf{A} = Ce^{j\theta}$$

may be written somewhat more concisely as

$$\mathbf{A} = C\underline{/\theta}$$

The complex number is now said to be expressed in *polar* form, a name which suggests the representation of a point in a (complex) plane through the use of polar coordinates.

It is apparent that transformation from rectangular to polar form or from polar form to rectangular form is basically the same as transformation between rectangular and exponential form. The same relationships exist between C, θ, a, and b.

The complex number

$$\mathbf{A} = -2 + j5$$

is thus written in exponential form as

$$\mathbf{A} = 5.39e^{j111.8°}$$

and in polar form as

$$\mathbf{A} = 5.39\underline{/111.8°}$$

In order to appreciate the utility of the exponential and polar forms, let us consider the multiplication and division of two complex numbers represented in exponential or polar form. If we are given

$$\mathbf{A} = 5\underline{/53.1°} \quad \text{and} \quad \mathbf{B} = 15\underline{/-36.9°}$$

then the expression of these two complex numbers in exponential form

$$\mathbf{A} = 5e^{j53.1°} \quad \text{and} \quad \mathbf{B} = 15e^{-j36.9°}$$

enables us to write the product as a complex number in exponential form whose amplitude is the product of the amplitudes and whose angle is the algebraic sum of the angles, in accordance with the normal rules for multiplying two exponential quantities:

$$(\mathbf{A})(\mathbf{B}) = (5)(15)e^{j(53.1°-36.9°)}$$

or

$$\mathbf{AB} = 75e^{j16.2°} = 75\underline{/16.2°}$$

From the definition of the polar form, it is evident that

$$\frac{\mathbf{A}}{\mathbf{B}} = 0.333\underline{/90°}$$

Addition and subtraction of complex numbers are accomplished most easily by operating on complex numbers in rectangular form, and the addition or subtraction of two complex numbers given in exponential or polar form should begin with the conversion of the two complex numbers to rectangular form. The reverse situation applies to multiplication and division; two numbers given in rectangular form should be transformed to polar form, unless the numbers happen to be small integers. For example, if we wish to multiply $(1 - j3)$ by $(2 + j1)$, it is easier to multiply them directly as they stand and obtain $(5 - j5)$. If the numbers can be multiplied mentally, then time is wasted in transforming them to polar form.

We should now endeavor to become familiar with the three different forms in which complex numbers may be expressed and with the rapid conversion from one form to another. The relationships among the three forms seem almost endless, and the following lengthy equation summarizes the various interrelationships

$$\mathbf{A} = a + jb = \text{Re}\{\mathbf{A}\} + j\text{Im}\{\mathbf{A}\} = Ce^{j\theta} = \sqrt{a^2 + b^2}\, e^{j\,\tan^{-1}(b/a)}$$
$$= \sqrt{a^2 + b^2}\,\underline{/\tan^{-1}(b/a)}$$

Most of the conversions from one form to another can be done quickly with the help of a calculator, and many calculators are equipped to solve linear equations with complex numbers.

We shall find that complex numbers are a convenient mathematical artifice which facilitates the analysis of real physical situations.

PRACTICE
●

A5.7 Express the result of each of these complex-number manipulations in polar form, using six significant figures just for the pure joy of calculating (a) $[2 - (1\underline{/-41°})]/(0.3\underline{/41°})$; (b) $50/(2.87\underline{/83.6°} + 5.16\underline{/63.2°})$; (c) $4\underline{/18°} - 6\underline{/-75°} + 5\underline{/28°}$.

A5.8 Find $\mathbf{Z}$ in rectangular form if (a) $\mathbf{Z} + j2 = 3/\mathbf{Z}$; (b) $\mathbf{Z} = 2\ln(2 - j3)$; (c) $\sin\mathbf{Z} = 3$.

Ans: A5.7: $4.69179\underline{/-13.2183°}$; $6.31833\underline{/-70.4626°}$; $11.5066\underline{/54.5969°}$.
A5.8: $\pm1.414 - j1$; $2.56 - j1.966$; $1.571 \pm j1.763$.

A BRIEF MATLAB® TUTORIAL

The intention of this tutorial is to provide a very brief introduction to some basic concepts required to use a powerful software package known as MATLAB. The use of MATLAB is a completely optional part of the material in this textbook, but as it is becoming an increasingly common tool in all areas of electrical engineering, we felt that it was worthwhile to provide students with the opportunity to begin exploring some of its features, particularly in plotting 2D and 3D functions, performing matrix operations, solving simultaneous equations, and manipulating algebraic expressions. Many institutions now provide the full version of MATLAB for their students, but at the time of this writing, a student version is available at reduced cost from The MathWorks, Inc. (http://www.mathworks.com/academia/student_version/).

Getting Started

MATLAB is launched by clicking on the program icon; the typical opening window is shown in Fig. A6.1. Programs may be run from files or by directly entering commands in the window. MATLAB also has extensive online help resources, useful for both beginners and advanced users alike. Typical MATLAB programs very much resemble programs written in C, although familiarity with this language is by no means required.

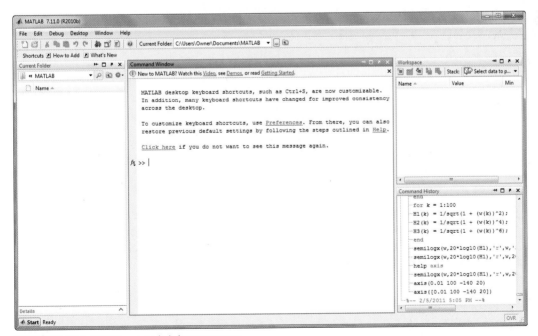

■ **FIGURE A6.1** MATLAB command window upon start-up.

Variables and Mathematical Operations

MATLAB makes a great deal more sense once the user realizes that all variables are matrices, even if simply 1×1 matrices. Variable names can be up to 19 characters in length, which is extremely useful in constructing programs with adequate readability. The first character must be a letter, but all remaining characters can be any letter or number; the underscore (_) character may also be used. Variable names in MATLAB are case-sensitive. MATLAB includes several predefined variables. Relevant predefined variables for the material presented in this text include:

eps	The machine accuracy
realmin	The smallest (positive) floating-point number handled by the computer
realmax	The largest floating-point number handled by the computer
inf	Infinity (defined as $1/0$)
NaN	Literally, "Not a Number." This includes situations such as 0/0
pi	π (3.14159...)
i, j	Both are initially defined as $\sqrt{-1}$. They may be assigned other values by the user

A complete list of currently defined variables can be obtained with the command *who*. Variables are assigned by using an equal sign (=). If the statement is terminated with a semicolon (;), then another prompt appears. If the statement is simply terminated by a carriage return (i.e., by pressing the Enter key), then the variable is repeated. For example,

EDU» input_voltage = 5;

EDU» input_current = 1e−3

input_current =

1.0000e−003

EDU»

Complex variables are easy to define in MATLAB: for example,

EDU» s = 9 + j*5;

creates a complex variable **s** with value $9 + j5$.

A matrix other than a 1×1 matrix is defined using brackets. For example, we would express the matrix $\mathbf{t} = \begin{bmatrix} 2 & -1 \\ 3 & 0 \end{bmatrix}$ in MATLAB as

EDU» t = [2   − 1; 3   0];

Note that the matrix elements are entered a row at a time; row elements are separated by a space, and rows are separated by a semicolon (;). The same arithmetic operations are available for matrices; so, for example, we may find $t + t$ as

EDU» t + t

ans =

   4   −2

   6   0

Arithmetic operators include:

∧	power	\	left division
*	multiplication	+	addition
/	right (ordinary) division	−	subtraction

The order of operations is important. The order of precedence is power, then multiplication and division, then addition and subtraction.

$$\text{EDU» } x = 1 + 5 \wedge 2 * 3$$

$$x =$$

$$76$$

The concept of left division may seem strange at first, but is very useful in matrix algebra. For example,

$$\text{EDU» } 1/5$$

$$\text{ans} =$$

$$0.2000$$

$$\text{EDU» } 1\backslash 5$$

$$\text{ans} =$$

$$5$$

$$\text{EDU» } 5\backslash 1$$

$$\text{ans} =$$

$$0.2000$$

And, in the case of the matrix equation $\mathbf{Ax} = \mathbf{B}$, where

$$\mathbf{A} = \begin{bmatrix} 2 & 4 \\ 1 & 6 \end{bmatrix} \quad \text{and} \quad \mathbf{B} = \begin{bmatrix} -1 \\ 2 \end{bmatrix}$$

we find $\mathbf{x}$ with

$$\text{EDU» } A = [2\ 4;\ 1\ 6];$$
$$\text{EDU» } B = [-1;\ 2];$$
$$\text{EDU» } x = A\backslash B$$

$$x =$$

$$-1.7500$$
$$0.6250$$

Alternatively, we can also write

$$\text{EDU» } x = A\wedge{-1}*B$$

$$x =$$

$$-1.7500$$
$$0.6250$$

or

$$\text{EDU}\gg \text{inv(A)*B}$$

$$\text{ans} =$$

$$-1.7500$$
$$0.6250$$

When in doubt, parentheses can help a great deal.

Some Useful Functions

Space requirements prevent us from listing every function contained in MATLAB. Some of the more basic ones include:

abs(x)	$\|x\|$	log 10(x)	$\log_{10} x$		
exp(x)	e^x	sin(x)	$\sin x$	asin(x)	$\sin^{-1} x$
sqrt(x)	$\sqrt{x}$	cos(x)	$\cos x$	acos(x)	$\cos^{-1} x$
log(x)	$\ln x$	tan(x)	$\tan x$	atan(x)	$\tan^{-1} x$

Functions useful for manipulating complex variables include:

real(s)	Re$\{s\}$
imag(s)	Im$\{s\}$
abs(s)	$\sqrt{a^2 + b^2}$, where $\mathbf{s} \equiv a + jb$
angle(s)	$\tan^{-1}(b/a)$, where $\mathbf{s} \equiv a + jb$
conj(s)	complex conjugate of $\mathbf{s}$

Another extremely useful command, often forgotten, is simply *help*.

Occasionally we require a vector, such as when we plan to create a plot. The command *linspace*(min, max, number of points) is invaluable in such instances:

$$\text{EDU}\gg \text{frequency} = \text{linspace(0,10,5)}$$

$$\text{frequency} =$$

$$0 \quad 2.5000 \quad 5.0000 \quad 7.5000 \quad 10.0000$$

A useful cousin is the command *logspace*().

Generating Plots

Plotting with MATLAB is extremely easy. For example, Fig. A6.2 shows the result of executing the following MATLAB program:

```
EDU» x = linspace(0,2*pi,100);
EDU» y = sin(x);
EDU» plot(x,y);
EDU» xlabel('Angle (radians)');
EDU» ylabel('f(x)');
```

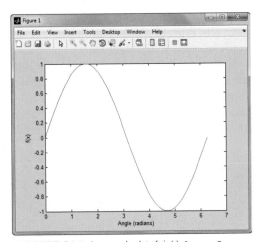

■ **FIGURE A6.2** An example plot of sin(x), $0 < x < 2\pi$, generated using MATLAB. The variable x is a vector comprised of 100 equally spaced elements.

Writing Programs

Although the MATLAB examples in this text are presented as lines typed into the Command Window, it is possible (and often prudent, if repetition is an issue) to write a program so that calculations are more convenient. This is accomplished in MATLAB by writing what is termed an m-file. This is simply a text file saved with a ".m" extension (for example, first_program.m). In a nod to Kernighan and Ritchie, we pull down **New Script** under the **File** menu, which opens up the editor. (Note that you can use another editor, for example, WordPad, if you prefer.)

We type in

$$r = \text{input('Hello, World')}$$

as shown in Fig. A6.3.

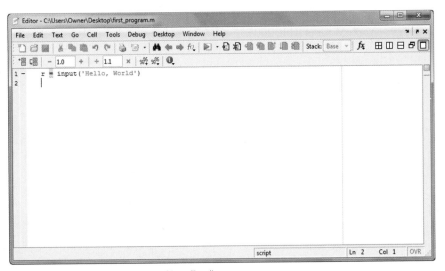

■ **FIGURE A6.3** Example m-file created in m-file editor.

We next save it as first_program in a convenient directory, taking care to select MATLAB Files (*.m) under **File Type.** Under the **File** menu, we select **Open,** and find first_program.m. This reopens the editor (so we could have skipped closing it earlier). We run our program by hitting f5 or selecting **Run** under the **Debug** menu. In the Command Window, we see our greeting; MATLAB is waiting for a keyboard response, so just hit the Enter key.

Let's expand a previous example to allow the magnitude to be user-selected as in Fig. A6.4. We are now allowed to enter an arbitrary amplitude for our plot.

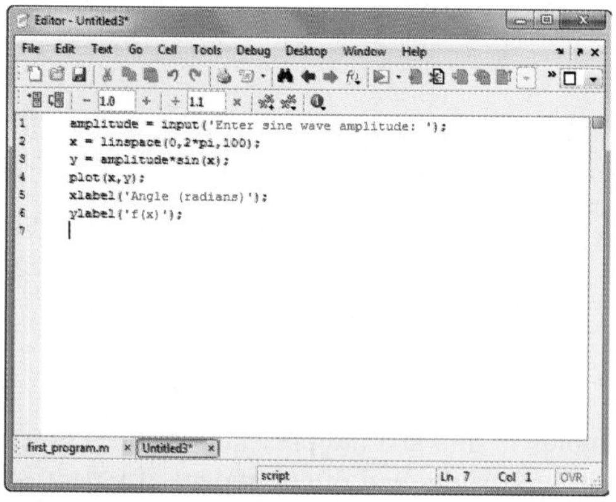

■ FIGURE A6.4 Example m-file named example1.m for generating sine wave plot.

We leave it to the reader to choose when to write a program/m-file and when to simply use the Command Window directly.

READING FURTHER

There are a large number of excellent MATLAB references available, with new titles appearing regularly. Two worth looking at are:

D. C. Hanselman and B. L. Littlefield, *Mastering MATLAB 7.* Upper Saddle River, N.J.: Prentice-Hall, 2005.

W. J. Palm, III, *Introduction to MATLAB 7 for Engineers,* 2nd ed. New York: McGraw-Hill, 2005.

ADDITIONAL LAPLACE TRANSFORM THEOREMS

In this appendix, we briefly present several Laplace transform theorems typically used in more advanced situations in addition to those described in Chap. 14.

Transforms of Periodic Time Functions

The time-shift theorem is very useful in evaluating the transform of periodic time functions. Suppose that $f(t)$ is periodic with a period T for positive values of t. The behavior of $f(t)$ for $t < 0$ has no effect on the (one-sided) Laplace transform, as we know. Thus, $f(t)$ can be written as

$$f(t) = f(t - nT) \qquad n = 0, 1, 2, \ldots$$

If we now define a new time function which is nonzero only in the first period of $f(t)$,

$$f_1(t) = [u(t) - u(t - T)]f(t)$$

then the original $f(t)$ can be represented as the sum of an infinite number of such functions, delayed by integral multiples of T. That is,

$$\begin{aligned}
f(t) &= [u(t) - u(t - T)]f(t) + [u(t - T) - u(t - 2T)]f(t) \\
&\quad + [u(t - 2T) - u(t - 3T)]f(t) + \cdots \\
&= f_1(t) + f_1(t - T) + f_1(t - 2T) + \cdots
\end{aligned}$$

or

$$f(t) = \sum_{n=0}^{\infty} f_1(t - nT)$$

The Laplace transform of this sum is just the sum of the transforms,

$$\mathbf{F}(\mathbf{s}) = \sum_{n=0}^{\infty} \mathcal{L}\{f_1(t - nT)\}$$

so that the time-shift theorem leads to

$$\mathbf{F}(\mathbf{s}) = \sum_{n=0}^{\infty} e^{-nT\mathbf{s}} \mathbf{F}_1(\mathbf{s})$$

where

$$\mathbf{F}_1(\mathbf{s}) = \mathcal{L}\{f_1(t)\} = \int_{0^-}^{T} e^{-\mathbf{s}t} f(t)\, dt$$

Since $\mathbf{F}_1(\mathbf{s})$ is not a function of n, it can be removed from the summation, and $\mathbf{F}(\mathbf{s})$ becomes

$$\mathbf{F}(\mathbf{s}) = \mathbf{F}_1(\mathbf{s})[1 + e^{-T\mathbf{s}} + e^{-2T\mathbf{s}} + \cdots]$$

When we apply the binomial theorem to the bracketed expression, it simplifies to $1/(1 - e^{-T\mathbf{s}})$. Thus, we conclude that the periodic function $f(t)$, with period T, has a Laplace transform expressed by

$$\mathbf{F}(\mathbf{s}) = \frac{\mathbf{F}_1(\mathbf{s})}{1 - e^{-T\mathbf{s}}} \qquad [1]$$

where

$$\mathbf{F}_1(\mathbf{s}) = \mathscr{L}\{[u(t) - u(t - T)]f(t)\} \qquad [2]$$

is the transform of the first period of the time function.

To illustrate the use of this transform theorem for periodic functions, let us apply it to the familiar rectangular pulse train, Fig. A7.1. We may describe this periodic function analytically:

$$v(t) = \sum_{n=0}^{\infty} V_0[u(t - nT) - u(t - nT - \tau)] \qquad t > 0$$

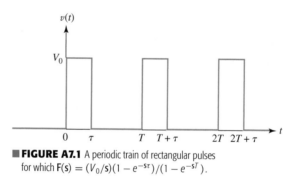

FIGURE A7.1 A periodic train of rectangular pulses for which $\mathbf{F}(\mathbf{s}) = (V_0/\mathbf{s})(1 - e^{-\mathbf{s}\tau})/(1 - e^{-\mathbf{s}T})$.

The function $\mathbf{V}_1(\mathbf{s})$ is simple to calculate:

$$\mathbf{V}_1(\mathbf{s}) = V_0 \int_{0^-}^{\tau} e^{-\mathbf{s}t}\, dt = \frac{V_0}{\mathbf{s}}(1 - e^{-\mathbf{s}\tau})$$

Now, to obtain the desired transform, we just divide by $(1 - e^{-\mathbf{s}T})$:

$$\mathbf{V}(\mathbf{s}) = \frac{V_0(1 - e^{-\mathbf{s}\tau})}{\mathbf{s}(1 - e^{-\mathbf{s}T})} \qquad [3]$$

We should note how several different theorems show up in the transform in Eq. [3]. The $(1 - e^{-\mathbf{s}T})$ factor in the denominator accounts for the periodicity of the function, the $e^{-\mathbf{s}\tau}$ term in the numerator arises from the time delay of the negative square wave that turns off the pulse, and the $V_0/\mathbf{s}$ factor is, of course, the transform of the step functions involved in $v(t)$.

EXAMPLE **A7.1**

Determine the transform of the periodic function of Fig. A7.2.

We begin by writing an equation which describes $f(t)$, a function composed of alternating positive and negative impulse functions.

$$f(t) = 2\delta(t-1) - 2\delta(t-3) + 2\delta(t-5) - 2\delta(t-7) + \cdots$$

Defining a new function f_1 and recognizing a period $T = 4$ s,

$$f_1(t) = 2[\delta(t-1) - \delta(t-3)]$$

we can make use of the time periodicity operation as listed in Table 14.2 to find $\mathbf{F}(\mathbf{s})$

$$\mathbf{F}(\mathbf{s}) = \frac{1}{1 - e^{-T\mathbf{s}}}\mathbf{F}_1(\mathbf{s}) \qquad [4]$$

where

$$\mathbf{F}_1(\mathbf{s}) = \int_{0^-}^{T} f(t)e^{-st}\, dt = \int_{0^-}^{4} f_1(t)e^{-st}\, dt$$

There are several ways to evaluate this integral. The easiest is to recognize that its value will remain the same if the upper limit is increased to ∞, allowing us to make use of the time-shift theorem. Thus,

$$\mathbf{F}_1(\mathbf{s}) = 2[e^{-s} - e^{-3s}] \qquad [5]$$

Our example is completed by multiplying Eq. [5] by the factor indicated in Eq. [4], so that

$$\mathbf{F}(\mathbf{s}) = \frac{2}{1 - e^{-4\mathbf{s}}}(e^{-s} - e^{-3s}) = \frac{2e^{-s}}{1 + e^{-2s}}$$

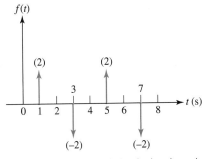

FIGURE A7.2 A periodic function based on unit-impulse functions.

PRACTICE

A7.1 Determine the Laplace transform of the periodic function shown in Fig. A7.3.

Ans: $\left(\dfrac{8}{s^2 + \pi^2/4}\right) \dfrac{s + (\pi/2)e^{-s} + (\pi/2)e^{-3s} - se^{-4s}}{1 - e^{-4s}}$.

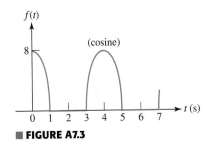

FIGURE A7.3

Frequency Shifting

The next new theorem establishes a relationship between $\mathbf{F}(\mathbf{s}) = \mathcal{L}\{f(t)\}$ and $\mathbf{F}(\mathbf{s} + a)$. We consider the Laplace transform of $e^{-at}f(t)$,

$$\mathcal{L}\{e^{-at}f(t)\} = \int_{0^-}^{\infty} e^{-st}e^{-at}f(t)\, dt = \int_{0^-}^{\infty} e^{-(s+a)t}f(t)\, dt$$

Looking carefully at this result, we note that the integral on the right is identical to that defining $\mathbf{F}(\mathbf{s})$ with one exception: $(\mathbf{s} + a)$ appears in place of $\mathbf{s}$. Thus,

$$e^{-at}f(t) \Leftrightarrow \mathbf{F}(\mathbf{s} + a) \qquad [6]$$

We conclude that replacing **s** by $(s + a)$ in the frequency domain corresponds to multiplication by e^{-at} in the time domain. This is known as the ***frequency-shift*** theorem. It can be put to immediate use in evaluating the transform of the exponentially damped cosine function that we used extensively in previous work. Beginning with the known transform of the cosine function,

$$\mathcal{L}\{\cos \omega_0 t\} = \mathbf{F}(\mathbf{s}) = \frac{\mathbf{s}}{\mathbf{s}^2 + \omega_0^2}$$

then the transform of $e^{-at} \cos \omega_0 t$ must be $\mathbf{F}(\mathbf{s} + a)$:

$$\mathcal{L}\{e^{-at} \cos \omega_0 t\} = \mathbf{F}(\mathbf{s} + a) = \frac{\mathbf{s} + a}{(\mathbf{s} + a)^2 + \omega_0^2} \qquad [7]$$

PRACTICE

A7.2 Find $\mathcal{L}\{e^{-2t} \sin(5t + 0.2\pi)u(t)\}$.

Ans: $(0.588s + 4.05)/(s^2 + 4s + 29)$.

Differentiation in the Frequency Domain

Next let us examine the consequences of differentiating $\mathbf{F}(\mathbf{s})$ with respect to **s**. The result is

$$\frac{d}{d\mathbf{s}}\mathbf{F}(\mathbf{s}) = \frac{d}{d\mathbf{s}} \int_{0^-}^{\infty} e^{-st} f(t)\, dt$$

$$= \int_{0^-}^{\infty} -t e^{-st} f(t)\, dt = \int_{0^-}^{\infty} e^{-st}[-t f(t)]\, dt$$

which is simply the Laplace transform of $[-t f(t)]$. We therefore conclude that differentiation with respect to **s** in the frequency domain results in multiplication by $-t$ in the time domain, or

$$-t f(t) \Leftrightarrow \frac{d}{d\mathbf{s}}\mathbf{F}(\mathbf{s}) \qquad [8]$$

Suppose now that $f(t)$ is the unit-ramp function $t u(t)$, whose transform we know is $1/s^2$. We can use our newly acquired frequency-differentiation theorem to determine the inverse transform of $1/s^3$ as follows:

$$\frac{d}{d\mathbf{s}}\left(\frac{1}{s^2}\right) = -\frac{2}{s^3} \Leftrightarrow -t\mathcal{L}^{-1}\left\{\frac{1}{s^2}\right\} = -t^2 u(t)$$

and

$$\frac{t^2 u(t)}{2} \Leftrightarrow \frac{1}{s^3} \qquad [9]$$

Continuing with the same procedure, we find

$$\frac{t^3}{3!}u(t) \Leftrightarrow \frac{1}{s^4} \qquad [10]$$

and in general

$$\frac{t^{(n-1)}}{(n-1)!}u(t) \Leftrightarrow \frac{1}{s^n} \qquad [11]$$

PRACTICE

A7.3 Find $\mathcal{L}\{t \sin(5t + 0.2\pi)u(t)\}$.

Ans: $(0.588s^2 + 8.09s - 14.69)/(s^2 + 25)^2$.

Integration in the Frequency Domain

The effect on $f(t)$ of integrating $\mathbf{F}(\mathbf{s})$ with respect to $\mathbf{s}$ may be shown by beginning with the definition once more,

$$\mathbf{F}(\mathbf{s}) = \int_{0^-}^{\infty} e^{-st} f(t)\, dt$$

performing the frequency integration from $\mathbf{s}$ to ∞,

$$\int_{\mathbf{s}}^{\infty} \mathbf{F}(\mathbf{s})\, d\mathbf{s} = \int_{\mathbf{s}}^{\infty} \left[\int_{0^-}^{\infty} e^{-st} f(t)\, dt \right] d\mathbf{s}$$

interchanging the order of integration,

$$\int_{\mathbf{s}}^{\infty} \mathbf{F}(\mathbf{s})\, d\mathbf{s} = \int_{0^-}^{\infty} \left[\int_{\mathbf{s}}^{\infty} e^{-st}\, d\mathbf{s} \right] f(t)\, dt$$

and performing the inner integration,

$$\int_{\mathbf{s}}^{\infty} \mathbf{F}(\mathbf{s})\, d\mathbf{s} = \int_{0^-}^{\infty} \left[-\frac{1}{t} e^{-st} \right]_{\mathbf{s}}^{\infty} f(t)\, dt = \int_{0^-}^{\infty} \frac{f(t)}{t} e^{-st}\, dt$$

Thus,

$$\frac{f(t)}{t} \Leftrightarrow \int_{\mathbf{s}}^{\infty} \mathbf{F}(\mathbf{s})\, d\mathbf{s} \qquad [12]$$

For example, we have already established the transform pair

$$\sin \omega_0 t\, u(t) \Leftrightarrow \frac{\omega_0}{s^2 + \omega_0^2}$$

Therefore,

$$\mathcal{L}\left\{ \frac{\sin \omega_0 t\, u(t)}{t} \right\} = \int_{\mathbf{s}}^{\infty} \frac{\omega_0\, d\mathbf{s}}{s^2 + \omega_0^2} = \tan^{-1} \frac{\mathbf{s}}{\omega_0} \bigg|_{\mathbf{s}}^{\infty}$$

and we have

$$\frac{\sin \omega_0 t\, u(t)}{t} \Leftrightarrow \frac{\pi}{2} - \tan^{-1} \frac{\mathbf{s}}{\omega_0} \qquad [13]$$

PRACTICE

A7.4 Find $\mathcal{L}\{\sin^2 5t u(t)/t\}$.

Ans: $\frac{1}{4} \ln[(s^2 + 100)/s^2]$.

The Time-Scaling Theorem

We next develop the time-scaling theorem of Laplace transform theory by evaluating the transform of $f(at)$, assuming that $\mathcal{L}\{f(t)\}$ is known. The procedure is very simple:

$$\mathcal{L}\{f(at)\} = \int_{0^-}^{\infty} e^{-st} f(at)\, dt = \frac{1}{a} \int_{0^-}^{\infty} e^{-(s/a)\lambda} f(\lambda)\, d\lambda$$

where the change of variable $at = \lambda$ has been employed. The last integral is recognizable as $1/a$ times the Laplace transform of $f(t)$, except that $\mathbf{s}$ is replaced by $\mathbf{s}/a$ in the transform. It follows that

$$f(at) \Leftrightarrow \frac{1}{a} \mathbf{F}\left(\frac{\mathbf{s}}{a}\right) \qquad [14]$$

As an elementary example of the use of this time-scaling theorem, consider the determination of the transform of a 1 kHz cosine wave. Assuming we know the transform of a 1 rad/s cosine wave,

$$\cos t\, u(t) \Leftrightarrow \frac{\mathbf{s}}{s^2 + 1}$$

the result is

$$\mathcal{L}\{\cos 2000\pi t\, u(t)\} = \frac{1}{2000\pi} \frac{\mathbf{s}/2000\pi}{(\mathbf{s}/2000\pi)^2 + 1} = \frac{\mathbf{s}}{s^2 + (2000\pi)^2}$$

PRACTICE

A7.5 Find $\mathcal{L}\{\sin^2 5t\, u(t)\}$.

Ans: $50/[\mathbf{s}(s^2 + 100)]$.

INDEX

A Short Table of Integrals

$$\int \sin^2 ax\, dx = \frac{x}{2} - \frac{\sin 2ax}{4a}$$

$$\int \cos^2 ax\, dx = \frac{x}{2} + \frac{\sin 2ax}{4a}$$

$$\int x \sin ax\, dx = \frac{1}{a^2}(\sin ax - ax \cos ax)$$

$$\int x^2 \sin ax\, dx = \frac{1}{a^3}(2ax \sin ax + 2\cos ax - a^2 x^2 \cos ax)$$

$$\int x \cos ax\, dx = \frac{1}{a^2}(\cos ax + ax \sin ax)$$

$$\int x^2 \cos ax\, dx = \frac{1}{a^3}(2ax \cos ax - 2\sin ax + a^2 x^2 \sin ax)$$

$$\int \sin ax \sin bx\, dx = \frac{\sin(a-b)x}{2(a-b)} - \frac{\sin(a+b)x}{2(a+b)}; a^2 \neq b^2$$

$$\int \sin ax \cos bx\, dx = -\frac{\cos(a-b)x}{2(a-b)} - \frac{\cos(a+b)x}{2(a+b)}; a^2 \neq b^2$$

$$\int \cos ax \cos bx\, dx = \frac{\sin(a-b)x}{2(a-b)} + \frac{\sin(a+b)x}{2(a+b)}; a^2 \neq b^2$$

$$\int xe^{ax}\, dx = \frac{e^{ax}}{a^2}(ax - 1)$$

$$\int x^2 e^{ax}\, dx = \frac{e^{ax}}{a^3}(a^2 x^2 - 2ax + 2)$$

$$\int e^{ax} \sin bx\, dx = \frac{e^{ax}}{a^2 + b^2}(a \sin bx - b \cos bx)$$

$$\int e^{ax} \cos bx\, dx = \frac{e^{ax}}{a^2 + b^2}(a \cos bx + b \sin bx)$$

$$\int \frac{dx}{a^2 + x^2} = \frac{1}{a} \tan^{-1} \frac{x}{a}$$